Wireless Sensor Networks

Current Status and Future Trends

OTHER TELECOMMUNICATIONS BOOKS FROM AUERBACH

Ad Hoc Mobile Wireless Networks: Principles, Protocols, and Applications
Subir Kumar Sarkar, T.G. Basavaraju, and C. Puttamadappa
ISBN 978-1-4200-6221-2

Communication and Networking in Smart Grids
Yang Xiao (Editor), ISBN 978-1-4398-7873-6

Decentralized Control and Filtering in Interconnected Dynamical Systems
Magdi S. Mahmoud
ISBN 978-1-4398-3814-3

Delay Tolerant Networks: Protocols and Applications
Athanasios V. Vasilakos, Yan Zhang, and Thrasyvoulos Spyropoulos
ISBN 978-1-4398-1108-5

Emerging Wireless Networks: Concepts, Techniques and Applications
Christian Makaya and Samuel Pierre (Editors)
ISBN 978-1-4398-2135-0

Game Theory in Communication Networks: Cooperative Resolution of Interactive Networking Scenarios
Josephina Antoniou and Andreas Pitsillides
ISBN 978-1-4398-4808-1

Green Mobile Devices and Networks: Energy Optimization and Scavenging Techniques
Hrishikesh Venkataraman and Gabriel-Miro Muntean (Editors)
ISBN 978-1-4398-5989-6

Handbook on Mobile Ad Hoc and Pervasive Communications
Laurence T. Yang, Xingang Liu, and Mieso K. Denko (Editors)
ISBN 978-1-4398-4616-2

IP Telephony Interconnection Reference: Challenges, Models, and Engineering
Mohamed Boucadair, Isabel Borges, Pedro Miguel Neves, and Olafur Pall Einarsson
ISBN 978-1-4398-5178-4

Measurement Data Modeling and Parameter Estimation
Zhengming Wang, Dongyun Yi, Xiaojun Duan, Jing Yao, and Defeng Gu
ISBN 978-1-4398-5378-8

Media Networks: Architectures, Applications, and Standards
Hassnaa Moustafa and Sherali Zeadally (Editors)
ISBN 978-1-4398-7728-9

Multimedia Communications and Networking
Mario Marques da Silva, ISBN 978-1-4398-7484-4

Near Field Communications Handbook
Syed A. Ahson and Mohammad Ilyas (Editors)
ISBN 978-1-4200-8814-4

Next-Generation Batteries and Fuel Cells for Commercial, Military, and Space Applications
A. R. Jha, ISBN 978-1-4398-5066-4

Physical Principles of Wireless Communications, Second Edition
Victor L. Granatstein, ISBN 978-1-4398-7897-2

Security of Mobile Communications
Noureddine Boudriga, ISBN 978-0-8493-7941-3

Smart Grid Security: An End-to-End View of Security in the New Electrical Grid
Gilbert N. Sorebo and Michael C. Echols
ISBN 978-1-4398-5587-4

Systems Evaluation: Methods, Models, and Applications
Sifeng Liu, Naiming Xie, Chaoqing Yuan, and Zhigeng Fang
ISBN 978-1-4200-8846-5

Transmission Techniques for Emergent Multicast and Broadcast Systems
Mario Marques da Silva, Americo Correia, Rui Dinis, Nuno Souto, and Joao Carlos Silva
ISBN 978-1-4398-1593-9

TV Content Analysis: Techniques and Applications
Yiannis Kompatsiaris, Bernard Merialdo, and Shiguo Lian (Editors)
ISBN 978-1-4398-5560-7

TV White Space Spectrum Technologies: Regulations, Standards, and Applications
Rashid Abdelhaleem Saeed and Stephen J. Shellhammer
ISBN 978-1-4398-4879-1

Wireless Sensor Networks: Principles and Practice
Fei Hu and Xiaojun Cao
ISBN 978-1-4200-9215-8

AUERBACH PUBLICATIONS
www.auerbach-publications.com
To Order Call: 1-800-272-7737 • Fax: 1-800-374-3401
E-mail: orders@crcpress.com

Wireless Sensor Networks

Current Status and Future Trends

Edited by Shafiullah Khan,
Al-Sakib Khan Pathan, and Nabil Ali Alrajeh

CRC Press
Taylor & Francis Group
Boca Raton London New York

CRC Press is an imprint of the
Taylor & Francis Group, an **informa** business

CRC Press
Taylor & Francis Group
6000 Broken Sound Parkway NW, Suite 300
Boca Raton, FL 33487-2742

First issued in paperback 2016

© 2013 by Taylor & Francis Group, LLC
CRC Press is an imprint of Taylor & Francis Group, an Informa business

No claim to original U.S. Government works

Version Date: 20120719

ISBN 13: 978-1-138-19987-3 (pbk)
ISBN 13: 978-1-4665-0606-0 (hbk)

Library of Congress Cataloging-in-Publication Data

Wireless sensor networks : current status and future trends / editors, Shafiullah Khan, Al-Sakib Khan Pathan, and Nabil Ali Alrajeh.
    p. cm.
Includes bibliographical references and index.
ISBN 978-1-4665-0606-0 (alk. paper)
    1. Wireless sensor networks. I. Khan, Shafiullah. II. Pathan, Al-Sakib Khan. III. Alrajeh, Nabil Ali.

TK7872.D48W579 2013
681'.2--dc23

2012018946

Visit the Taylor & Francis Web site at
http://www.taylorandfrancis.com

and the CRC Press Web site at
http://www.crcpress.com

Contents

Preface

Human sensing is one way of collecting information, acquiring knowledge, and making reliable decisions. Wireless sensor networks (WSNs) imitate this human intelligence capability, but on a wider distributed scale, with faster, cheaper, and more effective ways that can be used for different applications. Recognizing this fact, we decided to introduce this book on WSNs that documents the current state of these networks and their future trends.

As the subject is relatively recent and enjoying fast development, we thought that the book should be written by researchers and experts in the field. We called on those experts to contribute to the book by writing chapters in their areas of expertise. Our role was to organize the flow of knowledge through the various sections of the book and to ensure consistency on the one hand and clarity on the other. This book would be useful as a reference to all students, graduates, academics, researchers of computer science, and engineers, whether working in professional organizations or research institutions.

The book covers the various issues associated with WSNs, including their structure, activities, and applications. It consists of 5 sections divided into 17 chapters.

Wireless Sensor Networks: Storage Issues and Applications

Section I consists of Chapters 1 through 4. These chapters include a review of applications of wireless sensor networks, an elaboration on data-centric storage in wireless sensor networks, and environmental forest monitoring using wireless sensor networks. In addition, Chapter 4 describes the fundamentals of wireless body area networks.

Medium Access Control (MAC) Layer Issues

Section II consists of Chapters 5 through 7. Chapter 5 emphasizes mobile medium access control protocols for wireless sensor networks. In addition, Chapter 6, Cooperative Transmission Techniques and Protocols in Wireless Sensor Networks, presents cooperative diversity sensor systems from both the physical layer and MAC aspects. Chapter 7 details WSNs operating in IEEE 802.11 networks while adapting a multichannel assignment in its operation.

Position Estimation, Energy-Centric Simulation, and QoS Issues

Chapter 8 presents location and position estimation in WSNs by discussing the basic principles and techniques used in localization algorithms and categories of these algorithms and also focusing on a few of the representative localization schemes. Moreover, Chapter 9 describes energy-centric simulation and design space exploration for WSNs, and Chapter 10 presents the fundamentals of MAC protocols and explains the specific requirements and problems these protocols have to withstand for WSNs.

Protocols and Data Gathering Issues

Chapters 11 through 13 are devoted to issues related to protocols and data gathering. Chapter 11 covers all layers of WSN protocol stacks and presents theoretical aspects, an analytical evaluation, and a comparison of various major WSN protocols specified in the varied research literature under specific classifications or categories. Chapter 12 explains various phases of data gathering and data management protocols used in sensor networks and details the main features and attribute parameters for each phase. Chapter 13 presents a comprehensive description of two broad categories of data gathering algorithms for WSNs—the classical algorithms that are not energy aware and the modern energy aware data gathering algorithms. In addition, it presents an extensive simulation study that demonstrates the individual as well as the comparative performances of these data gathering algorithms.

Security Issues in Wireless Sensor Networks

Security is a challenging issue in WSNs; therefore, Chapters 14 through 17 present privacy and security issues in WSNs. Chapter 14 discusses different techniques and algorithms proposed to address the privacy issues in WSNs and investigates two relevant privacy preserving schemes in WSNs. Chapter 15 provides a comprehensive discussion on state-of-the-art security technologies for WSNs and a brief discussion on the future direction of research in WSN security. Chapter 16 reviews security solutions based on watermarking, which is widely considered to have lower processing and storage requirements in contrast to cryptography-based solutions. Finally, Chapter 17 surveys recently proposed works on intrusion detection systems (IDS) in WSNs and presents a comprehensive classification of various IDS approaches according to their employed detection techniques.

Acknowledgments

The editors would like to thank all of the experts who contributed to this book, making it of a comprehensive nature. We hope that the book will be useful reference for WSNs. We would also like to invite all of the readers to provide comments and suggestions for our future endeavors on related topics.

MATLAB® is a registered tradedmark of The MathWorks, Inc. For product information, please contact:

The MathWorks, Inc.
3 Apple Hill Drive
Natick, MA 01760-2098 USA
Tel: 508 647 7000
Fax: 508-647-7001
E-mail: info@mathworks.com
Web: www.mathworks.com

Editors

Shafiullah Khan earned a PhD in wireless networks in 2011 from Middlesex University, United Kingdom. He earned a BIT degree in information technology from Gomal University, Pakistan, in 2005. Dr. Khan is currently serving as an assistant professor at the Institute of Information Technology (IIT) at the Kohat University of Science and Technology (KUST), K.P.K, Pakistan. His research interests include wireless mesh networks, ad hoc networks, sensor networks, and wireless network security. He is currently serving as the editor-in-chief of the *International Journal of Communication Networks and Information Security* (*IJCNIS*).

Al-Sakib Khan Pathan earned a PhD in computer engineering in 2009 from Kyung Hee University, South Korea. He earned a BSc in computer science and information technology from the Islamic University of Technology (IUT), Bangladesh, in 2003. Dr. Pathan is currently an assistant professor in the Computer Science Department at the International Islamic University Malaysia (IIUM), Malaysia. He is also the founding head of the NDC lab at the Kulliyyah of Information and Communication Technology (KICT), IIUM. Until June 2010, he served as an assistant professor in the Computer Science and Engineering Department at BRAC University, Bangladesh. Prior to holding this position, he worked as a researcher in a networking lab at Kyung Hee University, South Korea, until August 2009. His research interests include wireless sensor networks, network security, and e-services technologies. He is a recipient of several best paper awards and has several publications in these areas.

Dr. Pathan has served as a chair, organizing committee member, and technical program committee member in numerous international conferences/workshops, such as HPCS, ICA3PP, IWCMC, VTC, HPCC, and IDCS. He is currently serving as the editor-in-chief of the *International Journal on Internet and Distributed Computing Systems* (*IJIDCS*); an area editor of *IJCNIS*; senior editor of the *International Journal of Computer Science and Engineering* (*IJCSE*); Inderscience, associate editor of the International Association of Science and Technology for Development (IASTED)/ ACTA Press *International Journal of Computer Applications* (*IJCA*); and *Circuits, Signals, and Systems* (*CCS*); guest editor of some special issues of top-ranked journals; and editor/author of four published books. He also serves as a referee of a few renowned journals such as *IEEE Transactions on Dependable and Secure Computing* (*IEEE TDSC*); *IEEE Transactions on Vehicular Technology* (*IEEE TVT*); *IEEE Communications Letters*; *Journal of Communications and Networks* (JCN); Elsevier's *Computer Communications, Computer Standards and Interfaces*; IOS Press's *Journal of High Speed Networks* (*JHSN*); and the *EURASIP Journal on Wireless Communications and Networking* (*EURASIP JWCN*). He is a member of the Institute of Electrical and Electronics Engineers (IEEE), Cambridge, Massachusetts, IEEE ComSoc Bangladesh Chapter, and several other international organizations.

Nabil Ali Alrajeh earned a PhD in biomedical informatics engineering from Vanderbilt University, Nashville, Tennessee. Currently, Dr. Alrajeh is an associate professor of medical informatics in the Biomedical Technology Department at King Saud University, Riyadh, Saudi Arabia.

Dr. Alrajeh worked as a senior advisor for the Ministry of Higher Education; his role was in implementing development programs including educational affairs, strategic planning, and research and innovation. Dr. Alrajeh's research interests include e-health applications, hospital information systems, telemedicine, intelligent tutoring systems, and WSNs.

Contributors

Abror Abduvaliyev is a graduate student at the National University of Singapore, Singapore. He earned his MSc in computer engineering from Kyung Hee University, South Korea, in 2010. He earned his BSc (Hons) in electronic commerce from the IT faculty of Tashkent University of Information Technologies (TUIT), Tashkent, Uzbekistan, in 2008. His research interests are in computer and network security, intrusion detection systems, and WSNs. He is a student member of IEEE, ACM, and IACSIT.

Khandakar Ahmed earned his BSc (Engg.) in computer science and engineering (CSE) from Shahjalal University of Science & Technology, Bangladesh, in 2006 and MSc in Erasmus Mundus Networking and e-Business Centered Computing (EMNeBCC) in 2011 under the joint consortia of the University of Reading, United Kingdom, Aristotle University of Thessaloniki, Greece, and Universidad Carlos III de Madrid, Spain. He is a PhD research candidate in the school of electrical and computer engineering, RMIT University. He is currently a lecturer in the Department of Computer Science and Engineering at Shahjalal University of Science and Technology, Bangladesh. His research interests include distributed computer systems with an emphasis on in-network data-centric storage of WSNs, peer-to-peer and content delivery networks, and cloud computing. He is a member of the IEEE and reviewer of Elsevier's *Journal of Parallel and Distributed Computing* (*JPDC*). His research works have been published in conferences and peer-reviewed book chapters.

Mohamed H. Ahmed earned a BSc and MSc in electronics and communications engineering from Ain Shams University, Cairo, Egypt, in 1990 and 1994, respectively. He earned a PhD in electrical engineering in 2001 from Carleton University, Ottawa, where he worked from 2001 to 2003 as a senior research associate. In 2003, he joined the Faculty of Engineering and Applied Science, Memorial University of Newfoundland, where he works currently as an associate professor. Dr. Ahmed serves as an editor for *IEEE Communication Surveys and Tutorials* and *EURASIP Journal on Wireless Communications and Networking* (*JWCN*), and as an associate editor for the *Wiley International Journal of Communication Systems*. He served as a guest editor of a special issue on fairness of radio resource allocation, *EURASIP JWCN*, in 2009, and as a guest editor of a special issue on radio resource management in wireless Internet in the *Wiley Wireless and Mobile Computing Journal* in 2003. Dr. Ahmed is a senior member of the IEEE. He served as a co-chair of the Transmission Technologies Track in VTC'10-Fall and the multimedia and signal processing symposium in CCECE'09, and as a TPC member in ICC'11, ICC'10, WCNC'10, Globecom'09, ICC'09, ICC'08, WCNC'08, VTC-F'06, Globecom'04, and others. Dr. Ahmed won the Ontario Graduate Scholarship for Science and Technology in 1997, the Ontario Graduate Scholarship in 1998, 1999, and 2000, and the Communication and Information Technology Ontario (CITO) graduate award in 2000. His research interests include radio resource management in wireless networks, smart antennas, multihop relaying, cooperative communication, and ad hoc and sensor networks. Dr. Ahmed's research is sponsored by NSERC, CFI, Bell/Aliant, and other governmental and industrial agencies. Dr. Ahmed is a registered Professional Engineer (PEng) in the province of Newfoundland, Canada.

Falah H. Ali is a reader in digital communications and director of the Communications Research Group at the University of Sussex. He earned a BSc in electrical and electronics engineering and an MSc in electronic systems from Cardiff University in 1984 and 1986, respectively, and a PhD in communications from the University of Warwick in 1992. From 1992–1994, he was a postdoctoral research fellow at the University of Lancaster. In 1994, he joined the University of Sussex as a lecturer in electronics engineering and in 2000 was promoted to senior lecturer at the same university. Dr. Ali has more than 20 years of research experience, has published numerous papers, and acted as the principal supervisor for many projects. His research interests are in the areas of mobile communications and wireless networks. He is a fellow of IET, senior member of IEEE, and chartered engineer.

Habib M. Ammari is an associate professor in the Department of Computer and Information Science at the University of Michigan-Dearborn and the founding director of the Wireless Sensor and Mobile Ad-hoc Networks (WiSeMAN) Research Lab at the University of Michigan-Dearborn. He earned his second PhD in computer science and engineering from the University of Texas at Arlington in May 2008, and his first PhD in computer science from the Faculty of Sciences of Tunis in December 1996. He has published his work in prestigious journals such as *IEEE TC*, *IEEE TPDS*, and *ACM TAAS*. He published his first Springer book, *Challenges and Opportunities of Connected k-Covered Wireless Sensor Networks: From Sensor Deployment to Data Gathering*, in August 2009. Dr. Ammari has received several prestigious awards, including the Lawrence A. Stessin Prize for Outstanding Scholarly Publication from Hofstra University in May 2010, the Best Paper Award at EWSN in 2008, and the Best Paper Award at the IEEE PerCom 2008 Google PhD Forum. He was the recipient of the Nortel Outstanding CSE Doctoral Dissertation Award in February 2009 and the John Steven Schuchman Award for 2006–2007 Outstanding Research by a PhD Student in February 2008. He received a three-year U.S. National Science Foundation (NSF) Research Grant Award in June 2009 and the U.S. NSF Faculty Early Career Development (CAREER) Award in January 2011. He serves as an associate editor of several international journals, including *ACM TOSN* and *IEEE TC*. He has also served as the program/publicity chair of numerous IEEE and ACM conferences and as a reviewer of several IEEE and ACM *Transactions* journals.

Muhammad Naeem Ayyaz earned his BSc in electrical engineering from the prestigious University of Engineering and Technology, Lahore, Pakistan, and an MS and PhD in electrical engineering with an emphasis on computer engineering from Syracuse University, New York. His research interests span diverse areas including embedded systems, bioinformatics, and computer networks. His research has been published in various respected journals.

Dr. Ayyaz has been part of the faculty of electrical engineering at the University of Engineering and Technology, Lahore, for more than 20 years where he holds the title of professor and is also chairman of the Department of Electrical Engineering. Apart from this, he holds a consultant position at the Al-Khawarizmi Institute of Computer Science, Lahore, Pakistan.

Mohammad Abdul Azim earned his PhD in electrical and information engineering at the University of Sydney. After completing his PhD, he worked in the Malaysian Institute of Microelcectronic Systems (MIMOS), Malaysia, Institut National de Recherche en Informatique et en Automatique (INRIA), France, as a researcher, and Memorial University of Newfoundland

(MUN), Canada as a postdoctoral fellow. His work involves energy-efficient routing, clustering, aggregation, localization, outlier-detection, and cooperative communications for WSNs, path selection algorithms for mobile multihop relays in Worldwide Interoperability for Microwave Access (WiMAX) networks, and wireless local area network (WLAN) security. Dr. Azim is actively involved as a member of the technical program committee of various international conferences such as IEEE WCNC, ICC, and PIMRC, and also a regular reviewer of various journals and conferences in the area of wireless networking and protocols.

Rabia Bilal is currently doing her PhD in biomedical engineering at the University of Sussex, United Kingdom. She earned a BS in electronic engineering and an MS in telecommunication with a specialization in electronic engineering in 2002 and 2005, respectively. She has published a number of papers and written a book chapter. Her research interests are in the areas of body area sensor networks, breast cancer detection systems, and image processing.

Carlene Campbell earned her PhD in the School of Engineering and Information Sciences at Middlesex University, United Kingdom. She completed her MSc in telecommunications and computer network engineering at London South Bank University, United Kingdom, her postgraduate certificate in higher education professional practice at Coventry University, United Kingdom, and her BSc in computer and management studies at the University of Technology, Jamaica. Dr. Campbell currently lectures on business information technology (BIT) and systems security at the Computing Department, Faculty of Engineering and Computing, Coventry University, United Kingdom. Her current research interests include wireless communications, computer/communications networks and security, and information systems.

Bhawani Shankar Chowdhry is presently working as a dean of Faculty of Electrical Electronics and Computer Engineering and has been serving as a chairman in the Electronics Department, MUET, for the past 18 years. He is a pioneer of two other departments, Telecommunication Engineering and Biomedical Engineering, in the same university. He also completed a 1-year postdoctoral fellowship from the School of Electronics and Computer Science, University of Southampton, United Kingdom. He participated in various workshops at ICTP, Trieste, Italy, as a regular associate of ICTP from 1996 to 2011 and has earned five PhDs (the first one was in electronics/ICT in Pakistan) and supervised more than 50 MPhil/master's theses in the area of ICT. His research has been published in over 60 national and international journals and IEEE, and ACM proceedings. Dr. Chowdhry was the recipient of the HEC University Best Teacher Award in 2001 (awarded by the Federal Minister of Education), the National Cultural Award in 2002 in recognition of achievements in the field of engineering (awarded by the Federal Minister of Culture, Tourism, and Minorities Affairs in September 23, 2002), and the Presidential Highest Academic Distinction Award (Izaz-e-Fazeelat) in 2009.

M. N. Doja is a professor in the Department of Computer Engineering, JMI Central University, Delhi, India. He earned his BSc (Engg.) from the Birla Institute of Technology, India, MTech from the Indian Institute of Technology, Delhi, India, and a PhD from JMI Central University, Delhi, India. He has more than two decades of academic, research, and training experience in the fields of computer science and information technology. His research areas include computer networks, mobile wireless networks, network security, artificial intelligence, and soft computing. Dr. Doja has guided research students and postgraduate students in the fields of communication networks, ad hoc networks, and soft computing technologies. He is the author of various books and has over 100 publications in different journals and conferences of national and international repute. He is the chair and co-chair for various technical conferences held at national and international levels.

Wan Du earned a MEng in communications and information systems from Beijing University of Aeronautics and Astronautics, China, and a PhD in integrated electronics from Ecole Centrale Lyon in the Lyon Institute of Nanotechnology Laboratory. Dr. Du's research interests include modeling and performance evaluations of WSNs, medium access control (MAC), routing protocols for WSNs, distributed embedded system development, and radio frequency identification (RFID) systems.

A. K. Dwivedi is currently a PhD research scholar at the School of Studies in Computer Science and I.T., Pandit Ravishankar Shukla University, Raipur, Chhattisgarh, India. He earned his MPhil from Annamalai University, Chidambaram, Tamil Nadu, India in computer science, and his MSc in computer science from M.C.R.P. University, Bhopal, Madhya Pradesh, India. He has published more than 15 research contributions. He is member of the International Association of Engineers (IAENG) and the Association of Computer Electronics and Electrical Engineers (ACEEE). His current research interests are in WSNs and next generation heterogeneous wireless networks.

Muhammad Farooq-i-Azam earned his BSc in electrical engineering from the prestigious University of Engineering and Technology, Lahore (Taxila Campus), Pakistan, and MSc in computer science from the University of the Punjab, Lahore, Pakistan. By serving in various engineering positions in reputed organizations, he has accumulated experience in diverse areas such as planning, design and administration of computer networks, and design and development of digital circuits. He also has extensive work experience with computer networks and UNIX-based systems, Solaris, VAX/VMS machines, and various distributions of Linux.

He is a member and project administrator of an open source project, IPGRAB (www.sourceforge .net), which is a lightweight packet sniffer that is distributed with Debian Linux and originally authored by Mike Borella. He founded an information and computer security company, ESecurity, and has organized an annual information security event, CHASE, in Pakistan for the past few years. He is also part of the faculty in the Department of Electrical Engineering, COMSATS Institute of Information Technology, Lahore, Pakistan.

A. C. M. Fong holds four degrees in electrical engineering and computer science from Imperial, Oxford, and Auckland. He is currently a professor in the School of Computing and Mathematical Sciences, Auckland University of Technology. Prior to that, he was an associate professor in the School of Computer Engineering, Nanyang Technological University. His research interests include information engineering and communications. He serves on the editorial board of five international journals and organizing committee of numerous international conferences, and is a chartered engineer registered in the United Kingdom.

Mihai Galos earned his MSc at the University of Polytechnics, Timisoara in 2009, from the Department of Computer Science, majoring in computer hardware. He is pursuing his PhD at Ecole Centrale de Lyon, France. He is interested in embedded compilers, instruction set simulators, network simulators, and WSNs.

Nicholas Gomes is a student at Hofstra University studying computer science. He aspires to be a software engineer working in the video game industry. During a summer research assistantship, he worked with a team of students to implement protocols on sensor nodes. His work is reflected in his scientific articles and book chapters.

Mark A. Gregory became a member of IEEE in 1982 and a senior member in 2006. He earned a PhD and a MEng from RMIT University, Melbourne, in 2008 and 1992, respectively, and a bachelor of engineering (Hons) from the University of New South Wales, Sydney, in 1984.

He is a senior lecturer in the School of Electrical and Computer Engineering, RMIT University, Melbourne. Research interests include fiber optic network design and operation, wireless networks, security, privacy, and technical risk.

Dr. Gregory is a fellow of the Institute of Engineers Australia, has reviewed journal papers for the IEEE Engineering Management Society, and is an associated editor of the *Australasian Journal of Engineering Education*.

William I. Grosky is currently a professor and chair of the Department of Computer and Information Science at the University of Michigan-Dearborn. Before joining UMD in 2001, he was a professor and chair of the Department of Computer Science at Wayne State University as well as an assistant professor of information and computer science at the Georgia Institute of Technology in Atlanta. His current research interests are multimedia information systems, text and image mining, and the semantic web. He is a founding member of Intelligent Media LLC, a Michigan-based company whose interests are in integrating the new media into information technologies.

Dr. Grosky earned his BS in mathematics from MIT in 1965, his MS in applied mathematics from Brown University in 1968, and his PhD from Yale University in 1971. He has given many short courses in the area of database management for local industries and has been invited to lecture on multimedia information systems worldwide. He also serves on many database and multimedia conference program committees, was an editor-in-chief of *IEEE Multimedia*, and is currently on the editorial boards of 18 journals.

Noreen Imran is a doctoral candidate at Auckland University of Technology, Auckland, New Zealand. She earned her MS in communication and networks from Iqra University, Karachi, Pakistan, and MS in computer science from Bahria University, Karachi, Pakistan in 2007 and 2002, respectively. She was employed at Federal Urdu University of Arts, Science and Technology, Gulshad-e-Iqbal Campus Karachi as an assistant professor. Prior to that, she worked for more than five years as a lecturer at the COMSATS Institute of Information Technology, Wah Campus, Pakistan. Her research interests include mobile ad-hoc networks, security in wireless multimedia sensor networks, distributed video coding, and image and video watermarking.

Matthew Jacques has attended Hofstra University as a computer science major since the fall of 2007. In the summer of 2011, he interned at WiSeMAN Research Lab, where he worked on research involving wireless sensor networks. Matthew continues to do research for the computer science department at Hofstra Univerity while working on his graduate degree.

Bilal Muhammad Khan is an associate lecturer and visiting research fellow at the University of Sussex, United Kingdom. He earned a BS in electronic engineering and MS in telecommunications with a specialization in electronic engineering in 2002 and 2005, respectively, and a PhD in wireless communication and controls from the University of Sussex United Kingdom in 2011. Since 2011, Dr. Khan has been involved in various projects on the design of computer networks, programmable logic controllers (PLC), field-programmable gate arrays (FPGA), microcontrollers, systems administration, and software training. He has published a number of papers and written many book chapters. His research interests are in the area of WSNs, wireless local area networks (WLANs), WiMAX, and PLCs.

Fatemeh M. Kiaie earned her BSc and MSc in electronics and communications engineering from Iran in 2004 and 2008, respectively. She is currently working toward a PhD in electrical and computer engineering at Memorial University of Newfoundland, Canada. Her research interests include WSNs, cooperation communication, collision minimization, and routing protocols.

M. Bala Krishna earned his bachelor of engineering (BE) in computer engineering from Delhi Institute of Technology (presently Netaji Subhas Institute of Technology), University of Delhi, India, and master of technology (MTech) in information technology from the University School of Information Technology, GGS Indraprastha University, Delhi, India. He is presently working as an assistant professor at the University School of Information Technology, GGS Indraprastha University, Delhi, India. He had earlier worked as a senior research associate and project associate at the Indian Institute Technology, Delhi, India, in the areas of digital systems and embedded systems. He also worked on the projects related to communication networks. His teaching and research areas include computer networks, wireless networking and communications, mobile computing, and embedded system design. Currently, he is working in the area of wireless ad hoc and sensor networks and mobile and ubiquitous computing.

Kok-Keong Loo (Jonathan Loo) earned his MSc in electronics (with distinction) from the University of Hertfordshire, United Kingdom in 1998 and his PhD in electronics and communications from the same university in 2003. Currently, he is a reader (associate professor) at the School of Engineering and Information Sciences, Middlesex University, United Kingdom. He leads a research team in the area of communication and networking. His research interests include network architecture, communication protocols, network security, embedded systems, video coding and transmission, wireless communications, digital signal processing, and optical networks. Dr. Loo has successfully graduated 11 PhDs as the director of studies, and is currently supervising 8 PhD students in the above specialist areas. To date, he has been published in over 145 guest editorials, book chapters, journals, and conferences.

Bruce Maxim is an associate professor of computer and information science at the University of Michigan-Dearborn. His research interests include software engineering, human–computer interaction, game design, artificial intelligence, and computer science education. He has published a number of papers on the animation of computer algorithms, game development, and educational computing applications. He is the coauthor of a best-selling introductory computer science text and web content to support the world's most popular software engineering text. His recent research activities have been in the area of serious game development.

Dr. Maxim is the architect of the ABET accredited computer science curriculum and the ABET accredited software engineering curriculum at the University of Michigan-Dearborn. He is the winner of both distinguished teaching and distinguished community service awards.

Natarajan Meghanathan is currently an associate professor of computer science at Jackson State University, Jackson, Mississippi. He graduated with a PhD in computer science from the University of Texas at Dallas in 2005. He has published more than 125 peer-reviewed articles in several international journals and conference proceedings, and his research has been funded through the U.S. National Science Foundation and the Army Research Lab. He serves as the editor-in-chief of two international journals and is an active member on the editorial boards of more than 10 journals as well as in the organizing and technical committees of several international conferences. His research interests are in the areas of wireless ad hoc networks, sensor networks, software security, and computational biology. He was recently recognized as the Best Faculty Honoree from Jackson State University by the Mississippi State Legislature at their annual HEADWAE luncheon for the academic year 2010–2011.

Fabien Mieyeville graduated from Ecole Centrale de Lyon, France, in 1998 and earned a PhD in integrated electronics at the same institution in 2001. Since 2002, Dr. Mieyeville has been an associate professor at the Institute of Nanotechnology of Lyon, a CNRS laboratory at Ecole Centrale de Lyon. His primary research interests include hierarchical design methodologies for hardware and software heterogeneous distributed systems, particularly WSNs.

Brian Moore is affiliated with Coventry University as a senior lecturer. He is currently the module leader for the ethical hacking and network security courses in the Computing Department, Faculty of Engineering and Computing. He received his BSc (Hons) in the School of Computing Science at the University of Glasgow, United Kingdom. His research interests are in ethical hacking, computer/information security, biometrics, and computer forensics.

David Navarro earned his PhD in 2003 in microelectronics and systems. He is currently an associate professor at Ecole Centrale Lyon in the Lyon Institute of Nanotechnology Laboratory. His research interests are electronic systems design and modeling. Dr. Navarro mainly focuses on WSNs, complementary metal–oxide–semiconductor (CMOS) image sensors, and embedded green computing.

Athanasios D. Panagopoulos earned his degree in electrical and computer engineering and PhD from the National Technical University of Athens (NTUA) in July 1997 and in April 2002, respectively. From May 2002 to July 2003, he served in the Greek Army in the Technical Corps. From September 2003 to December 2008, he was a part-time assistant professor in the Higher School of Pedagogical and Technological Education. From January 2005 to May 2008, he was head of the Wireless and Satellite Communications Department in the Hellenic Authority of Information Security and Communication Privacy. Since May 2008, he has been a lecturer in the School of Electrical and Computer Engineering of NTUA. Dr. Panagopoulos has published more than 200 papers in international journals, book chapters, and conference proceedings. He has been involved in numerous R&D projects funded by the European Union. His research interests include mobile computing technologies, radio communication systems design, and wireless and satellite communications networks. He is a senior member of IEEE, and serves as an associate editor in *IEEE Communication Letters* and *IEEE Transactions on Antennas and Propagation*.

Rodrigo Roman is a security researcher working at the I2R in Singapore. He also collaborates with the NICS security lab at the University of Malaga, Spain, where he earned his PhD in computer science in 2008. At present, his research interests are mainly focused on the secure integration of sensor networks with other infrastructures, such as critical infrastructures, cloud environments, and the Internet of Things. Dr. Roman is actively involved in the academic community, having published over 25 refereed papers at international conferences and journals, and having organized and chaired several workshops and conferences (e.g., ESORICS, ACNS, SecIoT). In addition, he has participated in various Spanish (ARES, SPRINT) and international (Feel@Home, SMEPP) research projects related to network and sensor networks security.

Vasileios K. Sakarellos earned a 5-year engineering degree in electrical and computer engineering from the Aristotle University of Thessaloniki, Greece, in 2004 and a PhD in wireless cooperative telecommunications from the National Technical University of Athens, Greece, in 2010.

His scientific interests are in the field of channel modeling, wireless link design, cooperative diversity techniques in fading channels, and wireless communication network analysis. Dr. Sakarellos has been awarded with the K. Karatheodoris Fund and the Fundamental Research

2009 Fund from the National Technical University of Athens, Greece. He has published 20 scientific articles in international refereed journals and proceedings of international conferences, and is a member of the IEEE and a member of the Technical Chamber of Greece.

Boon-Chong Seet earned his PhD in computer engineering from Nanyang Technological University, Singapore, in 2005. Upon graduation, he was employed as a research fellow under the Singapore–Massachusetts Institute of Technology Alliance (SMA) program at the National University of Singapore, School of Computing. In 2007, he was awarded a visiting scholarship to the Technical University of Madrid, Spain, to pursue research under an EU-funded project on multidisciplinary advanced research in user-centric wireless network enabling technologies (MADRINET). Since December 2007, Dr. Seet has been with the Auckland University of Technology, New Zealand, where he is currently a senior lecturer in its Department of Electrical and Electronic Engineering. He was also a visiting faculty at the University of British Columbia, Canada. He has served as a guest editor for special issues in *IEEE Wireless Communications Magazine* and the ACM/Springer *Journal of Personal and Ubiquitous Computing*. He is a member of ACM and a senior member of IEEE.

Jaydip Sen has 18 years of experience in the fields of networking, communication, and security. He has worked for reputed organizations such as Tata Consultancy Services, India, Oil and Natural Gas Corporation Ltd., India, Oracle India Pvt. Ltd., and Akamai Technology Pvt. Ltd. His research areas include security in wired and wireless networks, intrusion detection systems, secure routing protocols in wireless ad hoc and sensor networks, secure multicast and broadcast communication in next generation broadband wireless networks, trust- and reputation-based systems, QoS in multimedia communication in wireless networks and cross-layer optimization-based resource allocation algorithms in next generation wireless networks, sensor networks, and privacy issues in ubiquitous and pervasive communication. He has more than 90 publications in reputed international books, journals, and refereed conference proceedings. He is a member of ACM and IEEE.

Howard Senior received his bachelor of business administration (Hons) at the School of Business and Management at the University of Technology, Jamaica. He is currently an advertising executive and research analyst at OGM Integrated Communications company, one of Jamaica's leading advertising agencies. He has vast experience in various management capacities across a number of industries, working in civil service, banking and finance, tourism, hospitality, and marketing research. His current interests include the impact and implications of information and communication technologies, especially wireless communications, on businesses across industries.

Faisal Karim Shaikh earned his MEng from Pakistan and PhD from TU Darmstadt, Germany. He is currently working as an assistant professor at Mehran University of Engineering and Technology, Jamshoro, Pakistan. Dr. Shaikh served as TPC chair and TPC member for several national and international conferences. He is also an editorial board member of the *International Journal of Ubiquitous Computing*. His research interests include dependable WSNs, mobile ad hoc network (MANETs), VANETs, and body area networks.

Dimitrios Skraparlis received his PhD from the Department of Electrical and Computer Engineering, National Technical University of Athens, Greece, in 2009, MSc in communications and signal processing from University of Bristol, United Kingdom, in 2003, and the 5-year electrical and computer engineering degree from Aristotle University of Thessaloniki, Greece, in 2002.

He is currently with Nokia Siemens Networks, Athens, Greece, working on Packet Core solutions for 4G telecommunication systems. From June 2003 to October 2004, he was with Toshiba TREL, United Kingdom, working on multiple-input multiple-output (MIMO) communications technology. He has also completed internships with Infineon Technologies, Munich, Germany, and IBM Research, Zurich, Switzerland.

Dr. Skraparlis has filed six patent applications worldwide related to wireless communications technology. He has also published 20 scientific papers in international refereed journals and conference proceedings and as book chapters.

For his work, he has been awarded a sponsorship from Toshiba TREL, United Kingdom, the K. Karatheodoris Fund and the Fundamental Research 2009 Fund from the National Technical University of Athens, Greece. His current research interests include MIMO wireless communications and multiuser diversity techniques, applied statistics, signal processing and transceiver architectures, as well as applied cryptography.

Arijit Ukil is currently working at Innovation Labs, Tata Consultancy Services Ltd., Kolkata, as a scientist. He is primarily engaged with research activity on ubiquitous computing, security, and privacy. Before joining Tata Consultancy Services Ltd in 2007, he worked as a scientist in Deference Research and Development Organization (DRDO), India, for 4 years, where his primary focus areas were signal processing, embedded systems, and wireless communication for radar applications. He earned his BTech in electronics and telecommunication engineering in 2002 and is currently pursuing a PhD. He has published more than 30 conference and journal papers, and three book chapters in IGI-Global, Intech-web publishers. He has been a reviewer of a number of IEEE conferences, such as IEEE VTC and IEEE WCNC. He has been invited to and delivered keynote lectures and tutorials in ETCC'08, ICCET'09, NCERDM-IT'09, NCETAC2010-IT, IWCMC'10, and ICCC-2011. He is listed in 2010 Marquis' *Who's Who*.

O. P. Vyas earned his MTech from IIT Kharagpur, West Bengal, India, in computer science. He received his PhD from IIT Kharagpur in joint collaboration with Technical University of Kaiserslautern (Germany) in computer networks. Dr. Vyas is currently a professor at the Indian Institute of Information Technology, Allahabad, Uttar Pradesh, India, with the additional responsibilities of program coordinator (Software Engg.) and in-charge officer (doctoral section). He is an active researcher and has published more than 80 research papers and 3 books, and completed one Indo-German Project under DST-BMBF. His current research interests are in data mining and knowledge engineering, new generation networking, and social network analysis and mining.

Tabassum Waheed earned her bachelor of engineering degree in computer systems engineering from NED University of Engineering and Technology, Karachi. She did her master's of engineering in computer systems engineering from the same university in 2006. She is currently pursuing a PhD from Hamdard University, Karachi, in computer engineering. She has taught undergraduate/graduate engineering courses at leading engineering universities of Pakistan and is presently associated with the Usman Institute of Technology Computer of Engineering Department as an assistant professor. She is also a certified corporate trainer and has delivered many vendor-specific training sessions for IT professionals. Her research interests are in WSNs, vehicular ad hoc networks (VANETs), wireless body area networks (WBANs), network security, and quality of service (QoS) provisioning.

Wai-Choong Wong is a professor in the Department of Electrical and Computer Engineering, National University of Singapore (NUS). He is currently the deputy director at the Interactive and Digital Media Institute (IDMI) in NUS. He was previously the executive director of the I2R from November 2002 to November 2006. Since joining NUS in 1983, he has served in various positions at the department, faculty, and university levels, including head of the Department of Electrical and Computer Engineering from January 2008 to October 2009, director of the NUS Computer Centre from July 2000 to November 2002, and director of the Centre for Instructional Technology from January 1998 to June 2000. Prior to joining NUS in 1983, he was a member of the technical staff at AT&T Bell Laboratories, Crawford Hill Lab, New Jersey, from 1980 to 1983.

Dr. Wong received his BSc (1st class Hons) and PhD in electronic and electrical engineering from Loughborough University, United Kingdom. His research interests include wireless networks and systems, multimedia networks, and source-matched transmission techniques with over 200 publications and 4 patents in these areas. He is a coauthor of the book *Source-Matched Mobile Communications*. He received the IEEE Marconi Premium Award in 1989, NUS Teaching Excellence Award (1989), IEEE Millennium Award in 2000, the e-nnovator Awards 2000, Open Category, and Best Paper Award at the IEEE International Conference on Multimedia and Expo (ICME) 2006.

David Yoon is an associate professor in the Department of Computer and Information Science at the University of Michigan-Dearborn. His research interests include the integration of CAD and CAM, NC machining, and distributed computing. He served as an associate editor of the *International Journal of Modelling and Simulation* and currently serves on the editorial board of *Computer-Aided Design and Applications*.

Jianying Zhou is a senior scientist at the Institute for Infocomm Research (I2R) and heads the Network Security Group. He earned his PhD in information security from the University of London, MSc in computer science from the Chinese Academy of Sciences, and BSc in computer science from the University of Science and Technology of China. Dr. Zhou's research interests are in computer and network security, mobile, and wireless communications security. He is a founder of the International Conference on Applied Cryptography and Network Security (ACNS).

WIRELESS SENSOR NETWORKS: STORAGE ISSUES AND APPLICATIONS

I

Chapter 1

Review of Applications of Wireless Sensor Networks

Habib M. Ammari, Nicholas Gomes, William I. Grosky, Matthew Jacques, Bruce Maxim, and David Yoon

Contents

1.1 Introduction

Sensor technology has been widely used in a variety of domains dealing with monitoring, such as health monitoring, environmental monitoring, and seism monitoring; control, such as agriculture control; and surveillance, such as battlefield surveillance [1]. A wireless sensor network (WSN) is composed of tiny, battery-powered devices, called sensor nodes. A sensor node has two components. The first one, named mote, is responsible for storage, computation, and communication. The second component, called sensor, is responsible for sensing physical phenomena such as temperature, light, sound, and vibration, to name a few. A sensor is always attached to a mote. Sensor nodes collect data and may perform in-network processing on the collected data at intermediate nodes before forwarding it to a central collection point, called the sink (or base station), for further analysis and processing.

The design and implementation of WSNs face several challenges, mainly due to the scarce resources and limited capabilities of sensor nodes such as battery power (or energy), bandwidth, storage, processing, sensing, and communication. To accomplish their task successfully, the sensor nodes are required to communicate with each other and act as intermediate relays to forward data on behalf of others so that it reaches the sink in a timely manner. Depending on the nature of deployment field and the application requirements, the sensor nodes may be densely or sparsely deployed. In addition, they may be deterministically or randomly deployed. In this type of network, energy is the most critical factor for the effectiveness of the underlying WSN. In fact, in hostile environments, such as battlefields, it is difficult or even impossible to access the sensor nodes and recharge or renew their batteries. Thus, energy poses a serious problem for network designers, especially in this type of environment. Furthermore, the sensor nodes are fragile and unreliable, thus making them unable to function properly. To cope with these severe problems, the sensor nodes are in general densely deployed. Therefore, all algorithms and protocols developed for sensor nodes should be as energy efficient as possible to extend the lifetime of the individual sensors, thus prolonging the operational network lifetime as long as possible.

In this chapter, we review various WSN applications spanning different domains, namely health care, agriculture, environment, industry, and military. The remainder of this chapter is organized as follows: Section 1.2 discusses the problems of setting up WSNs for the health-care domain and describes the key applications. Section 1.3 presents some applications from the agriculture domain. Section 1.4 gives an overview of application from the environmental domain. Section 1.5 provides a summary of applications from the industry domain, while Section 1.6 focuses on applications from the military domains. Section 1.7 is devoted to discuss newly deployed and ongoing applications of WSNs. Section 1.8 concludes this chapter.

1.2 Health-Care Applications

The health-care field is always looking for more efficient ways to provide patients with the best and most comfortable care possible. Thus, it is no surprise that several health professionals are excited about the prospect of using WSNs to monitor and treat patients [2,3]. Although the specific technologies required to make efficient and reliable home health monitoring systems commonplace in the health-care industry are not yet available, there is much research being done into the subject. From home health monitoring to improving clinical trials, WSNs could soon be heavily integrated into the medical field. Not only can WSNs help the health-care industry provide a new level of care but they could also drastically cut costs. In fact, there are predictions that by 2012, the industry could save almost $25 billion if WSNs are integrated into the health-care system [2].

In this section, we examine exactly how WSNs can be used for health care. Also, we discuss why they can possibly revolutionize the way patients are cared for in the near future.

1.2.1 Wireless Sensor Network Possible Setup

There are many ways that an effective home health monitoring system can be set up using WSNs. As an example of how this can be achieved, we take a brief look at a network that can be built using a three-tier architecture, as shown in Figure 1.1. The first tier would be the actual sensors that the patient has attached to (or possibly one day implanted in) his or her body. The network will work in a basic master–slave configuration, where all the sensors (slave nodes) will report their data to one central node (master). The latter will help process all the raw data that is being collected. Ideally, the sensor nodes should be as small and convenient as possible. Moreover, they should have an exceptionally long battery lifetime. Thus, it may be best to deploy these sensor nodes (or slaves) in a way that they simply collect the data, while the central node (or master) is equipped with a permanent power source and is responsible for data processing and computing [4]. The second tier can preprocess the data received and send its results to the patient medical caregivers. There are two ways to design this tier. The first approach requires that this tier be implemented on a PC when

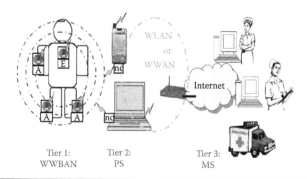

Figure 1.1 Telemedical system for health monitoring. (Data from Pan, J. et al., *Proceedings of the International Conference on Internet Computing Science and Engineering*, 160–5, 2008.)

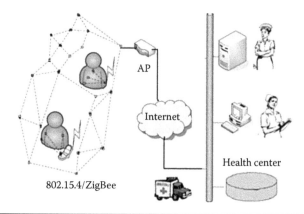

Figure 1.2 Architecture for in-community health monitoring system. (Data from Pan, J. et al., *Proceedings of the International Conference on Internet Computing Science and Engineering*, 160–5, 2008.)

there is a large amount of data and/or processing to be done. The second approach suggests the use of some sort of cellular devices that would allow the WSN to continue collecting and sending the data even if the patients leave their homes [4]. The third tier consists of the patient and the medical caregiver, and the data received has to be stored in the patient's records. This data can be analyzed by a medical professional or perhaps by a computer to look for abnormalities. Also, this tier can be built to directly contact emergency services if the incoming data indicates that a patient is in danger [4]. There are numerous ways in which this network can be built. Figure 1.2 shows an architecture for in-community health monitoring system. This is just a simple example showing how WSNs can be powerful tools for the health-care industry. Next, we look at some specific examples of how these networks can be used to improve health care.

One important thing to note about the server that will be collecting and analyzing this data is that to work effectively, it will have to be running some fairly complex software. Also, if this server is monitoring several patients, it will most likely need to be very powerful to manage all the data efficiently enough to indicate whether one of those patients under its "care" is in distress. In fact, this system should be able to tell with near absolute certainty if a patient needs medical attention. On the one hand, it would be useless if this system constantly contacts emergency medical services with claims that patients were in need of immediate attention, when in fact they were fine. On the other hand, if it is too rigid with its "red flags," then patients may die without medical personnel ever knowing anything was wrong. To prevent these two occurrences, it is necessary that the software be able to build up a medical history for the patients. Everyone has at least slightly different body chemistry, and vital signs that could look fatal to someone at first glance might be normal to another person. Because of this, the server must be aware of the patient's current medical trends. Also, instead of having a general list of "rules" as to when the patient needs medical attention, the software needs to make intelligent decisions based on trends it has seen as well as the patient's known history. As can be seen, this software, which will most likely be implemented on the central medical server, will be the most important and complex part of the system [4].

1.2.2 In-the-Body/Remote Monitoring

The medical field has also found many implementations for sensor networks. Applications range from monitoring equipment and patient information to remote monitoring over both short and long distances. Sensor networks have quite a few applications within a hospital. They can be attached to critical supplies, medications, and machinery to monitor depleted supply and location. This ensures that supplies and medications are never expended without ample notification and that supplies are not misplaced or sent to the wrong place. This also enables speedy access to instruments such as defibrillators that are generally only required during time-critical procedures such as resuscitation.

Sensors can also be affixed in the rooms of patients who are under close monitoring by doctors. Sensors can monitor movement and report on the movement patterns of the patient. They can also be attached to devices to report data such as heart rate, blood pressure, and medication distribution. New research is also affording ways to attach sensors directly to the patient's body to detect and report more vital signs in a faster manner. In addition to that, research is being done that will enable patients to wear a sensor on a wristband that reports the patient's full medical history, making it easier for doctors to administer treatments and forewarn them about allergies and bad drug combinations. Figure 1.3 shows the Pluto custom wearable mote [5].

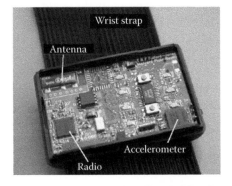

Figure 1.3   The Pluto custom wearable mote. (Data from Shnayder, V. et al., Technical Report TR-08-05, Division of Engineering and Applied Sciences, Harvard University, Cambridge, MA, 2005.)

1.2.3 Disaster Recovery

The pervasiveness of sensor networks continues into applications in disaster recovery. *CodeBlue* [5] is the most prominent recovery and response system developed by a research group from Harvard University. As its name indicates, *CodeBlue* is a network of sensors embedded in wristbands that can be attached to human beings. These sensors collect data pertinent to doctors and emergency responders of patients in need of intensive care. This data includes heart rate, pulse oximeter readings, EKG monitoring, motion capture, and location of person. These nodes can be attached to people being rescued from disasters to keep track of their vital signs and their locations.

CodeBlue was developed on three different types of motes: MICA2, MICAz, and Telos [6]. It was implemented in TinyOS and makes use of protocols for wireless networking and routing as well as end-user device support for laptops and personal digital assistants (PDAs). It operates on a completely open framework that uses a *publish/subscribe routing framework*. A pulse oximeter was mounted to the MICA2 mote and was equipped with an unobtrusive finger sensor attached to the DB9 connector to the mote. EKG monitoring circuitry was mounted to the Telos motes. This circuitry contained three small electrode leads. These leads were then attached to bodies to monitor heart activity. These sensors can collect, save, and transmit data wirelessly and can be used to provide quicker transfer from emergency responders to doctors by having all information readily available.

The most peculiar application of *CodeBlue* is the motion analysis sensor board. This sensor does so much more than monitoring location and movement. It is equipped with a three-axis accelerometer and a single-axis gyroscope to closely monitor muscle movement. This can be used to monitor muscle movements in patients with Parkinson's disease and help doctors prescribe medications and rehabilitation more tailored to an individual's need. It can also help detect when elderly people fall down and are unable to get up and report it to medics. This can be essential to saving a person's life.

1.2.4 Elderly/Chronically Ill

Among several beneficiaries from research in WSNs, there are two demographics, namely the elderly and the chronically ill, which could especially exploit the power and usefulness of this technology. These two categories of people tend to need their health constantly monitored. Because of this, several of them end up requiring prolonged hospital stays or moving into special facilities such as nursing homes [4]. This issue is becoming increasingly important, especially as the world population ages [7]. In fact, by 2025, the number of people in the world over the age of 65 is projected to be double the number in 1990 [8]. Providing proper monitoring can be expensive for their family and may force the patient to move out of their homes because living alone would be too much of a risk to their health. WSNs allow caregivers to keep a watchful eye on the patient's health with much lower cost and without forcing the patient into an unfamiliar environment. Furthermore, WSNs can be helpful to the elderly suffering from memory problems by providing them with advanced features such as helping them locate important objects in their homes or even something as simple as reminders as to when to take their medicine. There are numerous possible functions, such as these, that can be provided to increase the quality of life for elderly and chronically ill people who might not be able to live independently without this technology [7].

1.2.5 Better Diagnosis/Treatment

The field of medical diagnostics is another great potential area for using the sensor technology. If a patient believes they may be ill or experiences a symptom that comes and goes, these sensor nodes will be able to greatly speed up the process of diagnosis. They may even diagnose diseases earlier, leading to more effective treatment of the patients. WSNs allow patients to use various sensors to constantly monitor their vital signs and report the collected data back to the doctor. This will allow medical professionals to detect more subtle changes or trends in the patient's body. These sensors can provide the doctors with a large amount of data about the patients. If the doctors see their patients a few times to examine them, this will undoubtedly help them detect any illness [7]. Sometimes, it may happen that a patient experiences symptoms that come and go, such as sudden headaches or irregular heartbeats. In this case, the doctor will be able to send the patient home and just wait for the episode to occur again. The alternative would be either trying to diagnose the patient on description alone or holding the patient in the hospital until the symptom presented itself again [7].

1.2.6 Sleep Apnea

Home WSNs can be used to efficiently detect and treat one condition called *sleep apnea*. The National Heart Lung and Blood institute defines sleep apnea as "a common disorder in which you have one or more pauses in breathing or shallow breaths while you sleep" [9]. These pauses in breathing can be minutes long and can occur up to 30 times per hour [9]. Since sleep apnea occurs only during sleep, patients do not realize they have it. Once the technology is readily available, a doctor can provide a patient complaining of excessive tiredness with a small, unobtrusive mote to wear to bed. This mote records blood oxygen levels and helps confirm a diagnosis of sleep apnea. Also, it can wake the patient or contact emergency medical help if the patient's oxygen levels fall to a dangerous level. This method would be much less intrusive on a patient than requiring them to spend the night in a sleep lab to confirm the diagnosis [7].

1.2.7 Fall Detection

WSNs have also been tested for detecting fallen patients, primarily for use in nursing homes and hospitals. The system works through a series of omnidirectional cameras that transmit their images back to a central base station. The latter processes the image and attempts to detect if a patient has fallen in the camera's view radius with an accuracy of about 93% [10]. If someone has fallen, the network will alert the medical staff so that they can immediately assist the patient, preventing any further harm to the patient [10]. Although, at the moment, it is best suited for hospital and nursing home use, this technology shows a lot of promise. In fact, the cameras require well-lit conditions for the system to achieve high levels of accuracy [10]. However, this technology can eventually make its way into private homes, allowing elderly citizens to live more independently without worrying about falling and not being able to get up.

1.2.8 Cardiac

Wireless sensors would be a great leap forward for heart health as well. Typically, if a doctor believes that a patient may have occasional abnormal heart rhythms, they would test it by trying to actively stress the patient's body and heart until one of the arrhythmia occurred. This can be very uncomfortable and even possibly dangerous if the patient has serious health issues. Wireless EKG sensors, which can be worn by the patient at home, will constantly record the heart rhythm. Thus, if the arrhythmia occurs again, it will be recorded and sent to the patient's caregiver without any need to keep the patient in the hospital or subject them to a stress test [7].

WSNs have the potential to save many lives by monitoring patient's cardiac health. Having a network of cardiac sensors on patients' bodies can be the difference between life and death for several people, especially those at risk for fatal cardiac arrhythmia or heart attack. With these conditions, time is extremely important. Some arrhythmia can be fatal to the patient less than an hour after the first symptoms occur, and if the patient experiences a ventricular fibrillation, they can have less than 5 minutes to get medical help [8]. WSNs can detect the patient's distress and contact emergency services immediately [4]. Not only would this save valuable time that can be critical for the patient, it is also conceivable to imagine that the wireless network can respond to the call by sending the patient's medical records and real-time data about their current condition to the medical team as it is en route. This can alert them to issues, such as allergies to medications, and cut down on the time they would need to assess the patient's situation when they get to the scene. They can go straight into treatment because they already know the situation. It is obvious that the possible WSN applications in the health-care field are seemingly endless.

1.2.9 Security

Security is a very critical issue when deploying WSNs for monitoring a patient's health. Since the law requires that the patient's medical records be kept confidential, any home health-care system implemented on a WSN should meet at least a minimal level of security [11]. Also, for the system to be effective, doctors need to be able to rely on the fact that the data they are receiving on the patient is unaltered. If there was a large possibility that the data had been tampered with before registering in the medical file, the doctors may not take the medical histories and trends collected by the WSN seriously. For these reasons, a strong security protocol is absolutely necessary for WSNs deployed in the health-care field [8].

One of the more vulnerable areas in this system is the transmission of the data from the sensors to the sink. This is because, as we discussed earlier, one of the major issues with security in WSNs is finding a security protocol strong enough to prevent attack, yet light enough to not use up all of the resources of the sensor node on security. If the connection between the node and the sink is not properly secured, an attacker can send falsified information to the sink node, claiming to be one of the sensor nodes [8]. The attacker can cause the server to believe the patient is in distress and call an ambulance to their house for no reason. Even more maliciously, an attacker can even make the server believe the patient is fine when in fact they do need medical help. Since most models of home health-care networks involve a centralized server that collects and stores patient data for several different people, this server needs to have heavy security measures in place. A hacker gaining access to this server can change data around, potentially putting several people's lives in danger. At the very least, a hacker may have access to thousands of confidential medical records [8].

1.2.10 Conclusion

Although the concept of in-home health care by WSNs has not quite taken hold yet, it seems to inevitably be the future direction of the medical profession. When WSNs become commonplace in the home as electronic health aides, they will be beneficial to both the doctors and the patients. The doctors will be able to better perform their jobs with the extremely large pools of data collected on all of their patients. The latter will enjoy more freedom and privacy and more effective treatment. Also, the patients will benefit from lower costs compared to the scenario of having in-home health aides.

1.3 Agricultural Applications

The agriculture industry can benefit through the use of WSNs. Growing plants for a maximum crop yield means that farmers constantly need to monitor several different elements over large areas of land. There are numerous, important attributes, such as moisture, nutrient, sunlight, and temperature levels, to name a few, which can be monitored using sensors. It is essential to examine some agricultural applications of WSNs in order to get a broad view of the problems that they solve.

1.3.1 Data Collection for Farming

A specific sensor network that has already been developed is one that monitors every single plant in a vineyard. This project from Intel—*The Wireless Vineyard*—works by attaching Berkeley motes to each vine plant in a vineyard [12,13]. These motes are equipped with sensors to monitor things such as temperature, humidity, soil moisture, and the vital signs of each plant. They take readings and report back to the base station every 5 minutes. This enables farmers to understand each vine directly and make watering and lighting decisions to promote the best environment for each vine. Future implementations include plans to automate water and nutrient administrating for each vine directly based on the sensors' readings. Figure 1.4a shows a Ranch Systems node with a wind speed anemometer, while Figure 1.4b shows a Ranch Systems base station.

1.3.2 Moisture Tracking

Moisture tracking is a highly important aspect of agriculture that can easily be automated through the use of a WSN. In fact, well-established farms need accurate water readings to grow crops using the most cost-effective method possible. However, cost is not the only issue as agriculture consumes an estimated 70% of total worldwide water use [14]. Thus, creating efficient irrigation and water consumption is clearly a vital part of agriculture. Sensors that measure volumetric water data at different depths can be attached to motes so that they report their findings at timed intervals. This can help farmers find the source of any water problem to improve their crops' irrigation scheme. One project, summarized by *Francois Depienne*, describes an excellent example of a moisture tracking WSN [15]. In this project, a WSN is programmed and deployed in Bangalore, India, to assist local marginal farmers with irrigation management. By using buried sensors and elevated exposed motes, researchers are able to collect data on soil moisture content. Not only did the data collected help the locals, but it also provided researchers with data to better understand water management problems.

1.3.3 Nutrient Monitoring

Nutrient monitoring allows farmers to effectively spread their fertilizer and maintain crop yields. Similar to moisture tracking, a sensor would be buried in the ground in order to gather the volumetric nutrient data of the soil. The data that it collects would alert farmers about possible over- or underfertilization and reveal flaws in fertilization techniques. This will not only save farmers' money but also cut down on the environmental impact of overfertilization. It is worth mentioning that up-to-date nutrient data can also help farmers plan their crop rotations more efficiently, thus allowing the soil to renew quicker.

(a)

(b)

Figure 1.4   (a) Ranch Systems node with a wind speed anemometer and (b) Ranch Systems base station at Obsidian Ridge Vineyard in Lake County. (Data from Rieger, T., *Vineyard & Winery Management*, March/April:2–6, 2007, accessed from http://www.ranchsystems.com/ssite/FNL-24028%20Ranch%20Systems%20(E).pdf.)

1.3.4 Lighting Monitoring

Effective lighting is essential to growing crops efficiently. Monitoring sunlight or UV light exposure can help farmers ensure that crops are getting sufficient light for growth. This is important especially in farms with artificial lighting or with crops that are highly sensitive to the levels and spectrum of light they receive. In addition to helping with actively farmed plants, WSNs can help researching effective lighting strategies. Motes can track the intensities and spectrum of light that a plant receives. Using this data, researchers can find what strength and wavelength of light promotes the fastest growth for plants.

1.3.5 Temperature Monitoring

It is well known that plants grow optimally when multiple different environmental variables are correct. One of the most important of these variables is temperature. Proper tracking of temperature can mean the difference between bountiful yields and withered crops. WSNs can be deployed in planted fields to constantly measure soil and air temperatures throughout the day. Farmers and researchers can use this information to find out what temperature offers the best growth potential for crops and also where the best temperature spots are in a field.

1.3.6 Herd and Livestock Tracking

Using WSNs can help track herd and livestock activity. By setting up "zones" with the motes and attaching beacons to animals, farmers will be able to see herd movement across different zones. This can help with planning out grazing fields or keeping track of individual animals. In addition, researchers can use data to analyze animal behaviors. Perhaps, each type of animal prefers to feed under certain circumstances or in different environments. In addition, aggression or injury can be tracked allowing farmers to deal with time-sensitive events much more quickly and preventing loss [15].

1.4 Environmental Applications

Because of the growing concerns about human impact on the environment, it is important to analyze the use of WSNs in the environmental industry. WSNs are able to measure a large variety of environmental data for a huge number of applications. In particular, sensor nodes can be deployed to measure several attributes, such as temperatures, accelerations, magnetic fields, sound levels, and other scientific data. Based on this data, researchers can construct efficient, accurate models that can describe how an environment is working.

1.4.1 Habitat Monitoring

The ability to closely monitor the habitats of animals is something that has always been a problem for scientists. It can be almost impossible for researchers to get close enough to study animals without disturbing them, especially if they burrow underground. This has led the way to two applications for WSNs: the *Great Duck Island* network [11] and the *ZebraNet* [16].

Many animals are extremely sensitive to human interaction in their environments. Even the shortest time intervals of human presence can have major negative impacts on the existence of animals. Seabird colonies are especially sensitive to a disturbance in their environments. They often seek refuge on small islands to minimize their interactions with foreign species. One such island is Great Duck Island in Maine. Researchers have long been interested in the breeding patterns of storm petrels, but their concern of what their presence could cause has been a limiting factor. Using a sensor network, Polastre et al. have developed a nonintrusive manner of study.

The *Great Duck Island* network makes use of MICA motes. The motes were equipped with photoresistor, temperature, barometric pressure, humidity, and infrared sensors. These motes were placed in acrylic weatherproof containers and deployed both in and around storm petrel burrows. After deployment, the whole network was able to survive for 6 months and transmitted readings without a problem. Readings were transmitted to a base station on the edge of the island and then back to a laboratory in California through satellite. By using the differences between the ambient heat detected and the infrared heat detected, researchers were able to determine when the birds were occupying their burrows. They also found that humidity and temperature readings within the burrows remained constant while they varied greatly outside. Researchers concluded that the storm petrels remained burrowed for the summer month except for two 2-day excursions to the outside. Figure 1.5 shows a system architecture for habitat monitoring.

ZebraNet was created with the same motivations as the *Great Duck Island* network: researchers do not have knowledge about zebra migration patterns. Personal study would not be a viable option in the scenario either as it would disturb the zebras and possibly cause them to migrate to abnormal areas. Sensors, however, would be the perfect nonintrusive way to study zebras. *ZebraNet* was developed on MICA2 motes with equipped global positioning system (GPS) sensor. These sensors are mounted inside of a light collar that is placed around every zebra's neck. The goal is to take highly accurate position readings and report them back to researchers in a remote location.

The *ZebraNet* system is peculiar due to its unique requirements. Not only do the sensors need to be highly energy efficient due to the power requirements of GPS, but they also need to be able to handle mobility and the chance that they will be sparse in density. They also have to be fairly light in weight. To accommodate all of this, the sensors are equipped with a GPS that integrates with the mote's CPU and flash chip instead of having its own board, as well as a small rechargeable battery pack that scavenges solar energy through the use of small solar cells. The sensor only

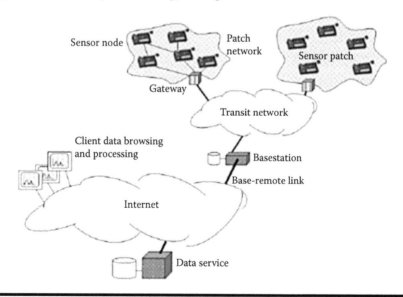

Figure 1.5 System architecture for habitat monitoring. (Data from Mainwaring, A. et al., *ACM International Workshop on Wireless Sensor Networks and Applications*, Atlanta, GA, 2002.)

senses once every 8 minutes and has a total alive time of less than 1 minute. It does not transmit this data right away, but rather saves it to the flash chip. Periodically, a mobile manned base station comes within the zebras' region and alerts the sensors. The sensors then transmit all of the data they have collected since the last encounter with a base station and then purge the flash RAM. When deployed, *ZebraNet* proved to provide a high data recovery rate with detailed and accurate positioning of the zebras, which was useful to researchers.

1.4.2 Conference Room Monitoring

In today's modern business world, many offices are laid out with cubicles. This construction promotes openness among employees and enables quick collaboration and accessibility. However, the biggest problem with this type of environment is that it hinders the possibility of holding an impromptu meeting or extensive discussion due to the disruptive potential it holds. To cater to the needs of employees to hold meetings, office buildings are generally equipped with private meeting rooms [17].

Although they can be efficient and useful, these private meeting rooms carry problems by nature. Generally speaking, a meeting room must be reserved weeks and sometimes months in advance of when the meeting actually takes place. For employees to hold on-the-spot discussions, they have to scour floors of the office building in the hopes of finding a free room. This creates a great inconvenience. However, Conner et al. have found a way to make this process exceptionally faster in a very peculiar way. One seemingly trifling aspect of most conference rooms is that they are equipped with motion sensors that automate the light fixtures in these rooms, as shown in Figure 1.6a. If they sense motion, the lights are turned on. Otherwise, the lights are turned off. Conner et al. [17] found a way to integrate these motion sensors with sensor nodes to create a simple yet powerful way to make full use of these meeting rooms.

This sensor network works as follows: when the motion sensor determines that the room is empty, it turns off the lights in the room. The sensor nodes sense when this happens and report the room as empty to the gateway node. When all the gateway nodes report to the sink, the sink compiles information on the occupancy status of all rooms. This is then posted on the Internet to disburse this information in a ubiquitous fashion. This information can then be accessed by all employees either by a computer or through PDA notifications, thus saving time and allowing unscheduled meetings to take place. These sensors also obtain reservation information from the

(a) (b)

Figure 1.6 **(a) Motion sensor node and (b) reservation status indicator. (Data from Conner, W. S. et al., in *Wireless Sensor Networks: A Systems Perspective*, ed. N. Bulusu and S. Jha, 289–307, Artech House, Inc., 2005.)**

main server. This information can be viewed directly from the sensor itself. It displays a series of LED lights to indicate top-hour, bottom-hour, and no reservation. It also saves trend information for both the specific meeting room and the specific people who hold meetings. It compiles this information and provides a trend prediction for room usage. Figure 1.6b shows a reservation status indicator, which allows mobile users to obtain the status of nearby conference rooms directly.

1.4.3 "Follow-Me" Application

Many large places, such as hospitals and office buildings, can become difficult to navigate. Although there may be maps and computer kiosks, people can still be lost while trying to find a specific room. Wang et al. have integrated a sensor application entitled the "Follow-Me" guidance system [38] in an attempt to tackle this problem. This system includes nodes with LED lights equipped to "light the way" for people. Figure 1.7 shows an example of "Follow-Me" deployment.

Sensor nodes are placed in a line on the wall in hallways in close vicinity to each other. They then detect each others' radio signal strength and decide the closest two nodes and designate them as neighbor nodes. They then link all the way back to the sink, which would generally be placed at the entrance of the building. From this sink, a person could choose a destination and the sensors would create a path between them and illuminate their LED lights to guide the person.

The biggest problem in implementing this application was making sure the nodes constructed a path that was *physically* possible to navigate. Being that the nodes used radio signal's strength to determine the shortest path, it is possible that they could create a path that would go through a wall or cut a corner. To deter this problem, Wang et al. constructed a *logical* topology that was used by the sensors alongside the radio connectivity to create a path. To create a logical topology, a deployment order was established. Nodes were turned off until all were placed in position. Then, they were switched on one by one. When one node was in discovery mode, and it found a node that was just switched on, it established that node as the nearest neighbor and therefore created a physically navigable path.

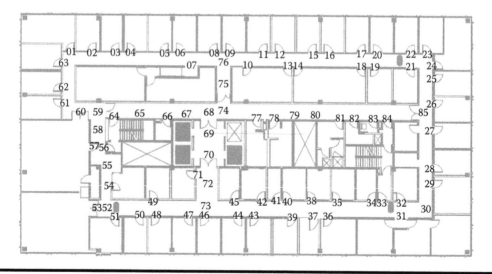

Figure 1.7 An example of "Follow-Me" deployment. (Data from Conner, W. S. et al., in *Wireless Sensor Networks: A Systems Perspective*, ed. N. Bulusu and S. Jha, 289–307, Artech House, Inc., 2005.)

1.4.4 Security

Camera systems and motion sensors are two widely used applications in modern-day security systems. Cameras are usually set up in public places and record video that carries a time stamp. This video is then usually archived for an amount of time. In places where privacy is an issue, motion sensors are installed in lieu of cameras. They are usually connected to alarm systems that are triggered when motion is detected in a room. While motion sensors alone are a plausible option for security, they do nothing in the way of providing a description of a perpetrator. Using sensor networks, Conner et al. [17] have found a way to couple motion sensors in private areas with camera systems in public areas to obtain images of individuals of interest.

The application of this starts by installing video cameras in a public place in a building, usually the lobby area. Then, sensor nodes with attached motion detectors are deployed in private areas of the building (dorm rooms, bathrooms, etc.). These sensors are all equipped with time settings that periodically update simultaneously to ensure that the time is accurate. When these nodes sense motion, they send a message with a time stamp back to the main server. This server keeps track of which sensors detected motion at which intervals of time. If a crime was committed in an area where privacy concerns made nonintrusive monitoring the only option, the server would be able to trace a path using the reported times from the sensor nodes. It would then suggest when the criminal made it back to the public area, and this time could be compared with footage from the security camera, thus providing an image of the criminal.

1.4.5 Weather Monitoring

One of the possible applications of WSNs in the environmental industry is weather monitoring. Tracking climate changes and weather shifts can be a difficult and challenging task. However, with the deployment of a large number of wireless sensor nodes, scientists can track weather changes more easily than setting up a complicated equipment. The sensors can sense a variety of phenomena from wind speed to barometric pressure changes in the field. Scientists and researchers can use the data gathered from WSNs to model weather and storm data [19].

1.4.6 Seism and Volcano Monitoring

Another challenging application of WSNs is measuring seismic and volcanic activities. This task can be difficult and dangerous [20]. With the use of wireless sensor nodes, scientists can have constant monitoring of different areas without the need for complicated equipment or rigs. Indeed, WSNs are able to conveniently and safely provide constant streaming data about temperatures, moment tensors, and vibrations in environments that are difficult to monitor otherwise. Using an accelerometer sensor, it is feasible to collect seismic response data [18]. Figure 1.8 shows the equipment setup, where sensors are placed on the specimen.

1.4.7 Pollution Monitoring

Pollution monitoring is another application that can see widespread use. In fact, industrial and governing bodies can use WSNs to track pollution levels in factories and cities. A chemical plant can distribute sensor nodes at different output areas and measure pollutants leaving the system. Cities and towns can place sensor nodes throughout high-traffic areas and see how much pollution is emitted into the air by vehicles. In addition, sensors could measure air quality in different areas of

Figure 1.8 Sensor placement on test specimen. (Data from Wong, J.-M. et al., *Proceedings of the Health Monitoring and Smart Nondestructive Evaluation of Structural and Biological Systems IV*, 2005.)

a city and alert officials when it drops so that they can plant trees and vegetation or redirect traffic. Measuring greenhouse emissions and air quality can help regulators keep pollution under control.

1.4.8 Energy Monitoring

Energy management is another ecological application of WSNs. Measuring energy use throughout systems can be simplified through the use of wireless sensors. Residencies and industries can place voltage sensors at key points in their power grid system to monitor current flow. They can use this data to view high flow areas or hotspots in their gird, thus allowing for possible optimizations. In addition, the sensor nodes may alert the monitors when they detect that tenants or machinery are using a significant amount of energy.

1.4.9 Water Quality Monitoring

Governing bodies can use sensors to measure water quality in reserves and in the wild. By placing motes in watertight containers and attaching aquatic sensors to them, researchers can gather data on different chemical levels in streams and rivers. Also, industrial plants can place sensors at outputs that feed into rivers to ensure that no harmful chemicals are distributed into the water. Moreover, residencies can use sensors to measure the quality of water they get from water tables or filters to ensure it is safe to distribute to tenants.

1.4.10 Early Fire Detection

One final place that sensor networks can be useful is in early fire detection. By monitoring things such as temperature, humidity, and wind speed, they can predict the occurrence of a brush fire.

They can also sense if a brush fire has occurred and can, together with satellite imagery, alert firefighters long before conventional methods do. This can be especially useful in places like California where there are miles of wooded area that could not possibly be monitored physically. Even when these places are monitored, the fires usually intensify to uncontrollable levels before people are alerted.

1.5 Industrial Applications

Private industry has been taking advantage of the amazing versatility of WSNs for a long time. They have developed a diverse set of sensor networks to provide a wide variety of unique functions and services. It is clear to see why industries would adopt wireless sensor technology so readily. Many industrial ventures need to gather data from remote or dangerous environments. The simple fact that no wires are required to link up the network of sensors saves money and allows for more complex networks in harder-to-reach spots [21,22]. In this section, we explore how different industries use WSNs to increase productivity, ensure safety, and cut down costs.

1.5.1 Common Uses

There are two most common uses for WSNs in industries, namely safety hazards detection and equipment failures detection. Since there are no wires, sensors can be placed inside equipment and measure attributes such as heat and certain movements or vibrations [21]. They can use these measurements to predict when a machine is likely to start malfunctioning so that preventative actions can be taken. Also, WSNs are used across many industries to ensure that regulations are being complied with and to keep their employees safe [23]. Furthermore, WSNs can detect almost all phenomena ranging from a major gas leak to a potential cave in mine tunnels.

Another interesting use of WSNs that is being researched for in industrial environments is mobile robot guidance. Indeed, some factories rely on mobile robots to transport materials and perform other similar tasks. In the past, they have relied on their own on-board sensors to guide themselves. But, this tactic was slow and not very effective. Now, the robots are able to interact with a network of wireless sensors set up around the factory that can guide it and send commands to it [22]. In fact, several companies are now integrating WSNs into their inventory management systems as well. This allows the company to track the materials all the way through the manufacturing to the final product. This process helps ensure that the inventory items do not get misplaced and that nothing is stolen [22].

1.5.2 Pharmaceutical Manufacturing

There is one particular industry in which WSNs have been readily adopted, namely the manufacturing of pharmaceuticals. Some critical conditions, such as humidity and temperature, must be kept at very specific levels during the manufacturing and storage of the product to ensure its purity and safety. This is a perfect application for WSN monitoring [24]. Even the transportation of the medicines must be carefully controlled, and wireless sensors are often packed in with the products as they are shipped so that the recipients can ensure that the drugs are kept within the required temperature and humidity levels for the whole trip [24]. Although several manufacturing plants use WSNs for these purposes, in some countries, where the industry is new, such as India, the critical readings within the plant are still recorded manually [24]. This can lead to errors and potentially unsafe conditions.

The major challenge for designers of WSNs in such strictly regulated industries, such as pharmaceuticals, is that the systems themselves must adhere to certain standards to ensure their readings are accurate and reliable [24]. A network of sensor nodes would not serve to help regulate the conditions in a manufacturing plant if the sensors themselves cannot provide accurate enough readings to ensure the safety and purity of the product being created.

1.5.3 Gas and Oil Industries

The oil and gas industries have greatly benefited from the technology of WSNs. In fact, in 2009, almost 25% of all wireless sensor motes deployed in private industrial ventures were used in this industry [21]. The workers face gas leaks and dangerous environments. Thus, without accurate sensors to alert them when something goes wrong, there would certainly be more job fatalities and disasters [21]. Wireless networks are especially appealing in this industry because of the scale over which the plants and production extend and the locations in which drilling sites and plants are located. Miles of pipelines must be kept safe from leakages. Also, the cost of wiring sensors over such distances can be very high [21].

The flexibility of WSNs helps the industry keep up with continually tightening standards. In addition, it can save the companies quite a bit of money in maintenance costs by alerting workers to systems that will need maintenance before the component actually goes offline or breaks down. Also, WSNs are invaluable in detecting the release of hazardous gasses. In fact, a properly deployed network of sensors can even calculate the exact location of the leak by analyzing the distribution of particles in the air [21]. These networks have kept many workers out of danger through early alerts to hazards, and their abilities to go where people cannot to monitor and take readings has allowed the workers to stay out of harmful ways.

1.5.4 Mining Industry

The mining industry makes similar use of sensor networks in the detection of releases of toxic gases in the mines. Also, WSNs can be used as early warnings to several other hazards that plague miners, such as underground explosions and deadly fire within the mines [25]. Sensor networks can even be programmed to detect subtle vibrations in the walls of the mines to alert the miners of potential cave-ins and other deadly occurrences. In a high-risk environment like a mine, it is important that the sensing systems are robust and every node is performing its job properly to ensure the safety of the miners. In situations like this, the sensor nodes on the network can be programmed to watch out not only for signs of hazards but also for other nodes in the system. If one node reports a hazardous condition, but finds that another nearby node does not make the same discovery, it will notify the controlling node in the system that its neighbor node may not be functioning properly so that it can be replaced [25].

The use of wireless networks is a very appealing solution in the mining industry because of their versatility and nonpermanent nature. A well-built and well-programmed network of wireless nodes can be constructed to allow constant shifting of nodes as needed and not lose any functionality in the system. As miners move farther into the mine and begin to work in different areas, the network can be added to or shifted around and still function properly. If the networks were linked together by wires, this would be much more difficult or even impossible to do. For reasons like these, the use of WSNs is rapidly spreading across several industries [25].

1.5.5 Railway Industry

Although we may not have noticed, in several of its areas, the railway industry is very dependent on wireless sensors to keep their trains running efficiently and safely. As trains move at high speeds with potentially dangerous momenta, keeping passengers and workers safe on the tracks has long been a problem. Also, the large distance over which the rails stretch makes it a perfect candidate for use of wireless networks of sensors. The use of WSNs allows each train to always know the location of other nearby trains. Such accurate readings allow for a higher volume of rail traffic and more productivity. When less accurate methods were commonly used, trains would need to stay farther away from each other on the tracks to ensure that no collisions would occur. However, if each train can be confident of the location and velocity of every other train in the area, then there is no reason to leave so much room between trains [26].

Like in other industries, the sensor nodes can easily be used to monitor the health of the equipment since there are no wires to get caught in the moving parts of the machinery. This allows the sensor nodes to be placed in locations where they can get the most accurate readings instead of simply where they can fit without getting in the way. WSNs on trains can prevent disasters, such as fires in the engine, and allow more accurate predictions as to when the machinery may malfunction or break. This would save the railway money with preventative maintenance [26]. Also, WSNs are used to measure the health of the rails themselves. When deployed at regular intervals along sections of track, the sensors can notify approaching trains of problems with the track that might be dangerous for them, such as twisted or deformed rails [26]. Railroad crossings are now starting to be equipped with sensors that can send back video data for analysis to detect people or cars that are in the crossing so that oncoming trains can be warned of the danger [26]. While they are clearly paramount to the safe and efficient running of such a large-scale operation as a rail line, the same technologies that keep trains from colliding can also be used to notify passengers at station platforms exactly where trains are in their schedule and if they will be arriving on time or not [26]. Figure 1.9 shows a typical scenario in which each railway wagon is equipped with multiple sensor nodes to monitor key parameters such as temperature, humidity, and smoke/fire, to name a few. In this architecture, the sensor nodes can act either as routers to forward packets to the sink node or as cluster heads in each wagon. The sink node and the cluster head send the collected data to the base station in the locomotive. Using an audiovisual system, the driver is able to monitor the sensor data and take appropriate decisions.

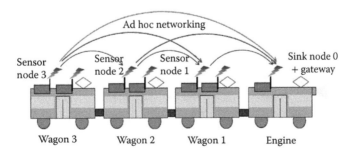

Figure 1.9 Typical scenario for rail wagons' health monitoring system. (Data from Shafiullah, G. M. et al., *Proceedings of the 2nd International Conference on Wireless Broadband and Ultra Wideband Communications*, 65, 2007.)

1.5.6 Security Issues

Security has always been a big and challenging issue in the industry. Also, it has been noticed that with the increasing reliance on WSNs to control and monitor plants and hazardous facilities, the increase of cyber attacks has risen drastically. Due to the nature of oil and gas plants, cyber criminals and terrorists may see them as easy targets. This means that the WSNs, which make up a very important part of the plant infrastructure, must be kept secure to prevent any damage to the plant and possible harm to workers or nearby citizens [27]. There are several types of attacks that the operators and designers of these high-security WSNs must be aware of. First of all, they need to ensure that all the data passing through the sensor network is encrypted. Someone with malicious intentions may try to eavesdrop on the sensor communication and gain information about the plant that way [27]. A simple attack, such as a denial-of-service attack, can cause a lot of trouble for the monitoring systems in a plant like this. If an attacker keeps sending the base station node in the network fake packets, the base station node will be unable to accept the real network traffic and the network will be unable to monitor the plant. Even just attempting to jam the signal by sending out constant broadcasts on the same frequency to disrupt network traffic can bring down the plant's monitoring systems. If the plant can no longer get data from these wireless sensors, then if there is a dangerous pressure buildup somewhere or some other dangerous event, it might not get detected quickly and may lead to a disaster [27]. If the security on the network is not robust enough, an attacker may also be able to introduce fake nodes into the system. They may even hijack the real ones and cause them to send improper data back to the base station node. This can make it appear that the plant is operating differently than it actually is [27].

As one can see, security is extremely important in any industry, not just gas and oil. Attackers can use unsecured WSNs to carry out many malicious actions from stealing company secrets and data to causing damage to a plant or machine by hijacking the sensor network monitoring it [27].

1.5.7 Conclusion

As technologies improve and sensor nodes become smaller, more accurate, more efficient, and more durable, their presence in all types of industry will increase rapidly. They are already used for all sorts of applications, from safety to convenience. These sensor nodes are cost-effective and cheap to install, easy to maintain, and update with changing industry regulations, and their uses appear to be almost limitless.

1.6 Military Applications

The same qualities that make the use of WSNs desirable for applications in the other fields, which we have discussed earlier, apply also to military applications. Given that information on the battlefield is important, WSNs are necessary to keep the commanders informed and allow them to make the best decisions to keep their soldiers safe [28]. The application of WSNs does not stop there though, as they also can be used for crucial tasks, such as supply management, equipment management, and damage assessment [29].

One of the more valuable and desirable qualities of WSNs that make them very convenient for military applications is their ability to spontaneously form networks and their ability for failure tolerance [29]. If one node goes down, for instance, gets destroyed by the enemy, the network can reconfigure itself and route information around the dead node, thus preserving the integrity of the network [29]. Also, their ability to withstand hazardous conditions is a major plus for use in combat [28]. One might even imagine planes flying overhead and scattering specialized motes across enemy territory or a battlefield. Once the motes hit the ground, they would be able to form up a network and almost immediately begin transmitting data about the terrain and the enemy back to the military commanders. Next, we discuss some uses of WSNs in the military domain.

1.6.1 Detection

Information availability is a major factor in deciding who wins or loses in a battle. With proper distribution of WSNs, military commanders would never be at a loss for information [30]. The sensor nodes are able to accurately detect enemy troop movement, thus giving the military commanders a great advantage in battle and preventing any sort of sneak attacks. In addition, the sensor nodes could even provide more useful services such as terrain mapping and environmental data from the battlefield [28]. Any information can help sway the scales, and lack of information can mean several lives lost.

The sensor nodes can be built and programmed as early warnings for chemical or biological attacks, thus allowing soldiers to protect themselves long before the threat even arrives. The use of WSNs as the "eyes and ears" of the battlefield can greatly increase the performance and efficiency of an army. Moreover, it can even prevent the excessive loss of lives during combat [30].

1.6.2 Soldier Health

As mentioned in Section 1.2, there have been a lot of research lately into networks of sensors that can be attached to the human body to monitor health and vital signs [32]. These body sensor networks (BSNs) can be invaluable for the military. BSNs can allow commanders to see the exact status of every soldier on the battlefield [31]. Based on injuries reported by the underlying BSN, the military leaders become aware of where to send medical personnel exactly. This can help medics triage the injured soldiers without even seeing them, thus allowing them to save more lives by heading to the more critically wounded soldiers first in a battle.

We should mention that equipping soldiers with BSNs can help soldiers localize the source of an explosion. Based on the reports of the explosive shock wave from each soldier's BSN, the base station can pinpoint the location of an explosion occurring nearby and send the information directly to the soldiers in the field. This allows the soldiers to react more quickly and effectively than if they tried to pinpoint the source of an explosion themselves [31]. Figure 1.10 shows how sensors, such as accelerometer, temperature, electroencephalography (EEG), and SpO_2 sensors, can be embedded within an advanced combat helmet worn by a soldier. This type of helmet helps continuously monitor the health status of each soldier. In the case of occurrence of any abnormalities, they should be detected by the underlying monitoring system, and the concerned soldier can be alerted.

1.6.3 Coordination

Because WSNs can be integrated with each other and with the other automated and technological aspects of the battlefield, this can allow commanders to more effectively coordinate all of their resources and soldiers to strike at the enemy more effectively. Combining all of the sensors and device controllers into one big interconnected network allows a better view of any activity on the battlefield. Also, it allows commanders to make better-informed decisions much more quickly compared to when all the systems remain separate.

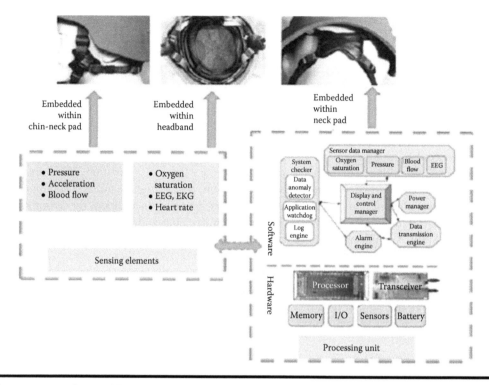

Figure 1.10 Advanced helmet for health monitoring. (Data from Lim, H. B. et al., *Proceedings of the International Workshop on Wearable and Implantable Body Sensor Networks*, 246–9, 2010.)

1.6.4 Problems

Although WSNs are indispensable in modern warfare, there are some problems with their use that need to be addressed. As we might imagine, the biggest problem facing WSNs in combat is security. Their communications must be encrypted so that they are not exploited by the enemy. What makes this even more difficult is that since the sensor nodes will very likely be deployed in disputed territory (if not behind enemy lines), the risk of enemy combatants discovering and capturing those sensor nodes is high. If the sensor node contains any data about how to encrypt or decrypt the messages it was sending, the enemies can listen to all of the WSNs' communications [28]. Worse yet, they can introduce fake sensor nodes into the system and feed the military leaders with false data. Although there are several proposals on how to address this issue, such as different encryption key schemes and methods for making WSNs hard to detect, the threat of enemies hacking into WSNs remains as one of the biggest threats to military use of WSNs [28,32].

1.6.5 Security

Security of military WSNs can be a bit more complicated to design and implement than security for other applications for a few reasons, which we discuss in the following. First of all, battlefields are a very dynamic environment and the swift and proper response from a WSN can literally mean the difference between life and death. Although there may be malicious parties who want to break into WSNs no matter where they are deployed or what they do, when these networks are deployed in military situations, the enemy will directly benefit from destroying or hacking our network and they will actively try to do so [28]. Not only does the security need to be engrained at the core of military sensor nodes, but also the nodes need to respond to each other quickly and continuously, keeping their users updated with the most current information. Every second wasted verifying other nodes and encrypting data cannot be afforded. So not only do these nodes need extra strong security, they also need it fast.

In a battlefield environment, there are other complications due to the fact that several sensor networks need to work together. Mobile nodes attached to soldiers or tanks need to constantly communicate with the stationary nodes to provide information on troop movement and battlefield conditions. When the mobile nodes move into the range of new static nodes, they need to be authenticated to ensure they are actually friendly nodes. This authentication process needs to be thorough and quick. Also, additional systems, such as unmanned aerial vehicles reporting enemy movement to soldiers on the ground, may be contacting those networks as well. All these interconnecting networks are constantly moving in and out of range of each other. This frequent mobility makes security and authentication a real, challenging problem [28].

There are some efficient solutions to these problems that have been reported in the literature. Here, we look at one security protocol, a "segmented key pool based scheme" [28]. It uses sets of preloaded keys in such a way as to allow a distribution of keys random enough to prevent a single captured node from compromising the network and yet fast enough to authenticate new nodes quickly so as to not interrupt normal operations of the networks [28]. Before the nodes are created, a large set of authentication keys is created. That set is then divided into several subsets. Each node on a stationary network (sensing nodes that will be deployed around the battlefield) randomly selects a number of keys from the subset that is given to that network. Each mobile node (nodes that will be on soldiers, tanks, and other equipment) randomly selects a number of keys from the entire large set. When a mobile node moves into the range of a stationary network and wants to connect, it will send out its list of authentication keys that were randomly selected from the entire set. The static nodes that receive these requests will compare the keys to the keys that they picked from their subset. When at least one node from the static network is able to match a key that the mobile node has broadcast, then that mobile node becomes authenticated to use the static network. This process takes a bit of time initially. However, when the mobile node finds a key that authenticates it on the static network in this location, it will save that key. Next time, it returns to the same location, it will know which key to broadcast, thus allowing for quick reconnecting to locations that have already been visited [28]. This method works well. But, it has its drawbacks. Security and connection speed are inversely related. The more random keys each node selects and stores before deployment, the faster the connections will occur. This is due to that fact that a new node entering the system will be more likely to share keys with the nodes it is communicating with. However, the trade-off is that increasing the number of keys in each node will also compromise more keys for each captured node. Despite these drawbacks, it shows how security and authentication might work on military WSNs [28].

1.6.6 Gunfire Origin Detection

The most prominent military application is a system to sense gunfire [33]. One of the scariest and most unknown things to the military is the opposition's deployment of snipers. Snipers are extremely discreet and generally vacate their position after firing only one shot. This leaves the military with only a small window of time to determine the sniper's position and take action. This inspired Ledeczi [33] from Vanderbilt University to develop the *PinPtr* application, which uses acoustic sensors to detect sound waves from muzzle blasts and shock waves from bullets. By analyzing these readings, the sensors can discern the position of the sniper within 1 m in a three-dimensional plane.

This system was developed on MICA2 motes running TinyOS with an attached custom acoustic sensor card. These were developed to be deployed both selectively and randomly with as little as 60 sensors and still be effective. Once deployed, these sensors automatically discover their positions, connect to each other over an ad hoc network, and sync their times. When a bullet is fired, the sensors detect the acoustics created by it and report the time of detection back to the base station. Using these times, the base station then makes an accurate prediction as to the location of the sniper.

Although there are some issues that need to be worked out, WSNs are already widely used by the military. During combat, having up-to-date information is crucial for success of the operation and safety of the soldiers. The cheap, versatile, and easily reconfigurable nature of WSNs makes them the perfect tool for the task.

1.7 Future Research Directions

There is little doubt that WSNs are the way of the future. In fact, *MIT Technology Review* says that WSNs will create a giant impact on our future lives [34]. In this section, we discuss some of the exciting applications of WSNs that are newly deployed or still in development.

1.7.1 Marine Deployments

Although there has been much research into WSNs over the past two decades, surprisingly, little of that research effort has been focused on underwater sensor networks. Marine networks would be very helpful in several fields such as military defense, weather forecasting, and biological or environmental study [35]. WSNs may be able one day to help us protect U.S. borders from drug smugglers, enemies, and other unauthorized ships. There is a lot of coastline in our country and there is no way that the coastguard can be everywhere at once. That is why WSNs are ideal for protecting our ocean boarders. Wireless sensor nodes can be placed on top of the water at regular intervals and can be programmed to distinguish between regular ocean waves and those that are created by passing ships [36]. Similar technologies can undoubtedly be used by the military to more efficiently track enemy submarines and other hostile ships that can be a possible threat to our national security [36].

Besides military applications, marine WSNs can also be used to measure the temperature and currents of the ocean through a combination of floating and underwater sensors for the purposes of study and weather prediction [35]. Moreover, WSNs deployed in the deep ocean can provide better detection and earlier warning for natural disasters such as tsunamis. Perhaps, the reason this field of study has lagged behind is because of the inherent challenges in marine deployment. First of all, even though most wireless nodes and sensors are designed for severe or hazardous conditions, nodes being fully submerged for long periods still present a challenge to designers [35]. The dynamically changing and sometimes violent ocean is a very challenging medium to deploy into. Particularly, it presents several problems that do not have to be addressed at all for networks deployed on land [35]. First of all, once deployed, the sensor nodes will be completely unreachable. WSNs are generally designed as if the sensor nodes will be unreachable after deployment anyways. But, no matter how they are deployed in the ocean, once they are under water, the sensor nodes would be completely on their own. This makes power efficiency very important, as there can be

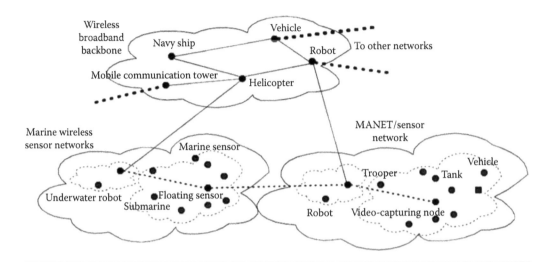

Figure 1.11 Marine wireless sensor networks in future digital battlefields. (Data from Mahdy, A. M, *Proceedings of the Seventh International Conference on Networking,* **530–5, 2008.)**

no battery or other equipment replacements. Also, depending on the depth at which the nodes are deployed, there may not even be enough sunlight to make solar power a viable option for these networks [35]. Figure 1.11 shows how marine WSNs can be used in future digital battlefields.

Networking and communication is also an issue in a submerged environment. Not only does the water affect the transmission of messages, but new routing protocols also would have to be established as well. In the traditional model for a marine network, there are some underwater nodes and some floating nodes to be able to get all the necessary measurements. When routing, the protocols will need to distinguish between the nodes that are above the water and the ones that are under the water to find the most efficient path [35].

Nowadays, terrestrial networks are receiving significant attention from both academia and industry. Despite these minor challenges, marine networks should also receive much more attention compared to their terrestrial counterparts. As traditional WSNs continue to secure a foothold as an indefensible new technology, we truly believe that there will be an increase in use of marine WSNs as well.

1.7.2 Smart Homes

Another exciting technology based on WSNs is the smart home. A smart home is a house with several integrated systems within it to make the home "aware" of what the occupants are doing and respond accordingly. Although most homes contain at least some pieces of technology that would make up a smart home, the cost and difficulty of installing fully integrated and fully automated systems has thus far prevented smart homes from becoming commonplace [37].

Smart homes are mostly geared toward convenience of the residents. But, the systems can of course be adapted to suit special needs and help the elderly and disabled live more independently. A future smart home will not only be able to tell when someone enters a room, but through use of small radio frequency identification chips that can be worn by occupants, the house can also tell exactly *who* has entered the room. Researchers are very excited as to the level of customization within the home that this technology can ultimately lead to. Room temperature, lighting, appliance operation, and even music play lists can be customized to fit the preferences of a particular occupant. In the future, when we walk into a room in our house, that room, through the use of several interconnected systems of WSNs, can determine who we are, turn the lights on automatically to our preferred level, adjust the temperature in the room so that we are comfortable, and turn on our favorite music play list [37]. This automation can ultimately help save power as well. Although it obviously would take power to run the smart home system, power would be saved by turning off everything in a room as soon as we leave. Lights and televisions would never again be accidentally left on [37].

The smart home can also implement an intelligent security system, thus making it much harder for intruders to enter the house undetected. Besides traditional security methods that would be present in the smart house system, there has been some preliminary research into behavior tracking so that the house can potentially recognize an intruder by their behavior. This means that even if someone was to steal the radio frequency identification chip from one of the smart houses occupants, the system can detect abnormalities in that occupant behavior and be alerted as a possible intruder [37]. Moreover, this behavior identification technology can be used to detect signs of distress or injury to the house's occupants. If the smart houses were to detect behavior that was indicative of someone having a heart attack, for example, the house could respond by automatically contacting emergency medical responders [37].

The smart house system would also make the jobs of emergency services much easier. In the future, police officers or medical workers arriving at a smart house can be aware of exactly where every occupant is inside the house. This would allow police to storm houses and extract their targets in a manner much safer for the officers and the suspects alike. When medical personnel receive a distress call from the house, they will know exactly which room to go to, thus saving valuable time that they need to care for the patient. Firefighters would know the exact location of anyone trapped in a burning building. In fact, the smart house may even be able to give those firefighters data on the structural integrity of the house so that they can avoid areas that may collapse, trapping them [37].

WSNs are ideal for implementing these smart homes because of their flexibility and the simple fact that the sensor nodes can be installed anywhere without the expense of running wires through the wall to connect all of the components of the network together. The roles that WSNs will play in our homes in the future are limited only by imagination. These networks will continue to appear in every industry, and in the coming years, we will rely on WSNs more and more. Wireless sensors will be a big part of the increasingly digital future of humanity [37].

1.8 Conclusion

In this chapter, we reviewed a sample of well-known applications of WSNs. More specifically, we considered five application domains, namely health care, agriculture, environment, industry, and military. For each application domain, we provided an overview of some useful WSN applications.

Given the usefulness of such sensor networking technology, we truly believe that several existing applications of WSNs will be commercialized and other new applications will emerge. In particular, multimedia, underwater, and underground WSN applications will receive more attention in the future from both academia and industry. Also, smart home and aware environments will see more progress and future development.

Acknowledgments

The authors gratefully acknowledge the insightful comments of the anonymous reviewers that helped improve the quality and presentation of this chapter significantly. The work of H.M. Ammari is partially supported by the U.S. National Foundation (NSF) grant 0917089 and a New Faculty Start-Up Research Grant from the University of Michigan-Dearborn, College of Engineering and Computer Science—Dean's Office.

References

1. Hof, H.-J. 2007. "Applications of Sensor Networks." In *Algorithms for Sensor and Ad Hoc Networks*, edited by D. Wagner and R. Wattenhofer, 1–20. LNCS 4621, Germany: Springer.
2. Omre, A. H. 2009. "Reducing Healthcare Costs with Wireless Technology." In *Proceedings of the Sixth International Workshop on Wearable and Implantable Body Sensor Networks (BSN)*, 65–70.
3. Chen, Y., W. Shen, H. Huo, and Y. Xu. 2010. "A Smart Gateway for Health Care System Using Wireless Sensor Network." In *Proceedings of the Fourth International Conference on Sensor Technologies and Applications*, 545–50.
4. Pan, J., S. Li, and Z. Wu. 2008. "Towards a Novel In-Community Healthcare Monitoring System over Wireless Sensor Networks." In *Proceedings of the International Conference on Internet Computing Science and Engineering*, 160–5.
5. Shnayder, V., B.-R. Chen, K. Lorincz, T. R. F. Fulford-Jones, and M. Welsh. 2005. "Sensor Networks for Medical Care." Technical Report TR-08-05, Division of Engineering and Applied Sciences, Harvard University, Cambridge, MA.
6. Moog Crossbow. Milpitas, CA. Accessed from www.xbow.com
7. Stankovic, J. A., Q. Cao, T. Doan, L. Fang, Z. He, R. Kiran, S. Lin, S. Son, R. Stoleru, and A. Wood. 2005. "Wireless Sensor Networks for In-Home Healthcare: Potential and Challenges." In *Proc. Workshop on High Confidence Medical Devices Software and Systems* (HCMDSS).
8. Oh, T. H., S. Lim, T. Lakshman, and Y. B. Choi. 2010. "Security Issues on Wireless Body Area Network for Remote Healthcare Monitoring." In *Proceedings of the International Conference on Sensor Networks, Ubiquitous, and Trustworthy Computing*, 327–32.
9. National Heart Lung and Blood Institute Diseases and Conditions Index. August 2010. "What is Sleep Apnea?" National Heart Lung and Blood Institute. http://www.nhlbi.nih.gov/health/dci/Diseases/SleepApnea/SleepApnea_WhatIs.html.
10. Wong, W. K., Y. C. Poh, C. K. Loo, and W. S. Lim. 2010. "Wireless Webcam Based Omnidirectional Health Care Surveillance System." In *Proceedings of the Second International Conference on Computer Research and Development*, 712–6.
11. Mainwaring, A., J. Polastre, R. Szewczyk, D. Culler, and J. Anderson. September 2002. "Wireless Sensor Networks for Habitat Monitoring." In *ACM International Workshop on Wireless Sensor Networks and Applications*. Atlanta, GA.
12. Rieger, T. 2007. "Wireless Vineyard Monitoring Technology." *Vineyard & Winery Management* March/April:2–6. Accessed from http://www.ranchsystems.com/ssite/FNL-24028%20Ranch%20Systems%20(E).pdf.
13. Camalie Vineyards. 2008. "Update Camalie Networks Wireless Sensing." Mount Veeder, CA. Accessed from camalie.com/WirelessSensing/WirelessSensors.htm.
14. World Business Council for Sustainable Development. 2005. *Water: Facts and Trends*. Geneva, Switzerland: World Business Council for Sustainable Development (reprinted 2009).
15. Depienne, F. 2007. "Wireless Sensor Networks Application for Agricultural Environment Sensing in Developing Countries." M.S. thesis, Ecole Polytechnique Federale De Lausanne, Lausanne, Switzerland.

16. Juang, P., H. Oki, Y. Wang, M. Martonosi, L. S. Peh, and D. Rubenstein. 2002. "Energy-Efficient Computing for Wildlife Tracking: Design Tradeoffs and Early Experiences with ZebraNet." *ACM Sigplan Notices* 37 (10): 96–107.

17. Conner, W. S., J. Heidemann, L. Krishnamurthy, X. Wang, and M. Yarvis. 2005. "Workplace Applications of Sensor Networks." In *Wireless Sensor Networks: A Systems Perspective*, edited by N. Bulusu and S. Jha, 289–307. Norwood, MA: Artech House, Inc.

18. Wong, J.-M., J. Goethals, and B. Stojadinovic. 2005. "Wireless Sensor Seismic Response Monitoring System Implemented on Top of NEESgrid."In *Proceedings of the Health Monitoring and Smart Nondestructive Evaluation of Structural and Biological Systems IV*.

19. Ituen, I., and G. Sohn, 2007. "The Environmental Applications of Wireless Sensor Networks." *International Journal of Contents* 3:1–7.

20. Tan, R., J. Chen, G. Xing, W.-Z. Song, and R. Huang. 2010. "Quality-Driven Volcanic Earthquake Detection Using Wireless Sensor Networks." In *Proceedings of the IEEE International Real-Time Systems Symposium (RTSS)*, 271–80.

21. Akhondi, M. R., A. Talevski, S. Carlsen, and S. Petersen. 2010. "Applications of Wireless Sensor Networks in the Oil, Gas and Resource Industries." In *Proceedings of the 24th IEEE International Conference on Advanced Information Networking and Applications (AINA)*, 941–8.

22. Low, K. S., W. N. N. Win, and M. J. Er. 2005. "Wireless Sensor Networks for Industrial Environments." In *Proceedings of the International Conference on Computational Intelligence for Modeling, Control and Automation and International Conference on Intelligent Agents, Web Technologies and Internet Commerce*, 271–6.

23. Ke, Z., L. Yang, X. Wang-hui, and S. Heejong. 2008. "The Application of a Wireless Sensor Network Design based on ZibBee in Petrochemical Industry Field." In *Proceedings of the International Conference on Intelligent Networks and Intelligent Systems (ICINIS)*, 284–7.

24. Potdar, M., A. Sharif, V. Potdar, and E. Chang. 2009. "Applications of Wireless Sensor Networks in Pharmaceutical Industry." In *Proceedings of the International Conference on Advanced Information Networking and Applications (AINA) Workshops*, 642–7.

25. Chehri, A., P. Fortier, and P.-M. Tardif. 2007. "Security Monitoring Using Wireless Sensor Networks." In *Proceedings of the Fifth Annual Conference on Communication Networks and Services Research (CNSR)*, 13–7.

26. Shafiullah, G. M., A. Gyasi-Agyei, and P. Wolfs. 2007. "Survey of Wireless Communications Applications in the Railway Industry." In *Proceedings of the the 2nd International Conference on Wireless Broadband and Ultra Wideband Communications*, 65.

27. Radmand, P., A. Talevski, S. Petersen, and S. Carlsen. 2010. "Taxonomy of Wireless Sensor Network Cyber Security Attacks in the Oil and Gas Industries." In *Proceedings of the 24th IEEE International Conference on Advanced Information Networking and Applications*, 949–57.

28. Durresi, A., M. Durresi, and L. Barolli. 2008. "Security of Mobile and Heterogeneous Wireless Networks in Battlefields." In *Proceedings of the International Conference on Parallel Processing (ICPP) Workshops*, 167–72.

29. Bri, D., M. Garcia, J. Lloret, and P. Dini. 2009. "Real Deployments of Wireless Sensor Networks." In *Proceedings of the Thrid International Conference on Sensor Technologies and Applications (SENSORCOMM)*, 415–23.

30. Durresi, A., M. Durresi, and L. Barolli. 2009. "Heterogeneous Multi Domain Network Architecture for Military Communication." In *Proceedings of the International Conference on Complex, Intelligent and Software Intensive Systems (CISIS)*, 382–7.

31. Lim, H. B., D. Ma, B. Wang, Z. Kalbarczyk, R. K. Iyer, and K. L. Watkin. 2010. "A Soldier Health Monitoring System for Military Applications." In *Proceedings of the International Workshop on Wearable and Implantable Body Sensor Networks*, 246–9.

32. Pinto, P. C., and M. Z. Win. 2006. "Design of Covert Military Networks: A Spectral Outage-based Approach." In *Proceedings of the IEEE Military Communications Conference*, 1–6.

33. Maroti, M., G. Simon, A. Ledeczi, and J. Sztipanovits. August 2004. "Shooter Localization in Urban Terrain." *IEEE Computer*, 37 (8): 60–1.

34. Feng, M. W. 2008. "Wireless Sensor Network Industrial View? What Will be the Killer Apps for Wireless Sensor Network?" In *Proceedings of the IEEE International Conference on Sensor Networks, Ubiquitous and Trustworthy Computing (SUTC)*, 270.
35. Mahdy, A. M. 2008. "Marine Wireless Sensor Networks: Challenges and Applications." In *Proceedings of the Seventh International Conference on Networking*, 530–5.
36. Tran, T. D., D. D. Do, H. V. Nguyen, Y. V. Vu, and N. X. Tran. 2011. "GPS-Based Wireless Ad Hoc Network for Marine Monitoring, Search and Rescue (MsnR)." In *Proceedings of the Second International Conference on Intelligent Systems, Modeling and Simulation,* 350–354.
37. Hussain, S., S. Schaffner, and D. Moseychuck. 2009. "Applications of Wireless Sensor Networks and RFID in a Smart Home Environment." In *Proceedings of the Seventh Annual Communication Networks and Services Research Conference (CNSR)*, 153–7.
38. Wang, X., F. Silva, and J. Heidemann. 2004. "Demo Abstract: Follow-Me Application—Active Visitor Guidance System." In *System Proc. 2nd ACM SenSys Conference*, Baltimore, MD.

Chapter 2

Data-Centric Storage in Wireless Sensor Networks

Khandakar Ahmed and Mark A. Gregory

Contents

2.1 Introduction

The wireless sensor network (WSN) is an emerging technology suitable for unattended monitoring of straddling infrastructures (e.g., industries, factories, and bridges) in a wide range of environments and even in humans. The WSN, along with mobile ad hoc networks, is one of the research trends in the field of infrastructureless wireless networks. Both mobile ad hoc networks and WSNs are highly infrastructureless with the dynamic nature of nodes and an unpredictable mobility pattern. However, unlike mobile ad hoc networks, limited resources and extremely large numbers of nodes are two major challenges in WSNs. A WSN is mainly application specific and data centric. The identity of a sensor is not as important as the data associated with it.

Data storage and retrieval methods in a WSN can be classified into three canonical approaches [1,2]: (1) External storage (ES), (2) local storage (LS), and (3) data-centric storage (DCS). In ES [3–6], nodes send data to a base station or gateway without waiting for a query. Nodes inside the network or any users outside the network send a query to the base station. This may create a bottleneck to the base station. This process also wastes energy when every single data is sent to the base station without aggregation. In LS [7–11], each node keeps its sensed data locally and uses flooding for query, consuming a significant amount of energy resources. In DCS [12–15], it is proposed to choose one (or more) rendezvous node(s), based on the event type, as the target node to store data. This reduces both storage and query costs.

The DCS is expedient in many scenarios, and two pragmatic cases are worthwhile to mention: One is the standard WSN structure comprising end nodes (sensors), which includes a central node referred to as a "sink node," a base station, and a gateway that connects the WSN to the external world. To make it efficient and robust, this kind of network can be amended where sensor nodes gather information from the environment and store it in the rendezvous nodes. The query, which usually comes from users who are external to the network, of a particular data type is forwarded via the gateway toward the relevant rendezvous node. Replies from the rendezvous nodes are then sent back to the consumer via the gateway.

On the other end of the spectrum, researchers have expanded their interest on WSN in the last few years to more intricate networks such as wireless sensor and actor networks (WSANs) [16,17]. A WSAN is very useful in an unattended WSN, in which an actor or actuator node can perform a certain action based on the information retrieved from one or more other sensors. An actuator node, in this case, plays the roles of both a sensor and an actor at the same time in the network. In WSAN, data collection from the environment, data storage, data processing, and performance of action all happen in a distributed fashion without any external intervention. Thus, DCS can be the suitable storage and information retrieval system in WSANs.

The first DCS is the geographic hash table (GHT) [1], which was proposed in 2002. Since then, a lot of research has been done on the topic targeting different challenges such as the nonuniformity of the network, multidimensional attribute (MDA), range query, data aggregation, and similarity search. However, none of the investigations cover all these issues together, although different proposals try to give solutions to different problems. In this chapter, most of the well-accepted proposed DCS techniques are presented, and a clear and in-depth study on them is provided to researchers and students in this field.

This chapter is organized as follows: Some major challenges relevant to DCS are discussed in Section 2.2. Section 2.3 briefly discusses most of the relevant DCS studies in the literature and classifies them into two major routing categories: (1) Point-to-point routing (Section 2.3.1) and (2) routing based on tree structures (Section 2.3.2). The subsections also present their main contribution and rationale based on the challenges mentioned in Section 2.2. Section 2.4

presents a future research direction while Section 2.5 concludes and provides a comparative table highlighting the salient features of each DCS mechanism.

2.2 Challenges in Data Storage

In providing data storage and search services for sensed events, a data storage scheme in a WSN faces many challenges. A few of them are challenges related to MDA, range queries, similarity search, data aggregation, nonuniformity of sensor network fields, multireplication, and load balancing of storage among sensor nodes. Sections 2.2.1 through 2.2.7 briefly introduce these challenges and refer to the research efforts that propose possible solutions.

2.2.1 MDA

A recent technological development in WSNs is the heterogeneity of sensors; sensors may have multiple capabilities in terms of computing, power supplies, communication, and sensing [18,19]. Hence, a heterogeneous network at present is able to detect multiple types of attributes such as humidity, temperature, level of a particular gas in the atmosphere, and so on. For example, in an air pollution–measuring application, the measurement data may be the fusion of several parameters such as temperature, level of carbon monoxide, level of smoke, and so on. In such an application, it is obvious that a storage and search mechanism for multidimensional queries must be in place.

Some authors propose multidimensional storage and query mechanisms for DCS [13,20,21].

2.2.2 Range versus Point Queries

Range query is another challenge for the DCS scheme. For example, a user may be interested in a particular range rather than a specific point value. For example, air pollution may be encountered if the level of carbon monoxide is in the range between 30 and 90 L/mol. So, a possible query is to find all the sensing points at which the level of carbon monoxide falls in the aforementioned range. Using range queries, users can drill down their search efficiently for events of interest. The aforementioned query illustrates this; an environmental scientist, in a particular forest, is presumably interested in discovering the aforementioned range of carbon monoxide and perhaps also wants to map it with some other parameters to take certain action or draw a conclusion [13].

Some authors [13,20–24] implement the range query mechanism in different formats. Chung et al. [23] mainly propose an efficient technique for similarity search in DCS, which works for both point and range queries. Li et al. [13] also show both point and range queries in their multidimensional DCS model. Shen et al. [21] describe a model called "distributed spatial–temporal similarity data storage," referred to as the SDS technique, which supports range query. A distributed index for features in sensor networks (DIFS) [22] performs data fusion based on the conveyance of data through the network. Routing is designed on top of the quad tree in a manner that balances communication load across the index, and the range is maintained along the hierarchy of sensor nodes.

2.2.3 Similarity Search

Due to the imprecise behavior of sensor hardware and the variation of environmental parameters, the similarity search problem in WSNs has received tremendous attention. Moreover, in certain applications or circumstances, in addition to searching for an exact match it is necessary to search within a specified similarity range. For example, a multidimensional query on attributes like temperature, carbon monoxide, forest name, location, and smoke level with values 100°| 150 L/mol| Melaleuca| North| 130 L/mol may also interest another similar set of values 90°| 150 L/mol| Melaleuca| South| 130 L/mol. Similarity search may be useful in many applications, such as finding similar flow patterns in oceans or wildlife activity patterns in habitat monitoring [21]. The traditional approach of similarity search is improper and inefficient for the energy-constrained sensor network. It is necessary to search for similar data that fits the query without collecting data from all sensors. Two DCS mechanisms implementing efficient techniques for a similarity search are described by Shen et al. [21] and Chung et al. [23]. One DCS mechanism [21] implements a similarity search with low energy consumption by assigning weight to different attributes based on their importance. In contrast, the other mechanism [23] maintains two variables for maintaining the range in the index node of each zone; a network is divided into zones based on the concept of Hilbert curve, and it uses three different functions referred to as backward, forward, and bidirectional probing to search for similar data.

2.2.4 Data Aggregation

Based on the basic data stored in sensors, DCS networks can facilitate data aggregation in a fully distributed way. By using data aggregation, the traffic generated by producer nodes can be reduced before answering consumer queries. Monitoring building integrity during earthquakes by engineers, habitat monitoring by biologists, monitoring temperature, and power usage in data centers by cluster computer administrators are examples of sensor applications that depend on the ability of sensors to extract summary (aggregate) data rather than raw data from networks.

In DCS, replicas can be ideally used for data aggregation as they (replica nodes) receive all the data from the producers in their surrounding area. However, there are many forms of data aggregation implementation. TinyDB [25], madwise [26], and TAG [9] are some of the aggregation mechanisms proposed mainly for WSN databases. They can be adapted to DCS networks as well.

Cuevas et al. [27] modeled data aggregation in multireplication DCS systems for WSANs. DIFS [22] and resilient DCS [24] in WSNs also provide data aggregation in DCS mechanisms.

2.2.5 Nonuniformity of Sensor Network Fields

In every deployment sensors may not be uniformly distributed, which means some sections/zones may be densely populated, whereas others may not be. Moreover, in mobile WSNs, sensors may move from one place to another, creating nonuniformity later. In the current state of the art, most DCS schemes are proposed on the consideration that sensors are uniformly distributed. This assumption leads to data losses in overburdened sensors.

Load-balanced DCS (LB-DCS) [28] deals with nonuniformity based on two mechanisms: First, it estimates network distribution, and then it exploits the data dissemination method based on the estimation.

2.2.6 Multireplication

Multireplication reduces the data loss that happens due to node failure or movement of data from one rendezvous zone to another. Furthermore, multireplication may facilitate data fusion and aggregation. The GHT [1] is the first mechanism that proposes multireplication using perimeter refresh protocol (PRP) to replicate data. It also extends GHT to structured replication–GHT

(SR-GHT) [1] to avoid the creation of a hotspot in the home node, where the number of replicas is $N_r = 4^d$, where d is a positive integer. However, the most interesting issue is that it never replicates data to all its replica nodes; rather, the producer node stores data only to its closest replica and a query node needs to query all the replica nodes. Tug-of-war (ToW) [29] also implements the same replication mechanism based on the 4^d grids. Similar to SR-GHT and ToW, quadratic adaptive replication (QAR) [30] uses the same grid-based replication mechanism. However, the number of replicas in QAR follows a more adaptable quadratic evolution ($N_r = d^2$), and it also allows the selection of the optimal number of replicas from a wider set of values.

2.2.7 Balancing the Storage

It is important to balance the storage load among the sensors to prevent the imposition of a storage load that is too high, which would create hotspots on some nodes. Load balancing is implemented in many research studies [21,28,31,32]. In SDS [21], a zone head estimates the zone's storage usage status using $\sum_{i=1}^{N} S_i/N$, where S denotes the percentage of a node's used storage and N is the number of total nodes in the zone. After receiving a storage request, each zone head first examines if its zone storage usage has reached its threshold, which is denoted by ϕ. If so, the head redirects the storage request to a neighboring zone and keeps a pointer to it. Dynamic load balancing (DLB) [31] uses a cover-up scheme to balance the load inside a zone. The LB-DCS [28] balances the load in a fashion similar to SDS [21] by estimating the zone density. However, unlike SDS, LB-DCS estimates zone density using the number of sensor nodes in a zone and then disseminates this density information throughout the network.

2.3 DCS Mechanisms

In the current state of the art, DCS can be classified based on two types of routing: (1) point-to-point routing and (2) routing that relies on tree construction techniques. The former deterministically maps the name of an event (e) to the routable address (i) associated with a particular node. In contrast, the latter relies on tree construction techniques. It divides the whole network into a tree structure and provides a map of data toward paths with minimal assumption about the underlying infrastructure.

2.3.1 DCS Mechanisms Based on Point-to-Point Routing

In point-to-point routing, the location or coordinate of a sensor node is used as its routing address. The routing methods of this category usually divide a network geographically and trace a sensor based on its geographic location. A few examples of point-to-point routings are greedy perimeter stateless routing (GPSR), logical stateless routing, carpooling, combing, and recursive hierarchical routing. Sections 2.3.1.1 through 2.3.1.8 briefly illustrate a few DCS methods that are formed on the top of point-to-point routing.

2.3.1.1 GHT

Ratnasamy et al. [1] are the pioneers who provided the noble concept of DCS in 2002. They were motivated to make an effective use of the vast amounts of data gathered by large-scale sensor networks using scalable, self-organizing, and energy-efficient data dissemination algorithms.

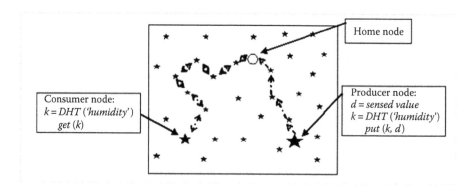

Figure 2.1 The GHT.

Authors use a distributed hash table [33–35] to map hash keys, generated from the event name or type, into geographic coordinates and store this event to sensor nodes in a geographic location close to these coordinates. The technique of GPSR [36] is used to store and/or retrieve data to or from a sensor node.

A GHT [1] uses a hash function over an event name, say, *humidity*, to find a corresponding key. When a sensor node senses an event, it is mapped to the corresponding key based on the event name. The GHT then uses the function *put (k, d)* in which *k* is a hash key used as the destined geographical location and *d* is the data, which forwards this packet to the location *k* using GPSR. The sensor node closest to the geographical location (*k*) is chosen as the home node, which stores the data, for this event type. Similarly, when a consumer wants to consume or query the data of an event type (for this particular case, *humidity*), GHT again maps the event type to the hash key (*k*) and uses the function *get (k)* to forward the query to the corresponding spatial/geographical location (*k*). The home node then replies by providing the stored data for that event type. In Figure 2.1, sensors are represented by "★". A producer node senses a value and forwards it to the home node denoted by "O". Another consumer node uses the same hash function in turn and retrieves the stored data from the home node. In the storing process, represented by "⋯→", the producer node sends data to the target node, whereas in the retrieval process, denoted by "↔", the query is first forwarded to the home node and replies are sent back to the consumer.

2.3.1.1.1 Routing Algorithm

GHT uses two types of routing algorithms: (1) GPSR and (2) PRP. GPSR is used to route packets geographically, whereas PRP is used to accomplish the replication of key–value pairs and consistent replacement at appropriate home nodes when the network topology changes.

There are two distinct algorithms that GPSR uses for routing: (1) A greedy forwarding algorithm [37] that moves packets progressively closer to the destination at each hop and (2) a perimeter-forwarding algorithm that forwards packets in cases where greedy forwarding is impossible. In the greedy forwarding rule, a node *S* forwards a packet to its neighbor *X*, which is closest to the destination (see Figure 2.2a). However, greedy forwarding fails when no neighbor is closer than *S* to the destination. The GPSR uses a perimeter routing algorithm to recover from this void condition. In Figure 2.2b, there is no neighbor in the range of *X* that is closer to the destination *D* than *X*. In such situations, perimeter forwarding uses a right-hand rule. Figure 2.2c illustrates the right-hand rule. The GPSR starts with the greedy forwarding rule, but switches to perimeter

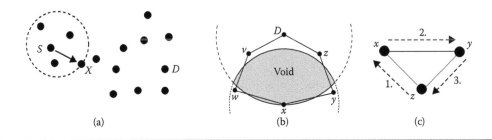

Figure 2.2 **(a) Greedy forwarding routing; (b) *x* has no neighbor closer to *D* than itself; (c) packets travel clockwise around the enclosed region. (Data from Ratnasamy, S. et al., *Proceedings of the 1st ACM International Workshop on Wireless Sensor Networks and Applications,* Atlanta, GA, 2002.)**

forwarding when the greedy rule fails. The GPSR again returns to greedy from perimeter mode when the packet reaches a node closest to destination than the node at which GPSR enters perimeter forwarding.

2.3.1.1.2 Evaluation

Ratnasamy et al. [1] use a simple analytical method to compare the performance of GHT with the performance of two other canonical methods, that is, ES and LS. The deployed sensor network has n sensor nodes that are able to detect T event types. The term D_{total} denotes the total number of events detected, Q denotes the number of event types for which a query is made, and D_q denotes the total number of successful responses for Q queries. So, the costs are as follows (*list* indicates that a full listing of events is returned, whereas *summary* indicates that only a summary of events is returned):

ES:

$$Total : D_{total}\sqrt{n}$$

$$Hotspot : D_{total}$$

LS:

$$Total : Q_n + D_q\sqrt{n}$$

$$Hotspot : Q + D_q$$

DCS:

$$Total : Q\sqrt{n} + D_{total}\sqrt{n} + D_q\sqrt{n} \ (list)$$

$$Total : Q\sqrt{n} + D_{total}\sqrt{n} + Q\sqrt{n} \ (summary)$$

$$Hotspot : Q + D_q(list) \quad or \quad 2Q \ (summary)$$

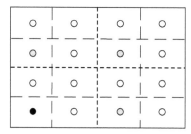

● Root point: (3, 3)

○ Level 1 point: (53, 3), (3, 53), (53, 53)

○ Level 2 mirror point: (28, 3), (3, 28), (28, 28), (78, 3), (53, 28), (78, 28), (3, 78),
(28, 53), (28, 78), (78, 53), (53, 78), (78, 78)

Figure 2.3 Figure showing SR. (Data from Ratnasamy, S. et al., *Proceedings of the 1st ACM*
International Workshop on Wireless Sensor Networks and Applications, **Atlanta, GA, 2002.)**

2.3.1.1.3 Multireplication

A home node becomes a hotspot when too many events with the same key are detected. To address
this issue, authors have extended GHT by employing structured replication (SR). SR introduces a
hierarchical scheme in which one can have $4^d - 1$ mirror images of the root home node.

Here, d denotes the depth of the hierarchy. As shown in Figure 2.3, a decomposition of two
levels ($d = 2$) is deployed with $4^2 - 1$ (15) mirror images at different levels. A producer node can
store the detected event to its closest mirror node reducing the storage cost to $O\left(\dfrac{\sqrt{n}}{2^d}\right)$ from
$O(\sqrt{n})$. In the retrieval phase, GHT needs to forward a query to all mirror images. The query
is first forwarded to the root node; it is then forwarded to all mirror images in level 1, and the
level-1 nodes forward this query to level-2 mirror images. This again increases the retrieval cost to
$O(2^d \sqrt{n})$ from $O\left(\dfrac{\sqrt{n}}{2^d}\right)$. Thus, for an event with D_i detected instances and Q_i queries, the total
cost of storing and retrieving this event is approximated as follows: $O\left(Q_i 2^d \sqrt{n} + D_i \left(\dfrac{\sqrt{n}}{2^d}\right)\right)$.

2.3.1.1.4 Discussion

A single home node that stores all the sensed data could become a hotspot for event queries for
the sensed data stored at the node and higher energy consumption than other nodes. However, to
solve this problem, local replication is done by means of PRP. This works fine to a certain extent as
several replicas can answer the consumer queries by balancing the load among them. However, for
popular events, many queries may reach the home perimeter as a result of which PRP creates routing hotspots around the home node. The GHT does not address multidimensional range queries,
nonuniformity, aggregation, and similarity search.

2.3.1.2 Data Storage and Range Query Mechanism for MDAs

Liao and Chen [20] propose an energy-efficient and scalable multidimensional range query mechanism (MDA) to retrieve data. It builds an in-network-distributed data structure by mapping

MDAs to their corresponding range spaces. Authors consider three assumptions prior to starting their work: (1) The sensors are uniformly and densely deployed, (2) each node can sense multiple events, and (3) each node maintains a neighbor table via the periodic exchange of beacon messages and knows its own geographic location.

A source node detects an event with a set of attribute values, $E = \{a_1, a_2 \dots a_n\}$. It is assumed that all attribute values are scalar and normalized to (0~1). An event is assigned with a k-bit code, where $k = 2m$. If $0 < a_i < 0.5$ the ith bit code is assigned to 0, otherwise it is assigned to 1. For $x_j = a_{2j} - 1$ and $y_j = a_{2j}$, attribute values of E are assigned a serial bit code $B = \{x_1, y_1, x_2, y_2 \dots\dots x_m, y_m\}$.

For example, an event $E = \{0.8, 0.7, 0.4, 0.3\}$ is assigned a four-bit code $B = \{x_1, y_1, x_2, y_2\} = \{1, 1, 0, 0\}$. Assignment of code B can be achieved as follows ($0 < a_i < 0.5$ and $0.5 < a_i < 1$ are generalized to $0 < 2a_i < 1$ and $1 < 2a_i < 2$, respectively):

$$\hat{B} = [a_1, a_2, \dots\dots, a_{2m}] 2 I_k$$

$$= [2a_1, 2a_2, \dots\dots, 2a_{2m}]$$

$$B = \left[\lfloor 2a_1 \rfloor, \lfloor 2a_2 \rfloor, \dots\dots \lfloor 2a_{2m} \rfloor \right]$$

$$= [x_1, y_1, x_2, y_2 \dots\dots, x_m, y_m]$$

This code B is mapped to a range space $R = [x_{\text{low}} - x_{\text{up}}, y_{\text{low}} - y_{\text{up}}]$. Hence, it is important to calculate $(x_{\text{low}}, y_{\text{low}})$ and $(x_{\text{up}}, y_{\text{up}})$ to find the range space R.

In code B, $x = x_1, x_2 \dots x_m$ are used to calculate X of R and $y = y_1, y_2 \dots y_m$ are used to calculate Y of R. In code B, if $x_1 = 1$ its value is between 0.5 and 1, and if $x_2 = 1$ its value is between 0.5 and $0.5 + (0.5)^2$. So, $X_{\text{low}} = x_1(0.5) + x_2(0.5)^2 + \dots\dots + x_m(0.5)^m$. Similarly, $Y_{\text{low}} = y_1(0.5) + y_2(0.5)^2 + \dots\dots + y_m(0.5)^m$. This is generalized as follows:

$$M_{k \times 2} = \begin{bmatrix} (0.5)^1 & 0 \\ 0 & (0.5)^1 \\ (0.5)^2 & 0 \\ 0 & (0.5)^2 \\ \vdots & \vdots \\ \vdots & \vdots \\ (0.5)^{k/2} & 0 \\ 0 & (0.5)^{k/2} \end{bmatrix}$$

Hence,

$$R_{\text{low}} = [x_{\text{low}}, y_{\text{low}}] = B \times M$$

Again, X_{up} is defined by

$$X_{\text{up}} = 1 - \left(x_1(0.5) + x_2(0.5)^2 + \dots\dots + x_m(0.5)^m \right)$$

So,

$$\ddot{B} = B \oplus J_k$$

$$J_k = [1,1,\ldots\ldots,1]_{1 \times k}$$

So,

$$\widehat{R}_{up} = \left[\widehat{x}_{up}, \widehat{y}_{up}\right] = \ddot{B} \times M$$

$$R_{up} = \left[x_{up}, y_{up}\right] = J_2 - \widehat{R}_{up}$$

So,

$$R = \left[x_{low} - x_{up}, y_{low} - y_{up}\right]$$

Example:

$$\widehat{B} = [0.8, 0.7, 0.4, 0.7, 0.3, 0.6] \times 2I_6$$

$$= [1.6, 1.4, 0.8, 1.4, 0.6, 1.2]$$

$$= \left[\lfloor 1.6 \rfloor, \lfloor 1.4 \rfloor, \lfloor 0.8 \rfloor, \lfloor 1.4 \rfloor, \lfloor 0.6 \rfloor, \lfloor 1.2 \rfloor\right]$$

$$= [1, 1, 0, 1, 0, 1]$$

$$R_{low} = B \times M_{6 \times 2} = [1,1,0,1,0,1] \times \begin{bmatrix} (0.5)^1 & 0 \\ 0 & (0.5)^1 \\ (0.5)^2 & 0 \\ 0 & (0.5)^2 \\ (0.5)^3 & 0 \\ 0 & (0.5)^3 \end{bmatrix}$$

$$= [0.5, 0.875]$$

$$\ddot{B} = B \oplus J_6 = [1,1,0,1,0,1] \oplus [1,1,1,1,1,1]$$

$$= [0,0,1,0,1,0]$$

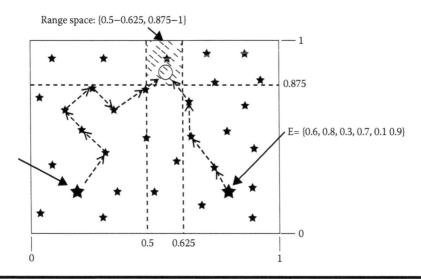

Range space: {0.5–0.625, 0.875–1}

Figure 2.4 Multidimensional range query mechanism (MDAs). (Data from Liao, W. H., and C. C. Chen, *IET Communications,* 4, 1799–808, 2010.)

$$\hat{R}_{up} = \ddot{B} \times M = [0.375, 0]$$

$$R_{up} = \left[x_{up}, y_{up} \right] = J_2 - \hat{R}_{up} = [1,1] - [0.375, 0] = [0.625, 1]$$

Hence, the range space is defined by $R = \{0.5\text{–}0.625, 0.875\text{–}1\}$. Figure 2.4 illustrates the MDAs storage and query process. For routing, GPSR (refer to Section 2.3.1.1.1) is used.

2.3.1.2.1 Simulation and Evaluation

The algorithm is simulated and compared with another range query scheme, that is, Distributed Index for Multi-dimensional Data (DIM) [13]. Authors evaluate MDAs against DIM, first, for the average message cost with insertion and queries and, second, for the number of dropped events in exchange for the quality of data. It is shown that in every case, MDAs outperform DIM.

2.3.1.2.2 Discussion

One of the major limitations of this DCS mechanism is the consideration of uniform distribution of sensor nodes. It is very likely to have node failure in large-scale deployed networks with a massive number of sensor nodes. This mechanism is highly inclined to data loss due to its no replication policy. It is also not suitable for range query, data aggregation, similarity search, and obviously nonuniform network distribution. However, simple and efficient rendezvous node selection procedures make this mechanism attractive among many other DCS mechanisms.

2.3.1.3 Distributed Spatial–Temporal Data Storage Scheme

Shen et al. [21] propose an efficient spatial–temporal SDS scheme for both static and dynamic WSNs. The efficient aggregation and query of data, similarity search for multiattribute data, and spatial–temporal search are identified by the authors as three major challenges faced by DCS schemes. The SDS is claimed to be unique by researchers in its spatial–temporal functionality and similarity search functionalities. It aims to reduce overhead, energy consumption, and searching latency.

In SDS, a deployed large-scale WSN field is considered a rectangular field. The entire field is divided into smaller rectangular zones. Each zone has a dedicated sensor node, named the zone head. It is assumed that each node in the network is configured with three basic pieces of information: (1) The number of zones, horizontally n_x and vertically n_y; (2) the zone ID assignments scheme (IDs are assigned to a zone sequentially from left to right); and (3) the ID and geographical location of its own zone. Each zone is assigned a zone ID. A node in a zone ID_i can calculate its Euclidean distance from another zone ID_j as $\left|ID_i, ID_j\right| = \sqrt{\delta x_{i,j}^2 + \delta y_{i,j}^2}$, where $\delta x_{i,j} = \left(ID_j - ID_i\right)\% n_x$ and $\delta y_{i,j} = \left(ID_j - ID_i\right)/n_x$.

A head node in a zone is responsible for communicating with other zones. All other nodes inside a zone are connected with the head zone.

2.3.1.3.1 Routing Algorithm

For routing a packet from one zone to another, SDS uses the carpooling algorithm. The basic idea of the carpooling algorithm is to select a neighbor zone of the source zone as the next hop, which is closest to the destination. The SDS proposes an improved version of the carpooling algorithm in which packets are combined and forwarded together to the neighbor zone until and unless their next hop is the same. This helps to reduce energy consumption to a certain extent.

2.3.1.3.2 Similarity Search

For similarity search, SDS proposes the following formula:

$$\text{Similarity} = \frac{\sum_{i=1}^{m} w_i \times B(v_i^{d1}, v_i^{d2})}{\sum_{i=1}^{m} w_i}$$

In this formula, m is the number of attributes, w_i is the weight of each attribute (each attribute is assigned a weight value based on its significance) (see Table 2.1), and $B\,(i, j)$ is a Boolean function returning 1 if $v_i^{d1} = v_i^{d2}$ and 0 otherwise.

2.3.1.3.3 Load Balancing

Shen et al. [21] introduce two types of load-balancing schemes in SDS: (1) Storage load balancing and (2) routing load balancing. In storage load balancing, the storage load is balanced among different zones. Every zone maintains a threshold, denoted by ϕ, of the percentage of the zone's used storage to indicate when the zone is at risk of being overloaded. When a zone's storage crosses ϕ, it starts forwarding all storage requests to a lightly loaded neighbor zone. Every zone calculates its storage status $\left(\sum_{i=1}^{N} S_i/N\right)$ and forwards this status to its neighbor zones periodically. In routing

Table 2.1   Example of Weight Settings

Attribute	Key Words	Weight
Object	Car, plane, truck, etc.	0.3
Model	F-16, F-17, etc.	0.2
Color	Red, purple, etc.	0.1
Direction	North, south, etc.	0.1
Division	Air Force, etc.	0.1
Pressure	Integer	0.1
Speed	Float	0.1

Source: Shen, H. et al., *IEEE Transactions on Mobile Computing,* 10, 982–96, 2011.

load balancing, instead of routing packets to one specific route, SDS calculates more than one shortest route between a source or relay and a destination. This helps to reduce congestion in any specific routing path. A tree-based data storage mechanism is also proposed to balance the load among sensors inside a zone. The SDS also provides a mechanism to deal with different situations in dynamic WSNs such as node departure, failure, and joining.

2.3.1.3.4 Simulation and Evaluation

After doing rigorous analysis on SDS, the authors simulate their scheme using "The ONE" [38] simulator to evaluate its performance. It is shown that SDS outperforms directed diffusion and GHT in spatial–temporal and range querying, single result querying, scalability, similarity searching, and performance in a dynamic WSN.

2.3.1.3.5 Discussion

The SDS efficiently implements range querying, spatial–temporal similarity search, and load balancing as discussed in previous subsections. It also modifies the carpooling routing algorithm to show considerable performance improvement. However, data aggregation is not addressed, although it is mentioned in the introduction of the paper [21] as one of the three major challenges. Furthermore, one of the major drawbacks of SDS is the single head node failure problem. Each zone is represented by a single head node, which is responsible for receiving data and giving responses to queries. No alternate mechanism is proposed in the case of head node failure. The SDS is also prone to data loss as it does not replicate its data inside or outside the zone.

2.3.1.4 Load-Balanced and Efficient Hierarchical DCS

Yao et al. [32] present a hierarchical Voronoi graph–based routing (HVGR) algorithm. Taking its inspiration from the Voronoi graph, HVGR constructs and maintains a virtual hierarchy of sensor nodes. This hierarchical phenomenon is self-organizing and doesn't need the use of a GPS or other geolocation devices. The network is divided into different levels of region based on landmarks. A landmark selection algorithm (see Section 2.3.1.4.2) is used to select at most m_i ($i = 1, 2, \ldots$)

landmarks in an $(i-1)$th level region. Then the network is divided into first-level subregions based on first-level landmarks. Each node in the first-level subregions broadcasts a *landmark* packet to the entire network. By receiving this packet, every node estimates its distance from all first-level subregion *landmarks*. Each node selects one *landmark* as its representative. Then each first-level subregion is again divided into second-level subregions with the second-level *landmark*. Each sensor node again chooses the closest *landmark* as their second representative. This process continues until the last-level subregions are small enough so that each node knows all other nodes in its subregion.

2.3.1.4.1 Routing

Figure 2.5 shows an example of the basic routing algorithm that is used to route packets from the source (S) to the destination (D). The first-level landmark of destination is L_1. The packet first moves hop by hop toward L_1 until it reaches the edge of the first-level region, say, R_1. Now, nodes in R_1 know the second-level landmark and the packet now moves hop by hop toward the second-level landmark, say L_2. Again, once it reaches node B of Region 2 in Figure 2.5, the packet starts moving toward L_3. Finally, the packet reaches the lowest-level subregion whose entry node knows the path to the destination and forwards the packet to the destination. The interesting feature of this algorithm is that it never overwhelms the landmark as the packets never go to the landmark; instead the packets move toward the landmark until they reach any node of that region. A series of hash functions H_i ($i = 1, 2, \ldots$) for each level of hierarchy is introduced. The term H_1 is used to enter the first-level region. Once the event enters the first-level regions, H_2 is used to enter the second-level region, and so on.

2.3.1.4.2 Landmark Selection

For landmark selection, authors use an optimized random landmark selection algorithm [39]. According to this algorithm, a random node, say, A, starts the landmark selection process. The node A becomes the master landmark of first-level landmark nodes. It selects $m_1 - 1$ nodes for m_1 first-level networks. Each first-level landmark then acts as master and continues the selection

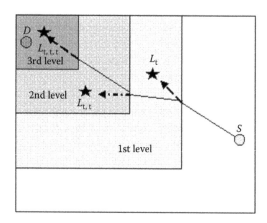

Figure 2.5   Region-oriented routing. (Data from Yao, Z., *Proceedings of the 5th Annual IEEE Communications Society Conference on Sensor, Mesh and Ad Hoc Communications and Networks—SECON '08,* **560–8, 2008.)**

process in its zone for selecting a second-level master node for the second level of the network. A master landmark node stops the landmark selection process once it finds that all sensor nodes in its region are within its communication range.

2.3.1.4.3 Load Balancing

For balancing the load among sensor nodes, the basic name-based routing is modified. During landmark selection, the network is divided unevenly. For example, an event E has a probability $1/m_1$ of being assigned to the first-level landmarks (L_1, L_2, …). If the network is divided unevenly ($L_1 > L_2$), L_2 is more likely to be overloaded earlier than L_1. Hence, the load is balanced by assigning tasks to regions in proportion to the sizes of the regions. An event is stored in a node in L_k's first-level region if and only if

$$\frac{1}{N}\sum_{i=1}^{k-1} N_i \leq H_1(E) \leq \frac{1}{N}\sum_{i=1}^{k} N_i$$

Here, N_i is the number of nodes in landmark L_i's first-level region, and $N = \sum_{i=1}^{m_1} N_i$.

2.3.1.4.4 Simulation and Evaluation

Yao et al. [32] also discuss path stretch reduction and handling of dynamic changes. The proposed HVGR technique is simulated extensively and compared with virtual ring routing (VRR) [40]. Using the result obtained from the simulation, it is showed that HVGR outperforms VRR and achieves the goals of good scalability, good efficiency, and good load balancing for routing and data storage.

2.3.1.4.5 Discussion

The HVGR is proposed focusing mainly on the construction of virtual hierarchical relations among sensor nodes. It does not consider range query, similarity search, data aggregation, non-uniformity, and multireplication. Due to the lack of replication, HVGR is susceptible to data loss.

2.3.1.5 Dynamic Load-Balancing Approach

Liao et al. [31] mention the unbalanced distribution of data among sensors as one of the major constraint in most of the DCS techniques. To address this issue, authors propose a grid-based DLB approach. The DLB approach relies on two schemes: (1) A cover-up scheme to deal with the problem of a storage node whose memory space is depleted and (2) multithreshold levels to achieve load balancing in each grid so that the load is balanced over all nodes.

The DLB divides the whole network in a grid with cells of the same size in such a way that all the nodes inside a cell are within one hop distance. Each grid is numbered using positive coordinates (X, Y) called grid IDs. A sensor node can calculate its grid ID (X, Y) using the following equation:

$$X = \lfloor (X_i - X_0)/d \rfloor \ \text{ and } \ Y = \lfloor (Y_1 - Y_0)/d \rfloor$$

Each node has a virtual grid ID and virtual coordinates, which are initially equal to the actual grid ID and coordinates. Initially, each node broadcasts a message within its grid by limited broadcast to exchange information regarding the *Grid_Node* table. A producer node uses the hash function on the event type, which is mapped into a grid and transformed into a grid ID using the aforementioned equation. The center of the grid is called a grid point. The node, after detecting an event, sends a *Put* packet to the grid ID and uses a geographic routing protocol (GPSR) to forward this packet to the node closest to the grid point.

2.3.1.5.1 Cover-Up Scheme for Load Balancing

To balance the load among sensors in a grid, the cover-up scheme is used. According to this scheme, every node of a grid has storage threshold levels. For instance, a sensor node has two levels of thresholds: (1) First level = 30 and (2) second level = 60. A grid point forwards and stores all packets in the node closest to the grid. When the node closest to the grid point reaches the first threshold level, it modifies its virtual coordinates to (∞, ∞) in order to "cover up" its original location. Hence, the geographic routing protocol forwarding an event data to a storage node will ignore the original closest node and find a new node that is the next closest node to the grid point. This process continues until all the nodes reach their first threshold level. The last node, the one farthest from the center, on reaching this threshold notes this fact and the second threshold level of storage is established.

However, by following this mechanism at some point all the nodes within a grid could become saturated. In such a case, Liao et al. [31] propose the use of an extended grid, which uses all the grids adjacent to the saturated one to select a new home node. When a query reaches a grid, all the sensor nodes of that grid receive this query and, using the data stored in them, reply back to the consumer.

2.3.1.5.2 Simulation and Evaluation

The proposed model is simulated in Java assuming the hierarchy depth $d = 1$. The total energy consumption simulated in this model includes energy consumed for both storing and retrieving an event. The energy consumption cost function is estimated based on the energy model [41]. The energy cost for sending a message S is determined by $E_{send} = E_{trans} \times s + E_{amp} \times d^2$, where E_{trans} is the energy cost of sending a bit, S is the message size, E_{amp} is the energy consumed in the amplifier, and d is the distance of message transmission. On the other hand, energy consumption for receiving a message is determined by the cost function $E_{send} = E_{rec} \times r$, where E_{rec} is the energy cost of receiving a bit and r is the message size. The DLB is evaluated against GHT for total energy consumption, performance of the hotspot storage space, standard deviation of storage space, average storage space for load balancing and, finally, the number of dropped events for the quality of data.

2.3.1.5.3 Discussion

This proposal uses a smart way of changing home nodes in order to keep the load balanced in a network. However, DLB does not focus on the recovery of data in the case of node failure and/or movement of a node from one zone to another. It also completely skips the replication process, which is a vital process in the case of DCS. According to the core design of DLB, it cannot be scaled or extended for similarity search, data aggregation, and range query.

2.3.1.6 LB-DCS

In the study by Albano et al. [28], an organic approach named LB-DCS is proposed to overcome the constraint of load imbalance in DCS-GHT [1]. LB-DCS functions on top of three mechanisms: (1) A density estimation protocol that is used to estimate network density f, which is included in *put* and *get* protocols; (2) a modified hashing function that includes f in its parameter list; and (3) a storage protocol enforcing quality of service (QoS) in the selection of a number of replicas for data storage.

Some authors [28,42] claim that depending on the event type the number of local replicas should be different. So, when a producer node produces any event, it also specifies a value in terms of the parameter q to specify the number of replicas. The *put* primitive takes q along with two other parameters: (1) Datum d and (2) metadatum k. On the basis of the q parameter, a home node selects q number of neighbor nodes using the *ball* method to replicate that event. This dispersal method is iterative. The home node, say, H (with coordinates X_H, Y_H), considers a ball with radius r (a randomly selected value). It sends a request for storage to all sensors within the range of this ball denoted by

$$B_{(X_H, Y_H)}(r) = \{sensors\ of\ coordinates\,(x, y) : |(x_H, y_H), (x, y)| < r\}$$

In turn, when a sensor receives a storage request, it acknowledges the request to H. The home node H calculates the number of acknowledgments (q') received from the nearest sensors. If $q' < q$, then H sends a storage request to $B_{(X_H, Y_H)}(2 \times r)$ sensors. This time, it considers only the sensors in $B_{(X_H, Y_H)}(2 \times r) - B_{(X_H, Y_H)}(r)$. This process continues until H receives q number of acknowledgments or goes out of the outermost perimeter.

Apart from ensuring QoS, authors also include nonuniform hashing, which can be used to balance the load even in a nonuniform distribution, say, Gaussian distribution, of sensor nodes in a network. For uniform hashing, they have used the *Rejection Method* [43].

2.3.1.6.1 Network Density Estimation

For handling dynamic networks, a hash function includes network density estimation f. For the purpose of estimating sensor density, a WSN is divided into $n \times n$ nonoverlapping square regions of side p. The point at the center is called the watch point. The sensor node closest to the watch point is called the sentinel. After selecting the sentinels, each sentinel broadcasts a request to its neighbors to count them. The number of neighbors is used as an estimation of the local density of a region. One proactive protocol (Broadcast) and two reactive protocols (Stripes and FatStripes) are used to deliver the estimate computed by sentinels to other sensor nodes. In Broadcast, each sentinel sends its density information to all the sensors once during initial setup. In Stripes, density estimation is stored in every sensor along the unicast back route from a sentinel. When a query for a sentinel arrives to a sensor, it first checks its cache. If record is not found, only then a request is forwarded to the target sentinel. The FatStripes protocol is the enhanced version of Stripes. As all the nodes in a transmission range spend the same energy for receiving an estimation of density, the estimation can be copied to all the sensors who receive it. This obviously increases the probability of a hit during querying. After collecting estimates, a sensor uses the formulas $d_{i,j}' = \dfrac{w_{ij}}{\sum_{ij} w_{ij}}$ and

$$d_{i,j}' = \frac{m \times d_{i,j}' + \sum_{i',j' \in N_{ij}} d_{i',j'}'}{2 \times m}$$ for calculating density in each region and final approximation based on the first computation, respectively.

2.3.1.6.2 Simulation and Evaluation

For performance evaluation, both LB-DCS and DCS-GHT are simulated in an NS-2 simulator. The number of data stored in different nodes is measured for different loads, and it is shown that

LB-DCS balances the load for both uniform and Gaussian distributions. The number of Medium Access Control (MAC) layer message exchanges during *get* and *put* functions is also measured for DCS-GHT and LB-DCS against different levels of density. The graph presented by Albano et al. [28] reports that for different values of density, LB-DCS exchanges a significantly lower number of MAC-level messages than does DCS-GHT. It also shows that the *put* operation exchanges more messages than the *get* operation due to the dispersion mechanism inside the zone during the *get* operation.

2.3.1.7 ToW

In ToW [29], authors propose a data-centric mechanism in which queries and events meet at a point that is selected based on the relative frequencies of events and queries. To minimize the communication cost, ToW adjusts the rendezvous point on the fly on an optimal basis. The ToW takes motivation from the SR mechanism in GHT. In SR, to alleviate a node's load, a detected event is stored in the nearest mirror image. This mechanism alleviates the storage cost; but as the query node has no idea which image node may have the required data, it needs to query all the images, thereby increasing the query cost. However, this mechanism is useful if event detection frequency is much higher than event query frequency. Similarly, if queries incur more cost than events, then the cost is reduced by letting a node send its query message to only its nearest image, as this event is stored in all mirror images.

2.3.1.7.1 Modes of Operation

The ToW operates in two modes, which are referred to as "write-one-query-all" and "write-all-query-one." The former mode allows a sensor to store an event in the nearest mirror image, but queries need to be disseminated to all mirror images. Conversely, in the latter operation mode, events must be stored in all mirror images to facilitate a query node to disseminate a query only to the nearest mirror image. In which mode a sensor node will operate depends on the resolution r. Resolution is determined based on the relative frequency of queries and events detection for a particular class of events.

2.3.1.7.2 Routing Algorithm

In ToW, authors also present a new routing algorithm, that is, "combing routing" to replace the hierarchical routing algorithm, which is presented in SR-GHT to route a message to the set of all level-r images of an event class. The cost related to hierarchical routing in SR-GHT is shown by

$$CR_{Hierarchical} \cong \delta(s,h) + \frac{\left(2+\sqrt{2}\right)\sqrt{n}}{2}(2^r - 1)$$

$$\cong \delta(s,h) + \left[\left(1+\sqrt{2}/2\right) \times 2^r - \left(1+\sqrt{2}/2\right)\right]\sqrt{n}$$

Here,

$$CR_{Hierarchical} = \delta(s,h) + S_1 + S_2 + \ldots\ldots + S_r$$

$\delta(s,h)$ is the distance between source node S and the root image h

S_i is the distance between each $(i-1)$-level image and the three associated i-level images

There are 4^{i-1} $(i-1)$-level images, and each $(i-1)$-level image is associated with three i-level images. The cost of communication between each $(i-1)$-level image and the three associated i-level images is $\sqrt{n}/2^i, \sqrt{n}/2^i, \sqrt{n}/2^i \times \sqrt{n}$.

Hence,

$$S_i = 4^{i-1} \times \left(\frac{2}{2^i} + \frac{\sqrt{2}}{2^i}\right)\sqrt{n} = \frac{\left(2+\sqrt{2}\right)\sqrt{n}}{2} \times 2^{i-1}$$

On the other hand, in the combing routing algorithm, the cost consists of a routing path from the message source to the nearest image and a horizontal path of length $\sqrt{n}\left(1-\dfrac{1}{2^r}\right)$ followed by 2^r vertical paths of the same length. The distance d from the source to the nearest image is $d = \delta(s,h)$ when $r=0$, and d is halved for every increment of r. So, $d = \dfrac{\delta(s,h)}{2^r}$. Hence, the total cost is defined by

$$CR_{Combing} \cong \frac{\delta\left(s,h\right)}{2^r} + \left(2^r+1\right)\left(1-\frac{1}{2^r}\right)\sqrt{n}$$

To optimize a sensor's mode of operation based on resolution, two quantities f_e and f_q referring to event frequency and query frequency, respectively, are defined. For write-one-query-all, the cost is defined by

$$C_{W_1Q_{all}} \cong 2\sqrt{2k\overline{\delta}}\sqrt{\frac{f_e}{f_q}}\sqrt{n}$$

Here, $\overline{\delta}\sqrt{n}$ is the average distance between a querying node and a home node of an event class and k is the number of nodes detecting events of a particular class denoted by C.

The cost of write-one-query-all is defined by

$$C_{W_{all}Q_1} \cong 2\sqrt{2k\overline{\delta}}\sqrt{\frac{f_q}{f_e}}\sqrt{n}$$

2.3.1.7.3 Evaluation and Comparison

The ToW mechanism is analytically compared with another very similar mechanism referred to as comb–needle (CN) [44], in which authors study push–pull event dissemination and query strategy in a grid network. Although both CN and ToW use dynamic strategies for information dissemination and gathering, there are some significant basic differences between them according to Joung and Huang [29]. First, ToW is based on DCS, whereas CN is not data centric; so a query source in CN has no prior knowledge about which specific node to be used to gather information. As a result, flooding is essentially a search technique in CN. Second, through rigorous mathematical evaluation, it has been shown that ToW has a smaller dominating constant in communication cost. As a result, on an average the query latency of ToW is half that of CN.

2.3.1.8 Similarity Search Algorithm

Chung et al. [23] propose the similarity search algorithm (SSA) based on the concept of a Hilbert curve over a DCS structure. The SSA is successful in searching similar data without collecting data from all the sensors in a network (see Section 2.3.1.8.2). Being motivated by the concept of Hilbert curve, authors divide the network recursively into 4^l square quadrants where l denotes the number of levels. The center (indexing node) of each square quadrant (cell) is denoted by I. It is important to select the proper number of indexing nodes (I) to avoid the performance degradation caused by the presence of too many indexing nodes and, on the other hand, the lack of storage space caused by too few indexing nodes. If the total memory space for storing data is A and the memory size of each sensor is z, then the number of indexing nodes n can be defined as $n \geq \lceil A/Z \rceil$. So, the number of levels l is defined by $l = \log_4 n \geq \lceil A/z \rceil$.

2.3.1.8.1 Network Overlay

The entire data range of an event is denoted by R, where R_L and R_U denote the lower and upper bounds of R, respectively. The data range R is divided into n equal subranges each equal to r, that is, $n \cdot r = R$. So the subrange of data for which I_{ID} is responsible is defined as $\left[R_L^{I_{ID}}, R_U^{I_{ID}} \right) = \left[R_L + (I_{ID} - 1) \cdot r, R_L + I_{ID} \cdot r \right)$. Figure 2.6 illustrates the Hilbert curve for level 1 and level 2, assuming that the whole data range R of an event is (0, 1).

Any detected event is mapped to a particular segment if the event falls in the range of that cell. Two parameters $\left(v_{min}^{ID}, v_{max}^{ID} \right)$ are used to record the minimum and maximum existing values of each segment. Initially, these two parameters are set to 0. When data is inserted in an index node, these two values are updated accordingly. For example, if a sensor detects an event with a value of 0.2, then the data is sent to I_0 as it belongs to the range [0, 0.25) and both the parameters $\left(v_{min}^{ID}, v_{max}^{ID} \right)$ of this cell are updated to 0.2 $\left(v_{min}^{ID} = 0.2, v_{max}^{ID} = 0.2 \right)$.

2.3.1.8.2 Similarity Search

The similarity search mechanism presented in [23] consists of two phases: (1) The similarity search query-resolving phase and (2) the query probing phase. The query-resolving phase determines an

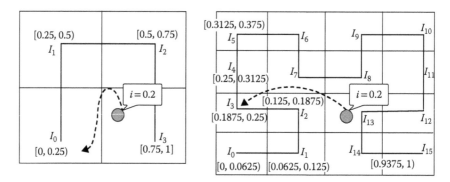

Figure 2.6 Level-1 and level-2 Hilbert curves. (Data from Chung, Y-C. et al., *Information Sciences*, 181, 284–307, 2011.)

indexing node that is most likely to provide an answer to the query. If the answer does not match exactly with the query, then the query-probing phase is initiated for finding the closest possible answer.

In the query-resolving phase, the following *locate*() function is used to find the *Target Node* I_T:

locate (V_q) = locate the indexing node I_{ID} such that $R_L + (I_{ID} - 1).r \leq V_q < R_L + I_{ID}.r$, where V_q is the search value given by the query.

The query is then forwarded to the *Target Node* I_T to retrieve data. If an exact match to V_q is found, then the query execution process is finished; otherwise, the query-probing phase becomes active.

In the query-probing phase, there can be two possible cases:

- **Case 1**: In this case, I_T is nonempty, and $v_s{}^{I_T}$ is the most similar local data in I_T.
- **Subcase 1**: If $v_s{}^{I_T}$ is larger than v_q, then all the data in I_{T+1} must be greater than v_q. But in I_{T-1}, there may be a $v_s{}^{I_{T-1}}$ that is closer to v_q.
- **Subcase 2**: If $v_s{}^{I_T}$ is smaller than v_q, then all the data in I_{T-1} must not be more similar to v_q than $v_s{}^{I_T}$ is. But in I_{T+1}, there may be a $v_s{}^{I_{T+1}}$, which is closer to v_q.
- **Case 2**: In this case, I_T is empty (i.e., no data is stored). The v_q has to be sent to both the neighbors (i.e., I_{T-1} and I_{T+1}) of I_T to find the most similar data.

Three functions are proposed, which are named backward probing, forward probing, and bidirectional probing. Backward probing and forward probing are used to deal with Case 1, whereas bidirectional probing is used to deal with Case 2.

2.3.1.8.3 Index Node Handoff and Failure

Two methods are proposed for workload sharing. One case is when an indexing node fails due to power shortage. In the first case, once the power level of a sensor goes beyond a certain threshold, the system can be defined or preinstalled and another sensor takes the responsibility. The basic idea is to transfer the current jobs and the data to a nearby sensor. In the second case, a node fails due to sudden unexpected damage. In this scenario, data have to be duplicated in the mirror indexing node. Neighbors of this victim indexing node are aware of this failure; hence, when a detected data/query is sent to this failed indexing node through these neighbors, they locate the mirrored indexing node by using the mirror mapping function and forward the data to that indexing node.

2.3.1.8.4 Range Query

Chung et al. [23] also extend their study to range query processing for both single-dimensional attributes and multi-dimensional attributes. A range query is divided into subqueries and the subqueries are forwarded to their corresponding indexing nodes. For example, the level of carbon monoxide in the air in a forest, the number of levels of the network 1, ranges from 50 to 150 L/mol. This range is split equally into four ($4^1 = 4$, four index nodes denoted by I_0, I_1, I_2, and I_3) subranges, that is, [50, 75), [75, 100), [100, 125), and [125, 150). Hence, if a range query is to find data within the subrange [60, 80), then the subranges of I_0 and I_1 are located as they cover the given query.

2.3.1.8.5 Simulation and Evaluation

The model SSA is simulated and evaluated against SR-GHT and a naive algorithm for network size, node density, node distribution, and the number of levels of a Hilbert curve. As communication forms the main part of energy consumption, the number of exchanged messages is used as the comparison metric for evaluation.

2.3.2 Data-Centric Conveyance Based on Tree Structures

In data-centric conveyance routing, sensor nodes communicate and route data based on reinforcement feedback from upstream nodes, resulting in a hierarchical tree structure. The position of a sensor node in a tree is used as its routing address. A few examples of this type of routing are path-based tree structure (pathDCS), HVGR, and virtual polar coordinate routing. Sections 2.3.2.1 through 2.3.2.2 briefly illustrate a couple of DCS methods that are formed on top of data-centric conveyance routing.

2.3.2.1 Distributed Index for Features in a Sensor Network

A DIFS [22] is designed to provide balance on communication load across an index keeping the search efficiency of a quad tree [45]. The DIFS also uses a geographic hashing mechanism such as GHT [1]. The DIFS constructs a multiply rooted hierarchical index that differs from traditional binary and quaternary trees. In DIFS, a nonroot tree can have multiple parents. Nodes are responsible for storing information for a specific range within a particular geographic region. A node covering a small area stores a wide range of values, whereas a node covering a large area stores a small range of values. The DIFS efficiently supports range queries, queries related to distribution of values in space, and so on.

The DIFS constructs a search hierarchy of histograms; unlike the single-tree hierarchies of SR and quad trees, each child in DIFS has *bfact* (*bfact* $= 2^i$, where $i > 1$) parents. The range of values maintained by a child in its histogram is *bfact* times the range of values maintained by its parents.

The range of values that an index node knows about is inversely related to the spatial extent covered by the node. This helps to ensure a balance in communication load over the network. A search in DIFS may originate from any node of the tree. A node, storing an event, forwards information first to the local index node having the narrowest spatial coverage but covering the widest value range. A histogram describing the values is then forwarded to a node having a wider spatial coverage but a narrower value range than the current index node. This process is continued. Unlike GHT, the DIFS function takes a string of characters, source locations, and a bounding box in its hash function to produce a hash location.

2.3.2.1.1 Simulation and Evaluation

In simulation and analysis, authors consider quad tree, diffusion, and SR using numerical simulations to compare with DIFS. Based on the simulation result provided by authors, SR [1] outperforms DIFS and quad tree [45] approaches in terms of communication for storage. The reason given is that the initial registration of an event with SR's local leaf-level mirror point needs no additional work. However, queries with sufficiently constrained search criteria, that is, DIFS and quad tree, lead to the pruning of branches in the search tree, consequently leading to the generation of lower search costs than SR. On the other hand, queries that request most of the entire range data result in no pruning; DIFS and quad tree do not perform better than SR. However, average bottleneck utilization is significantly reduced when the query range is sufficiently constrained. In quad trees, similar to DIFS, the bottleneck is always the root and its utilization is 1, whereas in SR every node of the tree is explored during every search. In terms of aggregate communication, cost of storage, and cost of search, quad tree outperforms DIFS, but it is not scalable to a large number of searches or stores. Using multiply rooted trees and a trade-off between geography and value coverage, DIFS scales well to a large extent.

2.3.2.2 Practical DCS

In pathDCS [46], paths between reference points called landmarks and the sensed data storage nodes called name locations are identified. The pathDCS algorithm defines tree structures in order to route messages. The novelty of this approach is that an event type is mapped to a path instead of a spatial location. The path is defined by an initial landmark and then followed by a set of procedural directions. The landmarks are also called beacon nodes, which are elected randomly or manually configured. Standard tree construction techniques are used to build trees rooted at each of these beacon nodes. This ensures all nodes know how to reach the beacons.

A path consists of a sequence of p beacons (b_i) and lengths l_i, where $i = 1, 2, \ldots p$. The packet is first routed to beacon b_1, then it is sent l_2 hops toward beacon b_2 using the tree rooted at b_2, and so on until it ends up at the previous $i - 1$ segment. It is then sent l_i hops toward the next beacon b_i. A beacon node b_i is the beacon whose identifier is closest to the hash function $h(k, i)$. In addition, the first segment length l_1 is equal to the distance to the first beacon b_1, whereas segment lengths for $i > 1$ are given by

$$l_i = h(k,i) \bmod hops(n_i, b_i).$$

The pathDCS algorithm has two parameters: (1) The total number of beacon nodes denoted by B and (2) the number of path segments denoted by P. The performance of pathDCS largely depends on these two parameters. Minimization of beacon nodes helps to minimize overhead. Data are also locally replicated by flooding within k-hops of the destination.

2.3.2.2.1 Beacon Handoff and Failure

The responsibilities of beacons are handed over to other nodes periodically. This happens so that node failure is avoided or the forwarding load on previously selected beacon nodes is reduced. Every beacon node maintains a threshold value measuring link quality. If link quality goes under the threshold value, all single hop nodes participate in the beacon selection process. The winning node in the beacon selection process kills the current beacon node and assumes the role of beacon henceforth. In the latter case, or deliberate handoff, a beacon randomly selects a neighbor node and exchanges its identifier with the neighbor.

2.3.2.2.2 Simulation and Evaluation

Performance evaluation is done through both high-level and packet-level simulations, as well as experiments on a sensor node test bed. The simulation is conducted over a TinyOS simulator

(TOSSIM). Authors correlate the percentage of total transmission (which increases exponentially) with the percentage of nodes. Success probability is also tested under failure and randomized parent selection for increasing network sizes. The pathDCS algorithm is also implemented in TinyOS and evaluated on the 100-node Intel Mirage MICAz testbed, as well as on 500 nodes in a TOSSIM packet-level emulator.

2.3.2.2.3 Discussion

It is noted that DCS methods are classified into generations in this chapter, that is, point-to-point and tree structures. Authors prefer the latter structure and point out different problems associated with the former routing method. The pathDCS algorithm mainly focuses on tree construction and, hence, query mechanism is left out of focus. However, authors also totally ignore other major challenges such as range query, similarity search, data aggregation, load balancing, and dealing with nonuniformity.

2.4 Future Research Direction

None of the routing protocols illustrated in this chapter are specifically suitable for DCS. The WSN has no Internet protocol (IP) addressing scheme in the network layer; hence, developing an efficient routing layer is challenging for researchers in this field. A deployed network can be divided into zones or sectors. The sector or zone number, referring to sectors or zones to which sensors belong, can be used to address the sensors logically. One, or more than one, sector head can act as a router for a particular zone or sector, which can also be referred to as a subnetwork. Moreover, there is no DCS scheme that addresses all the challenges revealed in Section 2.2. Additionally, most of the DCS schemes used GHT to hash event types or names over the geographic space. The efficiency of GHT decreases due to the skewness in storage and data routing when the network is nonuniformly distributed. Current DCS schemes also increase data storage cost in finding the destination of particular types of data, although they reduce query cost significantly. Hence, there is significant trade-off between data storage techniques and energy consumption, which needs to be optimized. Moreover, most of the DCS schemes discussed in this chapter are simulated in a simplified manner; they consider the number of messages exchanged as the prime cost metric and ignore the impact of short-range low-power communication layers.

2.5 Conclusion

Over the past two decades, a large amount of research work has been done on WSNs. However, the history of DCS goes back to less than one decade. Since sensor networks are becoming prime sources of data collection, research on DCS is of pivotal importance. The key DCS mechanisms have been classified in this chapter and are summarized in Table 2.2. It is clear that with the passage of time, researchers must address new challenges relevant to DCS and try to provide solutions as well. However, from Table 2.2 it is apparent that very few studies focus on the recently investigated important challenges such as similarity search, spatial–temporal search, dynamic load balancing, data aggregation, and dealing with nonuniformity. Aspiring researchers will find this chapter a promising source of information about previous and current research trends in the field of DCS and use their knowledge for the enrichment of this field.

Data-Centric Storage in Wireless Sensor Networks ■ 57

Table 2.2 Comparison of DCS Methods

	Title	Routing Category	Dimension (Attribute)	Range versus Point Query	Data Aggregation	Similarity Search	Multireplication	Load Balancing
1	GHT [1]	Point-to-point routing	Single	Point	No	No	No	No
2	Data storage and range query mechanism for MDAs [20]	Point-to-point routing	Multi	Range	No	No	No	No
3	Distributed spatial–temporal data storage scheme [21]	Point-to-point routing	Multi	Range	No	Yes	No	Yes
4	Load-balanced and efficient hierarchical DCS [32]	Point-to-point routing	Single	Point	No	No	No	Yes
5	DLB approach [31]	Point-to-point routing	Single	Point	No	No	No	Yes
6	LB-DCS [28]	Point-to-point routing	Single	Point	No	No	No	Yes
7	ToW [29]	Point-to-point routing	Single	Point	No	No	Yes	No
8	Efficient mechanism for similarity search [23]	Point-to-point routing	Both	Both	No	Yes	No	Yes
9	DIFS [22]	DCS based on tree structure	Single	Range	No	No	No	No
10	The pathDCS algorithm [46]	DCS based on tree structure	Single	Point	No	No	No	No

References

1. Ratnasamy, S., B. Karp, L. Yin, F. Yu, D. Estrin, R. Govindan, and S. Shenker. 2002. "GHT: A Geographic Hash Table for Data-Centric Storage." Presented at the Proceedings of the 1st ACM International Workshop on Wireless Sensor Networks and Applications, Atlanta, GA.
2. Campobello, G., A. Leonardi, and S. Palazzo. 2009. "A Novel Reliable and Energy-Saving Forwarding Technique for Wireless Sensor Networks." Presented at the Proceedings of the 10th ACM International Symposium on Mobile Ad hoc Networking and Computing, New York, NY, 269–78.
3. Pottie, G. J., and W. J. Kaiser. 2000. "Wireless Integrated Network Sensors." *Communications of the ACM* 43: 51–8.
4. Saroiu, S., P. K. Gummadi, and S. D. Gribble. 2002. "A Measurement Study of Peer-to-Peer File Sharing Systems." Presented at the International Conference on Multimedia Computing and Networking (MMCN), San Jose, USA, 152–66.
5. Yao, Y., X. Tang, and E. P. Lim. 2006. "In-Network Processing of Nearest Neighbor Queries for Wireless Sensor Networks." Presented at the Proceedings of the 11th International Conference on Database Systems for Advanced Applications, Springer-Verlag, Berlin, Heidelberg, 35–49.
6. Szewczyk, R., J. Polastre, A. Mainwaring, and D. Culler. 2004. "Lessons from a Sensor Network Expedition." *Wireless Sensor Networks* 2920: 307–22.
7. Intanagonwiwat, C., R. Govindan, and D. Estrin. 2000. "Directed Diffusion: A Scalable and Robust Communication Paradigm for Sensor Networks." Presented at the Proceedings of the 6th Annual International Conference on Mobile Computing and Networking, Boston, MA.
8. Wensheng, Z., C. Guohong, and T. La Porta. 2007. "Data Dissemination with Ring-Based Index for Wireless Sensor Networks." *IEEE Transactions on Mobile Computing* 6 (7): 832–47.
9. Madden, S., M. J. Franklin, J. M. Hellerstein, and W. Hong. 2002. "TAG: a Tiny AGgregation Service for Ad-hoc Sensor Networks." *SIGOPS Operating Systems Review* 36: 131–46.
10. Ye, F., G. Zhong, S. Lu, and L. Zhang. 2005. "Gradient Broadcast: A Robust Data Delivery Protocol for Large Scale Sensor Networks" *Wireless Networks* 11: 285–98.
11. Ye, F., H. Luo, J. Cheng, S. Lu, and L. Zhang. 2002. "A Two-Tier Data Dissemination Model for Large-Scale Wireless Sensor Networks." Presented at the Proceedings of the 8th Annual International Conference on Mobile Computing and Networking, Atlanta, GA.
12. Ratnasamy, S., B. Karp, S. Shenker, D. Estrin, R. Govindan, L. Yin, and F. Yu. 2003. "Data-Centric Storage in Sensornets with GHT, a Geographic Hash Table." *Mobile Networks and Applications* 8: 427–42.
13. Li, X., Y. J. Kim, R. Govindan, and W. Hong, 2003. "Multi-Dimensional Range Queries in Sensor Networks." Presented at the Proceedings of the 1st International Conference on Embedded Networked Sensor Systems, Los Angeles, CA.
14. Ganesan, D., D. Estrin, and J. Heidemann. 2003. "DIMENSIONS: Why Do We Need a New Data Handling Architecture for Sensor Networks?" *ACM SIGCOMM Computer Communication Review* 33: 143–48.
15. Ganesan, D., A. Cerpa, W. Ye, Y. Yu, J. Zhao, and D. Estrin. 2004. "Networking Issues in Wireless Sensor Networks." *Journal of Parallel and Distributed Computing* 64: 799–814.
16. Akyildiz, I. F., and I. H. Kasimoglu. 2004. "Wireless Sensor and Actor Networks: Research Challenges." *Ad hoc Networks* 2: 351–67.
17. Melodia, T., D. Pompili, V. C. Gungor, and I. F. Akyildiz. 2007. "Communication and Coordination in Wireless Sensor and Actor Networks." *IEEE Transactions on Mobile Computing* 6 (10): 1116–29.
18. Chatzigiannakis, I., A. Kinalis, and S. Nikoletseas. 2005. "An Adaptive Power Conservation Scheme for Heterogeneous Wireless Sensor Networks with Node Redeployment." Presented at the Proceedings of the 17th Annual ACM Symposium on Parallelism in Algorithms and Architectures, New York, NY, 96–105.
19. Shih, K. P., S. S. Wang, H. C. Chen, and P. H. Yang, 2008. "CollECT: Collaborative Event Detection and Tracking in Wireless Heterogeneous Sensor Networks." *Computer Communications* 31: 3124–36.
20. Liao, W. H., and C. C. Chen. 2010. "Data Storage and Range Query Mechanism for Multi-dimensional Attributes in Wireless Sensor Networks." *IET Communications* 4: 1799–808.

21. Shen, H., L. Zhao, and Z. Li. 2011. "A Distributed Spatial-Temporal Similarity Data Storage Scheme in Wireless Sensor Networks." *IEEE Transactions on Mobile Computing* 10: 982–96.
22. Greenstein, B., D. Estrin, R. Govindan, S. Ratnasamy, and S. Shenker. 2003. "DIFS: A Distributed Index for Features in Sensor Networks." Presented at the Proceedings of the First IEEE International Workshop on Sensor Network Protocols and Applications, 163–73.
23. Chung, Y-C., I. F. Su, and C. Lee. 2011. "An Efficient Mechanism for Processing Similarity Search Queries in Sensor Networks." *Information Sciences* 181: 284–307.
24. Ghose, A., J. Grossklags, and J. Chuang. 2003. "Resilient Data-Centric Storage in Wireless Ad-Hoc Sensor Networks." Presented at the Proceedings of the 4th International Conference on Mobile Data Management, Springer-Verlag, London, UK.
25. Madden, S. R., M. J. Franklin, J. M. Hellerstein, and W. Hong. 2005. "TinyDB: An Acquisitional Query Processing System for Sensor Networks." *ACM Transactions on Database Systems* 30: 122–73.
26. Amato, G., P. Baronti, and S. Chessa. 2005. "MaD-WiSe: Programming and Accessing Data in a Wireless Sensor Networks." Presented at the EUROCON 2005—The International Conference on Computer as a Tool, 1846–49.
27. Cuevas, M., Uruena, R. Cuevas, and R. Romeral, 2011. "Modelling Data-Aggregation in Multi-Replication Data Centric Storage Systems for Wireless Sensor and Actor Networks." *IET Communications* 5: 1669–81.
28. Albano, M., S. Chessa, F. Nidito, and S. Pelagatti. 2011. "Dealing with Nonuniformity in Data Centric Storage for Wireless Sensor Networks." *IEEE Transactions on Parallel and Distributed Systems* 22: 1398–406.
29. Joung, Y-J., and S-H. Huang. 2008. "*Tug-of-War: An Adaptive and Cost-Optimal Data Storage and Query Mechanism in Wireless Sensor Networks.*" Presented at the Proceedings of the 4th IEEE International Conference on Distributed Computing in Sensor Systems, Nikoletseas, S., B. Chlebus, D. Johnson, and B. Krishnamachari, Eds., Springer-Verlag, Berlin, Heidelberg, 5067, 237–51.
30. Rumín, Á. C., M. U. Pascual, R. R. Ortega, and D. L. López. 2010. "Data Centric Storage Technologies: Analysis and Enhancement." *Sensors* 10: 3023–56.
31. Liao, W-H., K-P. Shih, and W-C. Wu, 2010. "A Grid-based Dynamic Load Balancing Approach for Data-centric Storage in Wireless Sensor Networks." *Computers and Electrical Engineering* 36: 19–30.
32. Yao, Z., C. Yan, and S. Ratnasamy. 2008. "Load Balanced and Efficient Hierarchical Data-Centric Storage in Sensor Networks." Presented at the Proceedings of the 5th Annual IEEE Communications Society Conference on Sensor, Mesh and Ad Hoc Communications and Networks—SECON '08, 560–8.
33. Stoica, I., R. Morris, D. Karger, M. F. Kaashoek, and H. Balakrishnan. 2001. "Chord: A Scalable Peer-to-Peer Lookup Service for Internet Applications." *SIGCOMM Computer Communication Review* 31: 149–60.
34. Maymounkov, P., and D. Mazieres. 2002. "Kademlia: A Peer-to-Peer Information System Based on the Xor Metric." *Peer-to-Peer Systems* 53–65.
35. Rowstron, A., and P. Druschel. 2001. "*Pastry: Scalable, Decentralized Object Location, and Routing for Large-Scale Peer-to-Peer Systems Middleware 2001.*" Presented at the International Conference on Distributed Systems Platforms, Guerraoui R., Ed., Springer-Verlag, Berlin, Heidelberg, 2218, 329–50.
36. Karp, B., and H. T. Kung. 2000. "GPSR: Greedy Perimeter Stateless Routing for Wireless Networks." Presented at the Proceedings of the 6th Annual International Conference on Mobile Computing and Networking, Boston, MA.
37. Finn, G. G. 1987. *Routing and Addressing Problems in Large Metropolitan-Scale Internetworks.* DTIC Document, ISI/RR-87-180, Marina del Rey, CA, USC/Information Sciences Institute.
38. Keränen, A., and J. Ott. 2008. "Opportunistic Network Environment Simulator" [Special Assignment Report]. Department of Communications and Networking, Helsinki University of Technology. Available at http://www.netlab.tkk.fi/tutkimus/dtn/theone/pub/the_one.pdf. Last accessed January 2012.
39. Zhao, Y., L. Bo, Z. Qian, Y. Chen, and Z. Wenwu, 2005. "Efficient Hop ID Based Routing for Sparse Ad Hoc Networks." Presented at the Proceedings of the 13th IEEE International Conference on Network Protocols, Washington, DC, 179–190.

40. Caesar, M., M. Castro, E. B. Nightingale, G. O'Shea, and A. Rowstron. 2006. "Virtual Ring Routing: Network Routing Inspired by DHTs." *ACM SIGCOMM Computer Communication Review* 36: 351–62.
41. Heinzelman, W. R., A. Chandrakasan, and H. Balakrishnan. 2000. "Energy-Efficient Communication Protocol for Wireless Microsensor Networks." Presented at the Proceedings of the 33rd Annual Hawaii International Conference on System Sciences, Washington, DC, 2, 10.
42. Albano, M., S. Chessa, F. Nidito, and S. Pelagatti. 2007. "Q-NiGHT: Adding QoS to Data Centric Storage in Non-Uniform Sensor Networks." Presented at the Proceedings of the 2007 International Conference on Mobile Data Management, Washington, DC, 166–73.
43. Von Neumann, J. 1951. "Various Techniques Used in Connection with Random Digits." *Applied Math Series* 12: 1.
44. Liu, X., Q. Huang, and Y. Zhang. 2004. "Combs, Needles, Haystacks: Balancing Push and Pull for Discovery in Large-Scale Sensor Networks." Presented at the Proceedings of the 2nd International Conference on Embedded Networked Sensor Systems, Baltimore, MD.
45. Finkel, R. A., and J. L. Bentley. 1974. "Quad Trees a Data Structure for Retrieval on Composite Keys." *Acta Informatica* 4: 1–9.
46. Ee, C. T., and S. Ratnasamy. 2006. "Practical Data-Centric Storage." Presented at the Proceedings of the 3rd Conference on Networked Systems Design and Implementation, USENIX Association, Berkeley, CA.

Chapter 3

Environmental Forest Monitoring Using Wireless Sensor Networks

A Survey

Mohammad Abdul Azim, Fatemeh M. Kiaie,
and Mohamed H. Ahmed

Contents

3.1 Introduction

Wireless sensor networks (WSNs) are networks of tiny sensor nodes connected with wireless links. The fundamental purpose of deploying such networks is to monitor a wide variety of ambient conditions. A large number of diverse applications can be imagined using such networks. This includes environmental applications, wildlife monitoring, battlefield monitoring, target detection, rescue applications, health applications, agriculture and crop monitoring, consumption monitoring in retail management, and industrial applications.

Figure 3.1 shows a layout of a WSN where the users access the network data via the Internet. Sensor nodes consist of a processing unit along with storage, a wireless transceiver, a sensing unit with an ADC powered by the power unit connected to the battery (optionally attached to the power generator unit such as solar cell). Moreover, more complex and costly nodes can be equipped with a GPS for localization purposes. In the architectural perspective, the protocol stack used by the WSNs consists of the application, transport, network, data link, and physical layers.

Small to medium WSNs can be designed as flat networks where all nodes have the same functionality and accessibility. On the other hand, for larger WSNs, the network is divided into clusters where a subset of nodes (cluster heads) take the responsibility of managing other nodes, aggregating sensed data, and routing packets to the sink node. Such networks are known as hierarchical WSNs. Figure 3.2 depicts the structures of flat and hierarchical networks.

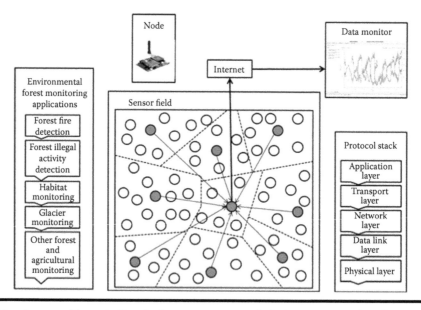

Figure 3.1   A general layout of a wireless sensor network (WSN) connected to the infrastructure-based network.

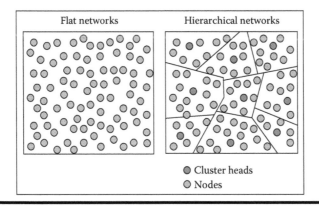

Figure 3.2   Flat versus hierarchical networks.

The specific class of WSNs deployed in the forest for real-time environmental monitoring is the main focus of this chapter. In fact, WSNs open up opportunities of remote real-time monitoring of the interested phenomenon of a desired environment without physical presence of people in the real field. Traditionally, an important phenomenon such as an event of forest fire can be monitored from ground patrol, areal patrol, observation towers, and low earth orbit (LEO) satellite monitoring systems. However, the former three are costly and involve manpower. As a result, the event detection is not efficient because of human errors and negligence. On the other hand, LEO satellite systems cannot provide continuous real-time event detection due to its orbital path. Moreover, the detection failure is inevitable due to the cloud cover.

For environmental disaster management, monitoring should avoid human presence in that area due to the hazardous conditions. For animal and habitat monitoring, it is desirable not to disturb the animal by the presence of the people. Human presence in the vicinity can excite animals, and make the creature not behaving normally, which is undesirable and can hamper the study to an objectionable extent. On the other hand, WSN can be deployed with little human presence on the habitat.

Due to the potential of the real-time environmental forest monitoring, a number of such deployments and studies have been carried out by the research community. Despite the suitability of WSNs for forest monitoring as described previously, a number of challenges have to be faced. The most important challenges are the following:

Large amount of nodes: The environment to be assessed or monitored by the network often needs to support a large area. To cover a large area, such as a forest, this can take a large number of sensor nodes, which makes it a costly option although it is predicted that the cost of the unit price of the sensor nodes will decrease significantly in future.

Scalability: In large WSNs, the network has to have the ability to handle an increasing number of nodes and associated tasks and functions. This capability (known as scalability) becomes one of the vital issues to be handled for the specific networks. Therefore, the design on all aspects must consider scalability before the real deployment in the environmental forest networks. Dividing the network into clusters is one of the approaches widely used in WSNs to handle scalability issues.

Battery and lifetime: WSN lifetime is often considered the battery life of the nodes as the battery replacement can be impossible or highly costly in an inaccessible terrain. Also, the uneven depletion of the battery power may cause the network to be inoperable prematurely. It is highly likely that the node next to the sink dies first due to their extra forwarding load compared to the far one. In that case, even though most of the far sensor nodes can be operable, the network as a whole becomes nonfunctional as no data can be reported to the monitoring center. Wang et al. [1] use Gaussian distribution for the node deployment, while Wu et al. [2] use an increasing number of nodes with a geometric proportion from the outer parts to the inner parts of the network to achieve longer lifetime compared to the uniform node distribution.

Unattendance: As some WSNs remain unattended during the operational time, nodes need to be self-configurable and the network needs to have self-healing capability. It is expected that the nodes can be destroyed due to the harsh environmental conditions. Moreover, the network needs to adapt according to the user requirements and remote configuration changes. Thus, an automated system network health should be monitored to cope up with the failure and sudden changes on the network.

The reminder of the chapter is organized as follows. Section 3.2 describes various forest monitoring applications including forest fired detection, forest illegal activities detection, habitat monitoring, and glacier monitoring applications, respectively. Some algorithms and techniques for environmental forest monitoring systems are discussed in Section 3.3. WSN modules for environmental forest monitoring applications are discussed in Section 3.4. Finally, we conclude the chapter by discussing open issues and future work for environmental forest monitoring applications in Section 3.5.

3.2 Forest Monitoring Applications

In this section, we discuss various applications of forest monitoring based on experimental studies reported in the literature.

3.2.1 Forest Fire Detection

WSN applications dealing with forest fire detection can play an important role in preserving precious forests from destruction. Such networks can provide an efficient and easy-to-use tool to monitor the interested zone in real time as discussed previously. Various studies of WSNs for forest fire detection are reported in the literature and most of them are discussed in this section.

Chaczkon and Ahmad [3] present a real-time technique to control and monitor bush firefighting, especially in the areas where mass causalities are likely to happen. The authors developed a small, integrated compact package to monitor temperature and humidity, as well as thermal radiation using MICA motes [4]. The device transmits periodic packets containing data on the temperature and humidity. Therefore, it can be used as an early warning fire detection system in the area of a bush fire or endangered public infrastructure.

Doolin and Sitar [5] describe the design of a system for wildfire monitoring incorporating wireless sensors and reported results from field testing during prescribed test burns near San Francisco, California. The system is composed of environmental sensors collecting temperature, relative humidity, and barometric pressure with an onboard GPS unit attached to a wireless networked mote. The motes communicate with a base station, which communicates the collected data to the software running on a database server. The data can be accessed using a browser-based Web application or by any other application capable of communicating with the database server. Sensors within the burning zone recorded the passage of the flame front before being scorched with temperature increasing, barometric pressure and humidity decreasing as the flame front advanced. Temperature gradients up to 5°C/s were recorded. The data also showed that the temperature slightly decreases and the relative humidity slightly increases from ambient values immediately preceding the flame front, indicating that locally significant weather conditions develop even during relatively cool slow moving grass fires. The specific goals of the project include developing and field testing of proof-of-concept wireless sensor technology for wildfire instrumentation, developing and field testing an asset tracking system for location and environmental monitoring of firefighting personnel and investigating possible applications in other domains such as monitoring of structural health, geologic hazards, and/or environment. In addition, successful field testing provides data that may be useful for firefighting and designing future generations of sensors and sensor platforms.

Lloret et al. [6] discuss the design, research, and development of a wireless multisensor network, which mixes sensors with IP cameras in a wireless network in order to detect and verify fire in rural and forest area. The study investigates the appropriate number of cameras, sensors, and access points for coverage of a specific area and scalability of the system. The multisensor network detects fire and sends alarm to a central server. The central server selects the closest wireless cameras to the multisensor, based on a software application, which are rotated to the sensor that triggers the alarm, and sends them a message in order to receive real-time images from the zone. The camera lets the fire fighters corroborate the existence of a fire and avoid false alarms. Figure 3.3 depicts such surveillance system.

Son et al. [7] developed a forest fire surveillance system (FFSS) in South Korea. The proposed architecture is composed of WSNs, a transceiver, middleware software, and a Web application. The nodes of this network gather measurements of temperature, humidity, and illumination from the environment. These data are collected at the sink node. This node sends the data to the transceiver (gateway) connected to the Internet. Then, a middleware program determines the forest-fire risk-level by this equation:

$$Y = 6.87 + 0.64P + 0.15EF + \left(\frac{1774.94}{CS}\right)$$

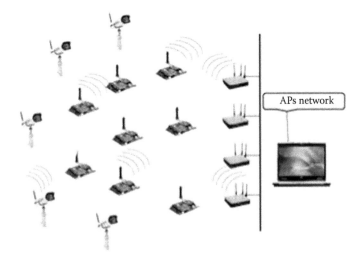

Figure 3.3 **Fire detection and verification design proposal.**

where *P*, *EF*, and *CS* are the effective humidity (%), solar radiation of the day (mJ/m²), and precipitation (mm), respectively. *EF* can be calculated as

$$EF = (1-r)\left(H_0 + \sum_1^4 r^n H_n \right),$$

where $r = 0.7$ and H_0 and H_n are the relative humidities of today and *n* days ago, respectively. If a fire is detected, FFSS automatically activates an alarm to facilitate an early extinguishing of the fire.

Zhang et al. [8] propose a forest fire detection technique using ZigBee WSNs. The environmental sensing parameters (temperature and humidity) are accounted to detect the event of fire. In the three-tier architecture of sensors, a cluster tree topology of network architecture is proposed. Data are collected and monitored by a remote host computer via the Internet gateway. Here, the three categories of the nodes are known as bottom nodes, cluster heads, and network coordinators. Clustered tree architecture of the network provides aggregation for superior energy consumption characteristics. Moreover, the hierarchical architecture provides scalability for a large forest network.

Kosucu et al. [9] try to combine the best of both simulation and test-bed worlds for forest fire detection. Authors constructed a number of outdoor test beds with randomly deployed unattended sensor nodes and collected actual sensor readings from this multihop sensor network. A fire simulator mimics the burning forest area with similar canopy characteristics, and collects the temperature readings that are mapped to the actual sensor positions in the test-bed configuration. Authors integrate the fire simulation results into outdoor test beds to test their fire detection algorithm. Experiments evaluate the success of the fire detection algorithm in terms of various system parameters including topology and physical configuration changes, link and node failures, wind direction, ignition point position, and sampling period variations for sensor readings.

Wang et al. [10] propose forest fire monitoring system based on the general packet radio service (GPRS) and ZigBee technology. First, the information of the natural phenomena such as temperature, relative humidity, atmospheric pressure, wind speed, and smoke concentration are collected by the ZigBee network through a cluster tree-based approach. The collected data at a coordinator node are then forwarded to the Internet through GPRS network. The FTP server collects the data from the Internet. Remote access is provided by the readily available GPRS networks. Obregon et al. [11] propose a similar idea, that is, a GPRS-based fire prevention system based on the phenomena such as temperature, relative humidity, rain gauge, and wind velocity and direction sensor systems. Power consumption results of various devices used in the experiments such as MICA, motherboard, GSM module, regulator, and station are reported through the paper.

3.2.2 Detection of Illegal Activities in the Forest

In addition to forest fire detection, another kind of forest monitoring application is the detection of deforestation and illegal animal hunting. Awang and Suhaimi [12] propose a forest monitoring system called RIMBAMON to monitor forests for preventing illegal logging and indiscriminate development activities as well as fire detection. The real-time data are captured, monitored, and displayed graphically in the system. This work uses two approaches: simulation approach and field testing. RIMBAMON consists of sensor nodes and a base station. The sensor nodes are deployed in the forest at a specific distance from one another depending on the line of sight and the functional communication range of the sensors. Each node captures information and may relay information from other nodes using multihop communication and transmits it to a monitoring node

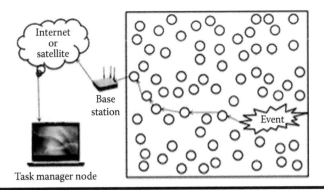

Figure 3.4 RIMBAMON scenario for a forest area monitoring.

known as the base station. The base station finally transmits the data to a task manager for further processing. Figure 3.4 shows the architecture of RIMBAMON.

3.2.3 Rainforest Monitoring

Wark et al. [13] report rainforest monitoring in Queensland, Australia, to observe the restoration of biodiversity. The rainforest deployment naturally poses challenge due to the limited solar energy and diverse and unpredictable radio environment. The deployment involves Fleck™-3 platform that consists of Atmega128 microcontroller and Nordic NRF905 radio transceiver operating in 915 MHz. The Fleck™-3 nodes are managed by Fleck operating system (FOS) and powered by NiMH batteries charged with a solar cell.

To support the unreliability of the wireless medium in the forest, the network employs link quality-based routing protocol, which is very similar to the collection tree protocol of TinyOS and uses the expected number of transmissions (ETX) as a routing metric. As receive signal strength indicator (RSSI) and link quality index (LQI) are not accessible in this specific platform, authors use packet reception rate (PRR) by snooping packets from neighbors. Besides, the deployment uses low-power listening (LPL) as medium access control (MAC) protocol that employs sleep scheduling and saves energy. A number of interfaces such as wind speed and direction, soil moisture, leaf wetness, temperature, relative humidity, video, and audio have been used to collect the data.

3.2.4 Habitat Monitoring

WSNs can provide close and detailed views of plant and animal habitats for researchers. Mainwaring et al. [14] report the monitoring of a seabird environment, while Hu et al. [15] report the monitoring of the population of a specific toad. The work by Hu et al. [15] presents a collection of requirements, constraints, and guidelines that serve as a basis for a general sensor network architecture for many applications. A tiered architecture is used for the deployment purpose. The lowest level consists of the sensor nodes that perform general purpose computing and networking in addition to application-specific sensing. The sensor nodes may be deployed in dense patches that are widely separated. Sensor nodes transmit data through the network to the gateway.

The gateway is responsible for transmitting sensor data from the sensor patch through a local transit network to the remote base station that provides wide area network (WAN) connectivity and data logging. The base station connects to database replicas across the Internet.

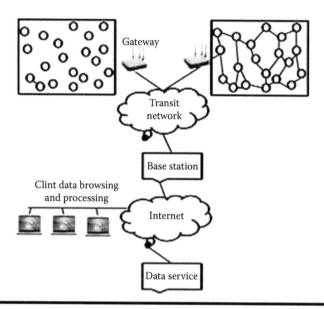

Figure 3.5   System architecture for habitat monitoring.

Finally, the data are displayed through a user interface. Figure 3.5 presents such a system architecture of the deployment.

Hu et al. [15] report development of the prototypes that recognize the vocalization of cane toad among nine different kinds of species found in northern Australia. The system incorporates high-frequency sampling of the toad's voice, compression, noise reduction of the signal, and recognition (by fast fourier transform [FFT] and machine learning procedures).

3.2.5 Glacier Monitoring

Martinez et al. [16] monitor the behavior of ice caps and glaciers, which is an important part of the earth climate. To accurately study this environment, the system must autonomously record glaciers over a reasonable geographic area and over a relatively long time. It also needs to be as noninvasive as possible to allow the sensor nodes (probes) to mimic the movement of stones and sediment under the ice. Figure 3.6 shows the architecture of the network.

3.2.6 Other Applications for Forest Monitoring

Burgess et al. [17] briefly describe the current state of wireless environmental monitoring from an end-user point of view and reports initial attempts to adapt this cutting-edge technology to the realm of forest research. The aim of the study is to develop and test a WSN suited to field experiments in forest and agricultural research. The key objective for this study is the design of WSNs that have (1) the ability to interface with a wide variety of environmental sensors as commonly used by agricultural and forestry researchers, (2) a sustainable power supply for long-term field operation of power hungry sensors in shaded forested environments, (3) radio and antenna performance suitable for intensive forest monitoring, (4) capability for in-field reprogramming by the end user to facilitate experimental needs, (5) real-time recovery of data from remote location via the cellular phone network, and (6) local data logging for backup and recovery of original readings.

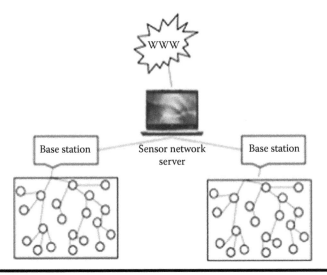

Figure 3.6   Generic sensor network architecture.

3.3 Algorithms and Protocols for Environmental Forest Monitoring

Besides the real deployments of the forest and environmental monitoring, we intend to identify, in this section, some useful techniques proposed to improve the performance of WSN-based forest monitoring.

Lopes and Ruiz [18] propose multitier, multimodal distributed video sensor networks for wildlife monitoring. This self-organizing network consists of three distinct tiers of sensor nodes: the first tier consists of infrared radiation sensors, while the second and third tiers consist of visual sensing devices. Figure 3.7 presents the tiered architecture of this network. Here, the first tier communicates preferably with the second tier and the second tier communicates preferably with the third tier. Communication that does not follow this rule is possible but unusual.

Sensors on the first and third tiers maintain three states, while sensors on the second tier maintain four states during the operation, as depicted in Figure 3.8. Sensors on the first tier in the first state awake and monitor the environment. They detect animals using the rough variation of the collected heat radiation from animals. If the variation is beyond a predefined threshold, a possible target is detected and the sensor goes into the second sate. It then triggers a wake-up call for the second tier devices and goes to the third state. In this state, it simply waits for a predefined time and goes back to the first state again.

Sensors in the second tier on their first state wait for the wake-up message from the first tier. In this state, nodes maintain all their components off except the transceivers. After receiving the wake-up call from the first tier, they go to the second state by turning on the visual sensor. In this state, nodes search for the moving objects for a predefined time. If an object is found, nodes go to the third state. On the other hand, if no object is detected in this period, nodes go to the first state again. In the third state, a sensor analyzes the object further for the color. If the color matches with the predefined color, that is, if a valid target is detected, the node goes to the fourth state, otherwise to the first state. From the fourth state, the device triggers a wake-up call for the third tier and goes back to the first state.

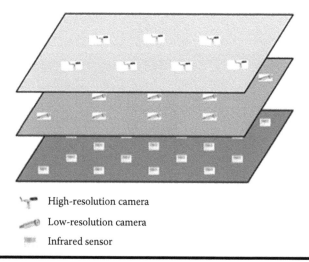

High-resolution camera

Low-resolution camera

Infrared sensor

Figure 3.7 Layout of the three-tier two-modality (visual, infrared) sensor networks.

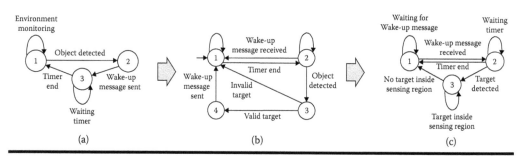

Figure 3.8 State transition diagram for (a) first tier, (b) second tier, and (c) third tier of monitoring application.

The third tier behaves as the second tier except that it does not have any upper tier; therefore, it only needs to maintain three states. For packet routing, authors use geographical and energy aware routing (GEAR) as a routing protocol for the deployment. The proposed system can save energy by not utilizing the high-energy-consuming visual sensor at all times. The high-energy nodes turn on based on the triggers by the lower layers. Experiments show a good number of successful target detection rates. We regard the 85% of detection accuracy as the trade-off of the system to have a less costly deployment.

Lan et al. [19] provide abstraction of environmental sensor networks, sensor network common interface between host and networks, and service-oriented architecture of the network. The paper describes the network management technique that provides intelligent interoperability and general messaging technique. In this work, Simple Object Access Protocol (SOAP) is used as message transmission protocol for information request and response. To make the network self-healing, the network is monitored by some specific nodes called snoopers. Based on the health of the network, the nodes adjust their behavior. Any failure detected by the snoopers is reported to the neighbors for necessary adjustments. Figures 3.9 and 3.10 describe such network abstraction and SOAP communication level, respectively.

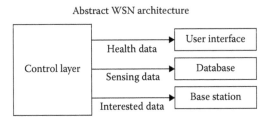

Figure 3.9   Abstract of sensor networks.

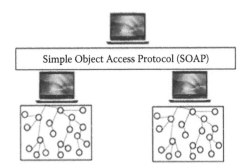

Figure 3.10   Simple Object Access Protocol architecture in the network.

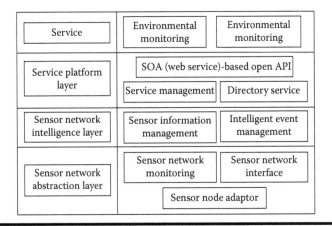

Figure 3.11   Service-oriented architecture of sensor networks.

As the environmental sensing applications have been developed independently, the interoperability issues arise. To address this issue, this work defines a service-oriented architecture for the sensor network comprising service platform layer, sensor network intelligence layer, and sensor network abstraction layer as depicted in Figure 3.11.

3.4 WSN Modules for Environmental Forest Monitoring

A large number of WSN-based environmental monitoring devices and modules from a number of vendors are available in the market. This section provides information regarding some of these modules.

Crossbow's MTS300 and MTS310CA [20] are compatible with MICA and MICA2 series motes. Both boards come with light, temperature, acoustic, and sound sensors, where MTS310CA has the additional sensors (two axis accelerometer and two axis magnetic sensors). Awang and Suhaimi [12] use MTS300CA types of sensors along with MICA2 nodes in their deployment. The base station node consists of a MICA2 mote attached to the MIB510CA programmer board, which uses serial communication for the data acquisition.

These sensor modules are compatible with IRIS, MICAz [21], and MICA2 sensor nodes (also known as MICA weather board or MICA environmental sensor board). The modules come with five sensors and one GPS sensor (only with MTS420) [22]. MTS420CA has a collection of five different types of sensors: (1) temperature and humidity sensor (Sensirion SHT11 [23]), (2) barometric pressure sensor from (Intersema), (3) GPS unit (LeadTek 9546), (4) accelerometer (ADXL 202 AE), and (5) light intensity sensor (Taos TLS 257) known as Fireboard [3,24].

Doolin and Sitar [5] use the Fireboard for their experiments along with the sensor node MICA2. It is to be noted that the sensor board can be connected to the MICA2 with a 52-pin connector. The architecture of the Fireboard is derived from the MICA weatherboard.

The Tmote Sky [25] comes with an optional suite of onboard low-powered sensors. The sensors include a digital temperature and humidity sensor (SHT15, Sensirion AG, Staefa ZH Switzerland) and analog total solar radiation and photosynthetically active radiation photodiodes (S1337-BQ and S1087, Hamamatsu Photonics K.K., Hamamatsu City, Japan). Burgess et al. [17] chose to include these sensors to provide basic meteorological measurements at each sensor node, but opted to position them off-board via a ribbon cable to permit better circulation of air around the temperature and humidity sensor.

Besides these wireless sensor boards, some other sensor devices are also used for forest monitoring. For instance, Martinez et al. [16] use 100 psi pressure sensors, two dual-axis microelectromechanical system (MEMS) tilt sensors, and a temperature sensor. Burgess et al. [17] use plant fluid (sap) flow sensors (HRM-30) and 4-channel thermocouple temperature (TT4) sensors of ICT International for their deployment. It is to be noted that the sap flow devices works on the heat ratio method.

For sensing devices, Zhang et al. [8] utilize SHT11 sensors consisting of integrated digital sensor for relative humidity and temperature with I2C bus. The device possesses the desirable characteristics of calibration-free alignment capability, autodormancy, and submersibility. Lopes and Ruiz [18] use different types of sensors in three different tiers: the first tier uses SDA02–54 Nicera passive infrared sensor, the second tier uses a CMUcam2 smart visual sensor, and the third tier uses the creative webcam instant camera.

Eko sensor systems are outdoor environmental real-time data collection devices with plug-and-play capability [26]. These systems comprise sensor nodes, gateway, and base station. Eko nodes support multiple sensors. The nodes operate on 2.405–2.480 GHz. Two kinds of Eko nodes are standard: Eko node and Long Range Eko nodes. The transmission powers of these two kinds (Eko and Long Range Eko) are +3 dBm and +18 dBm, respectively. With a receiver sensitivity of −101 dBm, the nodes can support 500–1500 ft and 2000 ft–2 miles under a line-of-site deployment.

Eko base stations comprise MEMSIC IRIS radio/processer, antenna, and USB processor board. Eko gateway runs on Debian Linux operating system. Gateways are responsible for the sensor network management and handled by XServ. The tool for viewing the real-time sensor data is known as EkoView. The EkoView comes with a browser, which is used to access the real-time data remotely from the Internet using computer or Smartphone. Besides the real-time viewing of data, users can set alert and run reports through the software.

Eko environmental monitoring systems come with various sensors such as soil moisture, ambient temperature, humidity, leaf wetness, soil water content, and solar radiation. Besides, Eko weather station is a collection of sensors and consists of rain collector, a temperature and humidity sensor with radiation shield, solar radiation sensor, barometric pressure sensor, and an anemometer in one package.

Campbell, Inc. provides proprietary sensor systems for real-time environmental monitoring [27]. Being non–plug and play, the Campbell system requires configuration for the deployment. The system comprises sensor nodes, a base station, and dataloggers. Non-web-based remote monitoring of Campbell dataloggers is possible through a range of modems. Campbell, Inc. provides three types of nodes: CWS220, CWS655, and CWS900. With the operation frequency range of 902–918 MHz (for the United States and Canada), the nodes support 50 frequency hopping spread spectrum (FHSS) channels. With 14 dBm transmitter power output and −110 dBm receiver sensitivity, the nodes support a transmission range of 1080 ft.

Campbell BS, named CWB100, provides communications between sensor nodes and Campbell dataloggers. Here, the communications from the sensor nodes are provided by the wireless radios. On the other hand, the communication between the base station and the dataloggers is provided by the data cable. The USB port is used for configuring the base station while connected directly to the computer. The dataloggers are the data storage for the collected sensed values. There are various options to access the data from the datalogger from a user device such as a PDA or a remote computer. Cellular or landline phone modems, narrowband UHF/VHF radios, or satellite transmitters can be used for distant access to the datalogger but of course with a high additional peripheral device cost. Apparently, the PDA access is a cost-effective solution compared to the other solutions stated earlier. LoggerNet is the Campbell software to manage datalogger network, data collection, network monitoring, troubleshooting, and data display.

A number of Campbell sensors are available for wireless nodes to collect data, including air temperature, relative humidity, barometric pressure, ice detector, leaf wetness, precipitation, snow measurement, soil moisture, water quality, wind speed and direction, and so on.

Waspmote from Libelium provides sensing services for the environmental monitoring [28]. Waspmote uses ATmega1281 microcontroller that operates on 8 MHz. Being modularized, the mote can use communication modules such as ZigBee, Bluetooth, and GPRS.

XBee-802.15.4, XBee-802.15.4 pro, XBee-ZB, and VBee-ZB pro are 2.6 GHz ZigBee modules that provide 500 to 7000 m of communication range. On the other hand, long-range modules such as XBee-868, XBee-900, and XBee-XSC operate on 868 and 900 MHz and provide ranges of 40, 10, and 24 km, respectively. The Bluetooth module provides a 30 m range. The GPRS module provides Internet connectivity via TCP/IP and UDP/IP sockets. The GPRS modules are capable of sending and receiving mails through SMTP and POP3 services. The uploading and downloading of files are handled by the FTP servers.

Waspmote gateway devices, named Meshlium, are equipped with WiFi, GPRS, ZigBee, and Bluetooth routers. Meshlium operates on linux systems and can be configured using web browsers. With a 32 GB capacity, Meshlium can store sensor data using database MySql and Postgre. To support different categories of sensing devices, it uses different types of sensor boards such as gas board, event detection board, and agricultural board.

3.5 Summary and Discussion

We have studied the utilization of WSNs for forest monitoring. We presented some algorithms and protocols proposed in the literature for this particular application. We also discussed various test beds and experimental work studies for forest monitoring applications. Based on the discussed work (in Section 3.4), we identify the following problems and challenges, which need further investigation.

3.5.1 Sensing

The success of WSN applications depends on choosing the right sensing parameters along with the frequency of sampling. For example, Zhang et al. [8] use only temperature and humidity as sensing parameters for the detection of fire, where the vital factor wind speed is neglected. Sensing with a very small frequency is undesirable, as shown by Awang and Suhaimi [12], where the network is not able to sample the environment of more than one sensor at a time. Sensing with high frequency gives a better view of the environment with a cost on the network lifetime. This is sometimes inevitable. For instance, Hu et al. [15] show that the WSN-based system requires a high sampling rate to detect the frog population. Unfortunately, processing that amount of data is highly expensive. Therefore, the system needs high-powered and costly nodes in the network. Besides, a reactive data sampling mechanism is often desirable where the sampling rate depends on the presence or absence of an event. For example, in an event-based sampling scenario, a forest fire monitoring system can maintain a low-rate sampling, where a high sampling rate only triggers when an excessive temperature is reported.

3.5.2 Withstanding Harsh Environment

The environment under surveillance can be a very harsh one depending on the application. Doolin and Sitar [5] report the total destruction of the node by fire. Designing flame and heat protection for motes and batteries in fire-prone areas should be considered to avoid motes being scorched by extensive heat. For bird monitoring applications [14], the placement of the parylene-sealed motes into the burrows was without enclosures. Using an environmental enclosure could make the deployment more robust.

As mentioned earlier, environmental condition must be considered for every deployment. A minute mistake or inattention may cause undesirable result or failure. For example, a Li-ion battery should not be charged in freezing temperature since charging such batteries may cause explosion. Similarly, other equipment may fail when deployed in extreme cold conditions. Barrenetxea et al. [29] report a failure of their disdrometer (sensor distinguishes between different kinds of rains) in the very first few days of the deployment.

Special care should be taken when the temperature drift is reasonably high because the crystals used in motes have an impact on the precision due to the temperature variation [30]. The colder the temperature, the slower the tiny-node crystals oscillate. To handle the rapid temperature shifts, algorithms and protocols should be designed with special care such that large drift of day–night temperature should not impact the performance of the protocols. An ultralow power MAC layer protocol often uses high-precision synchronization and may not be achievable in such deployment.

Outdoor deployment must care about the humidity; otherwise, it may cause permanent damage to the device due to short circuit. Barrenetxea et al. [29] report damage to SHT75 sensors deployed in highly humid environment due to poor packaging. Corrosion also may cause malfunction of the device and corruption of data. Simulating the environment may help identifying the problem before the real deployment.

3.5.3 Network Maintenance and Configuration

Things may turn wrong anytime by a device breakdown. Monitoring of the system and the data as much as possible is a good practice so that spatial measure can be taken just in case. To act efficiently, the system should be configurable and upgradable remotely. Therefore, acquiring remote control facilities as much as possible is highly recommended [29]. Deluge [31] is a good choice for TinyOS networks as it provides over the air programming mechanism to reprogram motes.

3.5.4 Link Quality

Geographical location has an impact on the reliability of a link. Dense foliage, rain humidity makes the link quality highly dynamic. Moreover, sensor network deployed in an environment in the presence of water bodies may become challenging as radio wave reflected from the water surface may interfere with the direct wave [32]; therefore, special care should be taken to deal with the interference. Routing techniques must take link quality as a metric to handle the dynamic link quality. In addition, the use of lower frequency is a better choice to handle this issue.

3.5.5 Localization

Position and time information is necessary to represent the data. Without this information, the reported data are not useful enough. For instance, in forest fire detection, the location of the fire is as important as the information about the existence of the forest fire itself. Moreover, routing, data aggregation, and MAC algorithms in WSNs often use location information to support their functionalities. Although a number of localization techniques have been developed and reported, localization techniques specifically designed for environmental sensor networks should be explored further due to the specific characteristics of the environment and propagation conditions. For example, a localization technique reports forest fire needs to take into account the channel characteristics such as the non-line-of-sight propagation due to the presence of trees and vegetation. An ordinary signal strength-based triangulation approach may not be appropriate for the case. Moreover, the accuracy of the localization may be different from the general localization case. A forest fire localization technique may not require pinpoint localization. Coarse grain location information of fire may be enough to send the firefighting machinery to the appropriate place. Le et al. [33] propose a forest fire localization technique where the network is divided into a number of grids. Based on this technique, the location of the sensors is obtained and represented by the grid that the sensor belongs to.

Wu et al. [34] propose a low-cost range-free localization technique that does not require any powerful anchor nodes, which can be a good candidate for a large-scale deployment in environmental sensor networks. However, with this localization technique, the irregularity of the radio pattern can cause inaccurate location estimation. Moreover, the nodes located at the edge of the sensor field derive less accurate location information, which makes the algorithm less suitable.

3.5.6 Application Specific Algorithms

A large number of algorithms are proposed in the area of WSNs. The efficiency of the algorithms often depends on the application itself. An MAC algorithm suitable for a small sensor network often is not suitable for a large-scale sensor network detecting forest fires. Therefore, using a MAC protocol developed specifically to forest fire detection ([35]) is more suitable than a general MAC protocol such as SMAC [36] or TMAC [37]. Similarly, for data aggregation, a well-accepted TAG algorithm may be less suitable than an adaptive weighted fusion algorithm [38] for fire detection network application. Choosing the right algorithm or technique can be based on the requirements of the sampling rate as well. Barrenetxea et al. [29] exclude network congestion management in their deployment as the low sampling rate is useless for specific monitoring.

3.5.7 Aggregation

Detection of driving events by observing readings of different sensors and using in-field power-aware decision fusion is desirable in environmental monitoring applications. Lu and Xue [38] provide a decision fusion algorithm for forest fire detection. However, providing the adaptation of the fusion algorithm based on the power availability of the node is often absent in such detection applications. This remaining power-based adaptation mechanism also poses a challenge for minimizing missing events and false alarm probabilities.

Data uncertainty is introduced not only in aggregation but also in data collection and propagation in resource constrain sensor environment. The data uncertainty may have a serious impact on critical environmental detection applications. Levis et al. [39] handle the uncertainty by providing rough set-based dynamic feature selection algorithm that allows the aggregation of data stream without altering the dynamic data semantics.

3.5.8 Ease of Use

WSN-based forest monitoring often includes sophisticated software and hardware. Burgess et al. [17] report extensive work on software writing to handle the end product with rudimentary features. The prime users of the environmental forest monitoring applications are usually nontechnical people. Therefore, it is expected that the system requires minimum or no configuration and developmental effort. A simple plug-and-play system is rather desirable for the general target users.

3.5.9 Compatibility

Available environmental monitoring systems are often incompatible with each other. Moreover, a specific sensor kit has its own strengths and weaknesses. As a result, the user often may not have a choice of deploying WSNs with multiple providers to exploit multiple strengths in a single deployment. Besides, if devices from different manufacturers are compatible, the user will have more flexibility in the design and implementation of new networks or expanding existing ones.

Acknowledgments

This survey is performed as a part of a project for WSN-based forest monitoring sponsored by the Department of Natural Resources (Forestry Section), Newfoundland and Labrador Government, Mathematics of Information Technology and Complex Systems (MITACS), and the Harris Center of Regional Policy and Development (Memorial University).

References

1. Wang, D., B. Xie, and D. P. Agrawal. 2008. "Coverage and Lifetime Optimization of Wireless Sensor Networks with Gaussian Distribution." *IEEE Transaction on Mobile Computing* 7 (12): 1444–58.
2. Wu, X., G. Chen, and S. K. Das. October 2006. "On the Energy Hole Problem of Nonuniform Node Distribution in Wireless Sensor Networks." In *Proceedings of the Third IEEE International Conference on Mobile Ad-hoc and Sensor Systems (MASS '06)*, 180–87. Vancouver, Canada.
3. Chaczkon, Z., and F. Ahmad. 2005. "Wireless Sensor Network Based System for Fire Endangered Areas." In *Proceedings of the IEEE International Conference on Information Technology and Applications (ICITA)*, Vol. 2: 203–07. Sydney.
4. Mica data sheet. Accessed February 4, 2012. http://stomach.v2.nl/docs/Hardware/DataSheets/Sensors/MICA_data_sheet.pdf.
5. Doolin, D., and N. Sitar. March 2005. "Wireless Sensors for Wildfire Monitoring." In *Proceedings of the SPIE Symposium on Smart Structures and Materials*. San Diego, CA.
6. Lloret, J., M. Garcia, D. Bri, and S. Sendra. 2009. "A Wireless Sensor Network Deployment for Rural and Forest Fire Detection and Verification." *MDPI Sensors* 9 (11): 8722–47.
7. Son, B., Y. Her, and J. Kim. September 2006. "A Design and Implementation of Forest-Fires Surveillance System Based on Wireless Sensor Networks for South Korea Mountains." *IJCSNS International Journal of Computer Science and Network Security* 6 (9B): 124–30.
8. Zhang, J., W. Li, N. Han, and J. Kan. 2008. "Forest Fire Detection System Based on a ZigBee Wireless Sensor Network." Springer-Verlag: *Frontiers of Forestry in China* 3 (3): 369–74.
9. Kosucu, B., K. Irgan, G. Kucuk, and S. Baydere. 2009. "FiresenseTB: A Wireless Sensor Networks Testbed for Forest Fire Detection." In *Proceedings of the IEEE International Conference on Wireless Communications and Mobile Computing*, 1173–77. Leipzing, Germany.
10. Wang, G., J. Zhang, W. Li, D. Cui, and Y. Jing. June 2010. "A Forest Fire Monitoring System Based on GPRS and ZigBee Wireless Sensor Network." In *Proceedings of the IEEE Industrial Electronics and Applications (ICIEA)*, 1859–62. Taiwan.
11. Obregon, P. D. P., S. Sondon, S. Sanudo, F. Masson, P. S. Mandolesi, and P. M. Julian. October 2009. "System Based on Sensor Networks for Application in Forest Fire Prevention." In *Proceedings of the IEEE Micro-Nanoelectronics, Technology and Applications (EAMTA)*, 61–65. Bariloche, Argentina.
12. Awang, A., and M. H. Suhaimi. 2007. "RIMBAMON: A Forest Monitoring System Using Wireless Sensor Networks." In *Proceedings of the IEEE International Conference on Intelligent and Advanced Systems (ICIAS)*, 1101–06. Kuala Lumpur, Malaysia.
13. Wark, T., W. Hu, P. Corke, J. Hodge, A. Keto, B. Mackey, G. Foley, P. Sikka, and M. Brunig. 2008. "Springbrook: Challenges in Developing a Long-Term, Rainforest Wireless Sensor Networks," In *Proceedings of the IEEE International Conference on Intelligent Sensors, Sensor Networks and Information Processing (ISSNIP)*, 599–604. Sydney.
14. Mainwaring, A., D. Culler, J. Polastre, R. Szewczyk, and J. Anderson. 2002. "Wireless Sensor Networks for Habitat Monitoring." In *Proceedings of the 1st ACM Int'l. Workshop on Wireless Sensor Networks and Applications (WSNA)*, 88–97. New York.
15. Hu, W., V. N. Tran, N. Bulusu, C. Chou, S. Jha, and A. Taylor. April 2005. "The Design and Evaluation of a Hybrid Sensor Network for Cane-toad Monitoring." In *Proceedings of the International Conference on Information Processing in Sensor Networks (IPSN)*, 503–08. Los Angeles, CA.
16. Martinez, K., J. K. Hart, and R. Ong. 2004. "Environmental Sensor Networks." *IEEE Computer* 37 (8): 50–56.
17. Burgess, S. S. O., M. L. Kranz, N. E. Turner, R. Cardell-Oliver, and T. E. Dawson. January 2010. "Harnessing Wireless Sensor Technologies to Advance Forest Ecology and Agricultural Research." *Elsevier Agricultural and Forest Metrology* 150 (1): 30–37.
18. Lopes, C. E. R., and L. B. Ruiz. November 2008. "On the Development of a Multi-tier, Multimodal Wireless Sensor Network for Wild Life Monitoring." In *Proceedings of the 1st IFIP/IEEE on Wireless Days*, 1–5. Dubai.
19. Lan, S., M. Qilong, and J. Du. December 2008. "Architecture of Wireless Sensor Networks for Environmental Monitoring," In *Proceedings of the IEEE International Workshop on Geoscience and Remote Sensing*, 579–82. Shanghai.

20. MTS300/310CA data sheet. Accessed February 4, 2012. http://www.memsic.com/support/documentation/wireless-sensor-networks/category/7- datasheets.html?download=168%3Amts-mda.

21. MicaZ data sheet. Accessed February 4, 2012. http://courses.ece.ubc.ca/494/files/MICAz_Datasheet.pdf.

22. MTS 400/420 data sheet. Accessed February 4, 2012. http://www.memsic.com/support/documentation/wireless-sensor-networks/category/7- datasheets.html?download=174%3Amts400_420.

23. Datasheet SHT1x (SHT10, SHT11, SHT15). Accessed February 4, 2012. http://www.sensirion.com/en/pdf/product_information/Datasheet-humidity-sensor-SHT1x.pdf.

24. The Fire Board Sensor Platform. Accessed February 4, 2012. http://firebug.sourceforge.net/fireboard.html.

25. Tmote Sky Data Sheet. Accessed February 4, 2012. http://www.bandwavetech.com/download/tmote-sky-datasheet.pdf.

26. Eko Environmental Systems. Accessed February 4, 2012. http://www.memsic.com/support/documentation/eko.html.

27. Campbell Scientific Documentations. Accessed February 4, 2012. http://www.campbellsci.ca/Products.html.

28. Libelium Waspmote Documentation. Accessed February 4, 2012. http://www.libelium.com/support/waspmote.

29. Barrenetxea, G., F. Ingelrest, G. Schaefer, and M. Vetterli. 2008. "The Hitchhiker's Guide to Successful Wireless Sensor Network Deployments." *In Proceedings of the ACM Conference on Embedded Networked Sensor Systems (SenSys)*, 43–56. NC, USA.

30. Werner-Allen, G., K. Lorincz, M. Ruiz, O. Marcillo, J. Johnson, J. Lees, and M. Welsh. 2006. "Deploying a Wireless Sensor Network on an Active Volcano." *IEEE Internet Computing* 10 (2): 18–25.

31. Hui, J., and D. Culler. November 2004. "The Dynamic Behavior of a Data Dissemination Protocol for Network Programming at a Scale." In *Proceedings of the ACM Conference on Embedded Networked Sensor Systems (SenSys)*, 43–56. NC, USA.

32. Corke, P., T. Wark, R. Jurdak, W. Hu, P. Valencia, and D. Morre. 2010. "Environmental Wireless Sensor Networks." *Proceedings of the IEEE* 98 (11): 1903–17.

33. Le, T. N., P. H. J. Chong, X. J. Li, and W. Y. Leong. February 2010. "A Simple Grid-Based Localization Technique in Wireless Sensor Networks for Forest Fire Detection," In *Proceedings of the 2nd International Conference on Communication Software and Networks (ICCSN)*, 93–98. Singapore.

34. Wu, K., C. Liu, and V. King. May 2005. "Very Low Cost Sensor Localization for Hostile Environments." In *Proceedings of the International Conference on Communications (ICC)*, 3197–201. Seoul, Korea.

35. Al-Habashneh, A. Y., M. H. Ahmed, and T. Hussain. May 2009. "Adaptive MAC Protocols for Forest Fire Detection using Wireless Sensor Networks." In *Proceedings of the IEEE Canadian Conference on Electrical and Computer Engineering*, 329–33. St. John's, NL, Canada.

36. Wei, Y., J. Heidemann, and D. Estrin. 2004. "Medium Access Control with Coordinated Adaptive Sleeping for Wireless Sensor Networks." *IEEE/ACM Transactions on Networking* 12 (3): 493–506.

37. Dam, T. V., and K. Langendoen. November 2003. "An Adaptive Energy-Efficient Mac Protocol for Wireless Sensor Networks." In *Proceedings of the 1st International Conference on Embedded Networked Sensor Systems (SenSys)*, 171–80. Los Angeles, CA.

38. Lu, G., and W. Xue. January 2010. "Adaptive Weighted Fusion Algorithm for Monitoring System of Forest Fire Based on Wireless Sensor Networks." In *Proceedings of the* IEEE *International Conference on Computer Modeling and Simulation*, 414–17. Sanya, China.

39. Levis, P., N. Lee, M. Welsh, and D. Culler. November 2003. "TOSSIM: Accurate and Scalable Simulation of Entire TinyOS Applications." *In Proceedings of the International Conference on Embedded Networked Sensor Systems (SenSys)*, 126–37. Los Angeles, CA.

Chapter 4

Fundamentals of Wireless Body Area Networks

Tabassum Waheed, Faisal Karim Shaikh, and
Bhawani Shankar Chowdhry

Contents

4.1 Introduction

We all like science fiction movies because the technology advancements shown in these movies give us a chance to use our imagination beyond boundaries and imagine things out of this world, but we also know that what we see is mostly fiction and has nothing to do with real life. For a fraction of a second, imagine that you are playing a baseball video game in which you are batting and to play the game you have to physically hit the ball. Your entire body movement is simulated as it is by the gaming character, giving you the ultimate gaming experience. Now this may seem something unbelievable, but thanks to phenomenal innovations in *wireless sensor networks* (WSNs) research, it is no longer a fantasy but a reality that makes interactive gaming an amazing phenomenon. These types of entertainment and commercial applications have become possible with the availability of wearable and implantable sensors that are parts of networks known as *wireless body area networks* (WBANs).

WBANs are new dimensions of WSNs and have evolved in recent years due to significant innovation in sensor miniaturization, embedded computing, and wireless technologies. WBANs have received tremendous interest from academia, industry, and consumer device manufacturers due to their novel defense and commercial applications. During the past decade, the drivers for WSN research were applications of surveillance, monitoring, tracking, and automation, but the drivers behind WBAN growth are applications that directly impact quality of human life, including telemedicine, ubiquitous health care, sports, and entertainment. Furthermore, because of a sedentary lifestyle and lack of exercise, the society is moving toward a disastrous situation, and life-threatening diseases are being diagnosed in almost every family. The traditional healthcare systems were not designed to cater to the huge flow of patients and are inadequate in current scenarios.

In simple words, a WBAN is a network of multiple interconnected low-power, lightweight, intelligent nodes that may be inside the body, on the body, or near the body, and perform sensing, processing, and communication. For this, a lot of research has been done in the areas of wired sensors, wireless sensors, and biochannels. With the emerging intriguing application of WBANs, they will have a significant impact on our lifestyle and will change the concept of how humans can use information technology for saving human life, improving its quality, and making health care affordable and accessible for everyone.

In this chapter, our contribution is twofold: (1) We have done a comprehensive survey of the evolution of WBANs with a special emphasis on the ongoing research on WBAN technologies and techniques, and (2) we have identified the pros, cons, and limitations of the present research and have proposed future directions in the field of WBANs. It is noteworthy that we assume cooperative WBAN entities and do not consider noncooperative behaviors such as selfishness and deliberate attacks.

The rest of the chapter is organized as follows: Section 4.2 discusses the major differences between WBANs and WSNs. Section 4.3 highlights the details of WBAN architecture, infrastructure design, various sensor types, node characteristics, and quality of service (QoS) design issues. In Section 4.4, we detail WBAN media access control (MAC) layer design issues and highlight the importance of the IEEE 802.15.6 standard. Section 4.5 deals with data dissemination strategies in WBANs. The importance of reliability and dependability mechanisms for WBANs are discussed in Section 4.6. Finally, Section 4.7 concludes the chapter and Section 4.8 provides future directions for WBANs.

4.2 Body Area Networks Versus Wireless Sensor Networks

The fundamental knowledge of WSNs can be used to develop WBANs, but they differ in many aspects. The potential applications of WSNs as identified by Chong and Srikanta [2] include military sensing, air traffic control, traffic surveillance, video surveillance, industrial automation, environment monitoring, and building monitoring. WSNs are very large scale, distributed, and self-organizing in nature, but WBANs are restricted to the human body only, and data acquisition is done at different times and aggregated in a different manner. The human body environment is different and poses different types of design challenges. As specified by Talha and Ahmed [9], the sensors/actuators of WBANs are unifunctional, their range is small, and their lifetime is less than a decade, whereas WSN devices are multifunctional with a large range and a lifetime of years. According to Chen and Gonzalez [6] and Honeine and Mourad [13], WBANs borrow many concepts from WSN, but the protocols, algorithms, and techniques of WSNs are not suited to WBANs because they differ in the following ways (Table 4.1):

- *Applications*: WSNs are used to monitor environment, whereas WBANs monitor interaction of human body with the environment.
- *Deployment and density*: WBANs consist of few nodes and there is no redundancy, whereas WSNs employ more nodes. In WSN, sensors are homogeneous in nature and insensitive to the placement. WBAN sensors are heterogeneous and need specific placement.
- *Data rate*: WSNs perform event-based monitoring, which is not periodic as events occur in an irregular manner. WBANs monitor periodic events so that data rate is stable.
- *Latency*: In case of WBANs, energy is an important factor for sustained operation and so it is traded for latency.
- *Mobility*: In case of WBANs, all nodes share the same mobility patterns.
- *Coverage*: The coverage area of WSNs is in kilometers, whereas WBANs are restricted only to human body.

Table 4.1 Differences between WSNs and WBANs

	Deployment	Density	Data Rate	Latency	Mobility	Coverage	Accuracy	Nodes
WSN	Random	High	Variable	Variable	No	Miles/km	Moderate	Homogenous
WBAN	Fixed	Low	Variable	High	Fixed	On body	High	Heterogeneous

■ *Energy harvesting*: In WSNs, energy is harvested through wind and sun, whereas in WBANs, it is done through human body motion. This is known as inertial power scavenging where human motion is converted into electrical energy, but this is challenging as the devices move with the body and topology keeps on changing. According to the Peltier–Seebeck effect, human temperature can also be converted into electrical energy.

Furthermore, Hang and Zhang [52] have highlighted the various challenges faced by WSNs and WBANs as follows:

■ *Context awareness*: In WSNs, environment is well defined and context awareness is not so important, whereas in WBANs, body physiology is very sensitive to context change.
■ *Access*: WSN sensors are replaceable or disposable, but implantable sensor replacement is difficult.
■ *Event detection*: In WSNs, early event detection is desirable but failure is reversible, whereas in WBANs, early event detection is crucial and failure is irreversible.
■ *Accuracy*: In WSNs, large node number compensate for accuracy, whereas in WBANs, each node must be accurate.
■ *Node function*: WSN nodes perform dedicated tasks, whereas WBAN nodes perform multiple tasks.

The advancement in embedded systems and implantable and wearable sensors will make WBAN viable for a variety of novel applications. WBANs will have a profound impact on how medical and health-care services are provided. They will make the availability of these services anywhere and everywhere all the time. Design of efficient, low-cost, low-power protocols/hardware is the primary research focus for WBANs. The ongoing research focus is on optimizing the physical (PHY) and the MAC layers, design of lightweight routing mechanisms, enhancing the cryptographic techniques for providing security, low-cost efficient sensor design, and developing killer applications for WBANs (Figure 4.1).

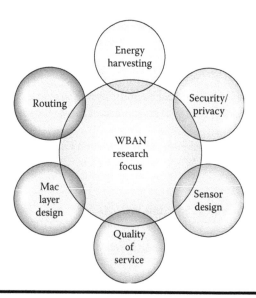

Figure 4.1   Current WBAN research trends.

4.3 WBAN Architecture

In this section, we elaborate the WBAN infrastructure in detail. Next, we detail the widely accepted WBAN system model and the associated traffic model. Following this, we present the typical WBAN topologies. Finally, we conclude this section by identifying the network requirements and design challenges of WBAN.

4.3.1 WBAN Infrastructure

There are one or more body sensor units (BSU), one body control unit (BCU), and a link to long-range network. The BSUs collect information and transmit to the BCU. There are two different infrastructures given by Onder and Cazalon [12]: managed and autonomous WBANs. The managed WBAN has a long-range connection, that is, a Global System for Mobile Communication (GSM) network such that the BCU sends an alert message to the closest hospital. On the contrary, in autonomous WBANs, the BCU has no connection with the long-range network; it is a more intelligent device and has been programmed to take decisions. For example, if the glucose level goes high, the actuator can inject insulin.

Furthermore, the basic types of WBANs can be categorized as stand-alone WBANs and interconnected WBANs. According to Wang and Pei [4], stand-alone WBAN is a network with sensor nodes in or in immediate vicinity of the bearer. The BSU information is recorded, processed, analyzed, and interpreted offline. The stand-alone WBAN can be connected to the Internet through BCU. The data can be raw or locally processed and sent in real time. The service provider or health-care provider can send feedback/alert/instructions accordingly. The user activities remain unhindered. For WBANs, various short-range communication standards are good candidates such as follows:

- *IEEE 802.11*: This standard is used in WLANs, having large form factor and is power hungry.
- *IEEE 802.15.1*: Bluetooth standard is less flexible and utilizes high-power consumption.
- *IEEE 802.15.4*: The ZigBee supports low-power devices and delivery of periodic and repetitive data. ZigBee is a fast, scalable, and low-power option for WBANs.
- *UWB*: The energy consumption of WBANs cannot be met by Bluetooth/ZigBee and so ultra wideband (UWB) offers a better low-cost integrated solution. The complexity is shifted in the receiver so that the transmitter is ultra low power.

WBANs are based on two types of sensors: wearable and implantable. The wearable sensors can be either woven into the fabric of clothes as an inseparable part of the garment or it can be an independent unit with transmission capabilities. The implantable sensors always incorporate a wireless module for transmission of measurements to an external device. They can be placed under the skin or in place of tooth or at the knee [15]. Generally, WBAN sensors are few, heterogeneous in nature, and require specific placement. Figure 4.2 shows normally utilized sensors for health-care applications.

4.3.2 WBAN System Model

A typical WBAN network is a network of on-body/in-body computing devices that is used in human-centric monitoring. The in-body network provides communication between implanted devices and a base station. On the contrary, on-body sensors allow communication between

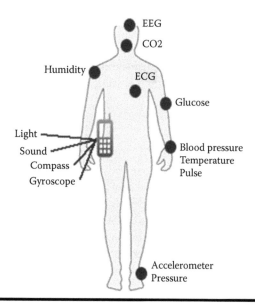

Figure 4.2 WBAN sensors.

wearable devices and the base station. The basic structure is a group of biosensors implanted in the human body and a BCU placed on the human body, which is connected to a base station. The biosensor has processor, memory, transceiver, and power, and it performs sensing, processing, and transmission/reception [8]. The BCU performs data aggregation and dissemination. It periodically collects the data and transmits it to the base station. A limitation is that the biosensors have limited power and less computational capabilities and not all nodes communicate with the BCU in a single hop. The idea behind WBAN is not only sensor connectivity but also includes many operational characteristics such as signal conditioning, on-node processing, low-power processing, power scavenging, and distributed inference [3]. The communication of traffic is limited to the body periphery for node-to-node transfer. Limiting the range decreases power consumption and interference and increases privacy. However, the signal gets attenuated due to body shadowing, path losses, and electrical properties of the body [10]. The dielectric constant increases the electrical length and the body tissue absorbs the signal. The antennas used include dipole, loop, and patch antennas. WBAN uses Wireless Medical Telemetry Services (WMTS)–licensed band, but it does not support voice or video. Accordingly, the unlicensed 2.4-GHz Industrial Scientific Medical (ISM) band or UWB can be utilized for WBANs.

4.3.3 WBAN Traffic Model

The traffic generated from medical sensors is diverse in nature and can be categorized [10] as follows:

- *On-demand*: Data transfer is initiated by the base station to acquire information. This can be continuous for surgical events or discontinuous for occasional information.
- *Emergency*: Data is transferred when a threshold value is exceeded at a node for certain parameter.
- *Normal*: Under noncritical conditions such as routine health monitoring, the transfer of data is periodic in nature.

4.3.4 WBAN Topology

The topology means the arrangement of the BSU in the network, which fulfills the application requirement. The WBAN topologies are categorized by Wang and Pei [4] as follows:

Point-to-Point: Where the two BSUs are connected directly and communicate with each other.

Star: In star topology, the BSUs forward the data to the aggregator, that is, BCUs and all devices are connected to a central node as shown in Figure 4.3.

Mesh: In mesh topology, the devices in each radio range can communicate directly or through multi-hopping as shown in Figure 4.4.

Star-Mesh: In this topology, each BSU can act as an aggregator and are connected to each other so that there is no single point of failure.

4.3.5 WBAN Communication Architecture

The communication architecture is needed to communicate in and around the body. Accordingly, the WBAN communication architectures are categorized [6] as follows.

4.3.5.1 Intra-BAN Tier 1

It is the tier 1 of the network and uses radio communication about 2 m from the body. Several architectures are available for intra-WBAN communication in which communication is between BSU and BCU as shown in Figures 4.3 and 4.4. Tier 1 can be infrastructure based or ad hoc in nature.

4.3.5.2 Inter-WBAN Tier 2

It involves communication between the BCU and the base station. It is the tier 2 of the network and bridges WBAN with the Internet and the cellular network as shown in Figure 4.5.

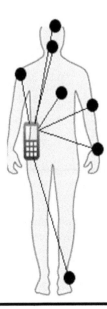

Figure 4.3   WBAN Star Topology.

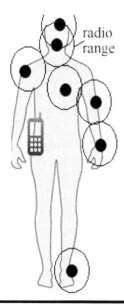

Figure 4.4   WBAN Mesh topology.

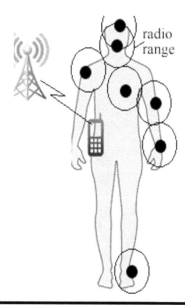

Figure 4.5   Inter-WBAN with infrastructure.

Alternatively, the two different BCUs can also communicate with each other to exchange information and to forward it to base station (Figure 4.6).

4.3.5.3 Beyond WBAN Tier 3

Tier 3 is used to extend the WBANs, enhance the coverage range, and connect it to the outside world and maintain the repository of all the gathered data from BSU and the actions taken on their basis.

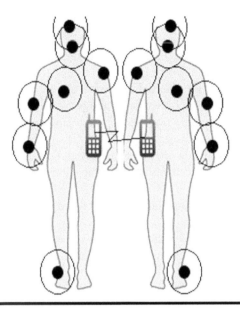

Figure 4.6   Ad hoc inter-WBAN.

4.3.6 WBAN Network Requirements

The networking requirement needs a minimalistic network such that the system's run time can be increased but at the cost of decreased QoS and the privacy. Such scenario is not appropriate for life-critical applications. The WBANs have specific network requirements [12] stated as follows:

- *Range*: For BSU to communicate in or around the same body, the radio range of 2–5 m is suggested.
- *Interference*: It should be suppressed for reliable communication between BSU of same or different applications.
- *Network density*: Different people can have different WBANs on them. The standards allow two to four networks per square meter.
- *Node density*: A maximum of 256 devices per network.
- *Data transmission*: Near real-time collection and data dissemination.
- *QoS*: The latency must be below 125 ms for medical applications.

4.3.7 WBAN Design Challenges

As WBANs are involved in multiparameter monitoring of the human vital signs, the primary requirement is that they should be unobtrusive, scalable, energy efficient, and secure. As Khan et al. [8] have identified, the challenges involved in the design of health-monitoring systems based on WBANs include the following:

- *Power*: In full active mode, the node operational life is very short because the batteries are drained quickly due to communication and processing. For applications such as a pacemaker, it is required that the device operates for years and this needs better scheduling and power management. Ongoing research is on micro fuel cells, biocatalytic fuel cells, and conversion of body temperature to fuel.

- *Computation*: Because of the limited power and memory, the biosensors have limited computational capabilities.
- *Comfort*: The devices should be part of the clothing or ornaments and should not cause discomfort to the users.
- *Security*: The physiological data collected is personal and so confidentiality and authenticity is needed. This is ensured through encryption and preventing unauthorized access.
- *Material constraints*: It is necessary that the shape, size, and material of the implanted sensors should be harmless to human tissues.
- *Robustness*: In harsh and hostile environment, device failure rate is high. Therefore, the protocol must have built-in mechanisms such that failure in one node does not cease the network.
- *Continuous operation*: Biosensors must operate for days and weeks without operator intervention.
- *Regulatory requirements*: The device should be designed with safety and should not harm the human body, and the power utilization of the device must be nonmalignant.

The definition of QoS varies from application to application and there are various perspectives. To date, no definition exists for the QoS of WSNs. Sometimes, it is referred as the optimization of the sensors sending data; for others, it is a generic term that measures quality of network characteristics. Usually, it relates to bandwidth, latency, and jitter. In case of WBANs, according to Ameen [11], it focuses on the collective services and can be specified in terms of reliability, timeliness, robustness, trust, and adaptability.

4.4 WBAN MAC Layers Design Issues

As discussed in Section 4.2 the ongoing research has established WBANs as a key technology that will impact human life tremendously by supporting innovative applications, services, and facilities. To provide these services, reliable transmission of data in real time is necessary in an energy-efficient manner. One of the ways of ensuring energy efficiency is through MAC layers, and so, this necessitates the design of an efficient MAC protocol that can cater a variety of application scenarios. Design of such a MAC layer mechanism is a hot research area, and tremendous research effort has been made to evaluate the existing MAC protocols, and a range of protocols has been suggested to support the variety of application traffic. The reason for focusing more on the MAC layer is that in WBANs, the radio frequency (RF) transceiver of the node takes the major part of the power, and MAC layer design has a significant impact in reducing the average power consumption of the sensor node [28]. For telehealth applications, a battery life of months or years is needed, and so, a good MAC protocol saves energy and should have a low duty cycle. Low duty cycle means nodes should not receive synchronization frames when no data is transmitted or received [28]. According to Gopalan and Park [23], the fundamental task of MAC is to prevent simultaneous transmission while it achieves maximum throughput and minimum delay and minimizes energy wastes. As identified by Ullah and Shen [16] and Gopalan and Park [23], energy efficiency can be optimized by controlling the energy waste, and the sources of energy waste are as follows:

- *Collisions*: Take place when more than one node transmits and receives at the same time and there is retransmission of data, which consumes extra energy.
- *Idle listening*: Nodes listen for an idle channel before transmission.

- *Overhearing*: Nodes receive sent packets for other destinations.
- *Traffic fluctuations*: The data rate keeps on varying.
- *Control packet overhead*: When payload and control information is transmitted, there is a transmission overhead as bandwidth gets wasted.

The current research has categorized MAC protocols as contention based and scheduled based. Contention based are the ones in which nodes contend for a channel. These are scalable, need no time synchronization, but have protocol overheads such as Carrier Sense Multiple Access (CSMA).The other type is scheduled-based protocol, in which channel is divided into time slots of fixed duration and each node transmits in a slot period. There are no issues of contention, idle listening, and overhearing and they conserve energy. In addition to energy efficiency, the other key feature requirements for MAC layer as specified by Ullah and Shen [16] are as follows:

- Must support transparent and simultaneous transmission on multiple PHY layers for on-body and in-body frequency bands.
- Must be adaptable to changes in topology, position, and density.
- Must handle diverse traffic of in-body/on-body nodes.
- Must adapt to changes in delay, bandwidth, and throughput.

Additionally, the MAC layer should also provide QoS for real-time applications. For example, in emergency situations, the nodes should be able to get quick access of the channel, thus medical events should be treated urgently [28]. WBANs carry correlated traffic in which one parameter fluctuation triggers other events such as rise in temperature may cause change in blood pressure, and so, in such type of applications, CSMA/CA experiences extra energy consumption. CSMA involves CCA, which is an issue in Medical Implant Communication Service (MICS) band due to tissue heating. Time Division Multiple Access (TDMA)-based protocols give better performance in correlated traffic due to low duty cycle, but they suffer from synchronization problem. A qualitative analysis of some of the existing standards is discussed here to identify their strengths and weaknesses.

4.4.1 IEEE 802.15.4

This has remained in focus for WBANs for a considerable period, but the ongoing research has suggested that it is not suitable for WBAN as it suffers interference [16]. This standard provides a low-power, low data rate solution and operates in two modes, that is, beacon-enabled mode and non-beacon-enabled mode. In beacon mode, the coordinator does the synchronization and the channel supports superframe structure, with active and inactive periods. The coordinator communicates with the nodes during active period and sleeps during inactive period. In non-beaconed mode, data is transmitted in periodic intervals using a slotted CSMA/CA protocol. Although it is widely suggested for WBANs, IEEE 802.1.5.4 performed poorly in terms of throughput and delay for high data rate application, but it gives a good lifetime for sensor for low data rate applications. Accordingly, Ghare and Kothari [51] have suggested certain modifications in IEEE 802.15.4 in superframe structure to support high data rate applications. The issue is that in WBAN scenario, the data rate of in-body and on-body nodes keeps on changing, which affects sensor lifetime.

4.4.2 Battery-Aware TDMA

A battery-aware TDMA protocol is also been suggested, which is based on battery discharge, channel models, and queuing techniques [52]. It uses the electrochemical properties of the batteries and the fading issues in the channel. The nodes receive beacons and transmit data in active periods, and no transmission takes place in inactive period. The protocol suffers from delay and packet dropping and does not cater emergency traffic.

4.4.3 Energy-Efficient TDMA

An energy-efficient TDMA protocol [53] is also proposed, which assumes a static topology and so no synchronization overhead is required. The network has a master node that coordinates transmission and a monitoring station to gather data. The protocol can support multiple PHY layers simultaneously and is energy efficient for streaming and short bursty traffic. However, since the WBAN topology is dynamic and keeps on changing, the protocol does not perform well and also lacks wake-up mechanism for on-demand events.

4.4.4 Heartbeat MAC

The heartbeat MAC protocol [54] is used to synchronize nodes, and collision-free transmission is guaranteed through dedicated time slots for each node. Thus, the protocol is energy efficient but does not support all types of traffic.

4.4.5 Reservation-Based Dynamic TDMA MAC

The reservation-based dynamic TDMA MAC protocol [16] is more suitable for periodic traffic, but is not efficient for emergency on-demand traffic. It follows a superframe structure with an uplink frame to support normal traffic and a downlink frame to support on-demand traffic.

4.4.6 IEEE Standard for WBANS: IEEE 802.15.6

Although WBANs are specialized than WSNs, the design requirements are very different in terms of energy efficiency, topology, scalability, variety of traffic, different frequency band, QoS, reliability, security, latency, bandwidth, and path losses. Keeping in view of these constraints, in 2007, the IEEE authorities initiated the working group for a communication standard, exclusively for WBANs known as IEEE 802.15.6 [31]. As specified by Astrin and Bang [33], this is a standard for short-range, wireless communication in the vicinity of or inside a human body. The idea behind is to cater both the medical (wearable and implanted) and the nonmedical applications of WBANs such as in entertainment and sports. As highlighted by [20], in IEEE 802.15.6 standard, various frequency bands are proposed including MICS and ISM and a variety of modulation schemes are also suggested to provide power and bandwidth efficiency. Kwak and Ameen [34] have highlighted that the key objective of the standard is to support low-complexity, low-power, highly reliable wireless communication outside/inside the human body. The idea behind IEEE 802.15.6 is to overcome the limitations of IEEE 802.15.1, 802.15.4 and to facilitate the deployment of WBANs. In April 2010, the first draft of IEEE 802.15.6 was presented and the specification defines three PHY layers—narrowband (NB), UWB, and human body communication (HBC)

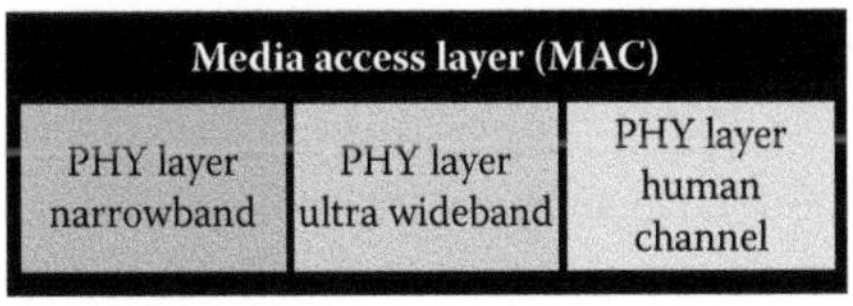

Media access layer (MAC)		
PHY layer narrowband	PHY layer ultra wideband	PHY layer human channel

Figure 4.7   IEEE 802.15.6 architecture.

layers—and one MAC layer (Figure 4.7). Ullah and Kwak [25] conducted a study to find the delay and throughput limits for IEEE 802.15.6 considering an ideal channel with no error using different frequencies.

4.4.6.1 IEEE 802.15.6 MAC Layer

This standard will support a MAC layer over several PHY layers. The draft operates with on-hop star and two-hop tree topologies. In the one-hop topology, frames are exchanged between nodes and hubs, and in two-hop, the hub and node use a relay agent. Time is divided into superframes, and each superframe is further divided into allocation slots. There are three access modes (Figure 4.8):

1. Beacon mode with beacon period superframe boundaries
2. Non-beacon mode with superframe boundaries
3. Non-beacon mode without superframe boundaries

As specified by Tachtatzis and di Franco [26], the Superframe structure starts with a beacon followed by two consecutive periods, each consisting of an Exclusive Access Phase (EAP), Random Access Phase (RAP), Type I/II access phase, and an optional B2 frame that precedes the Contention Access Phase (CAP). The length of these phases is variable given in number of allocation slots. The Superframe structure follows three type of access:

1. Random access: CSMA/CA for the NB and the UWB
2. Improvised and unscheduled access
3. Scheduled access-1 period, m-period

Figure 4.8   IEEE 802.15.6 MAC layer modes.

4.4.6.2 IEEE 802.15.6 PHY Layer

As WBANs can cater a variety of applications, several technologies are suggested for the physical layers such as on-off keying (OOK), frequency shift keying (FSK), and off-set quadrature phase shift keying (OQPSK). Research focus is more on UWB because of its low radiated power, large bandwidth, and scalability of data rates [17]. Various proposals were submitted for NB and UWB due to diversity of applications using in-body and on-body communication scenarios.

Narrow band: For NB, different frequencies are proposed, including MICS, wireless medical telemetry service (WMTS), and ISM bands, and a range of modulation schemes such as Gaussian frequency shift keying (GFSK), Gaussian minimum shift keying (GMSK), differential phase shift keying (DPSK), and OQPSK are suggested. To increase performance, other schemes such as direct sequence spread spectrum (DSSS), wake-up radio, and error correction are also explored [17].

UWB: Three signaling methods for UWB are proposed, including infra-red (IR), chirp, and frequency multiplexed ultra wide band (FM-UWB) by different companies, and it is expected that these proposals will merge with time.

4.5 Data Dissemination Strategies

The routing protocols used in WSNs cannot be directly applied to WBANs because the focus in WSNs is maximizing throughput and minimizing routing overhead, which is more important as compared to conserving energy. The idea behind energy-efficient protocols is to find routes that minimize energy but do not take into account other parameters such as QoS. In addition, protocols of WSNs are designed for static network that carries homogenous traffic, whereas WBANs carry diverse traffic and mobility is limited. Another important consideration in protocol design is the changing topology that affects routing design. So far, research has supported single-hop topology in WBANs, but multi-hop architectures are also being explored. As WBANs are highly distributed, networks and routing mechanisms must be energy efficient, reliable, and should ensure QoS. Routing algorithms of WSNs are not directly applicable to WBANs because the demands of WBANs are quite different. The design challenges as identified by Aboelaze and Aloul [1] and Hassanpour and Asadi [22] are as follows:

- Bandwidth variation due to fading, noise, and interference
- Limitation of energy source
- No redundancy in transmitted information
- Mobility of the nodes
- Multiple sinks
- Heterogeneous traffic and different computation capabilities of the node

As WBANs can be on-body as well as in-body, the design considerations are different for both the types. Efficient routing mechanisms are needed as some of the nodes are mobile and body movement introduces dynamic changes that conventional routing algorithms cannot cater. Routing algorithms for WBANs can be broadly classified as position-based routing, temperature-aware routing, cluster routing, and cross-layer routing (Figure 4.9).

According to Kim and Kim [21], because of the mobility issue, location information is also needed and so the design of routing algorithm should use position parameters. Some of the position-based routing mechanisms are group on earth observations (GEO) information position-assisted routing in which universal clock provides global synchronization of the nodes. Another is location aware routing (LAR), which is an on-demand routing protocol that limits the area of discovering nodes through location information. Geographical and Energy Aware Routing (GEAR)

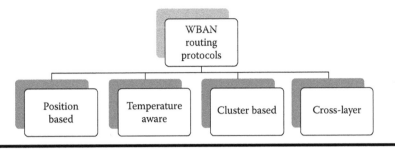

Figure 4.9 Routing strategies for WBAN.

protocol uses cost-based routing. Kim and Kim [21] have suggested an energy-efficient mechanism based on the intersection of circles.

A variety of routing mechanisms are suggested, including routing by cluster, which tries to maximize the system lifetime. One example is Low Energy Adaptive Clustering Hierarchy (LEACH), which is based on clustering. An extension of LEACH is AnyBody algorithm in which the cluster head aggregates the data and the cluster head keeps changing. Hybrid Indirect Transmissions (HIT) combines clustering with chaining, which improves efficiency, but it suffers from conflicts between the routes [28].

Implanted sensors are used in telemedicine applications such as in pacemakers, insulin pumps, and glucose monitors. For such type of applications, radiations and heating effects inside the human body are very important. When data is transmitted, there is a sudden rise in temperature, which is dangerous for human tissue, necessitating design of temperature-aware routing. Ullah and Higgins [28], Bag [62], Shio [63], and Takahashi [64] have identified various techniques to address this issue.

Thermal-Aware Routing Algorithm (TARA): Thermal-aware routing mechanisms include TARA, which routes packets by following a withdrawal strategy. This protocol does not take into account shortest path, and when a packet reaches a hot spot, it is sent back to the sender and sent again after cooling. Each node monitors neighbor packet count and calculates the communication radiation and power consumption to derive the current temperature of the neighbors. It gives better temperature distribution in the network, but it lacks in providing good network lifetime and better reliability.

Least Temperature Routing (LTR) chooses the neighbor with the lowest temperature as next hop and a defined hop count is used. In LTR, packet is dropped when number of hops increases beyond threshold. LTR avoids loops by maintaining lists of visited nodes. So the coolest nodes can be efficiently selected.

Adaptive Least Temperature Routing (ALTR) sends packet to least temperature nodes, but when maximum hop count is exceeded, shortest hop routing takes place.

In the abovementioned thermal-aware routing mechanisms, hot spots are a major [28] issue as temperature information maintenance causes overhead. TARA and ALTR share the same problem of not optimizing the energy efficiency, reliability, or delay and not suitable for large networks due to excessive packet dropping.

Least Total Route Temperature (LTRT) avoids hot spots by using least temperature route using Dijkstra's algorithm, but this has another type of overhead. The node temperature rises by 1 unit when a packet is received and decreases by 1 unit when a packet is not received during the given interval. It gives good performance as compared to other routing algorithms because of less packet dropping but suffers overhead duty to Dijkstra's algorithm [28].

Another mechanism proposed by Kamal and Rahman [35] is a Light Weight Temperature (LWT) mechanism, which is used in chemotherapy and incorporates a rendezvous node that keeps cluster information, and other nodes are publishers, brokers, or subscribers. In case of an event, the publisher stops the service, notifies broker, and broker notifies subscribers. The broker informs the rendezvous node, which stores the cluster information.

According to Honeine and Mourad [13], another design objective in WBANs is to reduce the absorption of power in human body known as Specific Absorption Rate (SAR), which is a measure of amount of power loss due to heat dissipation. In implanted WBANs, the absorption of power in the tissue is a big concern. Accordingly, concentrated efforts are made to calculate the SAR. The bioeffects caused by RFs are related to incident power density, network traffic, and tissue characteristics [28].

A cross-layer energy-efficient multi-hop protocol is designed especially for WBANs by the name of CICADA (Cascading Information Retrieval by controlling access with distributed slot assignment Protocol) [19]. It uses multi-hop TDMA, and same packets are used at both MAC and routing layers. The protocol uses spanning tree mechanism, and there is parent–child relation between nodes; however, CICADA lacks energy efficiency.

4.6 Reliability and Security Mechanisms

In this section, we first focus on different reliability mechanisms available for recovering data loss. Next, the security issues in WBAN are discussed.

4.6.1 End-to-End Reliability

The main objective is to assure and maximize the end-to-end (e2e) reliability of data transport. Generally, the data loss is due to poor links, perturbations, and congestion in the network. Table 4.2 compares the different strategies for e2e data transport in WSNs, which can also be applied to WBANs. Reliable Multi-Segment Transport (RMST) [55] protocol provides data transport reliability, which refers to the eventual transport of all messages from the sensor nodes to

Table 4.2 Reliability Mechanisms for WSN/WBAN

	Reliability		Congestion	
	MLD	*MLR*	*CD*	*CM*
RMST [56]	SNACK	HBH/e2e	—	—
Wisden [57]	NACK	HBH/e2e	—	—
ERTP [58]	IACK	HBH	—	—
CCF [59]	—	—	Buffer + MST	RA
PCCP [60]	—	—	MAT + MST	AIMD
TADR [61]	—	—	Buffer	Multipath
GIT [61]	Hybrid ACK	HBH	Buffer	Multipath/AIMD

the sink. RMST is a selective negative acknowledgment (SNACK) and timer-driven protocol and places responsibility for message loss detection (MLD) at the receivers (which can be intermediate nodes as well as the sink). The missing message requests are explicitly sent as a unicast from the sink toward the source node. RMST is only suitable for the applications that require large size data.

Xu et al. presented Wisden [56] a structural data transport system. Wisden periodically collects structural response data for monitoring damages. Wisden provides reliable data transport by using a hybrid message recovery scheme, that is, hop by hop (HBH) and e2e. Wisden uses HBH NACK scheme, where each source node stores the data in its Electrically Erasable Programmable Read-only Memory (EEPROM) and then transmits to the neighbor node toward the sink. Furthermore, Wisden also uses an e2e negative acknowledgment (NACK)–based scheme for message loss recovery (MLR). In Lea et al. [57] the authors have proposed energy-efficient and reliable data transport protocol (ERTP) for WSNs. ERTP uses implicit acknowledgment (IACK) as MLD. The retransmissions are dynamically calculated and carried out in HBH fashion to recover message losses. Ee et al. [58] presents congestion control and fairness (CCF) scheme for data transport in WSNs. CCF utilizes local congestion detection based on buffer occupancy and message service time (MST). The proposed rate control scheme is based on HBH implicit back-pressure mechanism. To mitigate congestion, CCF uses rate adaptation (RA). Priority-based Congestion Control Protocol (PCCP) [59] calculates a congestion degree as the ratio of message arrival time (MAT) and MST. PCCP uses implicit congestion notification (CN) by piggybacking the header of the forwarded messages, thus avoiding additional control messages and mitigating long-lived congestion efficiently. He et al. [60] present a solution for short-lived congestion by utilizing the idle or underloaded sensor nodes and propose a traffic-aware dynamic routing (TADR) algorithm. TADR routes messages around the congested areas and splits the flow along multiple paths. Shaikh et al. [61] have proposed generic information transport (GIT) framework, which supports tunable reliability using hybrid ACK and recovering data loss hop by hop. The congestion is detected by monitoring the sensor node buffers, and upon detection, it is controlled by adapting multipath and using additive increase multiplicative decrease (AIMD) data rates.

4.6.2 Security

WBANs is a promising technology for providing quality health-care facilities, ubiquitous monitoring, rehabilitation, and emergency response services for disaster management, and it is essential that special attention is given to secured storage and authorized access of patient-related information. As WBANs provide monitoring, processing, storage, and transmission of real-time data of patient, vital parameters with limited resources and conventional means of security are inadequate to satisfy the WBAN requirements. Thus, Patel and Wang [14] have suggested that new mechanisms need to be explored, which are lightweight and efficient. It means that to provide privacy, confidentiality, authentication, authorization, and integrity, innovative techniques of encryption and cryptography must be explored in depth such as elliptic curve cryptography as suggested by a current research. As the data is stored in a distributed manner, any breach in security leading to leakage of patient data can have serious consequences such as wrong treatment. On the contrary, accessibility of data is important and data should be easily available to caregivers, doctors, and insurance companies anytime and anywhere. A challenging situation arises that when security is enhanced, the availability is limited, making security implementation a conflicting goal. In addition, the WBAN environment is deployed in an environment open for eavesdropping, fabrication, and modification, and so, many vulnerabilities exist and threats are possible, such as compromised nodes, compromised database, and false data injection [30] has identified two important data security issues: secure and dependable distributed data storage, and fine-grained distributed data access control for private

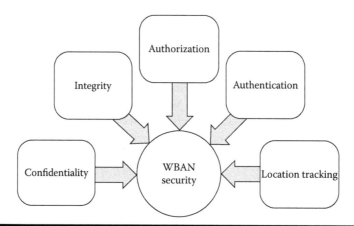

Figure 4.10   WBAN security requirements.

patient medical data. Many researchers have proposed identity-based encryption technique for providing security and privacy in WBANs. Chiu and Wang [18] have proposed a solution based on IBE(identity Based Encryption) which allows a public key to be generated from an arbitrary string. They have suggested that conventional IBE cannot be efficiently used for body sensors and IBE-Lite protocol proves to be a better option for privacy support in body sensor networks. The security requirements (Figure 4.10) as identified by Somasundaram and Sivakumar [44] are as follows:

Confidentiality: The patient information must not be disclosed to unauthorized person, and this can be ensured through symmetric key encryption.

Integrity: In monitoring of life-threatening diseases, the data should be protected from external modification.

Authentication: Identification of original source is necessary so that there is no fabricated data. One technique can be message authentication code.

Localization: Location tracking is necessary for tracking so that attacker cannot manipulate the location.

Availability: A node can be captured or disabled.

Secure management: The key distribution should be done in a coordinated manner.

4.6.2.1 Attacks

The sensor nodes have their vulnerabilities and attacks can be launched, which demand high protection such as variety of Distributed Denial Of Service (DDOS) attacks, including physical attacks of jamming and tampering of nodes; link attacks of collision, unfairness, and misdirection; and flooding attacks at the transport layer [44].

4.6.2.2 IEEE 802.15.6 Security

The IEEE 802.15.6 standard in addition to the details of the PHY and MAC layers has given special focus to the security implementation in WBANs and has defined three levels of security:

Level 0: Unsecured communication: It is the lowest level and data is transmitted in unsecured frames, and there is no consideration for authorization, integrity, confidentiality, and privacy.

Level 1: Authorization only: This is medium level of security and it has authorization but no encryption. It does not support any confidentiality or privacy.

Level 2: Authorization and encryption: This is the highest-level supporting authorization and encryption of data.

4.7 WBAN Applications

Research in WSN was driven by advancement in sensing, processing, and communication. The senor networks remained as a major component in warfare for both platform-centric and network-centric applications. With the increase in chip capacity, processor production capability has increased; there is a decrease in energy per bit processed. Now, one-chip sensors are available, which perform sensing, computing, and communication. Thus, there is an incredible enhancement in the potential applications of WSNs in monitoring events, objects, and interaction of objects with the environment.

The most prominent application difference between WSNs and WBANs is that WBANs monitor human body and interaction of human body with the environment, which is more challenging since the design constraints are different. The ongoing research is mostly concentrated on the telehealth applications as WBANs are gaining social acceptance in this area due to their ability to provide ubiquitous and pervasive monitoring of physiological and biochemical parameters in an environment in an unobtrusive manner. The reason for success of WBANs in patient-centric health care is due to the fact that aging population is increasing every day. In addition, sedentary lifestyle has introduced life-threatening diseases, and the existing means of diagnosis and treatment of these types of diseases are inadequate and costly. The current monitoring system cannot capture transient abnormalities and other momentary events. They are based on qualitative observation, whereas WBANs provide quantitative data collection through real-time sensing, extending health care beyond the confines of geographical limits. In traditional health care, the life-threatening disorders get undetected; however, WBANs will provide long-term trend analysis and handle transient abnormalities, emergency alerting, and improved comfort [14]. The emerging application domains of WBANs are shown in Figure 4.11.

4.7.1 Telehealth/Health-Care Applications

WBANs will be promising in ambulatory monitoring projects and will help in providing economical health-care facilities. For example, the HealthPAL from MedApps is a portable and dedicated remote-monitoring device that provides vital biometric data. It seamlessly gathers and sends patient readings to their electronic medical record for access by care providers. Another device is NanoSonic's new EKGear™ Sensor Shirt that measures heart rate and EKG and wirelessly transmits the information to a remote display and storage unit [37,38].

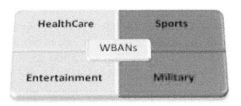

Figure 4.11   Application domains of WBANs.

WBANs provide an efficient solution to ubiquitous health care and provide long-term health-care monitoring and human-centric sensing. They can transmit vital signals such as blood pressure, heart rate, ECG, respiratory rate, temperature, and chest sounds so that caregivers can diagnose and react to adverse events. Researchers from computing, engineering, and medicine are collaborating to devise integrated one-button health-monitoring solutions. Industry giants like Philips have launched products that that will monitor physiological states without restricting movements such as the Telehealth Solutions [4,41], which specified the potential uses of BANs in health care. In case of remote patient monitoring system, WBAN can operate in ad hoc mode or in centralized mode where each sensor node transmits its own data to service node which sends the data to a record server connected with the hospital network [7] has used IEEE 802.15.4/ZIGBEE based sensor network to implement a WBAN for patient monitoring. After the Simulation study, they have suggested that WBAN should not be directly connected to the hospital network and a STAR configuration is more suitable.

WBANs will prove to be an indispensible technology for tailored health care and for preserving and increasing the span of human life. Some of the diseases that can be treated through specialized care are as follows:

- *Cardio diseases*: Prevents deaths, treatment can be prepared in advance and the medical service provider can get vital information on heart rate.
- *Cancer*: A sensor with the ability to detect nitric oxide (emitted by cancer cells) can be placed in suspected locations.
- *Alzheimer*: The elderly population is expected to get doubled and so homebound people can be remotely monitored and treated.
- *Glucose*: A biosensor can monitor glucose level and inject insulin.
- *Asthma*: Sensors can sense the allergic agents in the air.
- *Medical accidents*: Can maintain log and notify occurrence of an event.

4.7.2 Sports Applications

In sports, motion sensor can be used for feature extraction of the players or detect critical limits of the vital signs. For example, the Fitness Assistant from Fraunhofer, which is an intelligent tracksuit, provides a personal health and fitness program [39]. Another example is the Macaw designed for people who exercise regularly. Macaw [40] will be able to transform a smart phone into a personal health hub, connecting an individual's health application and wireless devices for tracking weight, fitness, and overall activity.

4.7.3 Entertainment Applications

Entertainment applications such as gaming, social networking, authentication, and personal information access will get a paradigm shift through WBANs. For example, in interactive gaming, the gaming players perform the actual actions through hands and body, which make the gaming an enjoyable experience. Social networking will allow users to exchange visiting cards through handshake, and personal information stored in WBANs can be used in shopping. Thus, the authentication will be more secure through physiological and behavioral biometrics. In addition, WBANs will pervade our lifestyle through killer applications such as wearable music system, navigation in car/walking, and infant monitoring.

4.7.4 Military Applications

WBANs will change the way military surveillance is done through soldier uniforms integrated with connected devices such as cameras and GPS, making a soldier-assisted system, and this will enhance the battlefield lethality and survivability. In disaster management, hazardous sites can be analyzed for firefighters and paramedics with the help of WBANs. Another important application is in aerospace where astronauts can use the wearable physiological monitoring system.

4.7.5 Worldwide WBAN Projects

A number of projects has been proposed, designed, and implemented to promote the use of WBANs in health care. These projects as highlighted by Al Ameen and Kwak [27] are supported by government agencies and private organizations and cover stress monitoring, cancer detection, elderly people monitoring, and so on. Some of these are given by Khan et al. [8], which are as follows:

- *Hip guard system*: This is for patients recovering from hip surgery. It uses embedded wireless sensors to monitor patient's leg and hip rotation [45].
- *Code blue*: This is a research project for prehospital and in hospital care, disaster response. The system has vital sign sensors, handheld devices, and location tracking tags [46].
- *MobiHealth*: It uses General Packet Radio Service/Universal Mobile Telecommunication System (GPRS/UMTS) communication for data transfer. It is for continuous monitoring of patients outside the hospital environment. The patient is not confined to hospital and can pursue other activities. It provides disease prevention, diagnosis, state monitoring, and clinical research [47].
- *UbiMon*: It uses wearable and implantable sensors for continuous monitoring to capture transient events [48].
- *eWatch*: Wearable sensing and notification built in a wrist watch. It is used in context-aware notification, elderly monitoring, and fall detection [49].
- *Artificial retina*: Retina prosthesis chips implanted within a human eye to assist blind people [50].

4.8 Open Ends: Future Directions

According to Jovanov et al. [3], despite of the maturity in technology and tremendous applications of WBANs in remote patient monitoring, WBANs will become commercially accepted and viable if research focus is on increasing the reliability of serial communication, extending the network coverage, type of payment policies, enhancing the battery life, provision of seamless integration, increased privacy/security, and standardization with interoperability. People who will benefit from WBANs and who will be the stakeholders of WBANs include users, caregivers, physicians, and researchers, but this also increases ownership and liability issues. Thus, as suggested by Jovanov et al. [3], to increase the opportunities of WBANs in telehealth, collaborative effort is needed between physicians, sociologists, technologists, and domain experts. Alongside, there are various challenges that need to be addressed to make WBANs popular in the consumer market. Some of these technical challenges identified by Wang and Pei [4] are as follows:

- Node size/form factor
- Low transmission power for radio interface
- Low electromagnetic pollution

- Node energy consumption should be less than energy scavenging limit
- Network energy efficiency
- Heterogeneous multi-hop links
- High density of node
- Data rate
- Lightweight protocols

WBANs will have a significant impact on our social life and their design needs careful analysis and consideration of various issues. In addition to the above mentioned challenges, Ashraf and Aboul [24] have highlighted some more open problems associated with the research of WBANs as

Physical challenges: These includes unobtrusiveness, sensitivity of sensors, energy of batteries, compatibility, bandwidth problems and effective data collection.

MAC layer challenges: These includes QoS requirements for emergency traffic, data prioritization and fairness schemes for vital signs monitoring.

Network layer challenges: These are load balancing routing strategies, thermal aware routing, congestion control and multipath routing.

Transport layer challenges: These include reliability at packet as well as at event level an flow control mechanisms.

Application layer: There should be a mechanism for translating meaningful information into knowledge.

Other challenges: Other issues are mobility, security, privacy, ease of deployment and scalability.

WBANs will impact our social life considerably. The social issues associated with WBANs as identified by Al Ameen and Kwak [27] are security, privacy, and legality. Not much work has been done so far to address the WBAN social issues. To diffuse in our lifestyle, they should be noninvasive; the characteristics needed are the number of nodes should be less; they should use small-size batteries; there should be a design trade-off between energy used and throughput, latency, and packaging; and strategic placement is required. As highlighted by Hanson and Powell [5], WBANs should provide as follows:

- *Value*: Must change the user quality of life
- *Safety*: Wearable and implanted sensors should not be obtrusive and painful
- *Security*: Should prevent unauthorized access
- *Privacy*: Sensitive information should not be revealed
- *Compatibility*: Nodes should interoperate and there should be seamless integration with other networks
- *Ease of use*: Nodes should be small and ergonomic
- *Interest protection*: Must protect the interest of all stakeholders

While developing protocols, standards, and applications, the communication model that is used as the basis is layered model, and each layer has its own functionality and interacts with its neighboring layer through defined interfaces. This type of approach provides interoperability between heterogeneous technologies but it trades off generality for efficiency. For WBANs achieving energy, efficiency is the primary design consideration, but the issue is that no matter how much effort is done to optimize the MAC layer, routing strategies, and transport layer, no significant change has been observed in the required efficiency. To meet the challenges, researchers have proposed a new approach that violates the layered architecture cutting the defined boundaries [36]. This approach is known as "cross-layer design," which increases the network performance by having interlayer cooperation. The cross-layer approach is the design of algorithms, protocols, or architecture that exploits or provides a set of interlayer interaction, which is the superset of the standard interface. The cross-layer approaches can be divided into two broad categories, that is, *design coupling approach* and *information sharing approach* [36]. The design coupling approaches ignore layer boundaries to propose algorithms that optimize network performance metrics by integrating functionalities from different layers. Information sharing approaches attempt to share information across layers while maintaining architectural protocol boundaries. The layers will share information with each other, for example, status of a node energy will be communicated to all nodes enabling each layer to adjust itself for better energy-efficient behavior and this will improve the performance. The communication can be between adjacent and nonadjacent layers as well as between unidirectional and bidirectional. Presently, no significant work has been done to support cross-layer designs for WBANs; thus, it is suggested that it is an open research area that can be explored in depth as there are various factors supporting it, such as the work by Jurdak [36]:

- Joint consideration of MAC and routing issues will influence the power consumption significantly.
- Mobility has its effect on physical, data link, network, and transport layers. Through cross-layer, it would be easy for a node to manage its resources.
- If the higher layers have knowledge of the physical layer, the routing strategies can be dynamically changed to support it.
- If the data link layer has knowledge of the physical layer, it can modify the error detection mechanisms.

So it implies that a focused effort must be done keeping in view of the constraints of WBANs to design cross-layer frameworks that will optimize the overall performance.

Another important WBAN research area is how wearable sensors on body can be used in the mobile body-to-body (B2B) networks. Mobile B2B networks lead to anytime, anywhere mobile connectivity, which ultimately reduces the number of base stations and will bring significant change in health-care provision and fitness monitoring [42]. The B2B networks are networks of interlinked wireless devices, carried, worn, or integrated into clothing and use the ISM band, with support for network capacity, better data rates, and green technology. The data will be routed from person to person before being received by the actual recipient. Thus, the design of B2B networks involves many challenges at PHY, MAC, and Routing layers. The B2B networks will conserve energy significantly because the signal propagation will take place over few meters to a maximum of 100 m, and frequency reuse can be achieved over much shorter distances, which is called "green" spectrum usage. Furthermore, the power amplifier will not operate at high output levels saving valuable battery energy [43].

4.9 Conclusion

WBANs are evolving every day and will become an indispensable technology in the near future and we can envision a "smart society," where we see a proliferation of smart devices interacting with each other and having ambient intelligence as an integral part. However, despite the tremendous research done presently on developing of standards, miniaturization of sensor, optimizing the wireless channel, design of efficient MAC mechanisms, security/privacy mechanisms, QoS provision, performance, energy harvesting/scavenging techniques, and system on chip (SoC)–embedded system design, there are many potential areas of WBANs that are still untouched and unexplored, such as network layer issues, transport layer congestion issues, provisioning of reliability for crucial real-time traffic, and seamless connectivity with other networks. To enhance the acceptance of WBANs, collaborative work must be done to develop killer applications that will maximize the productivity of WBANs and make them more beneficial for mankind.

The social acceptance of WBANs is gradually increasing with time and it can be anticipated that WBANs will find their place and will penetrate in our lives very soon through emerging applications. This has increased the research interest in various domains, including engineering and medicine. The industry is also collaborating to generate, explore, and discuss new ideas and avenues related to the design, deployment, and implementation of WBANs.

References

1. Aboelaze, Mokhtar, and Fadi Aloul. March 2005. "Current and Future Trends in Sensor Networks: A Survey." In *IEEE International Conference on Wireless and Optical Communications Networks (WOCN)*, 551–5. Dubai, UAE.
2. Chong, Chee-Yee, and P. Kumar Srikanta. August 2003. "Sensor Networks, Evolution, Opportunities, and Challenges." *Invited Paper Proceedings of IEEE* 91 (8): 1247–56.
3. Jovanov, E., C. Poon, Yang Guang-Zhong, and Zhang Y. T. November 2009. "Guest Editorial Body Sensor Networks: From Theory to Emerging Applications." *IEEE Transactions on Information Technology in Biomedicine* 13 (6): 859–63.
4. Wang, Bin, and Yang Pei. 2007. "Body Area Networks." In *Encyclopedia of Wireless and Mobile Communications*, edited by Borko Furht. Wright State University: Taylor & Francis.
5. Hanson, Mark A., and Harry C. Powell. 2009. "Body Area Sensor Networks: Challenges and Opportunities." *IEEE Computer* 42: 58–65.
6. Chen, Min, and Sergio Gonzalez. 2010. "Body Area Networks: A Survey." *Mobile Networks and Applications* 16 (2): 171–93.
7. Khan, Jamil Y. 2008. "Performance Evaluation of a Wireless Body Area Sensor Network for RPS." In *Engineering in Medicine and Biology Society, 2008, 30th Annual International Conference*, 20–25 August 2008, 1266–9. Vancouver, BC.
8. Khan, Pervez, Asdaque Hussain, and Kyung Sup Kwak. 2009. "Medical Applications of Wireless Body Area Networks." *JDCTA: International Journal of Digital Content Technology and its Applications* 3 (3): 185–93.
9. Umar Talha, M. and Jahanzeb Ahmed. 2010. "Body Area Networks—An Overview with Smart Sensors Based Telemedical Monitoring System." *1st International Conference on Business and Technology*, Islamabad.
10. Ullah, Sana, and Pervez Khan. November 2009. "A Review of Wireless Body Area Networks for Medical Applications." *International Journal Communication, Network and System Sciences* 2 (8): 797–803.
11. Ameen, M. A. "QoS Issues with Focus on Wireless Body Area Networks." In *IEEE Computer Society, Third 2008 International Conference on Convergence and Hybrid Information Technology*. INHA University, Korea.
12. Onder, Arif, and Loris Cazalon. "Body Area Networks." Course report SSY145, Chalmers University of Technology, Sweden.
13. Honeine, Paul, and Farah Mourad. "Wireless Sensor Networks in Biomedical: Body Area Networks, Systems, Signal Processing and their Applications XE." Applications \b\i 388–91.
14. Patel, Maulin, and Jianfeng Wang. February 2010. "Applications, Challenges and Prospective in Emerging Body Area Networking Technologies." *IEEE Wireless Communication* 80–7.
15. Konstantas, D. 2007. "An Overview of Wearable and Implantable Medical Sensors." IMIA Yearbook 2007: *Biomedical Informatics for Sustainable Health Systems* 1: 66–9.
16. Ullah, Sana, and Bin Shen. 2010. "A Study of Medium Access Control Protocols for Wireless Body Area Networks." *Sensors* 10 (1): 128–45.

17. Lee, Cheolhyo, and Jaehwan Kim. 2009. "Physical Layer Designs for WBAN Systems in IEEE 802.15.6 Proposals." In *2009 9th International Symposium on Communications and Information Technology*, 841–4. Icheon.
18. Tan, Chiu C., and Haodong Wang. 2008. "Body Sensor Network Security: An Identity Based Cryptographic Approach." In *WiSec '08 Proceedings of the First ACM Conference on Wireless Network Security*, 148–53. ACM, NY.
19. Latre, Benoit, and Bart Braem. January 2011. "A Survey on Wireless Body Area Networks." *Wireless Networks* 17 (1): 1–18. Kluwer Academic Publishers Hingham, MA, USA.
20. Kwak, Kyung Sup, and Sana Ullah. November 2010. "An Overview of IEEE 802.15.6 Standard." Invited paper, *Applied Sciences in Biomedical and Communication Technologies (ISABEL), 2010 3rd International Symposium*, 1–6. INHA University, Korea.
21. Kim, Kibyun, and Junhyung Kim. 2009. "An Efficient Routing Protocol Based on Position Information in Mobile Wireless Body Area Sensor Networks." In *Proceedings of the First International Conference on Networks & Communications*, 396–9. Kyungpook National University, Korea.
22. Hassanpour S., and B. Asadi. "Improving Reliability of Routing in Wireless Body Sensor Networks Using Genetic Algorithm." In *Computer Science and Automation Engineering (CSAE), 2011 IEEE*, 590–3. Islamic Azad University, Tabriz.
23. Gopalan, Sai Anand, and Jong Tae Park. 2010. "Energy-Efficient MAC Protocols for Wireless Body Area Networks: A Survey." *International Congress Ultra Modern Telecommunications and Control Systems and Workshops*, 18–20 October 2010, 739–44. Moscow.
24. Darwish, Ashraf, and Aboul Ella Hassanien. 2011. "Wearable and Implantable Wireless Sensor Network Solutions for Health care Monitoring." *Sensors* 11 (6): 5561–95.
25. Ullah, Sana, and Kyung Sup Kwak. 2011. "Throughput and Delay Limits of IEEE 802.15.6." In *IEEE WCNC*, 28–31 March 2011, 174–8. Cancun, Quintana Roo.
26. Tachtatzis, Christos, and Fbio Di Franco. 2010. "An Energy Analysis of IEEE 802.15.6 Scheduled Access Modes." In *IEEE Globecom Workshop on Mobile Computing*, 1270–5. Letter Kenny Institute of Technology, Ireland.
27. Al Ameen, Moshaddique, and Kyung Sup Kwak. 2006. "Social Issues in Wireless Sensor Networks with Healthcare Perspective." *BT Technology Journal* 24 (2): 138–44. Publisher: Springer Netherlands.
28. Ullah, Sana, and Henry Higgins. 2010. "A Comprehensive Survey of Wireless Body Area Networks." *Journal of Medical Systems* 1–30. Publisher: Springer Netherlands.
29. Bradal, Nourchene, and Samia Belhaj. October 2011. "Study of Medium Access Mechanisms Under IEEE 802.15.6 Standard." In *Wireless and Mobile Networking Conference (WMNC), 2011 4th Joint IFIP*, 26–28 October 2011, 1–6. Toulouse.
30. Li, Ming, and Wenjing Louu. February 2010. "Data Security and Privacy in Wireless Body Area Networks." *IEEE Wireless Communication* 17 (1): 51–8.
31. http://www.ieee802.org/15/pub/TG6.html, Accessed January 31st, 2012.
32. Martelli, Flavia, and Chiara Buratti. 2011. "On the Performance of IEEE 802.15.6 Wireless Body Area Network." In *WirelessConference2011-Sustainable WirelessTechnologies(EuropeanWireless)*, April 2011, 1–6. Austria.
33. Astrin, Arthur W., and Huan Bang. February 2009. "Standardization for Body Area Networks." *IEICE Transactions on Communications* 92-B (2): 366–72.
34. Kwak, Kyung Sup, and M. A. Ameen. "Current Status of the Proposed IEEE 802.15.6 WBAN MAC Standardization." UWB, Wireless Communication Research Centre, Inha University.
35. Kamal, Rossi Md., and Obaidur Rahman. 2011 "A Lightweight Temperature Scheduling Routing Algorithm for an Implant Sensor Network." *ICTC 2011 International Conference*, 396–400. Seoul.
36. Jurdak, Raja. 2007. "Wireless Adhoc and Sensor Networks-A Cross Layer Design Perspective." *Springer Signals and Communication Technology Series*.
37. http://medapps.net/healthPAL-HIW.html, Accessed February 02, 2012.
38. http://www.nanosonic.com/80/17/ekgearsensorshirt.html, Accessed February 02, 2012.
39. http://www.iis.fraunhofer.de/en/bf/med/sens/fitness/, Accessed February 02, 2012.
40. http://www.uspreventivemedicine.com/Macaw-App/Macaw-App-(1).aspx, Accessed February 02, 2012.
41. http://www.healthcare, Accessed February 02, 2012.

42. http://www.pcworld.com/article/210011/bodytobody_networking_the_next_big_thing.html, Accessed February 03, 2012.
43. http://www.microwavejournal.com/articles/12116-using-smart-people-to-form-future-mobile-wireless-networks, Accessed February 03, 2012.
44. Somasundaram, M., and R. Sivakumar. "Security in Wireless Body Area Networks: A Survey." *2011 International Conference on Advancements in Information Technology IPCSIT*, vol. 20.
45. http://www.ele.tut.fi/research/personalelectronics/kankaanpaa/projects/Puhvi/index.htm., Accessed February 07, 2012.
46. http://fiji.eecs.harvard.edu/CodeBlue, Accessed February 07, 2012.
47. http://www.hisp.uio.no/projects/mobihealth/, Accessed February 07, 2012.
48. http://www.ubimon.org/, Accessed February 07, 2012.
49. http://www.cmu.edu/qolt/ResearchXE"Research"\b\i, Accessed February 07, 2012.
50. http://artificialretina.energy.gov/, Accessed February 07, 2012.
51. Ghare, P. H., and A. G. Kothari. 2011. "Modification of Superframe Structure of 802.15.4 MAC for Body Area Networks." In *Communication Systems and Networks (COMSNETS), 2011 Third International Conference on*, 4–8 January, 1–4. Bangalore.
52. Hang, Su, and Zhang Xi. 2008. "Battery-Aware TDMA Scheduling Schemes for Wireless Sensor Networks." In *Global Telecommunications Conference, 2008. IEEE GLOBECOM 2008*. November 30–December 4, 1–5. New Orleans, LO.
53. Shafiullah, G. M., A. Thompson, P. J. Wolfs, and S. Ali. 2008. "Energy-Efficient TDMA MAC Protocol for Wireless Sensor Networks Applications." *Computer and Information Technology, 2008 ICCIT, 11th International Conference*, 24–27 December 2008. Khulna.
54. Li, Huaming, and Jindong Tan. January 2010. "Heartbeat-Driven Medium-Access Control for Body Sensor Networks." *IEEE Transactions on Information Technology in Biomedicine*, January 2010, 14 (1): 44–51.
55. Stann, F., and J. Heidemann. 2003. "RMST: Reliable Data Transport in Sensor Networks." In *Proceedings of the First International Workshop on Sensor Net Protocols and Applications*, 102–12.
56. Xu, N., S. Rangwala, K. K. Chintalapudi, D. Ganesan, A. Broad, R. Govindan, and D. Estrin. 2004. "A Wireless Sensor Network for Structural Monitoring." In *Proceedings of the 2nd International Conference on Embedded Networked Sensor Systems (SenSys)*, 13–24. New York.
57. Lea, T., W. Hub, P. Corkeb, and S. Jhaa. 2009. "ERTP: Energy-Efficient and Reliable Transport Protocol for Data Streaming in Wireless Sensor Networks." *Computer Communications*, 32 (7–10): 1154–71.
58. Ee, C. T., and R. Bajcsy. 2004. "Congestion Control and Fairness for Many-to-One Routing in Sensor Networks." In *Proceedings of the 2nd International Conference on Embedded Networked Sensor Systems (SenSys)*, 148–61. New York.
59. Wang, C., K. Sohraby, V. Lawrence, B. Li, and Y. Hu. 2007. "Upstream Congestion Control in Wireless Sensor Networks through Cross-Layer Optimization." *IEEE Journal on Selected Areas in Communications* 25 (4): 786–98.
60. He T., F. Ren, C. Lin, and S. Das. 2008. "Alleviating Congestion using Traffic-Aware Dynamic Routing in Wireless Sensor Networks." In *Conference on Sensor, Mesh and Ad Hoc Communications and Networks (SECON)*, 233–41. San Francisco, CA.
61. Shaikh, F. K., A. Khelil, B. Ayari, P. Szczytowski, and N. Suri. 2010. "Generic Information Transport for Wireless Sensor Networks." In *Proceedings of the third IEEE International Conference on Sensor Networks, Ubiquitous, and Trustworthy Computing (SUTC)*, 27–34. Newport Beach, CA, USA.
62. Bag, Anirban, and Bassiouni M. A. 2007. "Medium Access Control Protocol and Routing Algorithm for Wireless Sensor, Dissertation." University of Central Florida, ProQuest Information and Learning Company, Ann Arbor, MI.
63. Shio, Kumar Singh, M. P. Singh, and D. K. Singh. 2010. "A Survey of Energy-Efficient Hierarchical Cluster-Based Routing in Wireless Sensor Networks." *International Journal of Advanced Networking and Applications* 2 (2): 570–80.
64. Takahashi, Daisuke. 2008. "Temperature-Aware Routing for Telemedicine Applications in, Embedded Biomedical Sensor Networks." *EURASIP Journal on Wireless Communications and Networking*, 26: 1–26. Hindawi Publishing Corporation.

MEDIUM ACCESS CONTROL (MAC) LAYER ISSUES

Chapter 5

Mobile Medium Access Control Protocols for Wireless Sensor Networks

Bilal Muhammad Khan and Falah H. Ali

Contents

5.1 Introduction

Wireless sensor networks (WSNs) emerged as a new class in the arena of networks just 10 years back. This revolutionizes the way common sensing and actuation was performed and provides a breakthrough in wireless communication field. The network consists of tiny sensor nodes that interact with the environment by sensing. Since WSNs provide a faster, affordable, and flexible solution, they have become very popular and are being used in several real-world applications. With this flexibility and ease of deployment, they bring challenges and problems of their own. The main challenge is to use already limited bandwidth with such a scalable network, and for this innovative medium access, techniques are required to share the limited broadcast bandwidth in a fair and efficient manner as communication devices proliferate day by day.

Due to the nature of WSN, the nodes cannot be connected to a wired power supply; instead, they have to rely on the onboard small battery; almost majority of the medium access control (MAC) protocol designed and developed so far tackles the challenge of conserving the energy at the expense of quality of service (QoS). This chapter investigates and develops techniques to improve and enhance the QoS of MAC for the existing wireless industrial standard IEEE 802.15.4 in mobile application.

5.1.1 Mobile MAC Protocol Key Problems and Challenges

Mobility in WSNs raises new challenges and problems that occur at MAC level due to the existing link failure and new link formation. This in turn introduces delays, which are connected to establish connectivity. Moreover, due to mobility, neighborhood information is variable, which causes problems in scheduling. The mobile nodes may experience loss of connectivity in a network using schedule-based protocols. The reason is that as the node enters in the radio range of a new cluster, it has to wait for the next schedule synchronization point, and this leads to a temporary loss of connectivity and an increased latency for the packet that the mobile node wants to transmit. Moreover, mobility may also increase the chances of number of collisions in the network as the mobile node has no information about the neighboring nodes in the new cluster, thus can interfere in the ongoing transmission between the two nodes and causes collision in the network. This collision will lead to increase in number of retransmissions, resulting in increase in packet latency, high-energy consumption, and often resulting in packet drop.

Node mobility in WSN under certain scenarios can be beneficial, as in case of increasing network lifetime and network coverage [1–3]. However, as discussed in Section 5.1, mobility also imposes new challenges and problems on the design of MAC protocols. As the node moves away from the communicating node, frame loss and packet drop can occur due to the significant signal strength variations [4]. Moreover, due to hidden node and nonsynchronization with the new cluster, the mobile nodes contribute in the increase of number of collision, which plays a significant role in the degradation of QoS in wireless network. Especially in case of WSN where the resources are limited and energy conservation is significant, collision causes excessive loss of energy. Due to the loss of packet, the number of retransmission increases resulting in severe degradation in throughput, loss of energy, inefficient bandwidth utilization, and higher latency for the network.

Designing MAC protocol for mobile scenario is a challenging task, especially in case of WSNs, which are inherent resource constrained. The prime objective of such type of protocols in mobile scenario is to maintain connectivity and acceptable QoS while incurring less collision in the network. The protocols should be less complex and conserve energy. Moreover, the time associated for connectivity, neighborhood discovery, and synchronization should be minimized as these factors contribute significantly in the increase of latency.

5.2 Mobile MAC Protocol for Wireless Sensor Networks

The breaking of existing link and formation of new link due to one-hop neighbor moving out of the radio range due to mobility causes packet loss, increase in latency, increase in number of collision, and high-energy consumption [5–7]. Mobility causes rapid changes in topology of the network [8] and hence becomes increasingly challenging, especially for MAC protocols to adapt to such fast changes.

Mobility in WSN can be broadly classified into the following three main categories:

1. Nodes are mobile while the sinks are stationary.
2. Sinks are mobile and nodes are stationary.
3. Both nodes and sinks are mobile.

In a typical WSN, the sink is located centrally and is stationary; however, the area around the sink experiences high level of communication activities; due to this reason, the energy supplies of the node near the vicinity of sink deplete considerably fast. This may result in the disconnection of the sink with the entire network. To mitigate this problem, mobile sink concept is introduced. Gandham et al. [9] propose that sink relocation methods on the basis of energy consumption of individual sensors are evenly distributed and that overall network energy is minimized. Load balancing and data transmission is investigated by Luo and Hubaux [10]; the authors propose joint mobility and routing scheme to maximize the network lifetime. In this scheme, the mobile sink moves on a circular trajectory inside the area where the nodes are deployed. The sensing nodes that are inside this trajectory send the data to the mobile sink using the shortest possible path to conserve energy.

To address the problems of connectivity in sparse network where the numbers of sensing nodes are low, mobile data collector scheme is introduced [11]. In this scheme, the mobile node gathers all the buffer information stored in an individual node. Investigation over different mobility patterns is presented by Kansal et al. [12]. A random mobility pattern is introduced for mobile data collector by Shah et al. [13]. In this scheme, a mobile data collector randomly moves to collect the data buffered in the nodes of a sparse network. Since the movement of the data collector is random, the network experiences latency. Knowledge-based mobility or predictable mobility for mobile data collector is introduced to meet some latency bounds as well as save energy of the individual nodes [14]. The nodes in such scenario are aware of data transfer time and based on this go into sleep mode to save energy, hence increasing network lifetime [15]. In majority of WSNs, sensor nodes generate data at different rates. Data losses in sensor node occur if the buffer in the node is full and the contents of the buffer are not being transferred to the mobile data collector. To resolve such issues, a scheme referred to as controlled mobility [16,17] is proposed. To achieve controlled mobility, mobile element scheduling is proposed by Somasundara et al. [18]. The mobile element (ME) is scheduled in real time to visit the nodes before the buffers are full. To improve upon the protocol, two more proposals Earliest Deadline First (EDF) and Minimum Weight Sum First (MWSF) are proposed by Somasundara et al. [18].

In EDF, the decision of the next node to be visited by ME depends on the node having its buffer overflow deadline first. EDF scheme leads to significant amount of data loss due to the fact that nodes are far away as the network is sparse and two consecutive nodes can have the same deadline for buffer overflow. In MWSF, this problem is tackled by introducing the distance factor. The ME scheduling not only depends on the earliest deadline of the buffer overflow, but it also take into account the fact of how far the nodes are within the network. Even MWSF introduces the distance factor before deciding which node to visit, but the back and forth movement to reach faraway nodes is not completely avoided. Multiple mobile data collector with relay data collection algorithm is proposed by Jea et al. [19] to increase the scalability of the sparse network and to curb the problem of distance between the nodes. In this algorithm, nodes that are out of range from mobile data collector find the nearest sensor nodes to the MDC and relay their data to them using the shortest possible route.

Mobile MAC (MOB-MAC) is presented for mobile WSN scenarios by Raviraj et al. [4]. MOB-MAC uses an adaptive frame size predictor to significantly reduce the energy consumption. A smaller frame size is predicted when the signal characteristics become poor. However, the protocol incurs heavy delays due to the variable size in the frame and is not suitable for mobile real-time applications. Adaptive mobility MAC (AM-MAC) is presented in by Choi et al. [20]; it is the modification of Sensor-MAC (S-MAC) to make it more useful in mobile applications. In AM-MAC, as the mobile node reaches the border node of the second cluster copies and holds the schedule of the approaching virtual cluster as well as the current virtual cluster. By adopting this phenomenon, the protocol provides fast connection for the mobile node moving from one cluster and entering the other. The main drawback is that the node has to wake up according to both the schedules, but cannot neither transmit nor receive data packet during the wake-up schedule other than the current cluster. This contributes to significant delay and loss of energy due to idle wakeup.

Another variation of S-MAC is presented by the name of mobility sensor MAC (MS-MAC) [21]. It uses signal strength mechanism to facilitate fast connection between the mobile node and the new cluster. If the node experiences change in received signal strength, then it assumes that the transmitting node is mobile. In this case, the sender node sends not only the schedule but also the mobility information in the synchronous message. This information is used by the neighboring node to form an active zone around the mobile node so that whenever the node reaches this active zone, it may be able to update the schedule according to the new cluster. The drawback of this protocol is idle listening and loss of energy. Moreover, nodes in the so-called active zone spend most of the time receiving synchronous messages rather than actual data packets, thus resulting in low throughput and increase in latency. Another real-time mobile sensor protocol dynamic sensor MAC (DS-MAC) is presented [22]. In this protocol, according to the energy consumption and latency requirement, the duty cycle is doubled or half. If the value of energy consumption is lower than the threshold value (Te), the protocol doubles its wake-up duty cycle to transfer more data, thus increasing the throughput and decreasing the latency of the network. However, the protocol requires overheads, and during the process of doubling of wake-up duty cycle, if the value crosses Te, the protocol continues to transmit data using double duty cycle resulting in loss of energy.

MD-MAC is proposed by Hameed et al. [23], which is the extension of DS-MAC and MS-MAC. The protocol enforces Te value, and during any time if the value of energy consumption is doubled, then it halves the duty cycle. Moreover, the mobile node undergoes neighborhood discovery unlike other neighboring nodes that form an active zone as in the case of MS-MAC. MD-MAC is complex and requires high overheads.

5.3 IEEE 802.15.4 MAC Protocol for Mobile Sensor Network

IEEE 802.15.4 industrial standard carrier sense multiple access collision avoidance (CSMA/CA) MAC protocol is presented by Lam [24]. CSMA/CA uses random backoff values as collision resolution algorithm. Since WSN is resource constrained, the entire backoff phenomenon is performed blindly without the knowledge of the channel condition; this factor contributes in more number of collisions, especially when the number of active nodes is high. The IEEE 802.15.4 MAC protocol was mainly used in static scenarios, and initial research work and performance studies are done while considering static applications [25–28]. Recently, the use of mobile nodes in WSN became a very hot topic [17,29,30]. Introducing node mobility in WSN raises many challenges and problems, especially at MAC level. IEEE 802.15.4 MAC CSMA/CA is also used in mobile applications; however, the QoS is very poor [31–33]. The protocol incurs high latency over the mobile node as it requires long association process for the node leaving from one cluster to another cluster. If the node misses the beacon from the new cluster head consecutively for four times, then the node will go into orphan realignment process, hence inducing more delays. For the node to join the new cluster and not to miss the beacon from the new cluster head, it tends to wake up for long duration, which in turn significantly increases the energy consumption of the individual node. Moreover, as the mobile node succeeds in migrating from one cluster to another, it has to compete with the already available active nodes in the cluster, which contribute in significant number of collisions and packet drops, resulting in overall degradation of network QoS.

5.3.1 Association, Synchronization, and Orphan Scan in IEEE 802.15.4

The process of association and synchronization in IEEE 802.15.4 CSMA/CA MAC protocol is presented in Figure 5.1. The node association starts with an active scan procedure that scans all listed channels by sending beacon requests to all nearby coordinators. All the information received in a beacon frame will be recorded in a personal area network (PAN) descriptor. The results of the channel scan will be used to choose a suitable PAN. The node then sends a request to associate

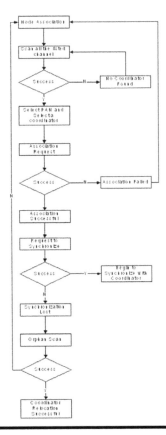

Figure 5.1   **Association and synchronization process in IEEE 802.15.4 CSMA/CA MAC protocol.**

with the chosen coordinator. The node updates its current channel and PAN id while waiting for an acknowledgment from the coordinator. Upon receiving an acknowledgment, the node then waits for the association results. The coordinator will take *aResponseWaitTime* symbols (*32*aBase SuperframeDuration, about 0.49 seconds*) to determine whether the current resources are available on the PAN to allow the node to associate. If sufficient resources are available, the coordinator then allocates a short address to the node and sends an association response command containing a new address and a status, indicating a successful association. If there are no sufficient resources, the node will receive an association response command with a failure status.

After the node associates with its coordinator, it will send a request to synchronize and start tracking the beacons regularly. If the node fails to receive a beacon *aMaxLostBeacons* times (equal to four times), it may conclude that it has been orphaned. The node then has the option to perform either the orphan device realignment procedure or the association procedure. If the node chooses to perform an orphan device realignment, it will do the orphan scanning by sending an orphan notification command to relocate its coordinator. The node waits for *aResponseWaitTime* symbols to receive a coordinator realignment command. The coordinator that receives the orphan notification command will search its list looking for the record of that node. If the coordinator finds the record, it will send a coordinator realignment command to the orphaned node together with its current PAN ID, MAC PAN ID, logical channel, and the orphaned node's short address. The process of searching the record and sending the coordinator realignment command takes within *aResponseWaitTime* symbols.

5.4 Case Study of IEEE 802.15.4 MAC Protocol

In WSNs, majority of the contention-based protocols use random backoff mechanism to avoid collision. The main drawback of using such protocols is that they rely on allocating random backoff delays to resolve the collisions, but as the number of nodes increases, the probability of nodes selecting the same backoff increases, thus resulting in access collision. In case of mobile sensor network, this problem becomes more serious as the mobile node enters from one cluster to another and tries to utilize the available resources that are already limited for the nodes in the cluster. This creates congestion, high latency, and low throughput in the network.

Section 5.4 fundamentally addresses the limitation of mobile nodes by implementing mobility adaptive (MA)-CSMA/CA The protocol enables existing CSMA/CA wireless standard used in WSN to be used in mobile applications efficiently. The proposed MAC is energy efficient while maintaining high goodput, low latency, and fairness among the nodes compared to the existing industrial standard CSMA/CA MAC protocol of IEEE 802.15.4. The energy efficiency is achieved by (1) minimizing the number of collisions and (2) transmitting the sleep schedule every time so that the node does not unnecessarily waste energy to perform backoff and sensing of the channel as the channel is being occupied by some other contending nodes. High goodput and fairness among the nodes is achieved by reallocating the guaranteed time slots (GTSs) to the nodes according to their priorities as well as data transmission frequency ensuring that every node will be able to access the shared communication link on fair bases.

Once the node joins the network, it contends normally as per CSMA/CA standard for the transmission of data; however, in case of the mobile node leaving one cluster and joining another cluster, the new cluster head allocates a GTS for the incoming node. This in turn minimizes the collisions in the new cluster because of the entrance of the new node and also by preallocating GTS to the mobile node the protocol saves valuable association and reassociation time for the nodes, which in turn improves the latency of the network and saves valuable energy of the mobile unlike the current CSMA/CA standard.

5.4.1 MA-CSMA/CA Main Stages

The transmission of each successful frame in MA-CSMA/CA MAC undergoes five main stages:

Stage 1: The contending node acquires backoff delay as per standard CSMA/CA protocol. In case of mobile node moving from one cluster to another, it has highest priority. The node constantly monitors the beacon strength received from different coordinators apart from the parent coordinator. At any stage if the beacon strength from one particular coordinator increases, the node assumes that it is moving toward the new coordinator and sends the request to the current cluster coordinator to allocate a GTS in the cluster toward which the node is moving. To maintain fairness among the nodes, the GTS resources are reserved once for the mobile node as they are about to enter the new cluster. After the first transaction, the node has to compete in the cluster like the other nodes. This GTS allocation before entering the new cluster significantly reduces the association time for the mobile node, which in normal circumstances any node using the standard CSMA/CA protocol undergoes. Moreover, as the node enters the new cluster, any ongoing transmission within the cluster will not be interrupted since it is preallocated with GTS. This in turn made MA-CSMA/CA more suitable for mobile scenario.

Stage 2: Nodes having data packet in the buffer undergo backoff delay before the transmission of packet.

Stage 3: Nodes perform clear channel access (CCA). If the channel is busy, then increment the backoff value.

Stage 4: If the channel is available after the second CCA, then the node transmits the data toward the intended receiver.

Stage 5: If the node is mobile and enters into the new cluster, then after the first transmission, GTS is de-allocated to maintain fairness in the cluster.

MA-CSMA/CA protocol is illustrated in Figure 5.2 for two nodes A and B. Node A is mobile and therefore it will continue to receive beacon from other nearby coordinators. As the signal strength of a particular coordinator increases in a consistent manner, the node assumes that it is moving in the direction of that coordinator and requests the current coordinator to send request

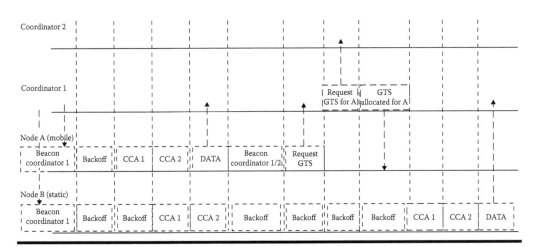

Figure 5.2   Illustration of MA-CSMA/CA MAC protocol.

for GTS allocation in the adjacent cluster for the mobile node. The coordinator upon receiving the request establishes the link with the adjacent cluster coordinator and requests for GTS for the mobile node.

In this way, the time of association is reduced considerably and the node will be able to transmit and receive the data packets normally. It is assumed that all the nodes A and B have data packets in their buffers. Therefore, according to the allocated delays, these nodes will perform the backoff. After the backoff period, node A performs CCA. After the two successful CCA, the node A sends data toward coordinator 1, and in the meantime, node B also starts performing CCA. As the channel was preoccupied by node A, node B increases its backoff value and repeats the whole procedure again until it transmits the buffered data packet.

5.5 MA-CSMA/CA Protocol

The proposed protocol and its working principle are presented in Figure 5.3. As the node joins the cluster, it receives beacon from the potential cluster heads. The node decides on the basis of the signal strength and chooses the cluster head having the strongest signal. In case of mobile node since it is constantly on the move, the proposed protocol uses the measure of signal strength from different cluster heads to its advantage. The node monitors the beacon signal strength from different cluster heads; when it reaches a point where the energy level of received beacon from the cluster head keeps on increasing consistently, the node assumes that it is moving toward the direction of this particular cluster head. The protocol allows the node to send a request to gain GTS in the upcoming cluster head even before joining the new cluster through its parent cluster head. Upon receiving the request of the mobile node, the current cluster head establishes the link with the potential new cluster head for the mobile node and receives the GTS for the node. This whole process is not in the conventional CSMA/CA WSN standard. Once the node leaves the cluster, it performs the association process, the process takes a long time, and if the nodes misses the beacon frame from the new coordinator four times, then it would go into orphan alignment process. This whole process of association going into new cluster increases the latency of the system. Moreover, after the association process, the new incoming node has to compete with the other nodes previously available in the cluster, which increase the probability of collisions, low throughput, high latency, and overall degradation of QoS of the network.

By allocating the GTS to the node even before it enters the new cluster through the old cluster head, MA-CSMA/CA significantly reduces the association time for the mobile node; moreover, due to GTS, the node will be able to successfully transmit the data in the new cluster without competing with the local nodes and not contributing to the number of collisions of the new cluster. To maintain fairness among the node, the GTS is allocated only once for every mobile node entering the cluster, and after that the node has to contend and perform the same procedure that other nodes follow to transmit the data toward destination

5.5.1 Pseudocode for MA-CSMA/CA Protocol

The pseudocode of the simulation loops is presented in Algorithm 1. The pseudocode matches the MA-CSMA/CA protocol as depicted in Figure 5.3. The system state during each slot is tracked using state vectors of dimension N, which is the number of nodes. Four vectors represent the values of number of backoff (NB), contention window (CW), backoff exponent (BE), and beacon strength (BS). The fifth vector (*delay*) represents both the states of the node (−1 if idle, 0 if

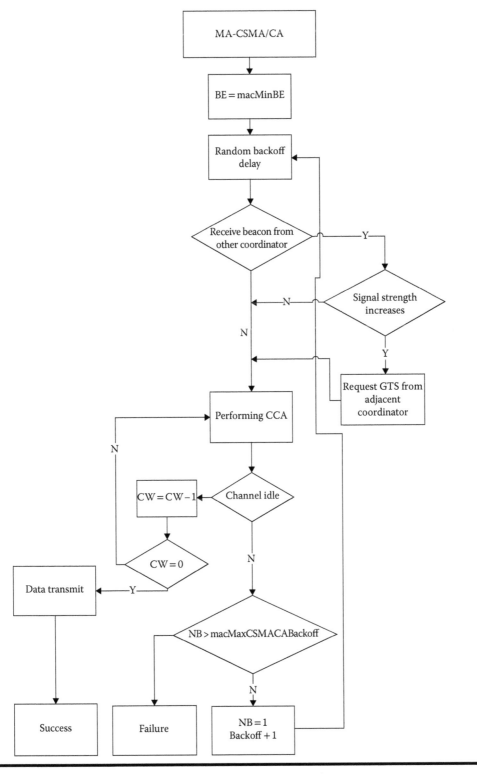

Figure 5.3   Block diagram representation of MA-CSMA protocol.

transmitting). When the node is in the transmitting state, the remaining transmission time in slots is maintained in the corresponding element of fifth vector (*busy for*). The number of transmission (*nbTransmission*), collisions (*nbCollision*), access failures (*nbFailure*), CCA's (*nbCCA*), and beacon threshold (*BTH*) are tracked for a simulation run of $T = 10^{10}$ slots.

Algorithm1. MA-CSMA/CA Protocol

```
%state vectors
NB(1:.N) = 0;
CW(1:.N) = 3;
BE(1:.N) = aMinBE;
delay(1:.N) = random number in [1....2^BE(i)];
BS(1:.N) = BTH_max
busyFor(1:.N) = 0;
Channel(1:.N) = Channel state [Idle = 0; busy = 1];

%Tracking Variables
nbTransmission;
nbCollision;
nbFailure;
nbCCA;
```

for T slots **do**
check nodes i for BS $==$ BTH$_{max;}$

if BS $==$ BTH$_{max}$
    select the node as the coordinator and perform association process;

end

check nodes i ready for CCA (IB $==$ 0) and update *nbCCA(i)*;

if BS<BTH$_{min}$

if *channel* idle **then**
    decrement CW for nodes i;
    check nodes j from i with *Channel* $= =0$;

    if more than one node j **then**
    increment *nbCollision(j)*;

    else
    increment *nbTransmission(j)*;
    set *busyFor(j)* to the transaction length in slots;
    end

else
    % Channel busy
    Check nodes j transmitting (*busyFor(j)>0*);

```
for node transmitting j do
    decrement busyFor(j);

    if busyFor(j == 0) then
        reset NB(j), BE(j), CW(j), and delay(j) to initial values for new transmission;

    end
end

for nodes i doing CCA do
    update CW(i), NB(i), BE(i);
    check nodes k from i with NB(k) == macMaxCSMABackoffs;

if maxBackoff reached for node k then
    increment nbFailure(k);
    reset NB(j), BE(j), CW(j), and delay(j) to initial values for new transmission;

end
end
end
```

elseif $BS >= BTH_{min}$
send request for GTS via current coordinator and save the ID for the interfering coordinator

```
if channel idle then
    decrement CW for nodes i;
    check nodes j from i with Channel ==0;

    if more than one node j then
    increment nbCollision(j);

    else
    increment nbTransmission(j);
    set busyFor(j) to the transaction length in slots;
    end

else
    % Channel busy
    Check nodes j transmitting (busyFor(j)>0);

    for node transmitting j do
    decrement busyFor(j);

    if busyFor(j == 0) then
        reset NB(j), BE(j), CW(j), and delay(j) to initial values for new transmission;

    end
end

for nodes i doing CCA do
```

update *CW(i)*, *NB(i)*, *BE(i)*;
check nodes k from i with *NB(k) = = macMaxCSMABackoffs;*

if *maxBackoff* reached for node k **then**
increment *nbFailure(k);*
reset *NB(j)*, *BE(j)*, *CW(j)*, and *delay(j)* to initial values for new transmission;

end
end
end

else BS = = BTH$_{max}$
update the coordinator ID;
wait for GTS allocated slot for the next beacon frame;
end

for nodes i in *delay* line **do**
delay(i) = delay(i) – 1;
end
end

5.6 Simulation and Analysis of MA-CSMA/CA

A Monte Carlo simulation of MA-CSMA/CA is performed. For this purpose, a vector-based simulator is developed in MATLAB®. The proposed protocol is also compared against the existing state-of-the-art CSMA/CA MAC protocol for the industrial standard of IEEE 802.15.4.

5.6.1 Network Model for Simulation

The protocols are compared in a typical multi-hop multi-cluster topology scenario. The network model for both the scenarios is presented in Figure 5.4. There are three types of nodes in WSN: (1) PAN coordinator, (2) FFD (full function device), and (3) RFD (reduced function device). The functions of PAN coordinator are to organize the network and to serve as the sink node. FFD can become the coordinator node for the cluster of nodes outside the range of the central PAN coordinator and serve as bridge node to the central coordinator. RFD is the device with limited capabilities. These devices are primarily used for sensing purposes.

5.6.2 Simulation Environment and Parameters

MATLAB is used as the simulation platform for the implementation of the new MA-CSMA/CA and the existing WSN industrial standard CSMA/CA. It is assumed that there are no retransmissions. Multi-hop network is considered and all nodes can send maximum of 1K of data, which is divided into a number of packets having a maximum of 143 data bits per packet. As slotted, CSMA/CA is therefore considered the size of beacon interval (*BI*), and superframe duration (*SD*) is calculated as follows:

$$BI = (aBaseSuperframeDuration)2^{BO} \qquad (5.1)$$

$$SD = (aBaseSuperframeDuration)2^{SO} \qquad (5.2)$$

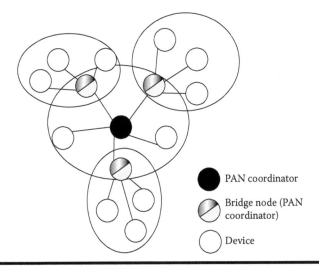

PAN coordinator

Bridge node (PAN coordinator)

Device

Figure 5.4   Multi-hop multi-cluster network.

In Equations 5.1 and 5.2, *BI* and *SD* are dependent on two parameters beacon order (*BO*) and superframe order (*SO*). The *aBaseSuperframeDuration* denotes the minimum duration of superframe, corresponding to *SO* = 0. This duration is fixed to 960 symbols where every symbol corresponds to 4 bits and the total duration of the frame equals to 15.36 ms, assuming 250 kbps in the 2.4-GHz frequency band; hence, each time slot has duration of 0.96 ms. In the simulation analysis, the abovementioned conditions are considered, and performance analyses are carried out by assigning same values to *SO* and *BO*. Moreover, to support mobility random waypoint model [34] is used. Rest of the simulation parameters are presented in Table 5.1.

The major performance matrices are simulated and the results of the proposed MA-CSMA/CA and existing CSMA/CA MAC protocols are compared in the following sections.

5.6.3 Setup Time

Setup time can also be referred to as response time. The response time is defined as the time required by any node in the network when the packet arrives in its buffer to the time the node about to transmit the packet. The setup time is mainly dependent on the MAC layer and its performance. The more efficient the MAC protocol is, lesser the setup time required by the contending node to access the shared communication link. The setup time is dependent on many parameters including control signal time (τ_{CS}), backoff exponential time (τ_{BE}) in case of CSMA/CA, association time (τ_{AS}) for MA-CSMA/CA, and radio turnaround time (τ_{RT}). The average setup time (τ_{sp}) for MA-CSMA/CA and CSMA/CA can be calculated from Equations 5.3 and 5.4 as follows:

$$\tau_{sp} = \sum_{m=1}^{M} \tau_{AS(m)} + \tau_{CS(m)} + \tau_{RT} \tag{5.3}$$

$$\tau_{sp} = \sum_{m=1}^{M} \tau_{BE(m)} + \tau_{CS(m)} + \tau_{RT} \tag{5.4}$$

Table 5.1   Simulation Parameters

Constants	Values
aBaseSlotDuration	60 symbols
aMaxSIFSFrameSize	18 octets
aNumSuperframeSlots	16
aUnitBackoffPeriod	20 symbols
macMinSIFSPeriod	12 symbols
CW	3
MacMinBE	0 to 3
maxCSMABackoff	5
Radio Turnaround Time	12 symbols
CCA detection time	8 symbols
N = for Multi-hop	36 nodes
E_{TX}	0.0100224 mJ
E_{RX}	0.0113472 mJ
E_{CCA}	0.0113472 mJ
E_{idle}	0.000056736 mJ

The value of aforementioned backoff exponent time is given by

$$\tau_{BE} = \left(2^{BE} - 1\right)\tau_{st}; \quad 0 \le BE \le 5 \tag{5.5}$$

$$\tau_{CS} = \left[\frac{T_{nb}}{T_{cs}}\right]\tau_{st} \tag{5.6}$$

where (τ_{st}) is the symbol time, (T_{nb}) is total number of control bits, and (T_{cs}) is total number of bits per symbol

$$\tau_{CS} = \left(ns\right)\tau_{st} \tag{5.7}$$

where (ns) is total number of symbols for which timer is activated.

Figure 5.5 shows the simulation results of average setup time for different number of active nodes N and traffic generated per node. The simulation results are obtained by choosing the values of backoff exponent (BE) equal to two values such as 3 and 2 for CSMA/CA. In case of CSMA/CA for ($BE = 3$), the setup time is much higher; this is due to the reason that the node before starting to send the packet undergoes a backoff delay, and after the backoff delay, it performs channel sensing; if the node finds that the channel is busy, then it undergoes a random delay. Due to this reason, the latency increases. In case of ($BE = 2$), the value of setup time is less than ($BE = 3$), but still considerably greater than the other techniques. In case of mobile nodes,

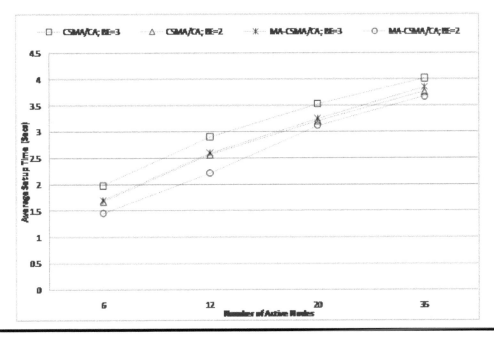

Figure 5.5 Multi-hop multi-cluster network scenario according to average network time.

CSMA/CA apart from initial backoff delay exhibits more delays due to initialization and orphan process. In case of CSMA/CA, if the node goes out of the range of the coordinator, then it starts the process of association. During this process, any packet sent toward the node is dropped. If the node gets acknowledgment during association process, then it waits for the beacon from the corresponding coordinator. In case the association process fails, then the node can go into orphan realignment procedure or perform the association procedure again. This whole procedure contributes to the latency of the network. The MA-CSMA/CA incurs low delays. The mechanism of requesting the GTS from the coordinator of other cluster before being disconnected from the parent coordinator saves considerable amount of time, which in the aforementioned protocols is wasted due to association process. Under variable and different number of active node conditions, the proposed protocol shows significant reduction in latency as compared to other protocols, which can be seen from the results.

5.6.4 Idle Channel Time

Idle channel time is one of the key factors in determining the efficiency of the MAC protocol. In WSNs, the resources, especially the bandwidth, are also scarce and hence the efficient utilization of the bandwidth becomes a priority for any MAC protocol. Figure 5.6 shows the comparison of the existing industrial standard MAC protocol and the proposed protocol that supports mobility. In case of CSMA/CA, it is clear from the results that the protocol shows worst bandwidth utilization. This is due to the fact that as the nodes collide or sense any activity over the channel, they back off randomly, and this random backoff increases exponentially, thus increasing the delay and decreasing the bandwidth utilization. It is because of this nature of CSMA/CA protocol that the nodes undergo long delays before transmission of data. In case of proposed MA-CSMA/CA MAC, nodes undergo initial backoff values, which

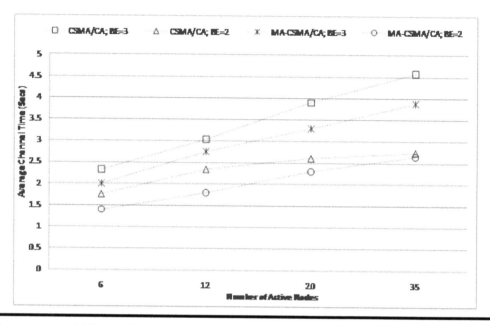

Figure 5.6 **Multi-hop multi-cluster network scenario according to idle channel time.**

are according to the standard; moreover, before the start of transmission, a feedback is provided to all the potential contending nodes about the duration of the current transmission. Therefore, the node freezes their backoff values and goes to sleep mode, as they wake up, the nodes start the remaining backoff count and access the channel. Hence, instead of providing a long random backoff delay, which results in excessive wait state with the probability that the channel may have been available during all the excessive random backoff duration, the proposed protocol provides initial backoff values along with the feedback of transmission so that nodes will know exactly when the channel is available for the transmission of data. It can be seen clearly from the results that the proposed MAC protocol significantly improves the channel utilization as compared to IEEE 802.15.4 industrial standard CSMA/CA in mobile application scenario.

5.6.5 Goodput

Goodput is the total number of bits received correctly at the destination without retransmissions and control signals. Figure 5.7 shows the comparison of the goodput of the proposed protocol and CSMA/CA in mobile application. The result shows significant degradation of goodput in case of CSMA/CA. There are several factors behind the degradation of the goodput for the CSMA/CA. From Section 5.6.4, it is clear that the channel utilization is very poor and the latency in terms of setup time is very high. The other influencing factor is the presence of random backoff and problem of collision. As the number of nodes in the network increases, it also increases the number of potential contending nodes, which causes higher collision, more random backoff, and more retransmission of data. Hence, more and more packets are collided and the overall QoS falls drastically. It can be seen that at lower value of (BE), the goodput of the network is decreased. This

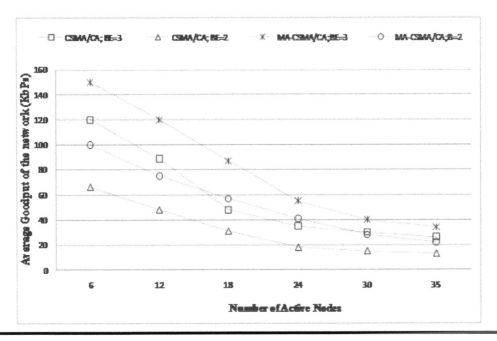

Figure 5.7 Multi-hop multi-cluster network scenario according to average goodput.

is due to the fact that more number of nodes will be ready to transmit the data with lower backoff delay causing considerable rise in collision and thus contributing to lower goodput results. In case of MA-CSMA/CA, collision is minimized by allocating GTS for the mobile node entering the new cluster. Moreover, the average setup time for the proposed protocol is very low and the channel utilization is also on the high side. Hence, the goodput of the proposed scheme increases not only under increase number of active nodes but also on variable data traffic.

5.6.6 Energy

Energy is also one of the key factors in WSNs. Since sensors are tiny devices, they are required to save energy. Figure 5.8 shows the comparison of energy for MA-CSMA/CA and CSMA/CA. It can be seen from the result that as the number of nodes increases, the energy consumption of CSMA/CA increases. The reason behind this is that CSMA/CA suffers heavily from collisions due to hidden nodes as well as contention collisions. As the packets of the nodes are collided, they have to be retransmitted. This retransmission of data packet along with the normal transmission of the packets over the network results in more collisions, which can be referred to as collision chain reaction. Due to retransmission and performing additional carrier sensing, nodes waste considerable amount of energy. Since the proposed protocol minimizes collision and all the nodes receive feedback from the coordinator about the ongoing transmission over the link, the nodes do not unnecessarily waste time for the channel to become idle and goes to sleep mode. Moreover, due to the introduction of mechanism by virtue of which mobile node can request slot to transmit the data in upcoming cluster even before joining the cluster saves significant time required for association, thus saving considerable amount of energy as compared to the conventional standards.

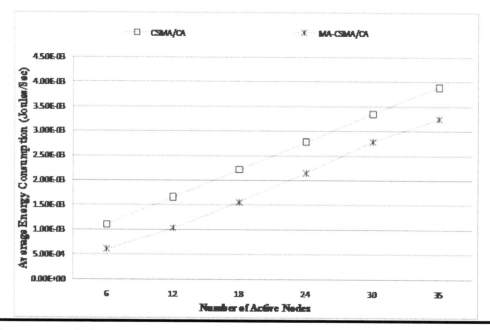

Figure 5.8 Multi-hop multi-cluster network scenario according to average energy consumption.

5.7 Future Work and Conclusion

In this chapter, MOB-MAC protocols are presented and compared. A careful analysis of these protocols reveals that the maintenance of QoS for mobile wireless scenario is a challenging task. There are several protocols available, which deal with this challenge; however, the protocols compromise complexity, latency, and energy efficiency to achieve these tasks. A future research direction can be optimization of the existing protocols so that they can maintain an acceptable QoS without compromising energy, latency, and complexity.

Moreover, the industrial wireless standard IEEE 802.15.4 is presented as a case study for mobile scenario. The protocol of IEEE 802.15.4 is adapted for mobile applications providing a contention based collision minimize transmission by employing mechanism of GTS allocation to the mobile nodes joining the new cluster which resolves the major issue of control message collisions as well. MA-CSMA/CA MAC solves the major problem of association for mobile node by introducing the request mechanism before entering a new cluster. Thus, in this way, the association time is almost negligible. MA-CSMA/CA works well for both static and mobile scenario without compromising the performance requirements. The proposed MAC is extensively compared with the current state-of-the-art industrial standard CSMA/CA for IEEE 802.15.4. The results clearly show that it outperforms the existing industrial standard of CSMA/CA protocol in all the performance aspects. There is considerable gain in the reduction of latency of about five times and the channel utilization is about six times better. A significant improvement of about 70–80% in goodput and consumption of energy is reduced by twofold. The results also show the superiority of the proposed protocol MA-CSMA/CA not only at low traffic load but also at high network load. As the number of active and mobile nodes in the network increases, the proposed protocol accomplishes far better performance than the existing IEEE 802.15.4 standard.

References

1. Grossglauser, M., and D. Tse. 2001. "Mobility Increases the Capacity of Adhoc Wireless Networks." In *IEEE Infocom 2001: The Conference on Computer Communications*, vol. 1–3, 1360–9. Anchorage, AK.
2. Luo, J., J. Panchard, M. Piorkowski, M. Grossglauser, and J. P. Hubaux. June 2006. "MobiRoute: Routing towards a Mobile Sink for Improving Lifetime in Sensor Networks." In *2nd IEEE/ACM International Conference on Distributed Computing in Sensor Systems*, 480–97. San Francisco, CA.
3. Vincze, Z., and R. Vida. October 2005. "Multi-Hop Wireless Sensor Networks with Mobile Sink." In *ACM Conference on Emerging Network Experiment and Technology*, 302–3. Toulouse, France.
4. Raviraj, P., H. Sharif, M. Hempel, and S. Ci. 2005. "MOBMAC: An Energy Efficient and Low Latency MAC for Mobile Wireless Sensor Networks." In *Proceedings of the 2005 Systems Communications*, August 14–17, 370–5. Montreal, Canada.
5. Munir, S. A., B. Ren, W. Jiao, B. Wang, D. Xie, and J. Ma. May 2007. "Mobile Wireless Sensor Network Architecrure and Enabling Technologies for Ubiquitous." In *Conference on Advanced Infonnation Networking and Applications Workshops (AINAW'07)*, 113–20.
6. Rahimi, M., H. Shah, G. S. Sukhatme, J. Heideman, and D. Estrin. May 2003. "Studying the Feasibility of Energy Harvesting in a Mobile Sensor Networks." In *Proceedings of the IEEE International Conference on Robotics and Automation*, vol. I, 19–24. Taipai.
7. Chakrabarti, A., A. Sabharwal, and B. Aazhang. April 2003. "Using Predictable Observer Mobility for Power Efficient Design of Sensor Networks." In *2nd International Workshop on Information Processing in Sensor Networks*, vol. 2634, 129–45. Palo Alto, CA.
8. Ali, M., T. Suleman, and Z. A. Uzmi. 2005. "MMAC: A Mobility-Adaptive, Collision-Free Mac Protocol for Wireless Sensor Networks." In *Proceedings of the 24th IEEE IPCCC'05*, 401–7. Phoenix, AZ.
9. Gandham, S. R. et al. 2003. "Energy Efficient Schemes for Wireless Sensor Networks with Multiple Mobile Base Stations." In *Proceedings of IEEE GLOBECOM*. San Francisco, CA.
10. Luo, J., and J.-P. Hubaux. 2005. "Joint Mobility and Routing for Lifetime Elongation in Wireless Sensor Networks." In *Proceedings of IEEE INFOCOM*. Miami, FL.
11. Ekici, E., Y. Gu, and D. Bozdag. July 2006. "Mobility-Based Communication in Wireless Sensor Networks." *IEEE Communications Magazine* 44 (7): 56–62.
12. Kansal, A. et al. 2004. "Intelligent Fluid Infrastructure for Embedded Networks." In *Proceedings of the 2nd International Conference on Mobile Systems, Applications, and Services*.
13. Shah, R. et al. 2003. "Data Mules: Modeling a Three-tier Architecture for Sparse Sensor Networks." In *Proceedings of the IEEE Workshop on Sensor Network Protocols and Applications*. Anchorage, AK.
14. Ghassemian, M., and H. Aghvami. June 2008. "An Investigation of the Impact of Mobility on the Protocol Performance in Wireless Sensor Networks." In *24th Biennial Symposium on Communications*, 310–5.
15. Narwaz, S., M. Hussain, S. Watson, N. Trigoni, and P. N. Green. 2009. "An Underwater Robotic Network for Monitoring Nuclear Waste Storage Pools." In *Sensors and Software Systems*, 236–255. Springer.
16. Pandya, A., A. Kansal, and G. Pottie. March 2008. "Goodput and Delay in Networks with Controlled Mobility." In *2008 IEEE Aerospace Conference*, 1323–30.
17. Dantu, K., M. Rahimi, H. Shah, S. Babel, A. Dhariwal, and G. S. Sukhatme. April 2005. "Robomote: Enabling Mobility in Sensor Networks." In *2005 Fourth International Symposium on Information Processing in Sensor Networks*, 404–9. Los Angeles, CA.
18. Somasundara, A. A., A. Ramamoorthy, and M. B. Srivastava. 2004. "Mobile Element Scheduling for Efficient Data Collection in Wireless Sensor Networks with Dynamic Deadlines." In *Proceedings of the 25th IEEE International Real-Time Systems Symposium*. Washington, DC.
19. Jea, D., A. A. Somasundara, and M. B. Srivastava. 2005. "Multiple Controlled Mobile Elements (Data Mules) for Data Collection in Sensor Networks." In *Proceedings of the IEEE/ACM International Conference on Distributed Computing in Sensor Systems*.
20. Choi, S.-C., J.-W. Lee, and Y. Kim. 2008. "An Adaptive Mobility-Supporting MAC protocol for Mobile Sensor Networks." In *IEEE Vehicular Technology Conference*, 168–72.

21. Pham, H., and S. Jha. 2004. "An Adaptive Mobility-Aware MAC Protocol for Sensor Networks (MS-MAC)." In *Proceedings of the IEEE International Conference on Mobile Ad-hoc and Sensor Systems (MASS)*, 214–6.

22. Lin, P., C. Qiao, and X. Wang. 2004. "Medium Access Control with a Dynamic Duty Cycle for Sensor Networks." In *Proceedings of the IEEE Wireless Communications and Networking Conference (WCNC)*, vol. 3, 1534–9.

23. Hameed, S. A., E. M. Shaaban, H. M. Faheem, and M. S. Ghoniemy. October 2009 "Mobility-Aware MAC Protocol for Delay Sensitive Wireless Sensor Networks." In *IEEE Ultra Modern Telecommunications … Workshops*, 1–8.

24. Lam, S. 1980. "A Carrier Sense Multiple Access Protocol for Local Networks." *Computer Networks* 4: 21–32.

25. Lu, G., B. Krishnamachari, and C. S. Raghavendra. April 2004. "Performance Evaluation of the IEEE 802.15.4 MAC for Low-Rate Wireless Networks." In *Proceedings of the IEEE International Performance Computing and Communication Conference (IPCCC'04)*, 701–6. Phoenix, AZ.

26. Woon, W. T. H., and T. C. Wan. 2008. "Performance Evaluation of IEEE 802.15.4 Wireless Multi-Hop Networks: Simulation and Testbed Approach." *International Journal of Ad-Hoc and Ubiquitous Computing* 3 (1): 57–66.

27. Zheng, J., and M. J. Lee. 2006. "A Comprehensive Performance Study of IEEE 802.15.4." In *Sensor Network Operations*, Chapter 4, 218–37. New York: IEEE Press, Wiley Interscience.

28. Koubaa, A., M. Alves, E. Tovar, and Y. Q. Song. April 2006. "On The Performance Limits of Slotted CSMA/CA in IEEE 802.15.4 for Broadcast Transmissions in Wireless Sensor Networks." IPP-HURRAY Technical Report, TR-060401.

29. Laibowitz, M., and J. A. Paradiso. May 2005. "Parasitic Mobility for Pervasive Networks." In *3rd International Conference on Pervasive Computing (PERVASIVE 2005)*. Munich, Germany.

30. Hu, L., and D. Evans. September 2004. "Localization for Mobile Sensor Networks." In *ACM, Mobi-Com 2004*. Philadelphia, PA.

31. Chen, C., and J. Ma. 2007. "Simulation Study of AODV performance over IEEE 802.15.4 MAC in WSN with Mobile Sinks." In *Proceedings of Advanced Information Networking and Applications Workshop 2007 (AINAW'07)*, 159–63. Beijing.

32. Attia, S. B., A. Cunha, A. Koubaa, and M. Alves. July 2007. "Fault Tolerance Mechanism for Zigbee Wireless Sensor Networks." In *19th Euromicro Conference on Real Time Systems (ECRTS'07)*. Pisa, Italy.

33. Zen, K., D. Habibi, A. Rassau, and I. Ahmed. 2008. "Performance Evaluation of IEEE 802.15.4 for Mobile Sensor Networks." In *5th International Conference on Wireless and Optical Communications Networks*. Surabaya, Indonesia.

34. Bettstetter, C., G. Resta, and P. Santi. July–September 2003. "The Node Distribution of the Random Waypoint Mobility Model for Wireless ad Hoc Networks." *IEEE Transactions on Mobile Computing* 2 (3): 257–69.

Chapter 6

Cooperative Transmission Techniques and Protocols in Wireless Sensor Networks

Vasileios K. Sakarellos, Dimitrios Skraparlis, and Athanasios D. Panagopoulos

Contents

6.1 Introduction to Cooperative Diversity Sensor Systems

Wireless sensor network technology has a vast potential for applications, which include energy network monitoring and management, efficient automation of industrial production, patient monitoring for better health care and assisted living, home monitoring and security, and environmental monitoring. In wireless sensor networks, nodes are usually interconnected in a wireless mesh arrangement, efficiently sharing information and resources through appropriately designed communication protocols [1–4].

The design of wireless sensor networks poses several challenges to a system designer; information should be moved with efficiency across the wireless sensor network and processing consumption as well as transmission power should be minimized as much as possible, yielding long battery life for each sensor node and increasing the overall efficiency of the wireless sensor system. Such design goals become especially important in the context of "green" communications, which require efficient usage of resources within the communication system [5–8].

Diversity has traditionally consisted of a key technique, which is used by system designers to achieve quality of service (QoS) and efficiency goals in a wide range of communication systems. The main idea behind diversity techniques is the creation of diverse paths of information; by utilizing appropriate preprocessing, encoding, and detection methods, reliability against fading channels and interference as well as increased system capacities and optimized energy efficiency can be achieved. Typical diversity techniques are antenna diversity, multiple antenna techniques (multiple-input multiple-output [MIMO]), and cooperation among nodes such as relaying and cooperative diversity.

Cooperation techniques therefore offer a useful tool, which achieves attractive and optimized trade-offs for the set of design parameters in wireless sensor networks. More specifically, cooperative diversity techniques [9, 10] have been proposed to improve the performance of wireless sensor networks exploiting the broadcast nature of the wireless medium. From a physical layer point of view, a cooperative sensor network consists of multiple nodes that share their resources, for example, energy and spectrum, to create multiple diversity channels that generally suffer from fading. The final destination receives multiple versions of the transmitted information signal through the cooperative relay nodes and combines them forming the final received signal. Therefore, the system performance is improved, typically in terms of availability, coverage range, and throughput. It should also be noted that a cooperative network can be considered as a virtual MIMO system, where each node corresponds to a virtual antenna of a MIMO system.

A general cooperative sensor system consisting of N relay nodes is shown in Figure 6.1. The source node S broadcasts the same information signal directly to the destination node D (sink sensor node), but also to relay nodes R_j, $j = 1, 2, ..., N$. These relay nodes retransmit the received signal to the destination node, which combines all the received signals and forms the final signal. The wireless channels among the nodes are characterized by fading, and in real life, these fading channels are correlated. Due to fading, errors may occur in the transmission of the information signal resulting in the decreased availability of the overall sensor network.

In propagation environments in general, the instant fading value of each channel is different, and as a result, the destination node receives multiple versions of the same information signal. This extra information is used by the destination to improve the estimation of the signal received by the

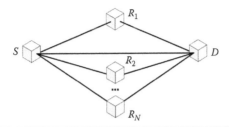

Figure 6.1 A general cooperative sensor network configuration.

source node through the direct link transmission. Therefore, cooperative diversity is introduced to a wireless sensor network by the multiple nodes that act as relays. Moreover, as can be seen in Figure 6.1, the source–destination direct link and the N relay channels form a virtual $(N + 1) \times (N + 1)$ MIMO system, where each relay node corresponds to a virtual MIMO antenna.

To study the performance of a complicated cooperative sensor network, the fundamental building blocks of this network should be taken into account. These fundamental systems are the relaying channel [11], the receive diversity combining system [12], and the fundamental cooperative system consisting of the source, the destination, and a single relay node [9].

6.2 Physical Layer Aspects of Cooperative Diversity Sensor Systems

6.2.1 Sensor Relay Types

6.2.1.1 Regenerative Sensor Relays

The nodes of a cooperative wireless sensor network operate as relays that forward an information signal to the final destination. Different relay types can be used in systems such as regenerative and non-regenerative relays. The regenerative sensor relays use the decode-and-forward (DF) technique. In the DF technique, the signal that the relay node receives from the previous hop is regenerated using the full receiver–transmitter processing chain, which contains the sequence of demodulation, channel decoding, encoding, and modulation. Finally, the information signal is forwarded to the next hop.

In a general cooperative sensor system with N serial sensor relays and $N + 1$ hops, it is possible to formulate the equivalent end-to-end signal-to-noise ratio (SNR) as

$$\text{SNR}_{\text{reg}} = \min(\text{SNR}_1, \text{SNR}_2, ..., \text{SNR}_N, \text{SNR}_{N+1}) \tag{6.1}$$

In Equation 6.1, SNR_j is the received SNR of the $j(j = 1, ..., N + 1)$ hop. It can be seen by Equation 6.1 that the minimum SNR characterizes the overall (end-to-end) multi-hop performance of regenerative senor relays against outages (in other words, SNR regions below a threshold that yield system outages).

6.2.1.2 Nonregenerative Sensor Relays

The nonregenerative sensor relays use the amplify-and-forward (AF) technique. In this case, the relay amplifies the received signal from the previous hop with either fixed or adaptive gain and forwards it to the next hop without decoding it. As a result, the additive noise from the previous hop is also amplified and forwarded to the next hop.

More specifically, there are two types of nonregenerative sensor relays: the adaptive gain relays and the fixed gain relays. The adaptive gain relays amplify the received signal with an adaptive gain in order that the transmitted power remains fixed. On the contrary, the fixed gain relays amplify the received signal with a fixed gain and as a result the transmitted power varies. These relays are less complex, but they cannot compensate for the fading of the first hop, which affects the transmitted power (amplifier's back-off effect).

In the following, SNR_j is defined as the received information SNR of hop j and TNR_j is defined as the total received power to noise ratio of hop j including the received information signal power plus the noise power of the previous hops that are also transmitted by the nonregenerative relay (as calculated by Sakarellos et al. [13] for a similar dual-hop problem). The total end-to-end SNR of an adaptive gain system with N serial relays and $N + 1$ hops can be calculated as [14]

$$SNR_{nonreg} = \left[\prod_{j=1}^{N+1} \left(1 + \frac{1}{TNR_j} \right) - 1 \right]^{-1} \tag{6.2}$$

In Equation 6.2, each relay estimates the combination of fading and noise, and it amplifies the received signal without separating the noise from the faded transmitted signal. If a more complex relay that is able to estimate the exact fading of the previous hop is used, this fading can be compensated. As a result, the received information SNR_j remains fixed instead of the total received TNR_j. Therefore, Equation 6.2 can be simplified (similarly to the analysis by Sakarellos et al. [13]) as

$$SNR_{nonreg}^{-1} = SNR_1^{-1} + SNR_2^{-1} + \cdots + SNR_N^{-1} + SNR_{N+1}^{-1} \tag{6.3}$$

In the fixed gain relay case, the SNR_j and the TNR_j also depend on the fading of the previous hops and Equations 6.2 and 6.3 need further calculation (similarly to the analysis by Sakarellos et al. [13]).

6.2.2 Combining Techniques

In a receive diversity system, the final destination (sink sensor) receives multiple versions of the same information signal and forms the final received signal by combining them. The most used combining techniques are maximal ratio combining (MRC), equal gain combining (EGC), and selection combining (SC) [12].

In the MRC technique, the final receiver amplifies the received signal with a gain that is relative to the SNR level of each signal. Then, the final signal is equal to the coherent sum of the received signals, as described by Stuber [12]. For the case of N received signals, the final SNR is given by

$$SNR_{MRC} = \sum_{j=1}^{N} SNR_j \tag{6.4}$$

In the EGC technique, the combiner amplifies the received signal with the same gain. The final signal is equal to the sum of the amplified signals. For the case of N received signals, the final instantaneous SNR is given by

$$SNR_{EGC} = \frac{\left(\sum_{j=1}^{N} \sqrt{SNR_j} \right)^2}{N} \tag{6.5}$$

Finally, in the SC technique, the combiner chooses the best received signal, yielding an instantaneous output SNR of

$$SNR_{SC} = \max\left(SNR_1, SNR_2, ..., SNR_N\right) \tag{6.6}$$

6.2.3 Cooperative Diversity Protocols in Sensor Networks

The cooperative diversity protocols describe the way that the relay nodes forward the information signal toward the final destination as well as the way that the destination combines the received signals. The main cooperative protocols are fixed relaying (FR), selection relaying (SR), and incremental relaying (IR). These protocols have been introduced by Laneman et al. [9].

6.2.3.1 Fixed Relaying Protocol

In the FR protocol, the sensor relays of a cooperative sensor triangle network always forward the received signals to the final destination. As a result, this protocol is limited by the condition of the sensor source–relay links. In a sensor network with regenerative sensor relays, if at least a sensor relay cannot successfully decode the received signal, then the final signal at the destination will also not be decoded. Therefore, a successful transmission requires that the destination as well as all the relays successfully decode the received signal. In a network with nonregenerative relays, if the SNR of the received signal of the source–relay links is low, the additive noise of these links will be amplified by the relays and will be forwarded to the final destination affecting and deteriorating the overall system performance. Therefore, in this case, a cooperative network is significantly limited by the source–relay links.

6.2.3.2 Selection Relaying Protocol

To overcome the disadvantage of the FR protocol, meaning the limitation of the overall system performance by the condition of the source–relay links, the SR protocol can be used. In this case, if the regenerative relays cannot decode the received signals or if the SNR of the received signals by the nonregenerative relays falls below a specified threshold, then these relays do not forward the received signal to the final destination. Therefore, not all the relays cooperate and the final destination combines the signals received by the rest of the relays. In this case, a cooperative system with a single relay falls back to direct link transmission. In the usual case where the sensor network uses the time division multiple access (TDMA) or the frequency division multiple access (FDMA) protocol and the destination combines the received signals with MRC or EGC, then the time or frequency slot that was initially assigned for the relay to retransmit can be used by the source to retransmit the same information signal. Finally, the final destination combines the two signals received by the source improving the system performance.

6.2.3.3 Incremental Relaying Protocol

In the IR protocol, the cooperative system initially uses the direct transmission from the source to the destination, and only if this transmission fails, the relays forward the information signal to the destination. The relay nodes can be informed by the destination that the signal has been successful or not decoded through a single bit transmission. This protocol optimizes the use of the energy and spectral resources of a cooperative network since the relays do not have to retransmit in the case of a successful direct transmission.

6.2.4 Outage Performance Analysis

In this section, the outage behavior of a fundamental cooperative sensor system is investigated. The considered cooperative system consists of the source node S, the destination node D, and a single relay node R. The SNRs of the S–D, S–R, and R–D hops are denoted as γ_j, where $j = 1, 2, 3$ respectively. In addition, the distances (link lengths) between S–D, S–R, and R–D are denoted as L_j, where $j = 1, 2, 3$, respectively.

The outage probability is defined as the fraction of time where the total end-to-end Shannon capacity $C\left(\text{bits}/(\text{Hz}\times\text{s})\right)$ at the destination D does not exceed a specified threshold C_{th}:

$$P_{\text{out}} = P\left(C < C_{\text{th}}\right) \tag{6.7}$$

By using the Shannon capacity equation, Equation 6.7 can be rewritten in terms of SNR:

$$P_{\text{out}} = P\left(\gamma_{\text{T}} < \gamma_{\text{th}}\right) \tag{6.8}$$

In Equation 6.8, γ_{T} is defined as the total end-to-end SNR describing the outage event. The SNR threshold γ_{th} depends on the spectral efficiency of the multiple access protocol.

If the TDMA or the FDMA are used, two degrees of freedom (DOF) are utilized: in the first time/frequency slot, the source S broadcasts a signal to both the relay R and the destination D, while in the second time/frequency slot, the relay R retransmits the received signal to the destination D. Both these protocols are based on half-duplex relay operation [9].

In the following, the outage probability of a cooperative sensor system is calculated and presented for various cooperative protocols. We present some formulas considering the following: firstly, the simple assumption that the fading channels are independent Rayleigh channels and secondly the usual scenario that the channels are considered correlated lognormal channels modeling large-scale fading such as shadowing [15,16]. The wireless sensors are usually in environments with obstacles; consequently, the correlated shadowing environment is the typical radio communication environment.

6.2.4.1 Outage Probability of Fixed Relaying Cooperative Sensor Network

6.2.4.1.1 Amplify-and-Forward

The total end-to-end capacity of a nonregenerative cooperative system using two time slots and MRC at the receiver is given [9] by

$$C = \frac{1}{2}\log_2\left(1 + \gamma_1 + \gamma_{23}\left(\gamma_2, \gamma_3\right)\right) \tag{6.9}$$

In Equation 6.9, $\gamma_{23}\left(\gamma_2, \gamma_3\right)$ is given by Equations 6.2 and 6.3 for the second hop and the third hop respectively. For Rayleigh fading, the SNR of each hop $j = 1, 2, 3$ is given as $\gamma_j = \text{SNR}\left|a_j\right|^2$, where SNR is the SNR without fading, while the fading of each hop is modeled by the independent and exponentially distributed random variables $\left|a_j\right|^2$ with parameters σ_j^2.

Taking Equations 6.7 through 6.9 into account, the outage event is calculated [9] by

$$|a_1|^2 + \frac{1}{\text{SNR}} \gamma_{23} \left(\text{SNR}|a_2|^2, \text{SNR}|a_3|^2 \right) < \gamma_{\text{th}}$$
(6.10)

In Equation 6.10, the SNR threshold is calculated as $\gamma_{\text{th}} = \left(2^{2C_{\text{th}}} - 1 \right)/\text{SNR}$. Alternatively, the spectral inefficiency of the two DOF (two time slots) can be incorporated in the SNR threshold using a rate-normalized SNR threshold [15,16]: $\gamma_{\text{th}} = \gamma_{\text{th.norm}}^2 + 2\gamma_{\text{th.norm}}$.

Finally, the high SNR behavior of the outage probability can be calculated [9] by

$$P_{\text{out}} \sim \left(\frac{1}{2\sigma_1^2} \frac{\sigma_2^2 + \sigma_3^2}{\sigma_2^2 \sigma_3^2} \right) \gamma_{\text{th}}^2$$
(6.11)

6.2.4.1.2 Decode-and-Forward

Using the same definitions as above, the end-to-end capacity of DF cooperative sensor network using MRC and suffering from independent Rayleigh fading is given as

$$C = \frac{1}{2} \min \left\{ \log_2 \left(1 + \gamma_2 \right), \log_2 \left(1 + \gamma_1 + \gamma_3 \right) \right\}$$
(6.12)

The outage event can be calculated by

$$\min \left\{ |a_2|^2, |a_1|^2 + |a_3|^2 \right\} < \gamma_{\text{th}}$$
(6.13)

Finally, the high SNR behavior of the outage probability can be calculated [9] by

$$P_{\text{out}} \sim \frac{1}{\sigma_2^2} \gamma_{\text{th}}$$
(6.14)

On the contrary, if shadowing fading is considered, modeled as correlated lognormal random variables, the outage probability of a cooperative system employing either MRC or SC can be calculated using the same analysis as above [15]:

$$P_{\text{out}} = 1 - \text{erfc} \frac{\left(u_{2,0} / \sqrt{2} \right)}{2} + \frac{1}{2} \int_{-\infty}^{u_{1,0}} du_1 \int_{-\infty}^{u_{3,0}} f_{u_1 u_3} \left(u_1, u_3 \right) \text{erfc} \left(\frac{u_{2,0} - \mu_{2/1,3}}{\sqrt{2}\sigma_{2/1,3}} \right) du_3$$
(6.15)

where

$$\begin{cases} u_{3,0} = \dfrac{\left(\ln \gamma_0 - \mu_3 \right)}{\sigma_3} \\[4mm] u_{j,0} = \dfrac{\left(\ln \gamma_{\text{th}} - \mu_j \right)}{\sigma_j}, \quad (j=1,2) \end{cases}$$
(6.16)

and

$$
\gamma_0 = \begin{cases} \gamma_{th}, & \text{for SC} \\ \gamma_{th} - \gamma_1, & \text{for MRC} \end{cases}
\tag{6.17}
$$

The parameters of the lognormal random variables γ_j are denoted as $\mu_j, \sigma_j \, (j = 1,2,3)$. The joint normal probability density function of the random variables u_1, u_3 is denoted as $f_{u_1 u_3}(u_1, u_3)$ [17]. The parameters $\mu_{i/j,k}$, $\sigma_{i/j,k} \, (i, j, k = 1,2,3, i \ne j \ne k)$ [17, Chapter 7] are given by

$$
\begin{cases} \mu_{i/j,k} = \dfrac{\rho_{ij} - \rho_{jk}\rho_{ik}}{1 - \rho_{jk}^2} u_j + \dfrac{\rho_{ik} - \rho_{jk}\rho_{ij}}{1 - \rho_{jk}^2} u_k \\[2mm] \sigma_{i/j,k} = \sqrt{\dfrac{1 - \rho_{jk}^2 - \rho_{ij}^2 - \rho_{ik}^2 + 2\rho_{jk}\rho_{ij}\rho_{ik}}{1 - \rho_{jk}^2}} \end{cases}
\tag{6.18}
$$

In Equation 6.18, the correlation coefficient between the lognormal variables (γ_i, γ_j), where $(i, j = 1,2,3, i \ne j)$, is denoted as ρ_{ij}, where $\rho_{ij} = \rho_{ji}$.

6.2.4.2 Outage Probability of Selection Relaying Cooperative Sensor Network

The total capacity of a SR cooperative system using MRC and the DF technique and suffering from independent Rayleigh fading is given [9] by

$$
C = \begin{cases} \dfrac{1}{2}\log_2(1 + 2\gamma_1), & \gamma_2 < \gamma_{th} \\[2mm] \dfrac{1}{2}\log_2(1 + \gamma_1 + \gamma_3), & \gamma_2 \ge \gamma_{th} \end{cases}
\tag{6.19}
$$

The high SNR behavior of the outage probability can be calculated [9] by

$$
P_{out} \sim \left(\frac{1}{2\sigma_1^2} \frac{\sigma_2^2 + \sigma_3^2}{\sigma_2^2 \sigma_3^2} \right) \gamma_{th}^2
\tag{6.20}
$$

Equation 6.20 is exactly the same as Equation 6.11 of the fixed AF technique.

On the contrary, for correlated lognormal channels and either the MRC or the SC techniques, the outage probability of a cooperative DF system is calculated by Sakarellos et al. [16]. In this work, the outage event is calculated using an alternative methodology than using Equation 6.19. The dual-hop part of the cooperative system is replaced by an equivalent system described by the same outage event. This equivalent system is then combined with the direct link to produce the final outage probability expression for SR. The numerical results of the two methodologies are exactly the same. More specifically, the outage probability can be calculated [16] by

$$P_{\text{out}} = \frac{1}{2}\int_{-\infty}^{u_{1,0}} du_1 \int_{-\infty}^{u_{2,0}} du_2 f_{u_1 u_2}(u_1, u_2)\, \text{erfc}\left(\frac{u_3(u_2) - \mu_{3/1,2}}{\sqrt{2}\sigma_{3/1,2}}\right)$$
$$+ \frac{1}{2}\int_{-\infty}^{u_{1,0}} du_1 \int_{-\infty}^{u_{3,0}} du_3 f_{u_1 u_3}(u_1, u_3)\, \text{erfc}\left(\frac{u_2(u_3) - \mu_{2/1,3}}{\sqrt{2}\sigma_{2/1,3}}\right) \tag{6.21}$$

In Equation 6.21:

$$\begin{cases} u_{1,0} = \dfrac{(\ln\gamma_{\text{th}} - \mu_1)}{\sigma_1} \\[2mm] u_{j,0} = \dfrac{(\ln\gamma_0 - \mu_j)}{\sigma_j}, \quad (j = 2,3) \\[2mm] u_2(u_3) = \dfrac{(\ln\gamma_3 - \mu_2)}{\sigma_2} \\[2mm] u_3(u_2) = \dfrac{(\ln\gamma_2 - \mu_3)}{\sigma_3} \end{cases} \tag{6.22}$$

Moreover, γ_0 is given by Equation 6.17 and $\mu_{i/j,k}$, $\sigma_{i/j,k}$ $(i, j, k = 1, 2, 3, i \neq j \neq k)$ are given by Equation 6.18, while the definitions of the other parameters are the same as those of expression Equation 6.15. Finally, the joint normal probability density function of the random variables u_i, u_j $(i, j = 1, 2, 3, i \neq j)$ is denoted as $f_{u_i u_j}(u_i, u_j)$ [17].

6.2.4.3 Outage Probability of Incremental Relaying

The IR uses one DOF when the *S–D* transmission is successful and two DOF when the relay repeats the source transmission. Thus, the expected spectral efficiency $\bar{C}_{\text{th}}$ is used in the outage probability of Equation 6.7. The high SNR behavior for independent Rayleigh fading is then calculated [9] by

$$P_{\text{out}} \sim \left(\frac{1}{2\sigma_1^2} \frac{\sigma_2^2 + \sigma_3^2}{\sigma_2^2 \sigma_3^2}\right)\left(\frac{2^{\bar{C}_{\text{th}}} - 1}{\text{SNR}}\right)^2 \tag{6.23}$$

6.3 Medium Access Control Aspects of Cooperative Diversity Sensor Systems

Due to the ad hoc and distributed nature of wireless sensor networks, it is a very important task for a system designer to deploy efficient protocols for medium access. Cooperative diversity techniques can in general be viewed as useful enhancements to the various sensor and ad hoc network protocol designs, improving on their weaknesses and offering more attractive trade-offs to a system designer.

Wireless sensor networks are characterized by a plethora of functional requirements that dictate the required optimizations of the several medium access protocol–related parameters. First, a wireless sensor system has to tackle collisions whenever nonduplexing protocols are deployed and also implement packet overhearing where nodes need to listen to packet preambles for scheduling purposes and also for exiting from sleep mode. Preambles and control packets incur additional inefficiencies to the system. Secondly, differing node capabilities is a challenging problem in wireless sensor networks and its solution is deployment specific. Geometrical and electromagnetic parameters are also a key factor and experienced as small-scale and large-scale (path loss and shadowing) fading, packet losses, and reduced coverage of individual or clustered sensor nodes. Lastly, circuit energy consumption as well as transmission energy consumption and energy management are key factors to the success of wireless sensor networks.

Since the wireless medium is shared among the sensor nodes, transmission collisions will degrade the reliability and energy efficiency of the wireless sensor system. Collisions result in extra energy consumption since unsuccessful packets need to be rescheduled and retransmitted. However, instead of discarding collided packets, it is possible to extract information from the corrupt packets. Network-assisted diversity multiple access (NDMA) [18] is a multiple-packet reception (MPR) method in which collided packets during a K-fold collision are stored in memory and later combined with retransmissions; colliding packets are assumed as interfering in the MIMO signal detection algorithm. However, this method can suffer in systems that experience fading since an outage of an acknowledgment caused by fading might be misinterpreted as a collision. In cases of high number of collisions, NDMA can also prove to be energy-inefficient since it requires K retransmissions from K colliding users immediately following a collision.

Cooperative diversity is deployed in the form of a nonregenerative relay that retransmits colliding signals [19]. Once a collision is detected, a node is selected to act as a nonregenerative relay, thus constructing a cooperative diversity system. However, in wireless sensor networks where nodes can also be in idle/sleeping state, NDMA methods can cause additional retransmissions. Ji and Zheng [20] propose focusing the control overhead on relay node selection and taking the channel conditions into account, thus providing more specific fairness guarantees to a wireless sensor network.

The use of extra signaling for relay selection has also appeared in several other contexts [21,22]. Mainaud et al. [23] propose deploying a smaller number of relays by identifying and randomly selecting groups and also basing the relaying decisions on the channel state.

The relay selection problem is addressed for wireless sensor systems with small packet sizes by also considering the residual energy of nodes [24]. In this work, nodes participate in relaying not by preselection but by using a contention slot and back-off mechanism only if they deduce that the channel quality exceeds a predefined threshold. Chen et al. [25] also tackled cooperation strategies in the context of ad hoc networks.

Finally, relay selection and comparison with noncooperation in networks that deploy energy harvesting techniques is tackled by Li et al. [26]. Cooperation is achieved through overheard packets from neighboring nodes of a sender [27]. Throughput and energy efficiency performance is quantified for multi-point-to-point scenarios in IEEE 802.11 networks with varying number of contending nodes.

Another important task especially important to the design of wireless sensor networks is node sleep management. Node sleep functionality dictates that nodes can enter a "standby" mode that conserves energy.

Managing the sleep periods for the sensors in a wireless system can potentially achieve strict energy constraints (low power listening [LPL]); sleep periods can either be scheduled over time (according to scheduling mechanisms, that is, a fixed wake-up mechanism [28] or receiver-initiated procedures [29]) or, alternatively, nodes can be configured in an idle listening mode, conserving energy, which can be "woken up" by unique packet transmissions (e.g., short preambles [30]) from neighboring nodes.

LPL with two variants of relay selection are analyzed, using shortened preamble packets for waking up nodes [31]. Some analytical results are presented by Ahmad et al. [32], regarding the control mechanisms in cooperating sensor nodes by analyzing the total energy consumption (according to the framework by Cui and Goldsmith [33]) and the packet latency.

6.4 QoS Support in Cooperative WSNs

Wireless sensor networks have some specific characteristics that should be taken into account for the imposition of the QoS requirements. QoS provisioning in WSNs is important in modern data applications and can be enhanced through the employing of cooperation strategies [34]. The subject of this section is firstly to present briefly the most common QoS performance indicators and the figures of merit for the performance of the WSNs. Secondly, this section focuses on the special demands to support the QoS in cooperative WSNs, and finally, the presentation of the technical challenges and the trade-offs of QoS provisioning in cooperative WSNs.

6.4.1 QoS in Wireless Sensor Networks

Before proceeding to the QoS of WSN [35,36], we should present some of the issues regarding traffic in WSNs. Traffic is quantified mainly by two metrics: the amount of traffic that is derived from the sensors and the disruption tolerance. The disruption tolerance is high for non-real-time applications and low for real-time applications. The traffic volume depends also on the type of data that are collected from the sensors. Many standardization groups (e.g., International Telecommunication Union [ITU]) [37–39] have made classification of multimedia traffic in audio, video, data, and other applications. The categorization of traffic in the WSNs should also taken into account the parameter of the error tolerance of the data.

To evaluate the general performance of wireless sensor networks, a set of measurements that are used to keep track of the WSN status over the time should be considered. These are the key performance indicators (KPIs) that are classified into two types, whether they describe the whole network resources or the QoS provisioned [35]. The main KPIs that can be measured in any data packet switched network are as follows:

1. *Latency or delay* is the required time for a data packet to be sent from a specific point to another. The round-trip delay is measured by the time needed for sending a packet that is returned to the sender. In wireless sensor networks, delay is also measured as the time needed for the sensors to send all the data to the sink nodes.
2. *Jitter* represents the delay variation of the received packets over time. Packets that are transmitted at a constant rate are not received necessarily at a constant rate due to the congestion of the network. Jitter may be evaluated with reference to the mean delay.
3. *Peak sensor node data throughput* is the maximum rate that is achieved during the data transmission in the network. This KPI is referred to as a single sensor node.
4. *Mean user throughput* is the measure of the average rate that is achieved during the data transmission in the network. This KPI is also referred to as a single sensor node. The calculation is usually made by comparing the size of the transmitted data with the time of the transmission for both downlink and uplink cases.
5. *Lifetime* of a wireless sensor network is defined as the time after which certain fraction of sensor nodes run out of their batteries, resulting in a routing hole within the network [40].

6. *Coverage/connectivity*: A wireless sensor network is defined as connected when every random node is able to communicate with every other node [41].

7. *Exposure* is also related to coverage capabilities of the wireless sensor network, and it is a measure of how well an object moving on a random path can be monitored by the WSN over a period of time [42].

The QoS performance of a wireless sensor network depends also on the scenarios and the application of the network under considerations, for example

- Surveillance and monitoring of remote areas is either an event-driven or a periodic data generation after query applications of WSN with medium delay-tolerant capabilities.
- Environmental monitoring is also an event-driven application with high delay-tolerant capabilities.
- Emergency communications is a multimedia communication and a periodic generation application after queries without delay-tolerant capabilities (real-time applications).
- Surveillance of critical infrastructures is also a periodic multimedia data-generation application after queries without delay-tolerant capabilities (real-time applications).

6.4.2 QoS Enhancement in WSN through Cooperation

While the QoS in WSNs may be enhanced through innovative computing methods and new sensor technologies, future trend should focus also on the sensor terminal/network cooperation and on the efficiency of resource reservation and allocation. Certain cooperation schemes should involve the dynamic formation of sensor nodes so that resources can be exchanged and common interests can be served. Instead of the traditional peer-to-peer communication between the sink and the sensor nodes using a direct link, where each sensor node is operating autonomously, cooperation among terminals is introduced. In such a scenario, the sensor terminals are communicating over a short-range communication in parallel to the sink communication. Such architecture offers virtual high data rate, lower energy consumption, and new applications and business solutions through WSNs. The concept of sensor node cooperation introduces a new form of diversity that results in an increased reliability of the communication, leading both to the extension of the coverage and to the minimization of the power consumption.

6.4.3 Technical Challenges for QoS Support in Cooperative WSNs

Cooperative WSNs have different needs, compared to the conventional WSNs and generally to the wireless data networks. In principle, WSNs need to optimize their overall performance of power, task distribution, and resource usage. Network services should aim to achieve such an optimization. An ideal cooperative WSN with the above QoS assumptions is networked and scalable, with a very little power consumption, auto-configurable and software programmable, capable of fast data acquisition, reliable and accurate, with low purchasing costs, and easy to install without requiring maintenance. In practical implementations of the cooperative WSNs, the above characteristics are often in conflict with each other, and then, trade-off solutions, tailored to the specific scenario, are needed. Thus, a sensor network design must take into account a large number of factors and technical challenges for QoS support [35,43]. The most important factors are briefly described as follows:

1. *Fault tolerance*: The loss of sensor nodes could have occurred due to various reasons—lack of power, physical damages, or environmental interference. It is important that the defect of a sensor should not affect the whole behavior and availability of the cooperative WSN. In this sense, the cooperative WSNs must be designed with fault tolerance to maintain the WSN's functionalities.

2. *Scalability*: Depending on the specific application, the number of sensor nodes may be on the order of hundreds or thousands in some particular cases. Consequently, transmission and communication cooperative protocols must be able to efficiently operate under an extremely high number of distributed nodes.

3. *Resource constraints*: In the WSNs [44], the most important constraints include battery life-time (energy), bandwidth, memory, buffer size, processing capability, and limited trans-mitted power. All these impose important technical requirements on the applications of cooperative routing protocols, computation algorithms, and so on. Low power consumption requirement is the most important one. This has a strong impact on the design/selection of the network cooperative protocols, which should have as a primary objective the maximiza-tion of the network lifetime at the cost of lower throughput or higher transmission delay. In the case that sensors are densely deployed, the sensors can be very close to each other so that short-range multi-hop cooperative communications are expected to consume less power than traditional single-hop (sensor node–sink) communications.

4. *Hardware constraints*: A typical sensor node consists of the following components—a sens-ing unit, a processing unit, a transceiver unit, and a power unit. Moreover, application-dependent components can also be integrated in a sensor node [44] to enhance the QoS in cooperative WSNs. Examples of additional components are location-finding systems, cooperative service systems, power generators, and mobilizers. The plurality of required components and functionality for each application thus poses complex requirements and constraints on the design, integration, management, and maintenance of WSN systems. Therefore, since hardware and software have constraints that need to be taken into account, the incorporation of new algorithms and protocols dedicated to cooperation can require additional hardware components (e.g., antennas and Radio Frequency [RF] chains for sup-porting diversity) as well as additional software implementations; such additional require-ments in hardware-constrained WSN nodes can be challenging to a system designer.

5. *Multiple sinks and traffic types*: In a WSN, there may exist multiple sinks and heterogeneous traffic types. This environment raises technical challenges for the design of efficient coopera-tive protocols to fulfill the requirements of the specific applications and scenarios.

6. *Synchronization*: Time synchronization is very important for the implementation of the cooperative WSNs and the application of cooperative routing. In fact, synchronized nodes can cooperate to transmit data in a tightly scheduled manner. In these conditions, collisions and retransmissions are expected to be drastically reduced, saving energy and improving the overall system lifetime [45].

7. *Compression, aggregation, and data fusion techniques*: Both compression and aggregation techniques reduce communication costs and increase reliability of data transfer between cooperative nodes in WSNs. The larger is the amount of produced data, the higher is the required data transmission efficiency. Data compression involves the reduction of the sensor data before transmission. Decompression of data usually occurs at the sink. Data aggrega-tion/fusion envisages the combination of data coming from different cooperative sensors

and various applications. These techniques increase the delay of the wireless sensor network, and this issue must be addressed in the QoS design of a cooperative WSN.

8. *Security*: Cooperative WSNs similarly with non-cooperative WSNs would suffer from threats and risks common to all the wireless systems. A malicious sensor node can compromise a cooperative sensor node, alter the data integrity, eavesdrop on messages, inject fake messages and waste network resources. The main constraints for the introduction of security services in WSNs concern limitations in storage, communication, computation, and processing capabilities. Consequently, the design of the security protocol in cooperative WSNs' design requires a full understanding of these limitations, while still allowing target performance required by the application.

9. *Cooperation incentives*: Some sensor nodes may refuse to cooperate to conserve their limited resources resulting in traffic disruption or overall QoS degradation in a cooperative WSN. Nodes exhibiting such behavior are termed selfish. Incentive mechanisms intend to provide a framework that forces players to cooperate for the best interest of all the participants. In other words, they provide a motive so that each individual sensor node prefers to work along with others, sometimes sacrificing their own resources and sometimes benefiting from the resources of others. The incentive mechanisms are usually distinguished into credit exchange systems and reputation-based systems [46].

Finally, the future trends are the design of "green" wireless sensor communication networks [5–8]. The objective of green communications is to decrease the energy consumption without deteriorating the QoS provision in the WSN. Summarizing the technical challenges for the support of QoS in the cooperative WSNs, they lead to the following optimization problems:

$$\min \ \{\text{Total cooperative WSN energy consumption}\} \tag{6.24}$$
$$\text{s.t.} \quad \text{QoS constraints}$$

or equivalently

$$\max \ \{\text{Total WSN lifetime}\} \tag{6.25}$$
$$\text{s.t.} \quad \text{QoS constraints}$$

6.5 Numerical Results and Discussion

In this section, numerical results for the performance of cooperative sensor networks located in indoor or on-body environments and suffering from large-scale correlated lognormal shadowing are presented. The outage probability is numerically evaluated for a fundamental cooperative system with a single relay node using Equations 6.15 [15] and 6.21 [16]. These expressions are easily calculated numerically and converge very fast due to the monotonically decreasing nature of the integrand functions. More specifically, replacing the infinite limit $-\infty$ with a small number such as -10 results in a numerical precision that greatly exceeds the accuracy of MATLAB®'s erfc computation in the integrant as well as the precision required for accurately representing outage probability values of practical interest. In the following numerical results, the integral in the final formula is evaluated using in-built MATLAB integration and verified with Gauss–Legendre integration in C.

The values for the system parameters are $\mu_j = 0\,$dB for all j. In addition, a model for shadowing correlation [47,48] is used:

$$\rho_{ij} = \left(\frac{\phi_T}{\phi_{ij}}\right)^\alpha \sqrt{\frac{L_i}{L_j}}, (i, j = 1, 2, 3, i \neq j) \tag{6.26}$$

where L_j are the path lengths, ϕ_{ij} is the angle between the paths i and j, $L_j \geq L_i, \phi_T = 2\sin^{-1}(L_c/2L_i)$, and $\alpha = 0.3$ is a parameterization exponent that depends on geometrical parameters. The decorrelation distance is $L_c = 1\,$m for indoor environments and $L_c = 0.2\,$m for on-body environments.

In Figure 6.3, a cooperative sensor network located in an indoor environment is considered. The system parameters are taken from Jeon et al. [49]: $\sigma_1 = \sigma_3 = 2.94\,$dB corresponding to non-line-of-sight (NLOS) conditions, $\sigma_2 = 1.39\,$dB corresponding to line-of-sight (LOS) conditions. The geometrical parameters, depicted in Figure 6.2, are $L_1 = L_2 = 5.6\,$m and $\phi_{23} = 30°$. The outage probability versus the rate-normalized SNR for the FR and the SR protocols and for the MRC and the SC techniques is plotted. The performance of a direct link as well as the performance of an energy-effective direct link is also plotted for comparison. The direct link transmits with the same energy as the source of the cooperative system, while the energy-effective direct link transmits with energy equal to the sum of the transmit energy of the source plus the transmit energy of the relay [15,16]. Therefore, the energy-effective direct link consumes the same energy as the cooperative system.

As can be seen from Figure 6.3, the SR protocol outperforms the FR protocol for high values of outage probability (especially for MRC), while both protocols have the same performance for low values of outage probability. The MRC technique outperforms the SC, especially for low values of outage probability. The SC technique has worse performance than the direct link, while the MRC technique outperforms the direct link except for high values of outage probability. All the cooperative protocols have worse performance than the energy-effective direct link throughout the whole range of SNR values. This is caused by the extra transmit power of the energy-effective direct link and the small values of standard deviation that correspond to channels that experience less severe fading conditions.

Figure 6.4 showcases the performance of a cooperative sensor network located in a more complex indoor environment, where system parameters are taken from Chrysikos et al. [50]: $\sigma_1 = \sigma_3 = 8.38\,$dB, $\sigma_2 = 6.04\,$dB, $L_1 = L_3 = 11.5\,$m, $L_2 = 8\,$m, and $\phi_{23} = 70°$ (see Equation 6.26). The SR is shown to outperform FR, but for very low values of outage probability, both protocols have the same performance. Both cooperative protocols outperform both the direct link and the energy-effective direct link, especially for low values of outage probability, for the MRC technique and the SC technique (not shown in the figure).

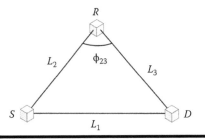

Figure 6.2   Cooperative sensor network system.

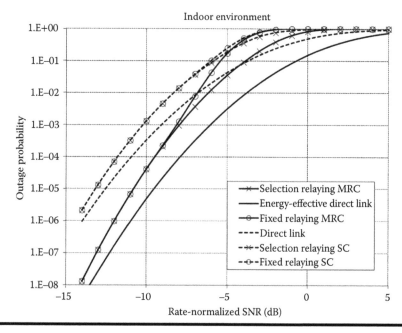

Figure 6.3 Comparison of different cooperative sensor network protocols in an indoor environment.

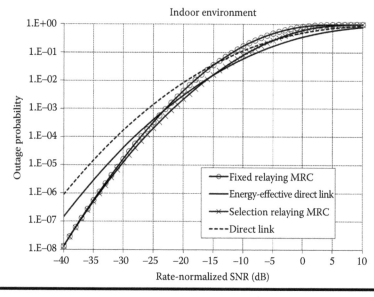

Figure 6.4 Comparison of different cooperative sensor network protocols in a complex indoor environment.

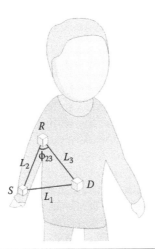

Figure 6.5 **On-body cooperative sensor network.**

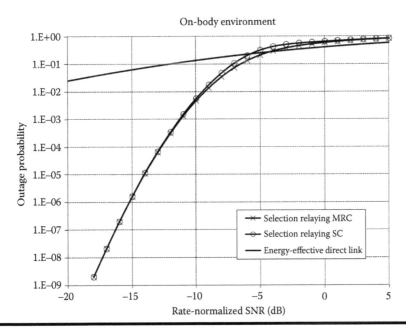

Figure 6.6 **Comparison of different cooperative sensor network protocols in an on-body environment.**

In Figure 6.5, a cooperative sensor network located in an on-body environment is considered. The system parameters are taken from IEEE P802.15-08-0780-09-0006 [51]: the source sensor S is considered to be at the right wrist, the destination sensor D at the navel, and the relay sensor R at the right upper arm of a walking person. This configuration corresponds to values: $\sigma_1 = 1177\,\mathrm{dB}$, $\sigma_2 = \sigma_3 = 2.49\,\mathrm{dB}$, $L_1 = 0.509\,\mathrm{m}$, $L_2 = L_3 = 0.36\,\mathrm{m}$, and $\phi_{23} = 90°$.

As can be seen by Figure 6.6, the SR cooperative protocol outperforms the energy-effective direct link throughout the whole range of the SNR values giving a great gain over noncooperation. The MRC technique has similar performance to the SC technique. Moreover, the FR protocol (not shown in the figure) has almost the same performance as the SR protocol.

6.6 Future Research Directions in Cooperative Sensor Networks

This chapter has presented several important aspects of physical layer and protocol design for cooperative wireless sensor networks. It has been demonstrated that the inclusion of cooperative diversity techniques in wireless sensor networks can have a significant impact on the efficiency and performance of wireless sensor networks and eventually facilitate and enable and invention of "killer applications" that would benefit real-world systems.

Following the analytical and simulation approaches presented in Sections 6.2 and 6.5, several questions still exist regarding the design and analysis of cooperative sensor networks. First of all, the investigation of more diverse channel models that are suitable for sensor network propagation analysis is an interesting research direction. While prior art has primarily focused on spatially uncorrelated Rayleigh channels, some recent efforts have attempted to tackle correlation and more analytically challenging distributions for the large-scale fading component such as the lognormal distribution; such examples have been covered in Sections 6.2 and 6.5. Nevertheless, more investigations regarding the physical layer aspects of sensor cooperation would shed some more light in the real-world applicability of cooperative diversity in the wireless sensor network domain.

Another important and interesting research area is the incorporation of physical layer and propagation characteristics into the analysis of protocols that enable node sleep management, fading prediction, and relay selection. Such extension of prior art would eventually offer sophisticated cross-layer models for highly optimized cooperative sensor network designs.

Moreover, one of the fundamental questions in cooperative diversity systems design is always current; does cooperation offer benefits over noncooperation? The previous section offers some paradigms; such analysis can be extended to more cases and sensor network applications.

The above examples consist only a subset of possible research directions in the field of cooperative sensor networks. It should be added that the study of performance indicators such as the ones described in Section 6.4 present unique opportunities for research. For example, it is a challenging task to provide simple and optimized solutions to complex systems that provide adequate fault tolerance, scalability, and throughput, given the resource and hardware constraints that are intrinsic in the nature of small cooperating sensor nodes.

6.7 Conclusions

Cooperative techniques offer the potential of new levels of management, scaling, and efficient use of resources in wireless sensor networks. By creating diverse paths of information flow and using appropriate preprocessing, encoding, and detection methods, reliability against fading channels and interference can be achieved as well as increased system capacities, higher QoS guarantees, and optimized energy efficiency.

Cooperation techniques become even more important in the context of "green communications," where the efficient design and use of resources in wireless sensor networks is seen as the primary design goal. Especially in cases of large-scale deployment of wireless sensor networks, cooperation can be used as a tool to optimize power consumption and battery life of the sensor systems.

This chapter has provided an overview of physical layer and protocol aspects of cooperative WSNs and has also provided an overview of real-world challenges to the system designer. Finally, some examples of WSN were given, which demonstrate various design trade-offs and choices regarding the cooperative techniques and show some performance regions where cooperation is of benefit compared to noncooperative WSN systems.

References

1. Yang, Z., and L. Tong. December 2005. "Cooperative Sensor Networks with Misinformed Nodes." *IEEE Transactions on Information Theory* 51 (12): 4118–33.
2. Akyildiz, I. F., W. Su, Y. Sankasubramaniam, and E. Cayirci. August 2002. "A Survey on Sensor Networks." *IEEE Communications Magazine* 40 (8): 102–14.
3. Gavrilovska, L., S. Krco, V. Milutinovic, I. Stojmenovic, and R. Trobec. 2010. *Application and Multidisciplinary Aspects of Wireless Sensor Networks: Concepts, Integration, and Case Studies.* New York: Springer.
4. Rajagopalan, R., and P. K. Varshney. Fourth Quarter 2006. "Data-Aggregation Techniques in Sensor Networks: A Survey." *IEEE Communications Surveys & Tutorials* 8 (4): 48–63.
5. Cheng, W., and X. Zhang. 2011. "On-Demand Based Wireless Resources Trading for Green Communications." In *IEEE INFOCOM Workshop on Green Communications and Networking*, 283–88. Shanghai, China: IEEE.
6. Yang, X.-S., and M. Ma. 2011. *Green Communications and Networks: Proceedings of the International Conference on Green Communications and Networks (GCN 2011).* Lecture notes in Electrical Engineering, Chongqing, China: Springer.
7. Chen, Y., S. Zhang, S. Xu, and G. Y. Li. June 2011. "Fundamental Trade-Offs on Green Wireless Networks." *IEEE Communications Magazine* 49 (6): 30–7.
8. Edler, T., and S. Lundberg. 2004. "Energy Efficiency Enhancements in Radio Access Networks." *Ericsson Review.* Accessed 2004, http://www.ericsson.com/ericsson/corpinfo/publications/review/200401/files/2004015.pdf.
9. Laneman, J. N., D.N.C. Tse, and G. W. Wornell. December 2004. "Cooperative Diversity in Wireless Networks: Efficient Protocols and Outage Behavior." *IEEE Transactions on Information Theory* 50 (12): 3062–80.
10. Sendonaris, A., E. Erkip, and B. Aazhang. November 2003. "User Cooperation Diversity (Two Parts)." *IEEE Transactions on Communications* 51 (11): 1927–48.
11. Van Der Meulen, E. C. Spring 1971. "Three-Terminal Communication Channels." *Advances in Applied Probability* 3 (1): 120–54.
12. Stuber, G. L. 2001. *Principles of Mobile Communications.* 2nd ed. Massachusetts: Kluwer Academic Publishers.
13. Sakarellos, V. K., D. Skraparlis, A. D. Panagopoulos, and J. D. Kanellopoulos. November 2010. "Outage Performance Analysis of a Dual-Hop Radio Relay System Operating at Frequencies above 10 GHz." *IEEE Transactions on Communications* 58 (11): 3104–9.
14. Hasna, M. O., and M.-S. Alouini. May 2003. "Outage Probability of Multihop Transmission over Nakagami Fading Channels." *IEEE Communications Letters* 7 (5): 216–8.
15. Skraparlis, D., V. K. Sakarellos, A. D. Panagopoulos, and J. D. Kanellopoulos. 2009. "Outage Performance Analysis of Cooperative Diversity with MRC and SC in Correlated Lognormal Channels." *EURASIP Wireless Communications & Networking* 2009. Article ID 707839: 1–7.
16. Sakarellos, V. K., D. Skraparlis, A. D. Panagopoulos, and J. D. Kanellopoulos. 2011. "Cooperative Diversity Performance of Selection Relaying over Correlated Shadowing." *Elsevier Physical Communication* 4: 182–89.
17. Papoulis, A., and S. U. Pillai. 2002. *Probability, Random Variables and Stochastic Processes.* 4th ed. New York: McGraw-Hill Science/Engineering/Math.
18. Tsatsanis, M. 2000. "Network-Assisted Diversity for Random Access Wireless Networks [J]." *IEEE Transaction on Signal Processing* 48 (3): 702–11.
19. Lin, R., and A. P. Petropulu. 2005. "A New Wireless Network Medium Access Protocol Based on Cooperation." *IEEE Transaction on Signal Processing* 53 (12): 4675–84.
20. Ji, W., and B. Y. Zheng. 2006. "A Novel Cooperative MAC Protocol for WSN Based on NDMA." In *Proceedings of the 8th International Conference on Signal Processing (ICSP'06)*, 16–20, Guilin, China.
21. Liu, P., Z. Tao, and S. Panwar. 2005. "A Cooperative MAC Protocol for Wireless Local Area Networks." In *Proceedings of the International Conference on Communications (ICC 2005)*, Vol. 5, 2962–8, Seoul, Korea.

22. Chou, C. T., J. Yang, and D. Wang. 2007. "Cooperative Mac Protocol with Automatic Relay Selection in Distributed Wireless Networks." In *Proceedings of the Fifth IEEE International Conference on Pervasive Computing and Communications Workshops (PERCOMW '07)*, 526–31, Washington, DC.

23. Mainaud, B., V. Gauthier, and H. Afifi. 24–27 November 2008. "Cooperative Communication for Wireless Sensors Network: A Mac Protocol Solution." In *Wireless Days, 2008. WD '08. 1st IFIP*, 1–5.

24. Nacef, A. B., S. Senouci, Y. Ghamri-Doudane, and A.-L. Beylot. 7–10 February 2011. "COSMIC: A Cooperative MAC Protocol for WSN with Minimal Control Messages." In *2011 4th IFIP International Conference on New Technologies, Mobility and Security (NTMS)*, 1–5, Paris, France.

25. Chen, Y., G. Yu, P. Qiu, and Z. Zhang. 11–14 2006. "Power-Aware Cooperative Relay Selection Strategies in Wireless Ad Hoc Networks." In *Personal, Indoor and Mobile Radio Communications, 2006 IEEE 17th International Symposium on*, 11–14 September, 1–5, Helsinki, Finland.

26. Li, H., N. Jaggi, and B. Sikdar. September 2011. "Relay Scheduling for Cooperative Communications in Sensor Networks with Energy Harvesting." *Wireless Communications, IEEE Transactions on* 10 (9): 2918–28.

27. Gokturk, M. S., and O. Gurbuz. 19–23 May 2008. "Cooperation in Wireless Sensor Networks: Design and Performance Analysis of a MAC Protocol." In *Communications, 2008. ICC '08. IEEE International Conference on*, 4284–89. Beijing, China.

28. El-Hoiydi, A., and J.-D. Decotignie. 2004. "Wisemac: An Ultra Low Power Mac Protocol for the Downlink of Infrastructure Wireless Sensor Networks." In *Computers and Communications, 2004. Proceedings. ISCC 2004. Ninth International Symposium on*, June–1 July, 1, 244–51, Washington, DC.

29. Sun, Y., O. Gurewitz, and D. B. Johnson. 2008 "Ri-Mac: A Receiver-Initiated Asynchronous Duty Cycle Mac Protocol for Dynamic Traffic Loads in Wireless Sensor Networks." In *SenSys '08: Proceedings of the 6th ACM Conference on Embedded Network Sensor Systems*, 1–14, New York, ACM.

30. Buettner, M., G. V. Yee, E. Anderson, and R. Han. 2006. "X-Mac: A Short Preamble Mac Protocol for Duty-Cycled Wireless Sensor Networks." In *SenSys '06: Proceedings of the 4th International Conference on Embedded Networked Sensor Systems*, 307–20, New York, ACM.

31. Nacef, A. B., S.-M. Senouci, Y. Ghamri-Doudane, and A.-L. Beylot. 5–9 June 2011. "A Cooperative Low Power Mac Protocol for Wireless Sensor Networks." In *Communications (ICC), 2011 IEEE International Conference on*, 1–6, Kyoto, Japan.

32. Ahmad, M. R., E. Dutkiewicz, and X. Huang. 21–23 October 2008. "MAC Protocol for Cooperative MIMO Transmissions in Asynchronous Wireless Sensor Networks." In *Communications and Information Technologies, 2008. ISCIT 2008. International Symposium on*, 580–5, Lao.

33. Cui, S., and A. Goldsmith. 2004. "Energy-Efficiency of MIMO and Cooperative MIMO Techniques in Sensor Networks." *IEEE Journal on Selected Areas in Communications*, 22: 1089.

34. Frattasi, S. S., B. Can, F. Fitzek., and R. Prasad. 2005. "Cooperative Services for 4G." In *Proceedings of the 14th IST Mobile & Wireless Communications*, Dresden, Germany.

35. Chakrabarti, S., and A. Mishra. February 2001. "QoS Issues in Ad Hoc Wireless Networks." *IEEE Communications Magazine*, 142–8.

36. Chen, D., and P. K. Varshney. June 2004. "QoS Support in Wireless Sensor Networks: A Survey." In *Proceedings of International Conference on Wireless Networks*, Las Vegas, NV.

37. *ITU-T Recommendation* G.1010. End-user Multimedia QoS Categories. Accessed November 2001, http://www.itu-t.org.

38. *ITU-T Recommendation* Y.1541. Network Performance Objectives for IP-based Services. Accessed December 2011, http://www.itu-t.org.

39. *ITU-T Recommendation* F.700. Framework Recommendation for Audio-visual/multimedia services. Accessed July 1996, http://www.itu-t.org.

40. Rai, V., and R. N. Mahapatra. 2005. "Lifetime Modeling of a Sensor Network." In *DATE '05, Proceedings of the Conference on Design, Automation and Test in Europe*, Washington, DC.

41. Meguerdichian, S., F. Koushanfar, M. Potkonjak, and M. B. Srivastava. 2001. "Coverage Problems in Wireless Ad-Hoc Sensor Networks." In *Proceedings of INFOCOM 2001*, 139–50, Anchorage, AK.

42. Meguerdichian, S., F. Koushanfar, G. Qu, and M. Potkonjak. 2011. "Exposure in Wireless Ad Hoc Sensor Networks." In *Mobile Computing and Networking*, 139–50. Las Vegas, NV.

43. Akyildiz, I. F., and M. C. Vuran. September 2010. *Wireless Sensor Networks*. New York: Wiley.

44. Bulut, E., Z. Wand, and B. K. Szymasnski. 2008. "A Cost-Quality Tradeoff in Cooperative Sensor networking." In *IEEE International Conference on Communication, ICC*, Beijing, China.
45. Elson, J., and K. Romer. January 2003. "Wireless Sensor Networks: A New Regime for Time Synchronization." *ACM SIGCOMM Computer Communication Review* 33 (1): 1–6.
46. Charilas, D. E., S. G. Vassaki, A. D. Panagopoulos, and P. Constantinou. 2011. "Cooperation Incentives in 4G Networks." In *Cooperative Wireless Communications*, edited by Y. Zhang, H.-H. Chen, and M. Guizani. New York: CRC Press.
47. IEEE 802.16j-06/013r3. Broadband Wireless Access Working Group. 19 February 2007. Multi-Hop Relay System Evaluation Methodology. http://ieee802.org/16.
48. Saunders, S. R. 1999. *Antennas and Propagation for Wireless Communication Systems*. New York: John Wiley & Sons, Inc., ISBN 0-471-98609-7.
49. Jeon, N.-R., K.-H. Kim, J.-H. Choi, and S.-C. Kim. 6–9 September 2010. "A Spatial Correlation Model for Shadow Fading in Indoor Multipath Propagation." In *Vehicular Technology Conference Fall (VTC 2010-Fall), 2010 IEEE 72nd*, 1–6, Ottawa, ON, Canada.
50. Chrysikos, T., G. Georgopoulos, and S. Kotsopoulos. 12–14 October 2009. "Empirical Calculation of Shadowing Deviation for Complex Indoor Propagation Topologies at 2.4 GHz." In *Ultra Modern Telecommunications & Workshops, 2009. ICUMT '09. International Conference on*, 1–6, St. Petersburg, Russia.
51. IEEE P802.15-08-0780-09-0006. Working Group for Wireless Personal Area Networks (WPANs). April 2009. Channel Model for Body Area Network (BAN).

Chapter 7

Adapting Multichannel Assignment and IEEE 802.11 Networks to Operate in Wireless Sensor Networks Environment

Carlene Campbell, Howard Senior,
Kok-Keong Loo, and Brian Moore

Contents

7.1 Introduction

Wireless sensor networks (WSNs) [1–9] are an emerging, fast-growing technology, and the growing interest can be attributed to new applications enabled by large-scale networks. The demand for using this medium is increasing with a wide range of deployment in monitoring and surveillance systems as well as for military, Internet, and scientific purposes. Packets from all nodes in a network converge at nodes near the sink, necessitating the prioritizing of the medium access control (MAC) protocol.

Challenges of WSNs include the widespread use of wireless applications, power consumption, lack of spectrum, end user acceptance, and interoperability. In fact, the complexity of mobility and traffic models, together with the dynamic topology and the unpredictability of link quality that characterize wireless networks, result in every application having different characteristics and requirements such as data type, rate of data transmission, and reliability.

This chapter focuses on the design of multichannel communication based on the 802.11 distributed coordination function (DCF) over a single radio for WSNs in order to improve its communication performance, that is, throughput, end-to-end delay, and channel access delay. Multichannel protocols utilize bandwidths better and, thus, they may perform favorably in applications demanding high data rates.

7.2 MAC Protocol

The MAC protocol [1,4,10,11] is responsible for reliable, error-free data transfer with minimum data retransmissions so that performance requirements such as control of bandwidth, power awareness, contention resolution, interference minimization, and collision avoidance are met.

Data collection in WSNs tends to suffer from heavy congestion especially in nodes nearer to the sink node, which gather, control, and store data collected by other sensor nodes. The MAC protocols proposed in the literature to combat these problems can be categorized into contention-free or contention-based protocols, although Kredo and Mohapatra [10] have classified these protocols as scheduled and unscheduled, or random, protocols. This chapter focuses on multichannel assignment MAC protocol for contention-based WSNs in order to efficiently utilize the medium by having node options to switch channels during congestion. This will aid future efforts in addressing most of the major limitations of WSNs associated with MAC and transport protocols with the use of the multichannel assignment. Multichannel assignment creates additional overhead in terms of switching delays, synchronization among nodes, extra control packets and, hence, more energy consumption. However, this chapter considers a WSN that streams high data rates and not the traditional WSN that periodically sends data to its sink node. This chapter explores multiple nonoverlapping channels with minimum overhead for increased capacity and minimum power usage.

7.3 Multichannel Approaches

Multiple nonoverlapping channels present in the IEEE 802.11 industrial, scientific, and medical (ISM) free frequency band have been exploited by mapping them to multiple radios to increase overall capacity and connectivity of the wireless mesh network's backbone. A centralized graph-based approach has been proposed by some researchers [12–14] in which links and nodes are considered as edges and vertices, respectively, of a graph and radio/channel assignment has been formulated by assigning edges to vertices. The limitation of these methods is that it is very difficult to capture network load information with a graph model. Network flow–based centralized approaches can be found in the literature [15–17] in which a multi-radio multichannel (MRMC) is modeled based on network flows that thereby overcomes the limitations associated with graph-based approaches. These approaches are not realistic as constant traffic sources are assumed all the time, whereas network traffic can be bursty in nature. A distributed gateway-centered MRMC approach has been developed [18,19] in which mesh gateways are considered the sink and source of data.

Although the MRMC approach increases network throughput, connectivity, robustness, and resilience enormously, it requires extra resources, for example, energy because the addition of extra radios consumes more power. Keeping these constraints in view, further investigation is needed for applying MRMC techniques directly to WSNs optimally. The effect of channel assignments is explored at the MAC layer with DCF [20], which is the basic and mandatory MAC mechanism of legacy IEEE 802.11 wireless local area networks (WLANs) that allows automatic medium sharing between compatible physical layers through the use of CSMA/CA and random backoff time following a busy medium condition.

7.4 Comparison of 802.11 and 802.15.4

A comparison of IEEE 802.15.4 and IEEE 802.11 is explored to determine the feasibility of WSNs in the 2.4 GHz frequency band as opposed to IEEE 802.15.4. Another factor to be considered is how effectively the IEEE 802.15.4 will be able to cope in the 2.4 GHz frequency band when the IEEE 802.11 becomes fully operational in the same 2.4 GHz frequency band. Will IEEE 802.15.4 be problematic, as at high traffic loads 802.11 uses a total bandwidth of up to 40 MHz leaving no channel for IEEE 802.15.4, and also will IEEE 802.15.4 be free from the channel interference of IEEE 802.11? The future of WSNs–which involve sending all data monitored to a sink–depends on channel assignment.

Both IEEE 802.15.4 and IEEE 802.11 use the CSMA/CA mechanism for contention-based networks. The slotted CSMA/CA mechanism adopted with the personal area network mode of IEEE 802.15.4 is different from the well-known IEEE 802.11 CSMA/CA scheme. The main differences involve the time slotted behavior, the backoff algorithm, and the clear channel assessment (CCA) procedure used to sense whether a channel is idle. The differences are outlined as follows:

■ In IEEE 802.15.4, each operation (channel access, backoff counting, CCA) can begin only at the boundary of time slots, which are termed "backoff periods." In IEEE 802.11, the notion of a slot exists only insofar as backoff counting is concerned.

- In IEEE 802.15.4, only when the backoff counter reaches zero does the node sense the channel (CCA).
- In IEEE 802.11, nodes are constantly sensing while in backoff, thereby incurring additional consumption of energy.
- In IEEE 802.15.4, the backoff counter of a node decreases regardless of whether the channel is idle or busy. In contrast, in IEEE 802.11 backoff counting pauses whenever the channel becomes busy.
- In IEEE 802.15.4, unlike in IEEE 802.11, the contention window size is reset to its minimum value at the beginning of each retransmission attempt.

When IEEE 802.15.4 and IEEE 802.11 use the same channels, their CSMA/CA functions enable them to share the same time slot. When the same channels are used by both, 802.15.4 suffers long delays as in most cases the 802.11 receive priority access due to its higher frequency range. An overlap between the two can adversely impact the operation of IEEE 802.15.4, since it is a low-power protocol using a small channel width compared to the transmitted power levels and channel width used by IEEE 802.11. The frequency bands in which these interference issues are critical for wireless networks include the 2.4 GHz ISM band. Figure 7.1 shows 802.11 and 802.15.4 channels in the 2.4 GHz ISM band.

The simulation model used is based on a network simulator 2 (NS2) [21] using the existing MAC protocol stack and the work done for cognitive radio cognitive network (CRCN) [22] graphical user interface (GUI) at the SNR lab at Michigan Technological University in Houghton, Michigan, and the Hyacinth model [23] for multichannel single-radio. This model has already provided many radio models including 802.11 and 802.15.4; this NS2 also incorporates different topologies and traffic generators, which enable the creation of different simulation scenarios. Different simulation scenarios are studied according to three different performance metrics: (1) Aggregate throughput, (2) delivery ratio, and (3) access delay. The sensor nodes are randomly placed in a 1000 × 1000 m² area. The number of nodes is 50 and the simulation is run for 300 seconds. Data is sent to a sink node. The DCF of IEEE 802.11 and IEEE 802.15.4 is used as the MAC layer. Large networks that are densely deployed are not assumed; in fact, we consider a sensor network with continuous streaming data that could be deployed for organizations, parks, and controlling vehicular traffic but not for remote monitoring. In this instance, nodes are always static and powered, and as such the depletion of battery life is not considered. The simulation of constant bit rate (CBR) traffic must be sent every 2 seconds to prevent buffer overflow and to replicate streaming data, and the effect of both 802.11 and 802.15.4 was investigated to analyze the effect with different data rates at different ranges.

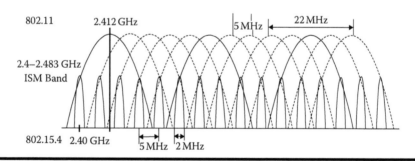

Figure 7.1   Channels comparison of 802.11 and 802.15.4.

7.4.1 Access Delay Comparisons

Figure 7.2 represents an access delay comparison between the 802.11 and the 802.15.4 networks. The access delay is the backoff time used in both the networks. Nodes transmit only to neighboring nodes within their range. In a particular scenario, nodes were placed at an interval of 10 m, the data rate was set at 100 kbps, and access delay was measured in units of seconds. This comparison was aimed at determining the efficiency of either network in relation to access delay based on distance between the nodes and varying data rates.

Both networks performed virtually similarly when transmitting data in this scenario up to 20 nodes. However, after 20 nodes both protocols started experiencing long delays in transmitting data packets. The comparatively higher delays experienced after 30 nodes at low data rates resulted from streaming data, which created buffer overflow and constant backing off as all the nodes were contending for the medium and the succession of data was not periodic. Even though 802.11 is designed for high data rates [20], the simulation result indicated that it can perform at lower data rates and short ranges. Both 802.11 and 802.15.4 utilize the CSMA/CA scheme when sending data. The protocol overheads that are associated with this scheme such as the contention process, interframe spacing, physical layer level headers (preamble + PLCP), and acknowledgment frames have a negative impact on small data size, consequently rendering 802.11 infeasible for operation at low data rates.

Figure 7.3 compares the access delay performance of both networks in a scenario where the interval between the nodes is increased from 10 m to 50 m and the data rate is increased from 100 kbps to 2 Mbps. When the distance between the nodes and the data rates is increased, a significant difference in access delay between the networks resulted. The result showed that 802.11 outperformed 802.15.4 by over 65% and that the 802.11 network had a significantly lower delay in packet transmission. The access delay in 802.11 gradually increase after 30 nodes. This is normal as all nodes are contending for the same medium.

The comparatively poor performance of 802.15.4 occurred because of the high data rate, streaming data, and distance to transmit data; these effects cause buffer overflow, data loss, and constant backing off of the medium. The 802.15.4 network does not have the ability to operate effectively under such severe constraints. The result is consistent with 802.15.4 network, which performs more effectively at short ranges between nodes and small data packet sizes [24]; therefore, it is inconsistent for the 802.15.4 network to operate efficiently with streaming data, which require high data rates.

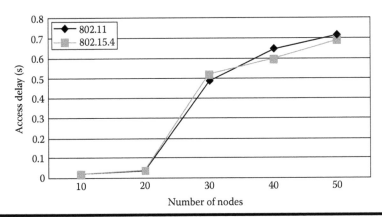

Figure 7.2 Delay comparison for 802.11 and 802.15.4 at a range of 10 m and data rate of 100 kbps.

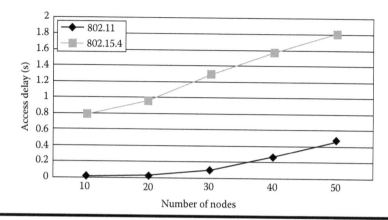

Figure 7.3   Delay comparison for 802.11 and 802.15.4 at a range of 50 m and data rate of 2 Mbps.

7.4.2 Packet Delivery Comparisons

Figure 7.4 represents a packet delivery ratio comparison between 802.11 and the 802.15.4 networks based on the distance between nodes and varying data rates. The delivery ratio is the ratio of the total number of packets received by the nodes to the total number of packets transmitted multiplied by the total number of receivers. The nodes were placed at an interval range of 10 m, and the data rate was set at 100 kbps. The performance of both networks followed the same basic pattern, that is, the packet delivery ratio decreased progressively as the number of nodes increased. The 802.11 network performed slightly better after 20 nodes than 802.15.4.

Figure 7.5 represents the packet delivery ratio comparison between the networks when the interval between nodes was increased from 10 m to 50 m and the data rates were increased from 100 kbps to 2 Mbps. Similar to the access delay, when the distance between the nodes and data rates is increased, the 802.11 network significantly outperformed the 802.15.4 network. This result indicates that 802.15.4 cannot perform well with streaming data even at low data rates and, thus, this network is not feasible for sensor networks with multimedia or surveillance systems that rely on the transmission of images and data over the wireless medium.

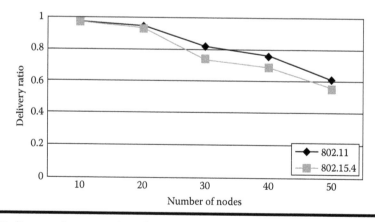

Figure 7.4   Delivery ratio comparison for 802.11 and 802.15.4 at a range of 10 m and data rate of 100 kbps.

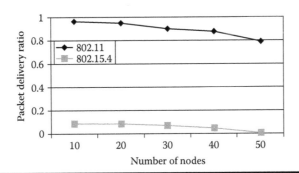

Figure 7.5   Delivery ratio comparison for 802.11 and 802.15.4 at a range of 50 m and data rate of 2 Mbps.

7.4.3 Aggregate MAC Throughput Comparisons

Figure 7.6 represents aggregate MAC throughput comparison between 802.11 and 802.15.4 networks. Aggregate MAC throughput is denoted by the total amount of data delivered to the sink per unit time by the MAC protocol, and it is measured in kilobits per second. The distance between nodes was 10 m and the data rate was set at 100 kbps. The result of the comparison indicated that as the number of nodes increased, aggregate throughput declined in both networks. The decline in throughput was greater in the 802.15.4 network compared to the 802.11 network.

Figure 7.7 represents aggregate MAC throughput comparison between the two networks when both distance between the nodes and the data rate are increased. The interval between the nodes was increased from 10 to 50 m and the data rate from 100 kbps to 2 Mbps.

As evident from the access delay and delivery ratio tests, significant difference in performance resulted in both networks when the distance between the nodes and data rate were increased.

The 802.11 network exhibited a comparatively higher aggregate throughput compared to the 802.15.4 network indicating the former's superior performance in a high-data-rate environment.

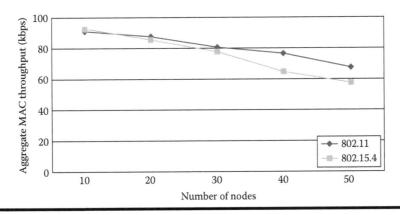

Figure 7.6   Throughput comparison for 802.11 and 802.15.4 at a range of 10 m and data rate of 100 kbps.

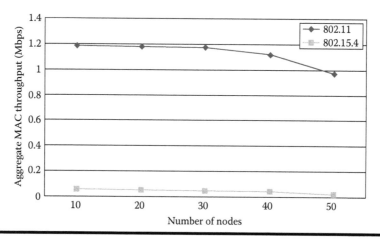

Figure 7.7 Throughput comparison for 802.11 and 802.15.4 at a range of 50 m and data rate of 2 Mbps.

On the other hand, the significantly poor performance of the 802.15.4 network in a high-data-rate wide-node-range environment makes it unsuitable for streaming data in a WSN.

The MAC sublayers for IEEE 802.15.4 and IEEE 802.11 MAC protocols were studied to give a comparative understanding of the 802.11 and 802.15.4 CSMA/CA scheme. The performance of both have been investigated and evaluated using simulation. The results obtained show that 802.15.4 is at a disadvantage when performing at long range with high-data-rate streaming and or at a low data rate with streaming data. The aggregate throughput, delivery ratio, and access delay performance metrics were used, which showed that 802.15.4 performed very poorly at high data rates and 802.11 performed slightly better after 20 nodes at low streaming data rates. Based on these outcomes, it is feasible to conclude that 802.15.4 is not a preferred choice for sensor multimedia or surveillance systems with streaming data for future multichannel multi-radio systems.

After investigating the performance of IEEE 802.11 and IEEE 802.15.4, it was found that the 802.11 contention-based protocols are feasible for use in multichannel assignment. The proposed design, a multichannel distributed coordinate function over single radio for WSNs, was implemented based on the fruitful result obtained with 802.11. The designed protocol was tested with simulation scenarios from NS2. The overall goal for such a design proposal was to utilize multichannel transmission in future 802.11 wireless sensor surveillance systems to process video data for automated real-time alerts and also to consider a more cost-effective solution for WSNs.

7.5 Multichannel Switching to a Single Radio

Multichannel assignments in 802.11 DCF over a single radio for WSNs, known as MC-DCF, will be used in NS2 to simulate various conditions. The customized MC-DCF protocol is designed by using the original random backoff timer that is invoked when a medium is busy to initiate channel switching based on a set of threshold criteria (see Figure 7.8). Nodes switch channels when the contention window of DCF reaches an assigned threshold. The sink node performs interface switching in order to receive data from channels coming from source nodes. Dynamic assignments have been utilized, in which nodes are not bound to a particular channel but have the option of switching between channels. This dynamic assignment provides significant performance benefits over a static approach as this can potentially utilize instantaneous traffic or interference

MC-DCF procedure

Channel is free immediate access	DIFS	Data
Busy channel	Back-off timer	Contention window
Differ access	Threshold	Channel switching

Figure 7.8 The multichannel distributed coordination function procedure.

information and reduce wastage of precious, already limited, bandwidth. WSNs cannot provide reliable and timely communication with high data rate requirements over single channels because of interference, radio collision, and limited bandwidth, since they are designed to mainly operate in remote areas that periodically send data to the host.

Why choose a multichannel over 802.11 DCF that uses a high data rate? The WSN is an emerging technology that is currently one of the fastest growing areas in the communication industry. The demand for using this medium is increasing due to its potentially wide range of deployment in monitoring and surveillance systems and other multimedia systems dealing with streaming or real-time data. With this in mind, the 802.11 standard is utilized more often, which uses a wide range of data rates.

The investigation and analysis of multichannel performance within a single hop (the link quality), that is, the packet reception rate, with different simulation scenarios according to three different performance metrics, that is, aggregate throughput, delivery ratio, and access delay, was performed as follows:

Sensor nodes are placed randomly in a 1000 × 1000 m² area, the radio range is set to 50 m, and the radio bandwidth is set to 2 Mbps. The number of nodes is 100 and the simulation time for each scenario is 500 seconds. The number of channels ranges from 3 to 10 since the spectral mask only defines power output restrictions up to ±11 MHz from the center frequency to be attenuated by 30 dB. It is often assumed that the energy of the channel extends no further than these limits.

The 802.11 channels are effectively 22 MHz wide; as a result, stations can use only every fourth or fifth channel without overlap, that is, typically 1, 6, and 11 in the Americas and, in theory, 1, 5, 9, and 13 in Europe, although 1, 6, and 11 are typical in Europe, too. However, if transmitters are closer together, overlap between channels may cause unacceptable degradation of signal quality and throughput. The MAC protocols are the IEEE 802.11 DCF and the customized MC-DCF.

Aggregate throughput is calculated as the total amount of data delivered to the sink per unit time by the MAC protocol, and it is computed as follows:

$$\text{Aggregate throughput} = \sum_{i=1}^{n} \left(R_i \times \frac{B}{t} \right) \tag{7.1}$$

where n is the number of receivers, and throughput is B/t where B is the bytes received by a receiver i in some duration of time, t, and $i = \{1, 2, 3, \ldots, n\}$.

Delivery ratio is the ratio of the total number of packets received by the nodes to the total number of packets transmitted multiplied by the number of receivers, and it is computed as follows:

$$\frac{\sum_{i=1}^{n} R_i}{\sum_{i=1}^{n} S_i} \tag{7.2}$$

where S_i is total data size of CBR packet node i sent, R_i means total data size of CBR packet node i received.

Access delay is the backoff time used in DCF [15]; the access delay can also be calculated as packet size multiplied by 8 (1 byte) divided by link size plus the propagation delay:

$$\frac{\text{Packet size} \times 8}{\text{Link size}} + \text{propagation delay}. \qquad (7.3)$$

Nodes only transmit to neighboring nodes within a range; transmitting over a wider range may consume more energy, which is not desired by WSNs [2].

7.5.1 Performance Analysis of 802.11 DCF and MC-DCF Using One, Two, and Three Channels

The performance of 802.11 DCF against MC-DCF using one, two, and three channels is analyzed and discussed in Figures 7.9 through 7.11. These comparative performances were measured within the context of the three metrics mentioned in Section 7.5 using CBR data streams.

A comparison between the 802.11 DCF protocol and the proposed MC-DCF protocol is illustrated in Figure 7.9. In this simulated evaluation, the channel access delay of CBR data stream on each protocol was measured. MC-DCF over three channels recorded the lowest level of channel access delay even when the CBR stream increased. Conversely, when transmitting over one channel the MC-DCF protocol recorded the highest level of channel access delay; this is similar in performance to the 802.11 DCF protocol, which was originally designed to operate over a single channel.

When the CBR stream was increased using a single channel, the outcome showed that 802.11 DCF and MC-DCF with one channel become saturated from backing off and buffer overflow. On the contrary, multiple channels resulted in a reduction in channel access delay.

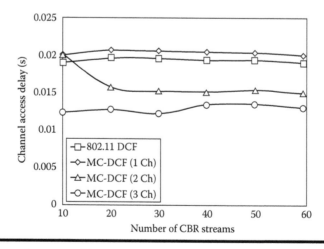

Figure 7.9 Impact of constant bit rate (CBR) streams on channel access delay for 802.11 distributed coordination function (DCF) and one-channel (1 Ch), two-channel (2 Ch), and three-channel (3 Ch) multichannel DCF (MC-DCF).

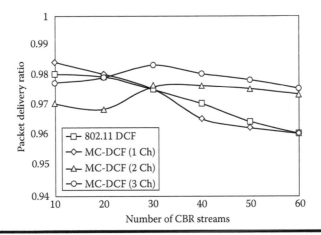

Figure 7.10 Impact of a constant bit rate (CBR) stream on packet delivery ratio for 802.11 distributed coordination function (DCF) and one-channel (1 Ch), two-channel (2 Ch), and three-channel (3 Ch) multichannel DCF (MC-DCF).

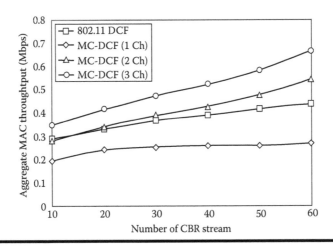

Figure 7.11 Impact of constant bit rate (CBR) streams on aggregate medium access control (MAC) throughput for 802.11 distributed coordination function (DCF) and one-channel (1 Ch), two-channel (2 Ch), and three-channel (3 Ch) multichannel DCF (MC-DCF).

The delivery ratio in Figure 7.10 shows that when using three channels, more packets were delivered compared to using one and two channels. Constant degradation occurred when the CBR stream increases in 802.11 DCF and MC-DCF with one channel (1 Ch). This degradation resulted in constant backing off where nodes were contending for the same channel, which gave rise to more packet loss.

In Figure 7.11, a similar trend is seen where MC-DCF with three channels has a better aggregate throughput, and more data are delivered to the receiving node. This showed that with the modification of the backoff algorithm, nodes have options to switch channels. If this procedure remained while using a single channel, backing off becomes more frequent as the threshold is reached more quickly. The MC-DCF with a single channel performs worse with more unsuccessful attempts and less data delivered to the receiving nodes. When using a single channel, the original 802.11 DCF performed better than the MC-DCF single channel.

7.5.2 Performance Analysis of the Impact of Node Density on 802.11 DCF and MC-DCF with One, Two, and Three Channels

Figures 7.12 through 7.14 show the impact of node density on the performance of 802.11 DCF and MC-DCF using one, two, and three channels by varying the number of nodes. This comparative performance was measured within the context of the three metrics mentioned in Section 7.5 by varying the number of nodes sending CBR streams every 2 seconds. It is noted that MC-DCF performed better when nodes have three channels on which to transmit simultaneously.

Figure 7.12 shows that 802.11 DCF and MC-DCF experienced the highest delays when more nodes transmitted more packets and the network became denser. When two or more channels are transmitting simultaneously, there is a relative improvement in delay. The MC-DCF with three channels recorded the lowest level of channel access delays with increase in the node density of the network.

In Figures 7.13 and 7.14, the packet delivery ratio and the aggregate throughput, respectively, show a comparatively better performance when two or more channels are used. Although there is a comparatively better performance over two or more channels, as the number of nodes transmitting packets through the network increases the performance of the network correspondingly degrades. This is not unusual as nodes will be switching channels, backing off, and entering the wait state, which is the norm in a contention-based network.

In Figure 7.13, the MC-DCF with three channels yielded the highest level of packet delivery ratio. As the density of the network increased, packet delivery ratio progressively declined, which was evident for all protocols in this simulating scenario. The single-channel protocols (802.11 DCF and MC-DCF [1 Ch]) recorded the lowest and the fastest declining packet delivery ratios of all protocols tested. An average of approximately 2.1% degradation of packet delivery ratio occurred compared with a total of 97% delivery ratio for the three channels.

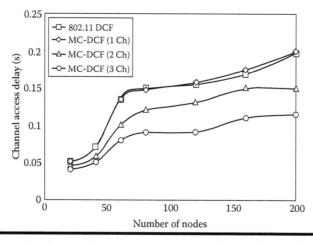

Figure 7.12 Impact of node density on channel access delay for 802.11 distributed coordination function (DCF) and one-channel (1 Ch), two-channel (2 Ch), and three-channel (3 Ch) multichannel DCF (MC-DCF).

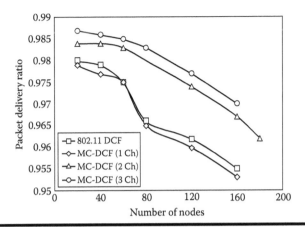

Figure 7.13 Impact of node density on packet delivery ratio for 802.11 distributed coordination function (DCF) and one-channel (1 Ch), two-channel (2 Ch), and three-channel (3 Ch) multichannel DCF (MC-DCF).

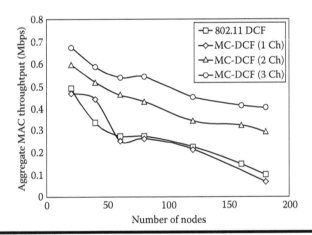

Figure 7.14 Impact of node density on aggregate medium access control (MAC) throughput for 802.11 distributed coordination function (DCF) and one-channel (1 Ch), two-channel (2 Ch), and three-channel (3 Ch) multichannel DCF (MC-DCF).

In Figure 7.14, MC-DCF with three channels recorded the highest level of aggregate MAC throughput even as the node density of the network increased. Similar to the aforementioned packet delivery ratio performance, 802.11 DCF and MC-DCF (1 Ch) recorded declines in performance with increases in the density of the network. In addition, all protocols recorded declines in aggregate MAC throughput with increases in the density of the network. The aggregate throughput as a function of the offered load for three channels showed a decrease in throughput of 14%.

7.5.3 Performance Analysis of a Sink Node with Single Radio

Figure 7.15 shows a sink node with a single radio switching between channels in order to receive data from more than one source node. This technique uses multichannel assignments in 802.11 DCF over a single radio for WSNs known as MC-DCF. All the nodes are aware of the channels in

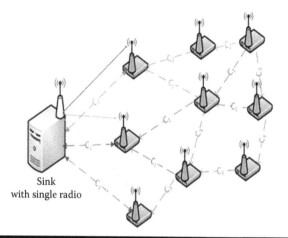

Figure 7.15 Sink node with single radio performing channel switching.

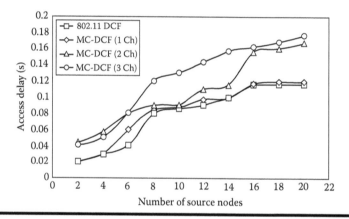

Figure 7.16 Impact of source node density on access delay at the sink for 802.11 distributed coordination function (DCF) and one-channel (1 Ch), two-channel (2 Ch), and three-channel (3 Ch) multichannel DCF (MC-DCF).

use, but each node interface can access only one channel at any given time. With only one radio, the sink node performs interface switching to another channel after receiving data from a source node that it is tuned into. This technique does not resolve the congestion and delay problems encountered by sink nodes in a contention-based network. However, from simulation results it is seen that there has been some improvement in performance at the sink node. Access delay and packet delivery ratio were measured at the sink node.

Figures 7.16 and 7.17 examine the performance of a sink node receiving data directly from source nodes within its range that are sending data. The more sources delivering to the sink the more delays encountered, and the packet delivery ratio decreases correspondingly. A sink node that must constantly switch channels in order to receive data incurs severe switching delays in addition to the time taken to accept data before switching. The findings indicate that 802.11 DCF and MC-DCF with a single channel performed better than the multichannel protocols. The sink node is operating in a single-channel mode with no extra overheads and switching delays occurring when receiving data.

The aggregate throughput degradation that has been observed in the previous simulation within two or more channels can be accounted for by the severe delays occurring mainly at the

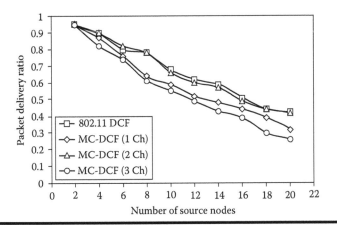

Figure 7.17 **Impact of source node density on packet delivery ratio at the sink for 802.11 distributed coordination function (DCF) and one-channel (1 Ch), two-channel (2 Ch), and three-channel (3 Ch) multichannel DCF (MC-DCF).**

sink node. This results in the dropping of packets. More work will be done in this area to improve the delivery of packets from the source to the sink in a multichannel environment.

In Figure 7.16, as the number of source nodes within the range of a sink increases, the level of access delay at the sink also increases correspondingly. This degradation in performance is consistent for all protocols, whether single- or multichannel protocols. The single-channel protocols (802.11 and MC-DCF [1 Ch]), however, outperformed the multichannel protocols at the sink. The sink node is operating in a single-channel mode with no extra overheads and switching delays occurring when receiving data.

In Figure 7.17, all protocols recorded declining levels of packet delivery ratio at the sink as the number of source nodes within the range of the sink increased. The highest rate of decline was evident in two or more channels.

The proposed MC-DCF, which is a backoff algorithm for multichannel access based on the 802.11 DCF protocols, was examined. This algorithm allows a node to access multiple nonoverlapping channels by accessing the channels dynamically through channel switching after a set threshold is met. The need for multichannel assignment in WSNs was analyzed and discussed, where in the future a sensor surveillance system with streaming data may find it difficult to operate in the 802.15.4 network due to congestion of the most frequently used 2.4 GHz frequency band. The simulation results proved futile for future development in this area for 802.11 networks. It was observed that better performance is achieved when using MC-DCF in analyzing the impact of WSN in the 802.11 network. Overall, MC-DCF exhibited prominent ability in utilizing multichannel transmission with 802.11 in wireless sensor surveillance systems that are of low cost, reliable, easy to manage, easy to deploy, and able to process video data for automated real-time alerts. Researchers will be able to achieve the goal of long-term, independently operating sensor networks under this constraint.

7.5.4 Multiple Sinks with Single Radio

Multiple sink nodes are used with single-radio interfaces. With only one radio, a sink node performs interface switching to another channel after receiving data from a source node into which it is tuned. However, multiple sinks yield a better performance within the network as nodes have options to send data to more than one sink node based on the traffic condition within the network.

If a sink node is experiencing congestion problems based on the backoff algorithm procedure of the DCF, the sending nodes will refrain from sending data to it and avail of the option of sending to another sink node.

Figure 7.18 shows delay impact from source nodes using multiple sink nodes. It has been shown that with three channels there is a 53% reduction in delay at the sink side compared to the high level of delay in the previous simulations in which only one sink node was used. In Figure 7.18, with two channels sending data from sources, there is approximately 32% delay improvement. Single-channel DCF and 802.11 DCF show little improvement. This indicates that single-channel performance does not improve with increasing number of sink nodes as decisions are made based on window size resetting, backing off, wait states, and the fact that all nodes are contending for the same medium. It is noted that MC-DCF with multiple channel switching and single-radio interfaces yields a better performance when using multiple sinks in comparison with single-channel protocols, which show a better performance in the previous results.

Figure 7.19 shows an improvement of over 41% for three channels in packet delivery ratio when the number of sinks is increased by three as compared to a single sink node. With two channels sending data from the sources to the sinks, there is an improvement by over 25% compared with the poor performance resulting with the single channel. Similarly, where the delay with the single channel shows no significant improvement, packet delivery ratio shows no major improvement.

The aggregate throughput in Figure 7.20 of the overall system with source nodes sending data to the sinks shows that with three channels 38% more data are delivered to the sink compared to data delivered with a single channel. Single channel does not show significant improvement with increasing number of sink nodes in receiving data from source nodes in all instances.

When analyzing the impact of MC-DCF with one, two, and three channels in comparison with the original 802.11 DCF, it was shown that increasing the number of sink nodes results in an improvement when two and three channels are used. There was little or no improvement when using a single channel or the original 802.11 DCF, which only operates on a single channel. The reason for this improvement is that each sink has less data to receive from the senders. The same amount of data simulated in the previous study was going to a single sink node. The improvement demonstrated that increasing the number of sink nodes obtained a better performance as the

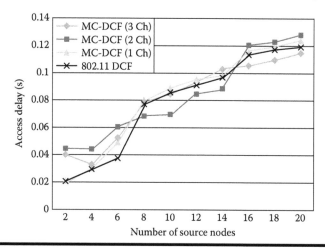

Figure 7.18 Delay impact from source nodes using multiple sinks with single radio interface for 802.11 distributed coordination function (DCF) and one-channel (1 Ch), two-channel (2 Ch), and three-channel (3 Ch) multichannel DCF (MC-DCF).

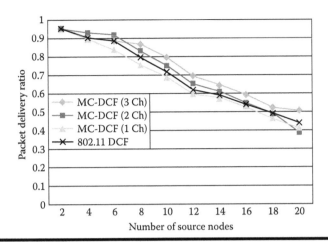

Figure 7.19 **Delivery ratio impact from sources using multiple sinks with single radio interface for 802.11 distributed coordination function (DCF) and one-channel (1 Ch), two-channel (2 Ch), and three-channel (3 Ch) multichannel DCF (MC-DCF).**

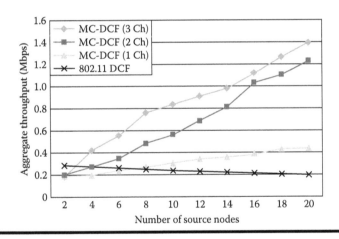

Figure 7.20 **Throughput of overall system using multiple sink nodes with single radio interface for 802.11 distributed coordination function (DCF) and one-channel (1 Ch), two-channel (2 Ch), and three-channel (3 Ch) multichannel DCF (MC-DCF).**

traffic load is split to be received by more sink nodes. Thus, channel switching by a single radio has less data to retrieve; therefore, less time is spent to switch between channels from senders and the queuing of packet data is reduced.

7.6 Multichannel, Multi-Radio Access at Sink Nodes in WSNs

7.6.1 Multiple Radios with Single Sink

A single sink node is used while increasing the number of sink radio interfaces. Figure 7.21 shows a sink node with three radio interfaces. Each interface is assigned to a channel and three sending nodes are assigned to each channel to create a one-to-one mapping against an interface. In this case, no channel switching is required. Each sender to the sink remains on the said channel

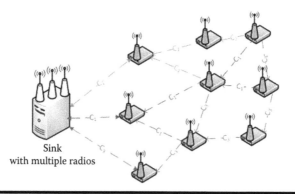

Figure 7.21 Single sink node with multiple radios.

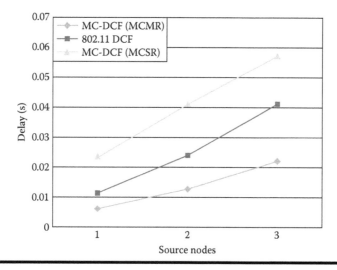

Figure 7.22 Delay impact with multichannel multiple radios communication at the sink node: In the figure, MC-DCF refers to multichannel distributed coordination function, MCMR refers to multichannel multi-radio, and MCSR refers to multichannel single radio.

throughout and no channel switching takes place. This allows a constant flow between the sending node and the radio interface. This technique does not fully alleviate the problem at the sink node; when there is congestion at the sending nodes, access delay increases as the option of switching channels is not permitted. This case can be applicable only when the network size is small.

Figure 7.22 shows the impact delay when MC-DCF uses a single sink with three radio interfaces to receive packet data, which creates a one-to-one mapping in receiving data from the sending nodes. The MC-DCF with a single radio from the previous simulations had to perform channel switching to receive data when two or more nonoverlapping channels were sending data to the sink. The previous results show that when two or more channels were used, there was poor performance; the repeat of this performance is shown in Figure 7.22, except that only three sources are assigned in this case to send data to the three interfaces, and each interface and each node is assigned to one of the nonoverlapping channels. When the one-to-one assignment was used, there was over 40% success in improvement of delay.

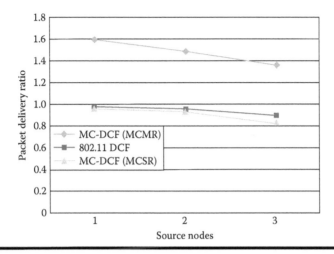

Figure 7.23 **Delivery ratio impact with multichannel multiple radios communication at the sink node: In the figure, MC-DCF refers to multichannel distributed coordination function, MCMR refers to multichannel multi-radio, and MCSR refers to multichannel single radio.**

This outcome indicates that if radio switching between channels is eliminated and data is constantly received from senders by the receiving radio interfaces, then performance at the sink can be improved. However, this would not be practical when network size increases, as the radio interfaces at the sink would also need to be increased. The limitation of non-overlapping channels would not make it feasible as there would not be enough channels to assign radio interfaces.

The packet delivery ratio in Figure 7.23 shows a similar improvement of approximately 46% for MC-DCF operating with multiple radios when compared with MC-DCF operating with a single radio in our previous study. Each interface on a sending node is assigned to different nonoverlapping channels. The 802.11 DCF showed little or no improvement as this protocol is designed to operate with a single channel only. As mentioned before, the one-to-one assignment is not ideal for large networks as it is not practical to have each radio interface assigned to a nonoverlapping channel from a sending node.

Figure 7.24 also shows an improvement of 53% in the one-to-one assignment with three non-overlapping channels for aggregate throughput. However, for small parks and building areas, this kind of implementation can be considered.

The aforementioned one-to-one scenario is not practical in all instances but depends on the size of the network and the number of nodes sending data directly to the sink. Ideally there are many nodes sending data to a sink, which creates a one-to-many assignment in which many nodes send data to the same radio interface. Sending nodes can be odd or even in number.

7.6.2 Multiple Sinks with Multiple Radios

Each scenario showed some level of improvement for MC-DCF when sink nodes obtain data from source nodes, compared with our previous results. In previous results the sink node with a single radio interface encountered severe degradation when receiving data from sources. This was due to the constant switching between source nodes interfaces.

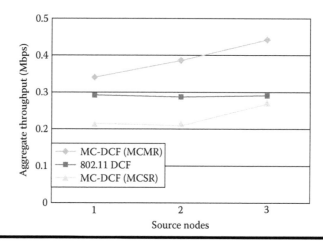

Figure 7.24   Throughput impact with multichannel multiple radios communication at the sink node: In the figure, MC-DCF refers to multichannel distributed coordination function, MCMR refers to multichannel multi-radio, and MCSR refers to multichannel single radio.

The present scenario analyzes the impact of sources sending data to three sink nodes. Each sink was equipped with three radio interfaces using the three nonoverlapping channels in IEEE 802.11.

Figure 7.25 shows a sensor network with multiple sink nodes, each having three radio interfaces.

Contention-based techniques are best resolved by preventive methods, but they are very difficult to predict as all nodes are always contending for a single channel. The multiple sink nodes and multiple radios shown in Figure 7.25 provide the required best overall practical access control in accessing the medium while reducing collision, delay, and the hidden node problem in a WSN. Data collection alleviates the heavy congestion that WSNs tend to suffer from especially in nodes nearer to the sink node that gather, control, and store the data collected by other sensor nodes. The idea is to achieve multiaccess, simultaneous transmission and maintain a good quality of communication, which can be obtained as long as distances between the sensor node and the sink node are short enough and the strength of the signals received is adequate.

The channel sensing and switching procedure within the backoff mechanism will also aid the elimination of unfair strategy on backing off for all nodes where nodes can switch channels and prevent severe delays and packet losses and increase throughput as constant traffic sources cannot be assumed at all times and traffic can be bursty in nature. The simulation results demonstrate the impact of transmitting streaming data within a sensor network with multiple sinks and radios.

Figure 7.26 shows the impact on delay caused by an increase in sink nodes and radio interfaces. It is shown from the simulation results that with three channels there is 96% reduction in delay at the sink side compared with the previous results in which the source node transmitting directly to the sink experiences high delays due to channel switching by the single radio interface. With two channels sending data from sources, there is approximately 87.4% improvement in delay. Single channel and 802.11 DCF show little improvement in delay. As mentioned in Section 7.5.4, single-channel performance does not improve with increasing number of sink nodes or radio interfaces as decisions are made based on window size resetting, backing off, wait states, and the fact that all nodes are contending for the same medium. It is noted that MC-DCF with multiple channel switching and multiple radio interfaces yields better performance when using multiple sinks in contrast to single-channel protocols and 802.11 DCF.

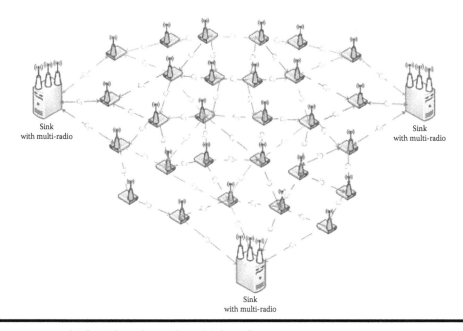

Figure 7.25   Multiple sink nodes with multiple radios.

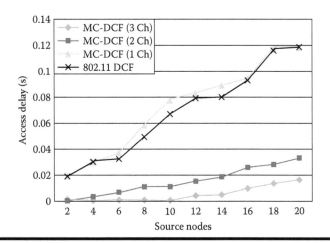

Figure 7.26   Delay impact from source nodes when using multiple sinks with multiple radio interfaces for 802.11 distributed coordination function (DCF) and one-channel (1 Ch), two-channel (2 Ch), and three-channel (3 Ch) multichannel DCF (MC-DCF).

Figure 7.27 shows an improvement of over 90% for three channels for packet delivery ratio when the number of sink nodes and radio interfaces increases by three as compared to the single sink node with a single radio interface considered in the previous simulation results. With two channels sending data from sources to sinks, packet delivery ratio improved by over 81% compared to the poor performance obtained with single channel. Similarly, in cases where the delay with single channel shows no significant improvement, packet delivery ratio also shows no major improvement.

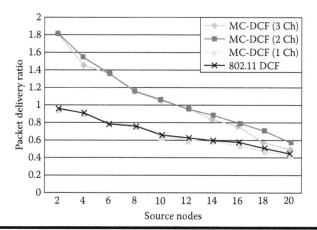

Figure 7.27 Delivery ratio impact from sources when using multiple sinks with multiple radio interfaces for 802.11 distributed coordination function (DCF) and one-channel (1 Ch), two-channel (2 Ch), and three-channel (3 Ch) multichannel DCF (MC-DCF).

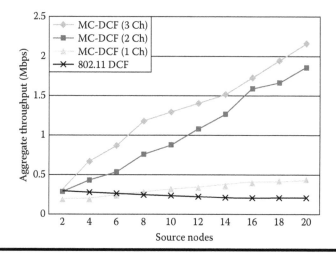

Figure 7.28 Throughput of overall system using multiple sink nodes with multiple radio interfaces for 802.11 distributed coordination function (DCF) and one-channel (1 Ch), two-channel (2 Ch), and three-channel (3 Ch) multichannel DCF (MC-DCF).

The aggregate throughput in Figure 7.28 of the overall system with source nodes sending data to sink nodes shows that with multiple sink nodes, channels, and radio interfaces, 92% more data is delivered to the sink compared to data delivered to a single sink with a single radio and a single channel. Single channel and 802.11 DCF in all instances has not shown any significant improvement with an increasing number of sink nodes receiving data from the source nodes.

From the simulation outcomes, it is proved that increasing the number of sink nodes and/or sink radio interfaces work more effectively. This improved the overall performance within the network. The network with multiple radio interfaces assignment in the sink node will be the network considered in the future, even though an increase in the number of sink nodes with single interfaces results in an improvement in performance. The simulation scenario with three sink nodes each equipped with three radio interfaces using the three nonoverlapping channels in IEEE 802.11 is the network to be considered in the future for static WSNs with streaming data. The simulation results show that an average of over 90% improvement in performance can be achieved. This kind of assignment can be made more cost-effective and energy efficient in the future.

7.7 Conclusion

This chapter determines the feasibility of considering 802.11 as a future medium for WSNs so that they operate in high-data-rate, multichannel environments with streaming data in the 2.4 GHz frequency band that requires timely and efficient data delivery. In addition, an original model was proposed that addresses the shortage of spectrum that limits current capability to introduce new wireless services and improve existing ones. An MRMC model was introduced that allows different wireless systems to share multiple channels and switch channels without causing excessive harmful interference to their neighbors. This system was proved using simulations to increase the amount of communications that can take place in a given network. This finding creates a framework in which the world of wireless services and applications may be revolutionized, resulting in less expensive networks that transmit at higher data rates than currently existing networks.

This chapter also shows that adapting multichannel assignments and IEEE 802.11 networks to operate in WSNs can be considered for use in future WSNs to operate at high data rates in multichannel environments with streaming data in the 2.4 GHz frequency band that requires timely and efficient delivery. Simulation outcomes have shown that increasing the number of sink nodes and/or the number of radio interfaces at the sink nodes results in better performance at the sink, which in turn results in overall performance improvement in the network. The multi-radio interfaces assignment at the sink node will be the more feasible network to consider in the future when compared to the multiple sink with single radio interfaces. The simulation scenario with three sink nodes, each equipped with three radio interfaces using the three nonoverlapping channels in IEEE 802.11, is the network to be considered in future static WSNs with streaming data. The simulation results show that an average of over 90% improvement in performance can be achieved. As such, this kind of assignment can be made more cost-effective and energy efficient in the future.

7.8 Future Research Directions

After experimenting with multichannel assignments and evaluating MC-DCF performance, it can be confidently said that the results are encouraging. However, these accomplishments need to be followed up by further development efforts to make such channel assignments a reality and apply MC-DCF in other contexts beyond multichannel assignments. The simulation experiments in this chapter encourage research on various interesting issues in various directions.

7.8.1 Overlapping Channels

Nonoverlapping channels were used to carry out the simulations discussed in this chapter. If transmitters are located close together, then overlapping channels may cause unacceptable degradation of signal quality and throughput. However, overlapping channels may be used under certain circumstances. By using overlapping channels, more channels are made available. The use of overlapping channels during medium access is an interesting and challenging future research direction.

7.8.2 Energy Efficiency

One of the most important issues faced in deploying WSNs is energy efficiency. Although this chapter uses nonoverlapping channels and assumes nodes are static and always powered, there is no certainty that multichannel communication helps to reduce energy consumption in WSNs. Evaluating the energy consumption of existing multichannel protocols, together with the impact of channel switching on energy consumption, can be a major future research direction.

7.8.3 Real-Time Constraints

In real-time applications, data is delay constrained and has a certain bandwidth requirement. It functions within a time frame that a user senses as immediate or current. For example, scheduling messages with deadlines is important to take appropriate actions in real time or set alerts that trigger critical activities.

However, due to the interference and contention on the wireless medium, this is a challenging task. Multichannel communication can help to reduce delays by increasing the number of parallel transmissions and help the network to achieve real-time guarantees. Reducing the delay in real-time applications can be an interesting research area in the future.

7.8.4 Multiple Applications Running on the Same Network

Latest operating systems for WSNs make it possible to have multiple applications running on the same network. This allows larger amounts of data to be transmitted in the network, and dealing with traffic using different priority levels is an energy-efficient way of avoiding collisions and interference, which can become a major issue. Multichannel communication is a topic that must be researched for solving the problems arising from running multiple applications in a network.

7.8.5 Cross-Layer Design

In wireless networks, users communicate using limited resources and changeable transmission media, which are prone to interference, weak signal strength, and other channel conditions. With these challenges, protocols find it difficult to function in isolation; this resulted in the invention of the cross-layer approach in order for protocols to develop and operate in unison. Cross-layer design is a design in which layers such as MAC and transport can exchange information between them in an intelligent way during communication to improve the performance of the system.

Some of the major challenges that WSNs need to overcome are as follows:

■ The constraints in computational energy and storage resources because of the limited energy of WSNs
■ Interference during transmission
■ Redundant information, since in most cases neighboring nodes often sense the same events from their environment and thus forward the same data to a base station
■ Topology changes due to node failure, even though most sensor nodes are usually stationary

Because of these challenges, protocols cannot develop in isolation and this results in the invention of the cross-layer approach. In IEEE 802.11, the available bandwidth cannot be estimated directly from the overall throughput achieved because of the following reasons:

- The maximum throughput is not constant for a given data rate; it is affected by the average packet length and the number of active contending nodes.
- The data rates of links are not the same as multiple rates are supported.

The layering approach to network design does not fit with the wireless network mentioned by Gavrilovska [25], who gives an in-depth analysis of cross-layering approaches for wireless ad hoc.

The transport layer, which provides the end-to-end communication service, mainly uses user datagram protocol (UDP) and transmission control protocol (TCP). A number of improved techniques are developed to improve reliable flow and congestion as well as error recovery in networks. However, the challenges to be overcome associated with WSNs are as follows:

- Sensor nodes are more constrained in terms of computational energy and storage resources; sensor nodes have limited energy which are usually provided by batteries and are difficult to replace when consumed.
- Interference during transmission, since more nodes are deployed in a sensor network, up to a hundred or thousand nodes, than in other wireless networks.

The transport layer using TCP for wireless transmission creates additional challenges as TCP assumes that packet losses are due to congestion. In wireless networks, a number of issues may cause packet losses:

- Bit error rate (BER), which is usually high based on changes within the environment.
- Bandwidth limitation.
- Round trip time (RTT); the overall throughput and delay will be affected because of longer latency within the wireless medium.

Another interesting issue that can be addressed by cross-layer design is the security problems of IEEE 802.11 wireless networks. This area has become a hot topic, especially when a simple MAC-based access control list is employed. With the MAC layer in WSNs utilizing multiple channels, there will be far more data streaming and more applications utilizing the medium; as such, the network will become more prone to be compromised by intruders or hackers. In the future, will WSNs opt to incorporate wireless intrusion protection systems or honeynets for security? A cross-layer design between the MAC layer and the network layer with stringent security measures in place may be a better solution. These are real issues to be addressed in the deployment of WSNs.

Therefore, cross-layer interactions form a technique to boost the performance by effectively adapts to the dynamic environment that can interactively communicate with each layer simultaneously. This will prevent the major challenges that the wireless systems faces. Cross-layer design can become a major research area in the future.

7.8.6 Upper Layers Multichannel Communication

Network settings is not possible to find a simple rate region. If the rate region can be reduced, then the set of feasible rates can be used to control congestion. The rate region is studied by several researchers [26–28]. In WSNs, local channel contention and interference on the shared communication medium causes network congestion [29]. One study [30] proposes an interference-aware rate control technique for WSNs. If multichannel communications are used to eliminate interference, the effects of congestion can be alleviated and fair rate control made possible for nodes that suffer from interference. Further research is needed to develop a congestion control or rate control algorithm that utilizes multichannel communication in WSNs.

7.8.7 Test Beds

Test beds replicate theories, computational tools, and innovations. When compared to WSN simulators, WSN test beds enable more realistic and reliable experiments aimed at capturing the subtleties of underlying hardware, software, and dynamics of a WSN. Deployment of a WSN test bed is further enhanced through an increasing collaboration between academia and industry.

The WSN test beds are the basis for experimentation with WSNs in real-world settings; they are also used by many researchers to evaluate specific applications pertaining to specific areas. A WSN test bed typically consists of sensor nodes deployed in a controlled environment. The WSN test beds provide researchers with an efficient way to examine and evaluate their algorithms, protocols, and applications. A WSN test bed can be designed to support different features depending on the objective of the test bed. One of the important features of a WSN test bed is that it can be designed to remotely configure, run, and monitor experiments. Another interesting feature is that a WSN test bed can be used for repeating experiments to produce similar results for analysis [31]. Selecting the appropriate level of abstraction in a simulation model is a complex problem. Thus, it is obvious that the accuracy of a simulator depends solely on its mathematical model. Accordingly, there is a trade-off between simulator accuracy and computational complexity. The more complex a simulation model the more computational resources and time required to execute it. As a result, designers of such simulation models tend to make them as simple as possible. It is impossible to take all the various aspects of a wireless channel into consideration when designing a simulation model [32]. Nonetheless, simulation tools are essential in providing affordable environments for the initial design and tuning of WSNs. This inherent difficulty in faithful modeling motivates many researchers to build their own WSN test beds. One of the advantages of a real WSN test bed over a simulator is that the former provides a realistic testing environment and allows users to get more precise testing results [31].

It is noted that WSN test bed monitoring is concerned with collecting information about a spectrum of parameters including node states (battery level, communication power), network topology, wireless bandwidth, link states, coverage bounds, and exposure bounds. Based on the collected network states, a variety of management control tasks can be performed. As emphasized by Willig [33], it would be helpful to have a repository of standard models for not only simulation codes but also implementation details on the test beds. However, experimenting with the simulations discussed in this chapter on real test beds and workloads from a set of different applications is important to continuously improve MC-DCF design through results and feedbacks arising from real, running scenarios.

Acknowledgment

The research effort detailed in this chapter was supported by National Agenda Project (NAP) of Korea Research Council of Fundamental Science and Technology, Seoul, Korea.

References

1. Akyildiz, F., T. Melodia, and K. R. Chodhury. 2007. "A Survey on Wireless Multimedia Sensor Networks." *Computer Networks* 51: 921–60.
2. Callaway, E. H. Jr. 2004. *Wireless Sensor Networks Architectures and Protocols*. Boca Raton, FL: Auerbach.
3. Campbell, C. E. -A., I. A. Shah, and K. K. Loo. 2010. "Medium Access Control and Transport protocol for Wireless Sensor Networks: An overview." *International Journal of Applied Research on Information Technology and Computing (IJARITAC)* 1 (1): 79–92.
4. Misra, S., M. Reisslein, and G. Xue. 2008. "A Survey of Multimedia Streaming in Wireless Sensor Networks." *IEEE Communications Survey … Tutorials* 10 (4): 1553–877.
5. Laboid, H. 2008. *Wireless Ad Hoc and Sensor Networks*. Hoboken, NJ: Wiley ISTE.
6. Cordeiro, C., and D. Agrawal. 2006. *Ad Hoc … Sensor Networks: Theory and Applications*. Hackensac, NJ: World Scientific Publishing Co.
7. Oh, D. -C., and Y. -H. Lee. 2009. "Energy Detection Based Spectrum Sensing for Sensing Error Minimization in Cognitive Radio Networks." *International Journal of Communication Networks and Information Security (IJCNIS)* 1 (1): 1–5.
8. Sen, J. 2009. "A Survey on Wireless Sensor Network Security." *International Journal of Communication Networks and Information Security (IJCNIS)* 1 (2): 59–82.
9. Garcia, M., J. Lloret, S. Sendra, and R. Lacuesta. 2010. "Secure Communications in Group-based Wireless Sensor Networks." *International Journal of Communication Networks and Information Security (IJCNIS)* 2 (1): 8–14.
10. Kredo, K. II, and P. Mohapatra. 2007. "Medium Access Control in Wireless Sensor Networks." *Computer Networks* 51: 961–94.
11. Kumar, S., V. Raghavan, and J. Deng. "Medium Access Control Protocols for Ad Hoc Wireless Networks: A survey." *Ad Hoc Networks* 4: 326–58.
12. Marina, M., and S. Das. 2005. "A Topology Control Approach for Utilizing Multiple Channels in Multi-Radio Wireless Mesh Networks." In *2nd International Conference on Broadband Networks (BROADNETS 2005)*, 381–90. October 3–7, 2005, Boston, MA.
13. Tang, J., G. Xue, and W. Zhang. 2005. "Interference-Aware Topology Control and QoS Routing in Multi-Channel Wireless Mesh Networks." In *Proceedings of the 6th ACM International Symposium on Mobile Ad Hoc Networking and Computing (MobiHoc'05)*, 68–77. May 25–27, 2005, Urbana-Champaign, IL.
14. Ramachandran, K., E. Belding, K. Almeroth, and M. Buddhikot. 2006. "Interference-Aware Channel Assignment in Multi-Radio Wireless Mesh Networks." In *IEEE INFOCOM*, 1–12. 25th IEEE International Conference on Computer Communications. April 2006, Barcelona, Spain.
15. Raniwala, A., K. Gopalan, and T. Chiueh. 2004. "Centralized Channel Assignment and Routing Algorithms for Multi-Channel Wireless Mesh Networks." *ACM SIGMOBILE Mobile Computing and Communications Review* 8: 50–65.
16. Kodialam, M., and T. Nandagopal. 2005. "Characterizing the Capacity Region in Multi-Radio Multi-Channel Wireless Mesh Networks." In *Proceedings of the 11th Annual International Conference on Mobile Computing and Networking (MobiCom'05)*, 73–87. August 28–September 2, 2005, Cologne, Germany.
17. Alicherry, M., R. Bhatia, and L. Li. 2005. "Joint Channel Assignment and Routing for Throughput Optimization in Multi-Radio Wireless Mesh Networks." In *Proc. ACM MobiCom*, 58–72.
18. Raniwala, A., and T. Chiueh. 2005. "Architecture and Algorithms for an IEEE 802.1 1-Based Multi-Channel Wireless Mesh Network." In *INFOCOM 2005. 24th Annual Joint Conference of the IEEE Computer and Communications Societies. Proceedings IEEE*. March 2005, Miami, FL.
19. Das, S. M., H. Pucha, D. Koutsonikolas, Y. C. Hu, and D. Peroulis. 2006. "DMesh: Incorporating Practical Directional Antennas in Multichannel Wireless Mesh Networks." *IEEE Journal on Selected Areas in Communications* 24: 2028.
20. Revision of IEEE Std 802.11-1999 … Part 11: Wireless LAN Medium Access Control (MAC) and Physical Layer (PHY). Approved 8 March 2007. IEEE-SA Standards Board.
21. Project, T. V., U. C. Berkeley, X. Parc, K. Fall, and E. K. Varadhan. 2009. "The ns Manual (formerly ns Notes and Documentation) 1," Facilities.

22. Network Simulator (NS2). Accessed November 27, 2010. http://www.cse.msu.edu/~wangbo1/ns2/nshowto8.html.

23. Accessed September 17, 2009. http://www.ecsl.cs.sunysb.edu/multichannel/

24. Yu, Q., J. Xing, and Y. Zhou. 2006. "Performance Research of the IEEE 802.15.4 Protocol in Wireless Sensor Networks." *2nd IEEE/ASME International Conference on Mechatronics and Embedded Systems and Applications*, 1–4. Proceedings of the 2nd IEEE/ASME International Conference. August 2006, Beijing.

25. Gavrilovska, L. 2006. "Cross-Layering Approaches in Wireless Ad Hoc Networks." *Springer Wireless Personal Communications* 37: 271–90.

26. Sarkar, S., and L. Tassiulas. 2003. "End-to-End Bandwidth Guarantees Through Fair Local Spectrum Share in Wireless Ad-Hoc Networks." In *Proceedings of the IEEE Conference on Decision … Control*, 564–69. Maui, HI.

27. Y. Yi, Y. S. and Shakkottai, 2004. "Hop-by-hop congestion control over a wireless multi-hop network." In *Proceedings of IEEE INFOCOM*, 2548–58. Hong Kong.

28. Xue, Y., B. Li, and K. Nahrstedt. 2003. "Price-Based Resource Allocation in Wireless Ad Hoc Networks." In *Proceedings of 11th International Workshop on Quality of Service*, vol. 2707, 79–96. New York, Monterey, CA: Springer-Verlag.

29. Gungor, V. C., M. C. Vuran, and O. B. Akan. 2007. "On the Cross-Layer Interactions Between Congestion and Contention in Wireless Sensor and Actor Networks." *Ad Hoc Networks* 5 (6): 897–909. ISSN 1570-8705.

30. Rangwala, S., R. Gummadi, R. Govindan, and K. Psounis. 2006. "Interference-Aware Fair Rate Control in Wireless Sensor Networks." In *SIGCOMM '06: Proceedings of the 2006 Conference on Applications, Technologies, Architectures, and Protocols for Computer Communications*, 63–74. New York, NY: ACM. ISBN 1-59593-308-5.

31. Yick, J., B. Mukherjee, and D. Ghosal. 2008. "Wireless Sensor Network Survey." *Computer Networks* 52: 2292–330.

32. De, P., A. Raniwala, S. Sharma, and T. Chiueh. 2005. "MiNT: A Miniaturized Network Testbed for Mobile Wireless Research." In *IEEE INFOCOM 2005*, March 13–17, 2005, City 2731–42.

33. Willig, A. 2006. "Wireless Sensor Networks: Concept, Challenges and Approaches." *Electronic Engineering and Information Technology* 123 (6): 224–231.

POSITION ESTIMATION, ENERGY-CENTRIC SIMULATION, AND QUALITY OF SERVICE ISSUES

III

Chapter 8

Location and Position Estimation in Wireless Sensor Networks

Muhammad Farooq-i-Azam and Muhammad Naeem Ayyaz

Contents

8.1 Introduction

A wireless sensor network is a network of tiny sensor nodes that communicate with each other through a wireless communication link. Each sensor node typically consists of a processing device, a small amount of memory, a battery, and a radio transceiver for communication. These sensor nodes obtain data, for example, temperature, pressure, and humidity; do some local processing; and transmit the data to a neighbor node or a beacon node, which in turn may be connected to a central computer where major processing is performed. As is evident, this central computer may be part of a bigger computer network so that the information can be communicated from this central computer to other computers that are part of the bigger network.

Wireless sensor networks can be used in diverse applications in both industrial and commercial environments. Some of the most common applications of wireless sensor networks include object tracking, habitat monitoring, fire detection, traffic monitoring, and area monitoring. Some of the typical characteristics of sensor networks are small nodes, mobile nodes, a dynamic network topology, harsh operating environments, and limited energy or power resources that these nodes should utilize efficiently as they may remain in an area for years without more energy being available.

There are many challenges involved at various levels and stages in the development of wireless sensor networks as discussed by Akyildiz et al. [1]. For example, the physical layer design of a wireless sensor node, which should be very small and yet accommodate all the functions that it is required to perform, presents many challenges. Similarly, new algorithms and protocols from link layer to application layer need to be developed. New operating systems are needed to run on such tiny nodes, and to write these tiny operating systems there is a need to develop new programming languages and new programming paradigms. One of the important problems encountered is the localization of sensor nodes, that is, the determination of the positions of nodes in the sensor field. This is important due to various reasons. For example, the data collected by a sensor node must be ascribed to the location from where it was collected. The data would not be useful if the location to which it belongs is not known. The set of values of temperature and humidity, for instance, collected by the sensor nodes is not meaningful unless the respective position coordinates from where these values were recorded are known.

An example of a wireless sensor network application where location information is important is target tracking. Likewise, in a sensor network meant for earthquake disaster relief, the sensor positions must be known to ascertain the location of survivors buried somewhere in the rubble of a collapsed building. Similarly, one of the biggest challenges faced by sensor networks is the efficient utilization of energy resources, which are not easily available to sensor nodes.

In addition, one of the most energy-dependent operations is data transmission from sensor nodes to base stations, which should use some energy-efficient and energy-aware routing algorithm. One of the approaches being worked on today that holds great promise is geographic location–based routing, which is based on mathematical modeling of sensor positions instead of using IDs. Again, for a location-based approach to be possible, the locations of the nodes must be known.

Due to various constraints, existing localization systems, such as global positioning systems (GPS), cannot be used for the localization of wireless sensor nodes. Therefore, new strategies and algorithms for the localization of sensor nodes need to be designed and developed. The algorithms should be designed within the defined constraints and the characteristics of the sensor nodes and the sensor network. In wireless sensor networks using only static sensor nodes, localization algorithm usually runs only at the time of initial deployment of the nodes. However, in a sensor network using mobile sensor nodes, the localization algorithm is needed to determine the new positions of mobile nodes as they move in the sensor field. Hence, localization algorithms for mobile sensor nodes need more energy compared to algorithms designed for static sensor nodes.

8.2 Background

Certain applications of wireless sensor networks require sensor nodes to be aware of their positions relative to the sensor network. For it to be significant and of value, data such as temperature, humidity, and pressure gathered by sensor nodes must be ascribed to the relative position from where it was collected. For this to happen, the sensor nodes must be aware of their positions. The literature has come to term this problem of location or position estimation of sensor nodes simply as localization. The term localization has earlier been used in robotics where it is used to refer to the determination of location of a mobile robot in some coordinate system. Under certain circumstances, the nodes should be aware of not only their position but also their direction or orientation relative to the network [2].

In a sensor network, nodes may be categorized as follows:

Dumb node (D): It is the node that does not know its position and eventually finds its location and position from the output of the localization algorithm under investigation. Dumb nodes are also known as free or unknown nodes.

Settled node (S): A settled node is a node that was initially a dumb node but managed to find its position using the localization algorithm.

Beacon node (B): A beacon node is a node that knows its position from the very start and always knows its position afterward without the use of the localization algorithm. It has a mechanism other than the localization algorithm to find its position. For example, the beacon node may be equipped with a GPS device or it may be placed at a position with known coordinates. Beacon nodes are also called reference nodes, anchor nodes, or landmark nodes.

It should be noted that sensor nodes may have symmetric or asymmetric communication links. If two nodes u and v are symmetric, then u reaches v and v reaches u as well. In the case of asymmetric communication links, either u reaches v or v reaches u but both u and v do not reach each other simultaneously.

Let us now consider a sensor network that is symmetric, two-dimensional, and arranged in a square shape. Then, this sensor network can be represented as a graph $G(V, E)$ where the set of sensor nodes can be represented as a set of vertices as follows:

$$V = \{v_1, v_2, \ldots, v_n\}$$

The set of edges E in the graph $G(V, E)$ comprises all edges $e = (i, j) \in E$ if v_i reaches v_j, that is, the distance between v_i and v_j is less than r where r is the maximum distance between the two nodes after which communication between them ceases to exist or if the distance between two nodes is greater than r, no direct communication between them is possible. In other words, if the distance between two nodes is greater than r, the two nodes are not neighbor nodes. The distance between two neighbor nodes v_i and v_j is defined as the weight $w(e) \leq r$ of the edge $e = (i, j)$ between them.

It is to be noted that the problem of localization is usually solved only for two dimensions with the supposition that, when needed or deployed, it could be extended to three dimensions. It is for this reason we have stated graph $G(V, E)$ to be a two-dimensional structure. Therefore, it can be stated that G is a Euclidean graph in which every sensor node has a coordinate $(x_i, y_i) \in R^2$ in two-dimensional space. The coordinate (x_i, y_i) represents the location of a node i in the given sensor field.

The sensor node localization problem can now be stated as follows:

> Let there be a multihop sensor network represented by a graph $G = (V, E)$. The graph has a set of beacon nodes B with known positions given by (x_b, y_b) for all $b \in B$. The localization problem requires one to find the position set (x_d, y_d) of as many dumb nodes $d \in D$ as possible. Finding the location of a node implies finding its latitude, longitude, and altitude.

The problem of node localization and positioning in a sensor network can be solved if each node is equipped with a GPS device. However, in the case of sensor networks, this is not a feasible option for a number of reasons such as the following:

■ The GPS receiver and protocols used are not designed to be energy efficient or energy aware. In the case of sensor networks, energy is a scarce resource and sensor nodes may be deployed without replacing any sort of battery for many years. Therefore, GPS devices are not suitable for solving the localization problem in wireless sensor networks. It is, nevertheless, possible that beacon nodes, which constitute only a fraction of the total number of nodes, are equipped with a GPS device so that these can serve as reference nodes for other nodes to solve the problem of their position awareness using the localization algorithm.

■ GPS devices are quite expensive. If these are added to every sensor node in the network somehow, the cost of deployment may increase to such an extent so as to render the sensor network solution unfeasible for a particular problem.

■ One of the required properties of sensor nodes is that they should be very small. With the addition of a GPS device, the size of the sensor nodes would become quite large, which violates one of the primary requirements of a sensor node.

■ GPS devices depend on satellites for their functioning. In cases or under circumstances where no satellite link is available, GPS ceases to function. In certain applications, this really can be the case, for example, indoor applications and Mars exploration.

Due to the aforementioned reasons, GPS devices are normally used only in a fraction of nodes, which serve as reference nodes to solve localization problem of other nodes. Such nodes are also called beacon nodes. Alternatively, it is possible to avoid the use of a GPS altogether by positioning a few nodes at fixed points so that their positions are known a priori so that these nodes can serve as beacon nodes. The sensor field can then exploit inherent radio frequency (RF) capabilities of sensor nodes or some other techniques to determine their positions using a localization algorithm. The accuracy with which dumb nodes can determine their location depends on the transmission range of beacon nodes and the distance between two adjacent beacon nodes, that is, the density of beacon nodes.

Various algorithms for the localization of wireless sensor networks have been proposed, and this is currently a hot area of research with many recent publications such as those by Amundson and Koutsoukos [3]; Mao and Fidan [4]; Pal [5]; Wang and Xu [6]; Zhang, Foh, Seet, and Fong [7]; and Kim and Kim [8].

8.2.1 Classification of Localization Algorithms

The majority of the existing localization algorithms may be classified as range-based or range-free depending on whether an algorithm uses distance estimation or some other information for estimating node locations. Range-based algorithms usually use sensor field geometry information to determine node locations. Communication between beacon nodes and dumb nodes also helps in determining geometric information about their relative placement, for example, the distance between two nodes or the angles of a triangle formed by the beacon nodes. This information is then used to determine node location. When distance is used as a primary means to determine node location, the technique is termed lateration, and when angle information is used for localization it is known as angulation. For node localization in a plane, precise distance measurements from at least three beacon nodes are required and we use trilateration for the position estimation of a node. The intersection of three circles around three beacon nodes gives a single point as the position of the node, as shown in Figure 8.1.

The same technique can be extended to three-dimensional space by the addition of a fourth beacon node. However, in actual practice, distance measurements are seldom precise and the intersection of three circles may result in more than one point. The scheme may be improved by employing more than three beacon nodes for a plane, and we then use multilateration to calculate the node position.

In the angulation technique, angle information is used to deduce the position of a node. Two beacon nodes and a dumb node form the vertices of a triangle and the lines joining them form the sides of the triangle. As the positions of beacon nodes are known, the distance between them, that is, one side of the triangle, is known. If the two angles that the dumb node forms with the two

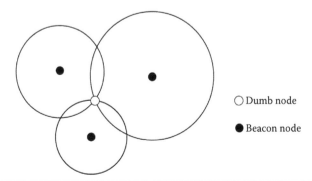

○ Dumb node

● Beacon node

Figure 8.1   Trilateration: Intersection of three circles around three beacon nodes gives position of the dumb node.

beacon nodes are measured, the location of the dumb node can be calculated as the third point of the triangle. This method of determining location of a node is termed triangulation.

In range-free localization algorithms, a node determines its position merely by finding the beacon nodes in its proximity and, for this reason, range-free algorithms are also termed proximity-based or connectivity-based algorithms. Such algorithms usually provide coarse-grained localization. However, with a sufficient number of beacon nodes with overlapping transmission regions, more accurate localization is possible. Range-free algorithms are robust against fluctuations of the wireless channel as the decision whether a node is in proximity of another node is based on connectivity information sampled over a long period of time. Hence, short and temporary variations in the wireless channel do not affect the accuracy of location estimation. A range-free algorithm calculates position without having to find the distance between the sensor nodes. A range-based localization algorithm is more accurate but has major computational cost and usually additional hardware and hence increased energy requirements. On the other hand, a range-free algorithm is less accurate but does not require additional hardware and has smaller computational overhead.

Localization algorithms may also be classified as either centralized or distributed algorithms depending on whether they use a central processor to estimate the positions of all nodes or nodes perform local processing to determine their positions. In the case of the centralized approach, node positions are calculated at a central processor. To calculate the positions of distant nodes, the central unit usually needs certain parameters from these nodes, which depend on the particular localization algorithm. These parameters from the distant nodes are sent to the central unit using the transceiver unit. The central unit computes positions of all nodes and sends the results back to them. As is evident, this approach may involve a lot of communication overhead for the dumb nodes, which in turn would require these nodes to have more battery power for sending and receiving the parameters. The centralized approach also introduces a single point of failure. If the central unit fails for some reason, the entire localization process fails. This approach is also prone to additional delays due to propagation delays involved in the transmission of parameters and results and the large amount of processing involved at the central unit. Furthermore, the central processing approach is not suited to the very nature of ad hoc networks.

In the decentralized or distributed approach, the dumb nodes determine their position themselves by performing local processing. The dumb nodes usually need information from neighbor nodes and beacon nodes to determine their position. For the decentralized approach to work, the dumb nodes must possess local processing capability. However, communication overhead is less compared to the centralized approach.

Localization algorithms may also be classified as fine-grained or coarse-grained. Some applications of sensor networks need to determine only the symbolic location of a sensor node, for example, whether a sensor node is inside room A or room B. This type of symbolic estimation of node position is termed as coarse-grained localization. In some other applications, we need to determine a more accurate estimate of node position, for example, node X is located at coordinates (x, y). This type of node localization is termed as fine-grained localization.

Another classification of localization algorithms depends on whether an algorithm is designed for the outdoor unconstrained environment or for the indoor constrained environment. The earliest localization algorithm [9] for the outdoor environment is attributed to Bulusu, Heidemann, and Estrin.

Yet another way of classifying localization algorithms is to consider whether an algorithm is designed for a sensor field in which the sensor nodes are fixed or for a sensor field comprising mobile sensor nodes. The majority of the applications of wireless sensor networks use static nodes. As a result, many of the localization algorithms consider sensor networks comprising entirely static nodes. However, a few applications of wireless sensor networks deploy mobile nodes and, hence, a few localization algorithms, such as the one proposed by Kim and Kim [8], take the mobile nature of the nodes into consideration.

8.2.2 Characteristics of Localization Algorithms

The main objective of a localization algorithm is to determine the position of a node. However, there are certain criteria that the algorithm must meet for it to be practicable. The criteria usually depend on the type of application for which the localization algorithm is designed. General design objectives or desired characteristics of an ideal localization algorithm are as follows:

- It is highly desirable that localization algorithms are RF based. The sensor nodes are equipped with a short-range RF transmitter. An efficient localization algorithm exploits this radio capability for localization in addition to its primary role of data communication.
- A wireless sensor network is ad hoc in nature. The localization algorithm should take the ad hoc nature of the network into consideration.
- The nodes in a network should be able to determine their positions in as little time as possible so that the localization algorithm has a low response time. This would enable sensor nodes to be deployed quickly.
- The position of the sensor node found by a localization algorithm should be accurate enough for the specific application for which this algorithm is being used.
- The algorithm must be robust so that it may work in adverse conditions.
- The algorithm should be scalable so that if sensor nodes are added or removed, it should still be able to work out the position of the nodes. Furthermore, the algorithm should produce acceptable results for sensor networks comprising small to large number of nodes.
- The localization algorithm should be energy efficient and, preferably, energy aware as well because sensor nodes are autonomous and normally do not have any external source of power.
- The localization algorithm should be adaptive to any change in the number of beacon nodes. If the number of available beacon nodes changes, the algorithm should still be able to provide location estimates. However, the accuracy of node estimates will change with the change in number of available beacon nodes. In general, with a higher number of beacon nodes, a localization algorithm is able to compute more accurate estimates of node positions.
- The algorithm should be efficient so that it is able to compute node locations with as small number of beacon nodes as possible.
- The algorithm should be universal so that it is able to compute node locations under all conditions of changing environments and weather. In particular, it should work in constrained environments such as indoors and unconstrained environments such as outdoors.

Only an ideal localization algorithm will be able to meet all the aforementioned goals. A localization algorithm in practice meets only a subset of these characteristics depending on the particular application for which it is designed.

8.3 Distance Estimation

Distance estimation between two nodes is an important function performed by range-based algorithms. A range-based algorithm estimates the position of a sensor node by using the distance information between the nodes, which in turn is calculated using some physical measured quantity.

The distances between dumb nodes and beacon nodes are usually determined by adding some additional hardware to the nodes or by using the existing radio communication facility on the sensor nodes. Certain characteristics of wireless communication between dumb and beacon nodes are determined by the distance between them. If these characteristics are quantified and measured at the receiving sensor node, these can be used to estimate the distance between the nodes. The characteristics generally used for this purpose are as follows:

- Received signal strength indicator (RSSI)
- Time of arrival (ToA)
- Time difference of arrival (TDoA)
- Angle of arrival (AoA)

8.3.1 Received Signal Strength Indicator

A signal is attenuated as it travels from a transmitter to a receiver. The longer the distance a signal has to travel, the greater the attenuation. Therefore, the strength of the received signal can be used to estimate the distance between the transmitter and the receiver. The distance can be calculated using the following information:

- Transmitted power of the signal
- Received power of the signal, that is, received signal strength (RSS)
- Path loss model

Using these three parameters, the power of the received signal, P_R^{ij}, transmitted by node i and received at node j at time t can be expressed as follows:

$$P_R^{ij}(t) = P_T^i - 10\eta \log(d_{ij}) + X_{ij}(t)$$

In this equation

$P_R^{ij}(t)$ is the power of the received signal at receiver node j, which is transmitted by node i at time t.
P_T^i is the transmitted power of the signal transmitted by node i.
η is attenuation constant, the value of which depends on the surroundings of the receiver node j.
d_{ij} is the distance between transmitter node i and receiver node j.
$X_{ij}(t)$ is the uncertainty factor or channel model whose value depends on multipath fading and shadowing.

The aforementioned equation can be solved for distance d_{ij} between beacon node i and receiver node j as all other parameters are known. Distance estimation using this technique is quite attractive due to the following factors:

- As the sensor nodes communicate with each other for data transfer, the solution to the problem of localization using RSSI values is only an added benefit. It does not need any additional hardware circuitry, and no size or weight is added to sensor nodes, which should remain as tiny as possible.
- Localization based on RSSI also saves power, which is otherwise needed by the additional hardware used for localization.

Apart from the merits of an RSSI-based localization scheme, there are some demerits as well. Some of these demerits are described as follows:

- The fundamental assumption in a localization scheme based on RSSI is that a signal suffers from the same amount of attenuation for the same distance traveled. However, in actual practice, this is not always the case due to factors such as multipath fading, fast fading, and shadowing. Savarese, Rabaey, and Langendoen [10] have reported that ranging errors on the order of ±50% are possible even when both transmitter and receiver nodes are stationary. This problem can be remedied, to some extent, by taking more measurements. In addition, statistical techniques may be used to filter incorrect values as suggested by Ward, Jones, and Hopper [11].
- If there are obstacles between a beacon node and a dumb node such that line-of-sight communication between them is not possible, the latter receives a signal transmitted by the beacon node after reflection from the surroundings. As a result, the signal suffers far greater attenuation than it does in the case of line-of-sight communication between the two nodes and, as a consequence, the RSSI value is not a true indicator of the distance vector (DV) between the two nodes and yields a distance estimate that is much greater than the actual distance [12]. The resultant error cannot be corrected by increasing the number of measurements since the additional measurements are still based on RSSI values of signals that are received after reflection from the surroundings.
- As the values of channel model X_{ij} and attenuation constant η depend on the surroundings, an estimation of these parameters is needed using calibration before sensor nodes are deployed in the sensor field. Under certain circumstances, such as deployment in a battlefield, this is not always possible.
- As suggested by Whitehouse and Culler [13], the transceiver units in all sensor nodes should be calibrated so that RSSI values correspond to the actual strength of the received signal.

8.3.2 *Time of Arrival*

The ToA technique of distance estimation uses the following relationship, which relates the distance traveled by a signal to the time taken provided that the speed of propagation is known:

$$d = v \times t$$

where d is distance, v is speed of the signal, and t is time taken by the signal to travel the distance d. Therefore, if time taken by a signal to propagate from a beacon node to a dumb node, which is called ToA or "time of flight," is measured and speed of propagation of the signal is known, the distance and hence position of the dumb node can be calculated.

There are two variations of the ToA technique:

1. One-way ToA
2. Two-way ToA

Furthermore, either of the aforementioned techniques may use an RF signal or an ultrasonic pulse for distance estimation.

8.3.2.1 One-Way Time of Arrival

In one-way ToA technique, the propagation time of one-way trip of the signal from the beacon node i to the dumb node j is measured. This is given by the difference between the sending time t_i at the beacon transmitting node and the receiving time t_j at the receiving dumb node. The distance d_{ij} between the two nodes i and j is then given as follows:

$$d_{ij} = v \times (t_j - t_i)$$

With this approach, the receiver node calculates its position in a secure manner without disclosing its location information to the transmitting node and, hence, this approach is also termed as "passive ToA" localization. Obviously, the transmitting node is usually a beacon node and the receiving node is a dumb node. As stated earlier in Section 8.3.2, an RF signal or an ultrasonic pulse can be used for distance estimation using this technique.

It is to be noted that for the one-way ToA technique to work, the receiver must know the time of transmission of the signal. In the case of an RF signal, the transmitting node can embed this information in the beacon signal that it sends to the receiver. However, RF signals travel at a very high speed, which is almost equal to the speed of light, that is, 3×10^8 m/s. Their use in distance estimation using time of flight requires extremely accurate and stable clocks and highly precise hardware for time measurement. For example, if sensor nodes are located 10 m apart and an RF signal is used for ranging, then the time taken by the signal to travel from the transmitting node to the receiving node is given as follows:

$$t = d/v = 10 / 3 \times 10^8 = 3.33 \times 10^{-8} = 33.3 \text{ nanoseconds}$$

If the transmitter and receiver clocks are not tightly synchronized and are out of sync by even 1 nanosecond, then it will result in an error on the order of 1 nanosecond in ToA measurement, which will be reflected in distance estimation error, $d_e = 1 \times 10^{-9} \times 3 \times 10^8 = 0.3$ m. The requirement of highly stable and accurate clocks and precise timing measurement means addition of complex and costly hardware to the sensor node and necessitates usage of a time synchronization algorithm along with the localization algorithm, which increases size and processing need and, thereby, energy consumption. As a result, one-way ToA measurements using RF signals are not considered a choice for sensor hardware existing today or in the future. However, in some media, other than air, for example, under water and earth, the propagation speed of signals is reduced and hence can be used for distance estimation using the time-of-flight technique. Furthermore, in many applications of wireless sensor networks, such as monitoring applications, we need to know not only the location where an event occurred but also the time when the event occurred, that is, the event's timing information. For example, if a sensor node samples temperatures in a sensor field, one might be interested in knowing both position and timing information of the temperature values. For a sensor network to be able to provide this timing information, the sensor nodes must be accurately synchronized. A localization algorithm can exploit this time synchronization to estimate node position. In fact, some proposed algorithms, such as the one proposed by Synapse by De Oliveira, Nakamura, Loureiro, and Boukerche [14], combine time synchronization and localization problems so as to propose a single solution of localization in time and space.

Ultrasonic waves have speeds far smaller than RF signals, for example, the speed of an ultrasonic wave in air at 20°C is 343.26 m/s. As a result, the hardware required for timing measurement is not complex. However, using ultrasonic pulses for distance estimation has its drawbacks, some of which are described as follows:

■ Use of ultrasonic pulses for distance estimation necessitates addition of extra hardware on sensor nodes for generation and reception of ultrasonic pulses. This will affect the size, cost, and energy efficiency of sensor nodes.
■ Speed of an ultrasonic wave changes with temperature. As a result, if there are large variations in temperature in the field of sensor network deployment, distance estimation will change with change in temperature. As a remedial measure to this problem, extra pieces of hardware need to be incorporated on sensor nodes, which again affects the size, cost, and energy efficiency of sensor nodes.

8.3.2.2 Two-Way Time of Arrival

In the two-way approach, the receiver node sends the signal back to the transmitter node, which measures the round-trip time for distance estimation between transmitter and receiver. Suppose sensor node i transmits a signal at its local time t_{i1}. It reaches sensor node j at its local time t_{j1}. After some delay, the sensor node j sends the signal back at its local time t_{j2}. The signal is received back at node i at its local time t_{i2}. This is illustrated in Figure 8.2.

Now,

$$\text{Total round-trip time including delay} = t_{i2} - t_{i1}$$

$$\text{Delay suffered at } j = t_{j2} - t_{j1}$$

$$\text{Actual round-trip time} = (t_{i2} - t_{i1}) - (t_{j2} - t_{j1})$$

$$\text{One-way time of flight} = \frac{(t_{i2} - t_{i1}) - (t_{j2} - t_{j1})}{2}$$

Hence, the distance d between the two nodes is given as follows:

$$d = \frac{(t_{i2} - t_{i1}) - (t_{j2} - t_{j1})}{2} \times v$$

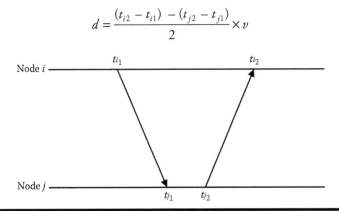

Figure 8.2 Signal propagation times in two-way time of arrival distance estimation technique.

It should be noted that node i subtracts its local time $t_{i2} - t_{i1}$ to get the total round-trip time. Similarly, node j subtracts its local time $t_{j2} - t_{j1}$ to get the processing delay. As both nodes have to process only their respective local times, no time synchronization between the nodes is necessary. This saves the extra hardware cost and energy required for time synchronization.

With the two-way approach, the beacon node, which is the transmitting node, has to carry out processing for distance and position estimation. The result is then sent back to the dumb node. As there are a large number of dumb nodes per beacon node, a particular beacon node may have to process localization information for many dumb nodes. This will increase processing overhead on the beacon nodes manifold. Furthermore, communication overhead is also involved as the beacon node has to send the result back to the dumb node.

With the one-way ToA approach, the beacon nodes do not have any processing overhead as localization processing is carried out by the dumb nodes. In this way, the processing task is uniformly distributed among all the dumb nodes, and beacon nodes are free to send timely beacons to the dumb nodes.

8.3.3 Time Difference of Arrival

Time difference between the receiving of two signals at a node is easier to measure compared to ToA of a signal. This time difference information can then be used to estimate the distance between two nodes. The advantage of using time difference instead of ToA for distance estimation is that errors in time difference measurement are tolerable and do not have a pronounced effect on the accuracy of estimation of distance between two nodes. As a result, the hardware required for time measurements is less complex and less costly and, hence, the method is also efficient in terms of energy consumption.

The TDoA techniques can be classified into two main categories [15]:

1. Multinode TDoA
2. Multisignal TDoA

8.3.3.1 Multinode Time Difference of Arrival

At least three beacon nodes B_1, B_2, and B_3 transmit signals at exactly the same time. Time differences among the arrival of these three signals at the receiving dumb node D are measured. The difference from a pair of beacon nodes, say, B_1 and B_2, defines a branch of a hyperbola on which the dumb node D is located. Similarly, difference from the pair of beacon nodes B_2 and B_3 will again give a branch of a second hyperbola. The receiving dumb node should lie on this second hyperbola as well. The point of intersection of the two hyperbolas gives the location of the dumb node D.

It should be noted that the nodes should be time synchronized with stable clocks for them to be able to transmit the beacon signals at exactly the same time. This type of TDoA is quite old and was used in classical long-range navigation systems such as LORAN.

8.3.3.2 Multisignal Time Difference of Arrival

In the ToA technique using RF signals, sophisticated hardware is needed for precise measurement of time and the nodes should also be time synchronized. One way to alleviate this problem is by the use of an ultrasonic signal along with an RF signal. The beacon transmitting node i transmits two signals simultaneously or after some fixed time interval, $t_{i2} - t_{i1}$, as shown in Figure 8.3.

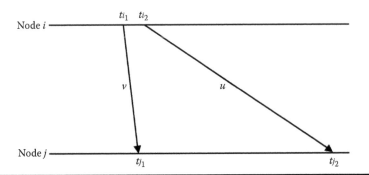

Figure 8.3   Signal propagation times in multisignal time difference of arrival distance estimation technique.

Due to a large difference in their propagation speeds, the RF signal is received first by the receiving dumb node j, which records the arrival time t_{j1} of the RF signal. As the speed of the ultrasonic signal is very small compared to the speed of the RF signal, the ToA of RF signal is treated as the time of transmission of the ultrasonic signal. After receiving the RF signal, the receiving dumb node prepares itself to receive the ultrasonic signal. The ToA, t_{j2}, of the ultrasonic signal is also recorded. If speed of the ultrasonic signal is u, an estimation of the distance d between the beacon transmitting node and the receiving dumb node is given as follows:

$$d \approx \left(t_{j2} - t_{j1}\right) \times u$$

It should be noted that sender and receiver clocks need not be synchronized because the RF signal provides for an indirect synchronization mechanism between the transmitting and receiving nodes. A disadvantage of this scheme is that the nodes must be equipped with separate pieces of hardware for transmitting, receiving, and processing two different kinds of signals. For ultrasonic signals, the nodes must be equipped with microphones and speakers. This results in increased cost, size, and energy consumption of the sensor nodes. Instead of RF and ultrasonic signals, any two signals with large difference in their speeds can be used for multisignal TDoA. An example of a localization system using multisignal TDoA is the Cricket location support system [16,17], which is described later in this chapter in Section 8.4.3.

Position estimation using multisignal TDoA is quite accurate compared to other ranging techniques. Priyantha, Miu, Balakrishnan, and Teller [18] have reported achieving accuracy to the order of a few centimeters. However, this is only possible under line-of-sight conditions. Under non-line-of-sight conditions, both RF and ultrasonic signals suffer from propagation delays and position estimation using multisignal TDoA is prone to error. In particular, ultrasonic waves may suffer from attenuation due to scattering and diffraction resulting from obstacles and atmospheric effects such as temperature, pressure, humidity, and turbulence.

8.3.4 Angle of Arrival

The direction of arrival of a signal at a dumb node can be used to estimate its position. The direction of a received signal can be determined by measuring the angle it makes with some reference direction or orientation. Alternatively, the angle between the dumb node and the beacon node may be measured. For the localization of a dumb node using this technique, AoAs from a minimum of three beacon nodes are measured. Position information of three or more beacon nodes along with the three angles of arrival can be used to estimate the location of the dumb node.

The AoA can be measured using directional antennas, a special configuration of antenna arrays, or a combination of both. When using directional antennas, these can be mounted on beacon nodes. To serve multiple dumb nodes, a directional antenna mounted on a beacon node rotates about its axis, thereby transmitting beacon signals in all directions. A dumb node may use a similar directional antenna configuration to receive the beacon signals. Alternatively, dumb nodes can also use special configuration of antenna arrays to receive and measure the AoA of a beacon signal. When an antenna array is used, antennas in the array are placed at known separations. The difference of ToA of the wave front at different antennas is used to estimate the direction from which the signal arrived.

Practical use of this technique is limited due to the complexities involved in deployment of special antennas. For example, mounting rotating directional antennas on tiny nodes is problematic and the rotating components are more prone to failure. Similarly, if an antenna array configuration is used, antennas in the array must be placed a specific distance apart, which is again a difficult proposition considering the tiny sizes of sensor nodes. Moreover, a greater accuracy of angle measurement is achieved only when separation distance between antennas in the array is small. However, with smaller separation distances, more sophisticated and precise hardware is needed for time difference measurements. Furthermore, shadowing, multipath fading, and non-line-of-sight conditions introduce a large amount of error in the estimated position, which is more than the same kind of errors introduced in other similar techniques, for example, RSSI, ToA, and TDoA. Due to these reasons, AoA is considered less of a choice for localization in sensor networks.

After distances have been estimated by using one of the aforementioned techniques, multilateration is used to estimate the position of a dumb node. Obviously, these techniques form the basis of range-based localization algorithms. Range-free algorithms do not use measurements to estimate distances and for localization. Instead, they analyze the connectivity information of neighbor nodes to deduce position information.

Apart from range-based and range-free methods, still another possible technique of localization is signal pattern matching. A database of unique signal signatures for all possible locations is created by using some property of radio signals. A dumb node localizes itself by comparing the pattern of received signals with the stored signal signatures. For example, a fixed number of beacon nodes may be deployed in the sensor field and RSSI values at each possible location may be calculated and stored in a database so as to serve as location signatures. This database can later be used for localization after actual sensor nodes are deployed in the sensor field. It should be noted that other signal characteristics instead of RSSI values, such as multipath pattern of a signal arriving at a given location, can also be used to create unique signatures for each location. Similarly, multiple signal characteristics may be combined to develop the signature database. This information can then be used to locate the position of a node. However, due to its very nature, this technique is not suitable for networks that are ad hoc and may have a dynamically changing topology, which is the case with sensor networks.

8.4 Single-Hop Localization Schemes

A number of localization schemes for single-hop networks have been developed that use the principles and techniques discussed in Sections 8.3.1 through 8.3.4. In a single-hop network, a dumb node has a direct, that is, single-hop, communication link to the beacon nodes. Some of the localization schemes for single-hop networks include "active badge," active office, Cricket, connectivity-based centroid algorithm, and approximate point in triangulation (APIT). Most of these single-hop localization algorithms were developed for context-aware computing before research on wireless sensor networks gained focus.

In a wireless sensor network, a dumb node may not always have a direct communication link with a beacon node. In other words, a wireless sensor network is usually multihop in nature. Therefore, localization schemes for single-hop networks are not suitable for large multihop wireless sensor networks. However, they provide insight into the problem of localization and provide basic principles, techniques, and foundation to develop algorithms for multihop sensor networks.

8.4.1 Active Badge

Active badge location system is one of the earliest location estimation systems proposed by Want, Falcao, and Gibbons [19]. It is meant to find location information of people in a building. Each person to be localized is supposed to wear a badge, which is termed active badge, because it emits an infrared (IR) beacon signal carrying a unique identification number every 15 seconds. The beacon signals transmitted by the active badges are collected by sensors placed at appropriate known locations throughout the building, for example, one such sensor may be installed in every room. The sensors in turn are connected to a central computer, which processes the information received from these sensors and determines the position of badges and thereby persons wearing them.

The active badge localization system is a range-free technique and uses centralized processing to determine locations of badges. The location information provided by the active badge scheme is symbolic, that is, it is a coarse-grained scheme that provides location information of a person as, say, in room 34.

Badges and sensors use pulse-width-modulated (PWM) IR waves for signaling. Unlike radio signals, IR signals do not penetrate walls and partitions in a building. Instead these are reflected and are not directional inside small rooms. Due to this property, location of a badge inside a certain room can be determined. A disadvantage of using IR is that the badges have to be worn by the persons to be localized.

Variations of the active badge scheme have been proposed for other applications. For example, Harter and Hopper have suggested the use of an "equipment tag" [20] similar to an active badge for the location estimation of equipment and devices for context-aware applications.

8.4.2 Active Office

The active office localization technique is a range-based technique, which uses central processing and provides fine-grained localization. It was proposed by Ward, Jones, and Hopper [11] to determine positions of devices for context-aware applications. The scheme uses multisignal TDoA to estimate distances. Each device to be localized carries a wireless transceiver unit consisting of a microprocessor, a 418-MHz radio transceiver, and an array of ultrasonic transducers. Each of these units has a unique 16-bit address. On the ceiling of the room in which positions of the devices are to be determined, a matrix of ultrasonic receivers is mounted. These receivers are also connected to a central computer through a serial link as shown in Figure 8.4. The central computer also controls an RF transmitter unit, which broadcasts a radio message consisting of a preamble and a 16-bit address every 200 milliseconds. The central computer chooses the device whose 16-bit address is to be sent in the next message. At the time a radio message is broadcast to the devices to be localized, a reset signal is also sent to the ultrasonic receivers mounted on the ceiling through the serial link.

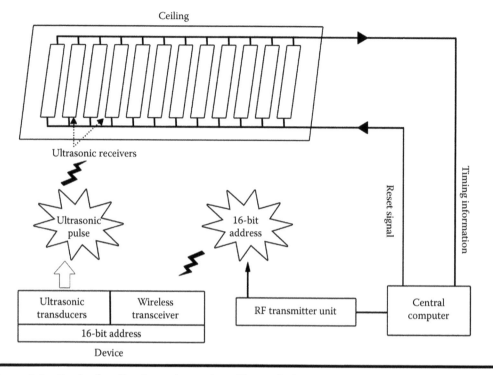

Figure 8.4   Device localization using active office technique.

All the devices receive the broadcasted radio message. The device whose address matches the address in the message transmits a short ultrasonic pulse, which is received by the ultrasonic receivers mounted on the ceiling. The receivers measure time of flight of the ultrasonic pulse using the difference between the time when it received the reset signal and the time when it received the ultrasonic pulse. The timing information is then used to estimate distance of the device from the ultrasonic receivers. As positions of receivers are already known, distance information from these receivers can be used to solve a multilateration problem to localize the device.

A very good accuracy of at least 95% of averaged readings within 8 cm of true position has been reported. Authors claim that if enough transducers are mounted on the device, the accuracy is high enough to provide information relating to the orientation of the device as well.

8.4.3 Cricket System

In the active badge and active office location estimation systems described in Sections 8.4.1 and 8.4.2, location is calculated by a central computer. This information is not communicated to the badge or device itself, and it is supposed to be used by the central computer for context-aware applications. In the Cricket location support system [17] proposed by Priyantha, Chakraborty, and Balakrishnan, devices calculate their own positions instead of a central computer. The design goals of Cricket include user privacy, decentralized administration, network heterogeneity, low cost, and fine-grained localization. Similar to the active office system, Cricket uses multisignal TDoA with a combination of radio and ultrasound signals for distance estimation and thereby position determination. The reported location granularity achieved by the Cricket system is 4×4 ft^2.

Rather than mounting it on the devices to be localized as in the active office scheme, the Cricket system places an RF transmitter and an ultrasonic transducer in the ceiling of the room so as to serve as beacons. The device to be localized serves as the listener and performs local processing to determine its own position. Multiple beacons are placed in the ceiling of the room of interest. Each beacon transmits an RF signal and an ultrasonic signal concurrently. The device receives the RF signal first and then waits for the ultrasonic pulse to arrive. Time difference between the arrivals of the RF and ultrasonic signals is used to estimate the distance from the beacon node. Distance and position information from three or more beacon nodes enables the device to determine its own position by solving a multilateration problem.

In the Cricket localization scheme, the nodes are not time synchronized. As a result, RF transmissions from different beacon nodes may collide. Furthermore, a listening device may correlate an RF signal from one beacon node with an ultrasonic pulse from another beacon node. This problem can be solved by using a carrier sense–style channel access protocol. However, it increases cost and energy consumption. An alternative is to use a fixed or deterministic transmission schedule for the beacon nodes. However, this is also prone to problems due to clock drifts. Instead of this, the Cricket system uses randomization to alleviate the problem of beacon collisions. It chooses the beacon transmission times in a random manner with uniform distribution within a given interval. As a result, broadcasts from different beacons are independent so as to avoid repeated synchronization and, at the same time, prevent persistent collisions.

8.4.4 Bulusu's Algorithm

Unlike the single-hop localization schemes discussed in Sections 8.4.1 through 8.4.3, which were basically designed for context-aware applications, Bulusu's algorithm was designed for the localization of nodes in sensor networks. This single-hop, range-free localization algorithm proposed by Bulusu, Heidemann, and Estrin [9] assumes an idealized radio model in which all sensor nodes possess the same transmission power so that their transmission ranges are identical. It is further assumed that all sensor nodes transmit in an idealized spherical fashion. Each node transmits periodic beacon signals every T seconds containing its position information. The neighbor beacon nodes are synchronized in time such that their beacon signals do not overlap in time and, in any time interval T, each beacon node transmits exactly one beacon signal. However, beacon nodes have an overlapping region of transmission.

Each dumb node j keeps a count of the number of beacon signals received from a particular beacon node i in some fixed time interval t. Knowing the time period T after which a beacon signal is transmitted by the beacon node i, the dumb node j can also compute the total number of beacon signals transmitted by the beacon node i in time interval t. Using both these parameters, the dumb node j can compute a connectivity metric for a particular beacon node. The connectivity metric is given by the percentage of beacon signals received by the dumb node j, which were transmitted by the beacon node i. If

$N_s(i, t)$ = Number of beacons transmitted by beacon node B_i in time t. Subscript s in $N_s(i, t)$ refers to *sender* node.

$N_r(i, t)$ = Number of beacons transmitted by beacon node B_i and received by dumb node D_j in time t. Subscript r in $N_r(i, t)$ refers to *receiver* node.

CM_{ij} = Connectivity metric for beacon node B_i at dumb node D_j

Then

$$CM_{ij} = \frac{N_r(i, t)}{N_s(i,t)} \times 100$$

Higher the value of CM_{ij} greater the number of B_i beacon signals received by the dumb node D_j and smaller the distance between B_i and D_j. The dumb node D_j calculates this connectivity metric for all the beacon nodes in the set S comprising all beacons nodes from which it receives beacons. From this set S of beacon nodes, it selects a subset N of those neighbor beacon nodes for which the connectivity metric exceeds a certain threshold, say, 90%. The dumb node D_j then localizes itself by determining the centroid of the selected beacon nodes. If the number of beacon nodes selected in subset N is k and their positions are given by (x_{i1}, y_{i1}), (x_{i2}, y_{i2}), … (x_{ik}, y_{ik}), then the estimated location of dumb node D_j is given as follows:

$$(x_j, y_j) = \left(\frac{x_{i1} + \ldots + x_{ik}}{k}, \frac{y_{i1} + \ldots + y_{ik}}{k} \right)$$

According to experimental results obtained by Bulusu, Heidemann, and Estrin [9], the localization error falls within 30% of the separation distance between two adjacent beacon nodes. One major advantage of Bulusu's algorithm and all other proximity-based and range-free algorithms is that no additional hardware is required to measure timing information so as to calculate the distance between a beacon node and a dumb node. Without this additional hardware, size of the node remains tiny and energy, which is a scarce resource in sensor networks, is also conserved.

The disadvantage of Bulusu's algorithm is that it requires a high density of beacon nodes in the sensor field so that each dumb node has at least three neighbor beacon nodes. Furthermore, the algorithm works only for single-hop sensor networks and is not suitable for sensor networks, which are multihop in nature.

8.4.5 Approximate Point In Triangulation

The APIT is an area-based range-free localization scheme proposed by He, Huang, Blum, Stankovic, and Abdelzaher [21]. The authors of this scheme proclaim that it performs best in a sensor network with an irregular radio pattern and random node placement and has a low communication overhead. Rather than context-aware applications, APIT algorithm is designed for sensor networks. The APIT uses beacon nodes and RSSI information from neighbor nodes of a dumb node to solve the problem of localization. It employs distributed processing, and each dumb node determines its position by locally processing the available information.

At first, three beacon nodes at a single hop from the dumb node are chosen and the dumb node determines whether it is inside or outside the triangle formed by the three beacon nodes. The decision whether the dumb node is inside or outside the triangle is made by using the RSSI information from neighbor nodes of the dumb node. Next, a different set of beacon nodes is selected and again the dumb node determines whether it is inside or outside the triangle formed by the new set of beacon nodes. After repeated iterations of this process, the dumb node is able to estimate its position by narrowing down the estimate of the area of intersection of the triangles for which it was found to be inside and excluding the area of triangles for which it was found to be outside. The process of repeated selection of beacon nodes and triangle test is carried out until all the audible beacon nodes are exhausted or the required level of accuracy of location information is achieved.

The decision whether a node is inside or outside a triangle is based on, as the authors of the scheme call it, the point-in-triangulation (PIT) test. The PIT test is further categorized as perfect PIT test and APIT test. According to the perfect PIT test theory, if a node M is inside a triangle, when M is moved in any direction it must be nearer to or farther from at least one vertex of the triangle. Similarly, if the node M is outside the triangle, when M is moved there must be a direction in which the new position of M is closer to or farther from all the three vertices of the triangle. The concept is illustrated in Figure 8.5.

In a sensor network comprising static sensor nodes, it is not possible to move the nodes. Therefore, an approximation of the perfect PIT test called APIT is used instead. In the APIT test, a dumb node uses information from other neighbor dumb nodes to determine whether it is inside or outside a triangle. If no neighbor of a node M is farther from or closer to all the three vertices of a triangle simultaneously when compared to node M, then node M assumes that it is inside the triangle; otherwise, it assumes that it is located outside the triangle, as illustrated in Figure 8.6.

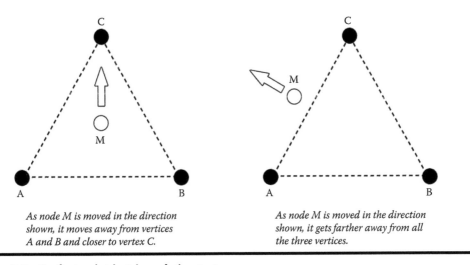

As node M is moved in the direction shown, it moves away from vertices A and B and closer to vertex C.

As node M is moved in the direction shown, it gets farther away from all the three vertices.

Figure 8.5   Perfect point-in-triangulation test.

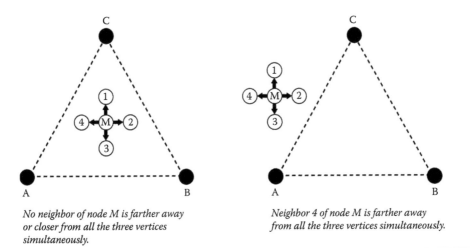

No neighbor of node M is farther away or closer from all the three vertices simultaneously.

Neighbor 4 of node M is farther away from all the three vertices simultaneously.

Figure 8.6   Approximate point-in-triangulation test.

A dumb node can decide whether a neighbor node is farther from or closer to all the three vertices of a triangle than itself by comparing the RSSI information from the neighbor node with that of its own. A higher RSSI value of a neighbor dumb node than that of the host dumb node received from the same beacon node indicates that the neighbor node is closer to the beacon node than the host node.

The assumption that the distance of a dumb node from a beacon node is proportional to the RSSI value is a source of error in APIT because RSSI values are also affected by factors such as non-line-of-sight conditions, reflection, and scattering. As a result, a dumb node that is otherwise closer to a beacon node may have a smaller RSSI value due to obstacles. Similarly, RSSI values may also vary due to burst disturbance effects. However, error due to it can be minimized by taking a running average of RSSI values over time.

There are other possible sources of error in APIT as well. For example, consider a dumb node M, which lies inside a triangle closer to one of its edges. If a neighbor N of the dumb node M lies outside the triangle near the same edge at a greater distance, than it is farther away from all the three vertices (beacon nodes) of the triangle and, hence, node M can assume that it, too, is outside the triangle.

The severity of aforementioned errors can be reduced by making more PIT tests, which is possible only if there is high density of beacon and sensor nodes in the sensor network. Therefore, APIT performs better in dense sensor networks.

8.5 Multihop Localization Schemes

In practical applications of wireless sensor networks, majority of dumb nodes are located at a multihop distance from the beacon nodes. The fraction of dumb nodes located at single-hop or multihop distance from a particular number of beacon nodes depend on a number of factors such as density of beacon nodes, size of the sensor field, nature of the application for which sensors are deployed, and topology of the sensor network. The single-hop localization schemes discussed in Sections 8.4.1 through 8.4.5 are not applicable under these circumstances. Using the same principles and techniques of distance estimation and localization as discussed in the aforementioned sections, a number of localization algorithms for multihop wireless sensor networks have been developed. Some of these algorithms are only academic in nature, whereas some are more suited to practical applications. In Sections 8.5.1 through 8.5.5, we describe a few of the representative localization algorithms for wireless sensor networks.

8.5.1 Ad Hoc Positioning System

One of the earliest localization algorithms for multihop sensor networks that estimates node positions from mere connectivity information and, hence, is range-free is the ad hoc positioning system (APS), which was proposed by Niculescu and Nath [22].

The algorithm uses distributed processing and each dumb node determines its position using multilateration. To perform multilateration, range estimates to at least three beacon nodes are required. However, in a multihop sensor network, it is not possible for each dumb node to have a direct single-hop communication link with three beacon nodes. To resolve this problem, APS proposes to use connectivity information to estimate the range of a beacon node that is at a multihop distance from the dumb node. It suggests three basic techniques to perform range estimation of multihop distant beacon nodes via intermediate nodes:

1. The DV-hop propagation method
2. The DV-distance propagation method
3. Euclidean propagation method

All three techniques use flooding to propagate information in the network similar in nature to the operation of the DV routing protocol. Starting with each anchor, a node propagates information to its immediate first-hop neighbors only. The propagated information depends on which of the three variations of the APS algorithm is being used. This approach is well suited to nodes having limited bandwidth and power. It is to be noted that level and complexity of signaling depends on the total number of beacon nodes and the average degree of each node, that is, number of single-hop neighbors of a node.

Each of the three aforementioned techniques vary in the amount of signaling, power consumption, and the degree of position accuracy achieved and, hence, each technique may be suitable for a certain class of problems.

8.5.1.1 DV–Hop Propagation Method

This is the most basic form of the APS algorithm. Each node in the sensor field maintains a table having entries of the form $\{X_j, Y_j, h_j\}$, where (X_j, Y_j) is the location of a beacon node i and h_i is the number of hops between node maintaining the table and the beacon node i. The number of hops to a beacon node can be counted by incrementing a "count" field in a message at each hop as it is transmitted from a beacon node to its immediate neighbors. This table is also maintained by the beacon node i itself. When the beacon node i has obtained positions and hop counts for all other beacon nodes j, the average size of a hop can be calculated as follows:

$$c_i = \frac{\sum \sqrt{\left(X_i - X_j\right)^2 + \left(Y_i - Y_j\right)^2}}{\sum h_i}$$

for all beacon nodes j where $i \neq j$.

This average size of one hop, c_i, computed by the beacon node i is treated as a correction factor and is propagated throughout the network using controlled flooding as discussed earlier in Section 8.5.1. Knowing the locations of beacon nodes and the correction factor c_i, a dumb node can use multilateration to estimate its own position. The steps involved in the APS DV-hop algorithm can be summarized as follows:

■ The algorithm starts from the beacon nodes, which propagate their position information to their immediate neighbors.
■ All other nodes work along the same technique as the DV algorithm and receive and propagate the position information of beacon nodes. Ultimately, all the nodes have position information of all the beacon nodes and also the number of hops to these beacon nodes.
■ Once a beacon node receives the position information of other beacon nodes and the number of hops to them, it can compute the average size of one hop.

- The average hop size is propagated as correction factor to all other nodes in the network using flooding in a controlled manner.
- All dumb nodes already have number of hops to the beacon nodes. Now, when a dumb node receives the correction factor, it can convert its distance to the beacon node from hops to units of length.
- Finally, multilateration is used to compute the node position.

It should be noted that the correction factor calculated by each beacon node will be different. Therefore, each dumb node will receive different correction factors from different beacon nodes. Niclescu and Bath suggest that a dumb node should use the first correction factor it receives to estimate its position and drop all subsequent correction factors. In general, such a policy ensures that a dumb node uses the correction factor received from its closest beacon node. In addition, if the network is large, a transistor–transistor logic (TTL) field can be set in the packets used for the distribution of correction factors so that amount of signaling and congestion in the network is reduced. This again ensures that the correction factor calculated by a beacon node is used by the dumb nodes in its immediate neighborhood. Limiting the propagation of correction factor by the TTL field also supplements the policy of usage of the first correction factor received.

Distance calculation and position estimation by using the correction factor in DV-hop propagation method is best explained with the help of an example: Consider the segment of a sensor network with three beacon nodes, B_1, B_2, and B_3, and six dumb nodes as shown in Figure 8.7. Each beacon node knows the Euclidean distance to other beacon nodes as given in the figure. Now, the correction factors computed by B_1, B_2, and B_3 are as follows:

$$c_{B1} = \frac{60+150}{2+5} = 30 \text{ m}$$

$$c_{B2} = \frac{60+100}{2+4} = 26.667 \text{ m}$$

$$c_{B3} = \frac{150+100}{5+4} = 27.778 \text{ m}$$

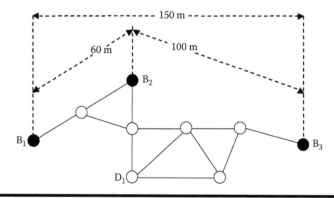

Figure 8.7   Node localization using dv–hop propagation method.

Let us now consider dumb node D_1 and find the distances estimated by it to the three beacon nodes. Node D_1 is at a distance of 3 hop counts from the beacon node B_1, 2 hop counts from B_2, and 3 hop counts from B_3. With 2 hop counts, B_2 is the nearest beacon and, most likely, it will receive the first correction factor, that is, 26.667 m, from this beacon node. As a result, the dumb node D_1 can now estimate distances from the three beacon nodes as follows:

Distance of D_1 from B_1 = 26.667 × 3 = 80 m
Distance of D_1 from B_2 = 26.667 × 2 = 53.334 m
Distance of D_1 from B_3 = 26.667 × 3 = 80 m

The dumb node D_1 already knows the position information of the three beacon nodes. Using the additional information of distances to these beacon nodes, it can now employ trilateration to estimate its own position.

The main advantages of the DV-hop propagation method are simplicity, small amount of processing required, low cost as no additional equipment on sensor nodes is required, low energy cost, and the fact that measurement errors are avoided as it does not use any measurement technique for range estimation. Its major drawback is that it is most suitable only for isotropic networks where hop length is uniform across all segments of the network. The algorithm may not produce desired results for anisotropic networks, where node density and hence hop length in certain segments of the network do not correspond well to the correction factor calculated by the beacon node.

8.5.1.2 DV–Distance Propagation Method

In this variation of the APS algorithm, range between neighbor nodes is estimated using RSSI measurements and this distance information, instead of average size of hop, is propagated in the network. In other words, the DV algorithm now uses cumulative distance instead of hop counts. As a result, a dumb node now knows its distance from the beacon nodes by receiving this distance information through DV exchange instead of computing it by multiplying hop counts with average size of a hop as in the DV-hop algorithm. Knowing the distance information and positions of beacon nodes, a dumb node uses multilateration to estimate its own position.

As the distance estimates are better in this technique, the DV-distance method provides better position information than DV-hop, that is, position information calculated by DV-distance is less coarse. Furthermore, the DV-distance method is suited to anisotropic networks as well, because the variation in the size of hops between various nodes does not affect the actual measured distances between the nodes as is the case with the DV-hop propagation method. However, this method has its own drawbacks, for example, measurement errors can produce incorrect results.

8.5.1.3 Euclidean Propagation Method

The Euclidean propagation method provides better and fine-grained position information compared to DV-hop and DV-distance methods. It works by propagating true Euclidean distance to the beacon nodes. Combining this distance information with the positioning information of beacon nodes, a dumb node is able to estimate its own fine-grained position using multilateration.

For a dumb node D to be able to calculate its Euclidean distance from a beacon node B, it must have at least two neighbors A and C, which have distance estimates for the beacon node B. In Figure 8.8, the neighbors of D, that is, A and C have distance estimates AB and CB. The dumb node D also has estimates for AD, CD, and AC. With this available information, the dumb node

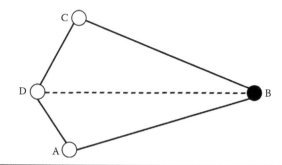

Figure 8.8   Node localization using Euclidean propagation method.

D can calculate the Euclidean distance BD from the beacon node B. In a similar manner, when node D has calculated Euclidean distances to three or more beacon nodes, it can use multilateration to estimate its own position.

This method is also suitable for anisotropic networks. However, it requires a higher ratio of beacon nodes compared to DV-hop and DV-distance methods, which work well with smaller numbers of beacon nodes.

8.5.2 Convex Position Estimation

The convex position estimation (CPE) algorithm was proposed by Doherty, Pister, and Ghaoui [23]. It treats the problem of sensor node localization as an optimization problem and applies linear programming to arrive at the solution for the localization problem.

A convex set is a set of points such that any two points in the set can be connected with a path comprising points contained within the set. The set of all the nodes in a sensor network forms a convex set as a path between any two nodes comprises only nodes that are within the set of the sensor nodes. There is a path P between nodes M and N if either M or N communicates with the other or both M and N communicate with each other. The path comprises nodes M and N and all other intermediate nodes that participate in the communication between them.

A linear program (LP) is a mathematical optimization technique to find out the best possible solution to a problem satisfying a set of linear constraint relationships. For example, an LP problem can be formulated to find the minimum labor needed to complete a given task under certain constraints. Formally, an LP problem can, in general, be represented as follows:

Minimize $\mathbf{c}^T\mathbf{x}$
Subject to $\mathbf{A}\mathbf{x} \le \mathbf{b}$

where $\mathbf{x}$ is the vector of variables to be determined, and $\mathbf{c}$ and $\mathbf{b}$ are vectors of known values. It is noted that $\mathbf{A}$ is a matrix of known values. The expression to be optimized, that is, $\mathbf{c}^T\mathbf{x}$ is called the "object function." The relational expression $\mathbf{A}\mathbf{x} \le \mathbf{b}$ specifies the constraints that should be satisfied by the optimized solution.

The CPE algorithm provides an optimization framework and formulates the problem of position determination of dumb nodes as an LP. For a two-dimensional sensor field, each node's position can be described in rectangular coordinates as (x, y). For the LP problem, a single vector $\mathbf{x}$ with positions of all the nodes is formed as follows:

$$\mathbf{x} = \left[x_1 y_1 \dots x_m y_m, x_{m+1} y_{m+1} \dots x_n y_n \right]$$

The first m entries are known positions of beacon nodes and the remaining $n - m$ entries are determined by the algorithm.

As is clear, locations of a subset of nodes, that is, beacon nodes are known. Furthermore, connectivity information of beacon and dumb nodes is also known, which is modeled as a set of geometric constraints on the node positions. For example, if a node M can transmit in a radius of 10 m and another node N can receive communication from node M, then it means that N lies within a 10-m radius of node M. Similarly, another node P, which does not receive communication from node M, lies outside this 10-m radius around node M. This connectivity information serves as a constraint for the LP problem to be solved for determining node positions. Such connectivity constraints limit the set of estimated positions of dumb nodes. In general, the localization problem is then formulated as follows:

Given: Positions of beacon nodes
Find: Positions of dumb nodes
Subject to: Connectivity constraints of beacon and dumb nodes

Each node in the sensor field maintains a table of connectivity information by exchanging messages with neighbor nodes. The table maintains a list of neighbor nodes from which the node receives messages. A receiver node can receive messages from only that transmitter node whose transmission radius encompasses the receiver node. This provides a natural bound on the distance between the transmitter and receiver nodes.

Each node sends its table of connectivity information, that is, list of neighbor nodes from which it receives messages to a central processing device. The central processor also has the list of all beacon nodes along with their position information. The central processor then solves the LP problem to determine positions of the dumb nodes. Using connectivity constraints and knowledge of positions of beacon nodes, the algorithm narrows down the area and generates a bounding box in which a dumb node may be located. The bounding box is treated as a set of all possible locations where the dumb node may be located. The algorithm then selects the center of the bounding box as the best possible estimate of the node position.

Higher the number of beacon nodes smaller the size of the bounding box, thereby giving a better estimate of the node position. If the number of beacon nodes used in defining the bounding box is small, the "settled nodes"—the dumb nodes whose positions have already been determined—can also be used along with the beacon nodes to find positions of remaining dumb nodes. However, location estimates of the settled nodes can have errors, which may propagate in subsequent computations and result in inaccurate position estimates of dumb nodes. To overcome this problem, authors of the algorithm suggest the use of eight settled nodes to ensure the same certainty of position estimation as given by a single beacon node.

Additional constraints, such as AoA, can also be used along with connectivity information to generate a smaller bounding box; as a result, a better estimate of the node position can be obtained. As the CPE algorithm does not require range estimation between nodes, it is a range-free algorithm. As a result, it does not require any additional hardware on the sensor nodes. However, it is a centralized algorithm and all the nodes must send connectivity information to the central processor. The central processor, after processing and determination of node positions, sends position information back to each node. This entails communication overhead resulting in high energy cost. Security of node position information may also be compromised during processing at the central node or during transmission from central processor to the dumb node. As in all centralized algorithms, the central processor can also be a single point of failure.

Due to these problems, CPE algorithm may not be a viable option of localization in current practical applications of sensor networks. However, it provides additional insight and a method to approach the solution to the localization problem and can serve as a platform on which to develop and optimize better localization algorithms.

8.5.3 Iterative and Collaborative Multilateration

Once a dumb node has estimated its position with the help of beacon nodes, it can be used as a beacon node by other dumb nodes to estimate their positions. This basic scheme of position estimation is the basis of iterative location estimation [24,25]. The algorithm is essentially a two-step process. As proposed by Savarese, Rabaey, and Beutel [24]; and Savarese, Rabaey, and Langendoen [10], all the dumb nodes compute a rough estimate of their positions in the first phase using an algorithm called "hop-terrain." This is referred to as the "start-up phase." In the second phase, called the "refinement phase," the computed position estimates are refined using an iterative process in which the immediate neighbor nodes of a dumb node take part in the process.

A similar algorithm has been proposed by Savvides, Han, and Srivastava [25]. The authors call this algorithm "ad hoc localization system" (AHLoS). The algorithm builds on the idea of "atomic multilateration." In the case where a dumb node is at a distance of one hop from at least three noncollinear beacon nodes, the dumb node can perform multilateration to determine its position. This is termed atomic multilateration. Iterative multilateration uses atomic multilateration as its main primitive. The algorithm supports central, distributed, or cluster-based processing. If the algorithm is deployed using distributed processing, first the dumb nodes with a distance of one hop from at least three beacon nodes estimate their position using atomic multilateration. These dumb nodes, after their positions have been estimated, are immediately available as beacon nodes and the other dumb nodes can utilize the position information of these beacon nodes for their own position estimation using iterative multilateration.

If the algorithm uses centralized processing, the central node has the global knowledge of the sensor network. The network is represented by a graph G on which the algorithm operates. The graph has weighted edges to denote separations between two adjacent nodes. At first, the positions of the dumb nodes that are at a distance of a single hop from at least three beacon nodes are determined using atomic multilateration. With central processing, the algorithm first determines positions of those dumb nodes that have connectivity with maximum number of beacon nodes. This results in better accuracy and faster convergence. As soon as a dumb node has settled, that is, it has estimated its position, it becomes a beacon node so that remaining dumb nodes can use it for multilateration. This iterative procedure repeats until the positions of all the dumb nodes have been estimated.

A drawback of iterative multilateration is the propagation and accumulation of error, which results from the use of settled nodes as beacon nodes. As the position estimates of these nodes are not as precise as those of beacon nodes, multilateration involving these nodes may accumulate such errors.

It is also possible that the algorithm does not converge due to accumulation of error and the accumulated error increases with time if the algorithm keeps running. As a countermeasure, it has been suggested to add confidence weights to position estimates of settled nodes being used as beacon nodes and then solve a modified optimization problem [10]. With this approach, the algorithm converges for almost all cases of location estimation and also produces better position estimates. Furthermore, the convergence problem can be avoided by using better and high-precision ranging techniques.

For iterative multilateration to be possible, at least three beacon or settled nodes must be in the neighborhood of the dumb node for which position is to be estimated. However, this is not always the case. Not all the nodes in the sensor field have three neighbor nodes with known position estimates. Under such a situation, it is not possible to find the position of a dumb node by mere atomic or iterative multilateration. Another technique called "collaborative multilateration" [10,25] may be used for position estimation of such nodes. Collaborative multilateration is a technique in which a dumb node uses multihop information from a neighbor dumb node and solves a joint set of location estimation functions to determine positions of both dumb nodes simultaneously. The solution obtained may or may not be unique. A unique solution to the problem of location estimation using collaborative multilateration is reached only when a "participating node" has at least three participating neighbors. A participating node is a sensor node that is either a beacon node or a dumb node with at least three participating neighbors.

Collaborative multilateration may be used to enhance iterative multilateration. As a result of collaborative multilateration, the number of dumb nodes reduces and the number of settled nodes increases. These settled nodes can further be used in another round of iterative multilateration to determine locations of more nodes and to improve the accuracy of location estimates.

8.5.4 Multidimensional Scaling

Multidimensional analysis is a data analysis technique used primarily in psychometrics and psychophysics. Shang, Ruml, Zhang, and Fromherz [26] applied this technique to propose a range-free localization algorithm, MDS-MAP, which can estimate node positions using mere connectivity information. However, the algorithm can also make use of additional information, for example, position information of beacon nodes or estimated distances between neighbor nodes. With this additional information, the algorithm is able to provide better and accurate results. The algorithm works even if no beacon nodes are available in which case it generates relative coordinates for the dumb nodes.

The underlying theory and concept of multidimensional scaling (MDS) is best explained by an example described in the literature [26]. Consider a small cloud of colored beads suspended in the air. The distances between each pair of beads are measured. Now, if the beads fall down on the ground, the earlier configuration of the beads in the air can be reconstructed by using the distance measurements recorded earlier. Position of each bead in the reconstructed arrangement is chosen such that the distances in the new arrangement match the distances in the original layout. It is this problem of reconstructing the arrangement that is solved by MDS. It should be noted that the reconstructed arrangement is an arbitrarily rotated and flipped version of the original layout because the new layout is constructed without using absolute information.

MDS-MAP executes in three steps:

Step 1: Distance between each possible pair of nodes is estimated using an all-pairs shortest path algorithm, for example, Dijkstra's or Floyd's algorithm.

Step 2: The MDS is used to estimate node positions such that the estimated positions satisfy the distances estimated in step 1. This step generates a relative map of sensor nodes such that the nodes have the same neighbor relationship as the underlying network. Relative map is generated in the absence of any beacon nodes and is well suited to sensor networks in which powerful sensors and expensive infrastructure using beacon nodes cannot be installed. Relative positioning information is sufficient for some applications such as direction-based routing algorithms.

Step 3: Resulting coordinates are normalized by taking into account the known positions of beacon nodes. This step generates absolute position information of dumb nodes.

It is noted that MDS-MAP is a centralized algorithm and is not much suited to practical applications of wireless sensor networks. An improved version of the MDS-based localization algorithm, called MDS-MAP(P), which is short for MDS-MAP using patches of relative maps, has also been proposed [27,28]. Main theme of the improved MDS-MAP(P) is to build local maps of the neighbor nodes at each individual node and merge all the local maps to form a global map. In this way, computation of shortest path distances between faraway nodes is avoided. Furthermore, the local maps of nearby nodes are usually more accurate.

The original MDS-MAP works well with sensor networks in which nodes are uniformly distributed and the shortest path between two nodes correlates with the true Euclidean distance between them. However, in the case of anisotropic and irregular networks, it suffers from performance degradation and the estimated node positions are prone to error. On the other hand, MDS-MAP(P) works well with both uniform and irregular networks. As it uses local maps, a good local map of faraway nodes can be constructed and merged with the global map thereby providing better results for anisotropic networks. Another advantage of MDS-MAP(P) is that it offers a partially distributed processing approach as local maps are constructed at individual nodes. However, a central processor is still required where the global map is constructed and node positions are estimated.

8.5.5 Miscellaneous Algorithms

In addition to the algorithms described in Sections 8.5.1 through 8.5.4, a number of localization algorithms have been proposed such as recursive position estimation (RPE) [29] and directed position estimation (DPE) [2]. The RPE, as the name suggests, utilizes recursive steps to estimate and refine position estimation. Although its communication cost is low, it needs more beacon nodes. The DPE algorithm is similar to RPE, but reduces the required number of beacon nodes using directed recursion.

Examples of some recent localization algorithms are the ones proposed by Chang, Hung, Lin, and Li [30]; Wang, Wu, and Shu [31]; and Rencheng, Lisha, Teng, and Liding [32].

Wang, Wu, and Shu have proposed a decentralized positioning system that utilizes a particle filter for more accurate measurement of RSS and then uses least-mean-squares (LMS) algorithm to find the position of a node [31]. It is to be noted that accuracy of location estimation depends on how accurately RSS is measured and on factors such as reflection, scattering, and other similar phenomena that affect the recorded measurement. Therefore, the algorithm suggests the use of a particle filter to improve RSS measurement with subsequent application of the LMS algorithm to estimate the path loss factor of the distance-dependent path loss model. The algorithm is a range-based decentralized algorithm and hence requires less transmission energy when compared with a centralized approach.

Another low-cost strategy for localizing wireless sensor network nodes is the bounding-box algorithm [33]. Whereas DV-hop and range-based algorithms require static anchor, the bounding-box algorithm does not require any fixed anchor and has low hardware requirements. This algorithm uses a mobile beacon node that is always aware of its position and broadcasts its current location while moving in the sensor field. The node receiving this beacon signal can then know the bounding-box region where it is inside and thus find its own position. A problem with the bounding-box algorithm is that it is unable to locate relative positions of neighbor nodes. Due to this problem, the bounding-box algorithm is not suitable for certain applications such as location-aware routing. Chang, Hung, Lin, and Li [30] have suggested a mechanism, which they call "distinguishing relative locations" or simply DRL, to remedy this problem. The DRL uses mobile beacon nodes that transmit beacons to define the bounding-box region and tone signals to help distinguish relative locations of any two neighbor nodes. Unlike beacon signals, tone signals do not carry any position information and hence they consume less amounts of energy.

Another recent algorithm [32] combines two approaches, that is, iterative and collaborative multilateration approaches, using maximum likelihood estimation (MLE) [34] and MAX-MIN [35] localization algorithms. For small errors in distance measurement, the MLE algorithm provides location estimation with reduced error. However, a drawback of the MLE algorithm is that it requires a lot of floating-point computations.

The MAX-MIN algorithm is based on the bounding-box algorithm. After calculating the distance between the blind node and the beacon nodes, the MAX-MIN algorithm builds bounding boxes. The beacon nodes are taken as reference and circles are drawn with a reference node taken as the center. Radius of a circle is equal to the distance between the reference node and the blind node. Further, squares are circumscribed against each circle. The center of region of intersection of these squares is designated as the estimated location of the dumb node. We need at least three beacon nodes to locate a dumb node. The MAX-MIN algorithm does not need complex computations, and only simple additions and subtractions are required to estimate the position of a node. However, unlike MLE, the position accuracy of MAX-MIN is good enough for large errors in distance. If the distance error is below 7%, MLE estimates the position of the blind node more accurately than the MAX-MIN algorithm. However, as the distance error increases beyond 7%, the MAX-MIN algorithm provides far more superior results than MLE.

The hybrid algorithm proposed by Rencheng, Lisha, Teng, and Liding combines the two algorithms to produce an optimum result. The hybrid algorithm first estimates the location of a dumb node by using both MLE and MAX-MIN. Let (x_1, y_1) be the coordinates of a dumb node estimated using MLE and let (x_2, y_2) be the coordinates calculated using MAX-MIN. Then, the hybrid algorithm is based on choosing parameters α and β so as to calculate the revised location with reduced error using the following formula:

$$\begin{bmatrix} x \\ y \end{bmatrix} = \alpha \begin{bmatrix} x_1 \\ y_1 \end{bmatrix} + \beta \begin{bmatrix} x_2 \\ y_2 \end{bmatrix}$$

The aforementioned formula combines the results of both MLE and MAX-MIN to produce a location estimation that is better than results produced by either of the algorithms used alone.

8.6 Future Research Directions

Wireless sensor networks form a new and emerging area of research. Practical applications of wireless sensor networks are still being conceived and developed. Along the way, new techniques and algorithms are being proposed and developed for various layers in sensor networks. So is the case with node localization. As discussed in Sections 8.1 and 8.2, majority applications of wireless sensor networks need to know the position information from where data was collected. This is possible only if the sensor node collecting and transmitting the data knows its position by using a localization algorithm.

Majority of proposed localization algorithms are generic in nature and do not take any particular application of sensor networks into consideration. However, it is possible that a single localization algorithm may not be suited to the entire spectrum of wireless sensor network applications. For example, if an algorithm is suitable for the location awareness of sensor nodes in a body sensor network that monitors the physiological activity of a living being, it is likely that this algorithm is not suitable for a sensor network that is used for the surveillance of a certain area, such as a battlefield. Therefore, work needs to be done to determine the suitability of proposed localization algorithms for various applications, and if no current algorithm is suited to a particular problem new algorithms might need to be developed and tested for it.

All the localization algorithms depend on some kind of measurement, such as RSSI or timing, which is made by the underlying sensor hardware. However, these measurements are prone to error due to practical limitations of the hardware and may result in poor localization accuracy. This problem can be alleviated by careful calibration of sensor nodes or by making the algorithm robust against measurement errors by using techniques to detect and either reject or correct these errors. Therefore, work is needed to determine the frequency of measurement errors, their impact on localization, and how sensor nodes are to be calibrated so as to alleviate this problem. Similarly, new techniques need to be developed so that a localization algorithm is able to detect an improbable measurement. This can be accomplished, for example, by using consistency checks, such as symmetry and geometric constraints. In a similar manner, statistical models may be used to filter measurement errors.

Most of the current work in the area of node localization focuses on static sensor nodes as is the case with wireless sensor network applications. However, future applications will use mobile nodes as well. A mobile node will be able to move to the area that needs sensing coverage. The localization algorithm should be able to detect this movement and determine the new position of the node. Similarly, a node may move more frequently, for example, to collect data from or transmit information to other nodes in the sensor field. The localization algorithm, in this case, should be able to determine the location of the moving node in real time. Major work needs to be done to resolve issues in the localization of such mobile nodes.

Some current and future applications involve a dynamically changing topology of wireless sensor networks with sensor nodes being added to or removed from the sensor field. The localization algorithm should be scalable to either a very small or a very large number of sensor nodes and should provide the desired level of accuracy in both these extreme cases. Not many current localization algorithms take the scalability factor into consideration. So, additional work is needed so that the localization algorithms are scalable and work well with hundreds and thousands of sensor nodes.

Majority of node localization algorithms use a set of beacon nodes whose position information is known through a GPS device or a similar device or by positioning the nodes at locations with known coordinates. The beacon nodes know their positions throughout the period of their deployment. Work needs to be done to determine the minimum and optimum number of beacon nodes that will result in the desired accuracy of a localization algorithm. Similarly, some proposed localization schemes, such as localization using particle dynamics [36], use a large number of beacon nodes. By employing the principles used by these algorithms, new and efficient techniques that use fewer number of beacon nodes can be built.

Security of sensor nodes is an important issue in wireless sensor networks. Various types of attacks can be mounted against a sensor node at various layers. For example, a sensor node may be forced to provide wrong information by feeding it wrong input through a malicious sensor node planted by an adversary. Similarly, a dumb node may estimate wrong position for itself if an attacker is able to take control of a beacon node or is able to plant malicious beacon nodes in the sensor field. The compromised or planted malicious beacon node sends beacons signals with wrong position information to the dumb node. If the dumb node uses multilateration with this wrong position information as the input, the calculated position for the dumb node will be in error and, as a result, the data collected by this node will be ascribed to a wrong location by the sensor field. The problem can be alleviated if the nodes in the sensor field have a reliable mechanism to trust and authenticate each other. This issue and similar security issues need to be addressed in localization algorithms.

Real-world applications of sensor networks are deployed in a three-dimensional space. However, current work on localization algorithms focuses on two-dimensional planar space. Therefore, the current two-dimensional localization algorithms need to be extended to three-dimensional space. Further experiments and simulations need to be carried out to test and verify their suitability for deployment in actual three-dimensional space.

Error accumulation and propagation is a severe problem in some range-free and range-based localization algorithms, such as iterative and collaborative multilateration approaches. A localization algorithm may not converge due to error propagation or it may result in unacceptable errors in location estimation. New techniques need to be developed to limit the accumulation and propagation of errors in localization so that the accuracy of localization can be increased.

8.7 Conclusion

In this chapter, we define the problem of localization of nodes in wireless sensor networks, describe the desired characteristics of a localization algorithm, and discuss how localization algorithms can be classified. We also discuss a few of the representative localization algorithms. None of the current localization algorithms is suitable for the entire class of applications of wireless sensor networks. For example, a localization algorithm that is suitable for static sensor nodes may not work well with mobile sensor nodes. It should be noted that these algorithms describe only the basic principles and techniques that may be used for localization of sensor nodes. A complete localization application and framework for a practical wireless sensor network can be built using a combination of these techniques.

Each of these localization algorithms provides a different level of accuracy of estimated positions and is directly related to the number of beacon nodes deployed in the sensor field. A better level of position accuracy can be achieved if a higher number of beacon nodes are deployed. Range-based localization schemes provide fine-grained position estimates. However, these techniques usually require extra pieces of hardware to be added to sensor nodes for measurement of various parameters, which adds to size, weight, and energy costs. If size of integrated circuits and hardware keeps getting smaller following Moore's law, range-based schemes might be able to use extra hardware without much additional cost and, therefore, may hold promise in future sensor networks. Range-free schemes, on the other hand, approximate the position of a dumb node by using mere connectivity and proximity information. Therefore, these schemes do not need extra hardware, thereby saving cost in terms of weight, size, and energy. However, location estimates provided by range-free schemes are coarse-grained estimates.

Similarly, some localization algorithms, such as MDS and MDS-MAP, may estimate only relative positions of nodes, and other algorithms, such as APIT and APS, may localize using absolute coordinates. Localization algorithms that provide only relative positions do not need beacon nodes. However, if beacon nodes are added to the network, these algorithms will determine absolute positions as well.

Algorithms with a centralized approach may have large communication overheads and are not suited to the ad hoc nature of wireless sensor networks. However, the accuracy of estimated positions is better with such algorithms because the central node has global knowledge of the sensor network. On the other hand, localization algorithms using distributed processing are easily scalable to small or large number of nodes in the sensor network.

Localization algorithms employing iterative and collaborative multilateration techniques hold promise as a large number of dumb nodes can be localized using fewer beacon nodes. If errors are detected and are prevented from being propagated, a good level of accuracy can also be achieved.

Acknowledgment

We thank Mansoor Sheraz, software developer at Information Technology Center, COMSATS Institute of Information Technology, Lahore, Pakistan, for his help in drawing the figures of this chapter.

References

1. Akyildiz, I. F., W. Su, Y. Sankarasubramaniam, and E. Cayirci. August 2002. "A Survey on Sensor Networks." *IEEE Communications Magazine* 40 (8): 102–14. IEEE Communications Society.
2. De Oliveira, H. A. B. F., F. E. Nakamura, A. A. Loureiro, and A. Boukerche. October 2005. "Directed Position Estimation: A Recursive Localization Approach for Wireless Sensor Networks." In *Proceedings of the 14th International Conference on Computer Communications and Networks, ICCCN '05*, 557–62. Piscataway, NY: IEEE.
3. Amundson, I., and X. Koutsoukos. 2009. "A Survey on Localization for Mobile Wireless Sensor Networks." In *Proceedings of the 2nd International Conference on Mobile Entity Localization and Tracking in GPS-Less Environments, MELT '09*, 235–54. Berlin, Heildelberg: Springer-Verlag.
4. Mao, G., and B. Fidan, ed. 2009. *Localization Algorithms and Strategies for Wireless Sensor Networks*. New York: Information Science Reference, IGI Global.
5. Pal, A. 2010. "Localization Algorithms in Wireless Sensor Networks: Current Approaches and Future Challenges." *Network Protocols and Algorithms* 2 (1): 45–74. Macrothink Institute.
6. Wang, L., and Q. Xu. June 2010. "GPS-Free Localization Algorithm for Wireless Sensor Networks." *Sensors* 10 (6): 5899–926.
7. Zhang, Q., C. H. Foh, B. C. Seet, and A. C. M. Fong. May 2010. "Location Estimation in Wireless Sensor Networks Using Spring-Relaxation Technique." *Sensors* 10 (5): 5171–92.
8. Kim, E., and K. Kim. June 2010. "Distance Estimation With Weighted Least Squares for Mobile Beacon-Based Localization in Wireless Sensor Networks." *IEEE Signal Processing Letters*, 7 (6): 559–62. IEEE Signal Processing Society.
9. Bulusu, N., J. Heidemann, and D. Estrin. October 2000. "GPS-Less Low-Cost Outdoor Localization for Very Small Devices." *IEEE Personal Communications* 7 (5): 28–34. IEEE Communications Society.
10. Savarese, C., J. M. Rabaey, and K. Langendoen. 2002. "Robust Positioning Algorithms for Distributed Ad-Hoc Wireless Sensor Networks." In *Proceedings of the General Track: 2002 USENIX Annual Technical Conference*, 317–27. Berkeley, CA: USENIX Association.
11. Ward, A., A. Jones, and A. Hopper. October 1997. "A New Location Technique for the Active Office." *IEEE Personal Communications* 4 (5): 42–47.
12. Bulusu, N., D. Estrin, L. Girod, and J. Heidemann. July 2001. "Scalable Coordination for Wireless Sensor Networks: Self-Configuring Localization Systems." In *Proceedings of the Sixth International Symposium on Communication Theory and Applications*. Ambleside, Lake District, United Kingdom.
13. Whitehouse, K., and D. Culler. September 2002. "Calibration as Parameter Estimation in Sensor Networks." In *Proceedings of the 1st ACM International Workshop on Sensor Networks and Applications, WSNA '02*, Atlanta, GA.
14. De Oliveira, H. A. B. F., E. F. Nakamura, A. A. F. Loureiro, and A. Boukerche. May 2007. "Localization in Time and Space for Sensor Networks." In *Proceedings of 21st International Conference on Advanced Information Networking and Applications Workshops, AINAW'07*, Niagra Falls, ON, 539–46. IEEE Computer Society.
15. Akyildiz, I. F., and M. C. Vuran. 2010. *Wireless Sensor Networks*. Chichester: John Wiley & Sons.
16. Priyantha, N. B. 2005. *The Cricket Indoor Location System*. Doctoral Dissertation, Massachusetts Institute of Technology, Cambridge, MA.
17. Priyantha, N. B., A. Chakraborty, and H. Balakrishnan. 2000. "The Cricket Location-Support System." In *Proceedings of the 6th Annual International Conference on Mobile Computing and Networking, Mobicom '00*, 32–43. NY: ACM.

18. Priyantha, N. B., A. K. Miu, H. Balakrishnan, and S. Teller. July 2001. "The Cricket Compass for Context-Aware Mobile Applications." In *Proceedings of the 7th Annual International Conference on Mobile Computing and Networking, Mobicom '01*, 1–14. NY: ACM.
19. Want, R., A. Hopper, V. Falcao, and J. Gibbons. January 1992. "The Active Badge Location System." *ACM Transactions on Information Systems* 10 (1): 91–102.
20. Harter, A., and A. Hopper. January 1994. "A Distributed Location System for the Active Office." *IEEE Network* 8 (1): 62–70.
21. He, T., C. Huang, B. Blum, J. A. Stankovic, and T. Abdelzaher. 2003. "Range-Free Localization Schemes for Large Scale Sensor Networks." In *Proceedings of the 9th Annual International Conference on Mobile Computing and Networking, Mobicom '03*, 81–95. NY: ACM.
22. Niculescu, D., and B. Nath. 2001. "Ad Hoc Positioning System." In *Proceedings of the IEEE Global Telecommunications Conference, GLOBECOM '01*, vol. 5, 2926–31. IEEE.
23. Doherty, L., K. S. J. Pister, and L. E. Ghaoui. April 2001. "Convex Position Estimation in Wireless Sensor Networks." In *Proceedings of Twentieth Annual Joint Conference of the IEEE Computer and Communications Societies, INFOCOM 2001*, Anchorage, AK, vol. 3, 1655–63. IEEE Computer Society.
24. Savarese, C., J. M. Rabaey, and J. Beutel. May 2001. "Locationing in Distributed Ad-Hoc Wireless Sensor Networks." In *Proceedings of the International Conference on Acoustics, Speech and Signal Processing, ICASSP 2001*, 2037–40. Salt Lake City, UT.
25. Savvides, A., C. C. Han, and M. B. Srivastava. July 2001. "Dynamic Fine-Grained Localization in Ad-Hoc Networks of Sensors." In *Proceedings of the 7th Annual International Conference on Mobile Computing and Networking, Mobicom '01*, 166–79. NY: ACM.
26. Shang, Y., W. Ruml, Y. Zhang, and M. P. J. Fromherz. June 2003. "Localization from Mere Connectivity." In *Proceedings of the 4th ACM International Symposium on Mobile Ad Hoc Networking & Computing, MobiHoc '03*, Annapolis, MD, 201–12. ACM Press.
27. Shang, Y., and W. Ruml. March 2004. "Improved MDS-Based Localization." In *Proceedings of Twenty Third Annual Joint Conference of the IEEE Computer and Communications Societies, IEEE INFOCOM 2004*, vol. 4, 2640–51.
28. Shang, Y., W. Ruml, Y. Zhang, and M. Fromherz. October 2004. "Localization from Connectivity in Sensor Networks." *IEEE Transactions on Parallel and Distributed Systems* 15 (11): 961–74.
29. Albowicz, J., A. Chen, and L. Zhang. 2001. "Recursive Position Estimation in Sensor Networks." In *Proceedings of The 9th International Conference on Network Protocols, ICNP 2001*, 35–41. IEEE Computer Society.
30. Chang, C. Y., L. L. Hung, C. Y. Lin, and M. H. Li. May 2010. "On Distinguishing Relative Locations with Busy Tones for Wireless Sensor Networks." In *Proceedings of The 2010 IEEE International Conference on Communication, IEEE ICC*, 1–5. IEEE.
31. Wang, C. L., D. S. Wu, and F. F. Shu. April 2010. "Design and Implementation of Decentralized Positioning System for Wireless Sensor Networks." In *Proceedings of The 2010 IEEE Wireless Communications and Networking Conference, IEEE WCNC*, 1–6. IEEE.
32. Rencheng, J., M. Lisha, G. Teng, and W. Liding. March 2010. "A New Hybrid Localization Technology of Wireless Sensor Networks." In *Proceedings of The 2010 International Conference on Measuring Technology and Mechatronics Automation, ICMTMA '10*, 185–88. IEEE.
33. Galstyan, A., B. Krishnamachari, K. Lerman, and S. Pattem. April 2004. "Distributed Online Localization in Sensor Networks Using a Moving Target." In *Proceedings of Third International Symposium on Information Processing in Sensor Networks, IPSN '04*, Berkeley, CA, 61–70.
34. Savvides, A., H. Park, and M. B. Srivastava. September 2002. "The Bits and Flops of the N-hop Multilateration Primitive For Node Localization Problems." In *Proceedings of the 1st ACM International Workshop on Wireless Sensor Networks and Applications, WSNA '02*, Atlanta, GA, 112–21. ACM Press.
35. Simic, S. N., and S. Sastry. 2002. "Distributed Localization in Wireless Ad Hoc Networks." Memorandum No. UCB/ERL M02/26, University of California, Berkeley, CA.
36. Zhao, W., D. Liu, and Y. Jiang. 2006. "Positioning Algorithm of Wireless Sensor Network Nodes." In *Proceedings of the 2006 International Conference on Intelligent Information Hiding and Multimedia Signal Processing, IIH-MSP'06*, Pasadena, CA, 271–73. IEEE Computer Society.

Additional Reading

1. Akyildiz, I. F., and X. Wang. September 2005. "A Survey on Wireless Mesh Networks." *IEEE Communications Magazine*, 43 (9): S23–30. IEEE Communications Society.

2. Al-Karaki, J. N., and A. E. Kamal. December 2004. "Routing Techniques in Wireless Sensor Networks: A Survey." *IEEE Wireless Communications* 11 (6): 6–28. IEEE Communications Society.

3. Anjum, F., S. Pandey, and P. Agrawal. November 2005. "Secure Localization in Sensor Networks Using Transmission Range Variation." In *Proceedings of IEEE International Conference on Mobile Adhoc and Sensor Systems, MASS'05*, Washington DC, 195–203.

4. Aspnes, J., T. Eren, D. K. Goldenberg, A. S. Morse, W. Whiteley, Y. R. Yang, B. D. O. Anderson, and P. N. Belhumeur. December 2006. "A Theory of Network Localization." *IEEE Transactions on Mobile Computing* 5 (12): 1663–78. IEEE Computer Society.

5. Battelli, M., and S. Basagni. 2007. "Localization for Wireless Sensor Networks: Protocols and Perspectives." In *Proceedings of Canadian Conference on Electrical and Computer Engineering 2007*, Vancouver, BC, 1074–77.

6. Blumenthal, J., F. Reichenbach, and D. Timmermann. 2005. "Position Estimation in Ad hoc Wireless Sensor Networks with Low Complexity." In *Proceedings of Joint 2nd Workshop on Positioning, Navigation and Communication 2005 & 1st Ultra-Wideband Expert Talk 2005*, 41–49. Germany, WPCN.

7. Bulusu, N., J. Heidmann, D. Estrin, and T. Tran. February 2004. "Self-Configuring Localization Systems: Design and Experimental Evaluation." *ACM Transactions on Embedded Computing Systems (TECS)* 3 (1): 24–60.

8. Bulusu, N., and S. Jha, ed. 2005. *Wireless Sensor Networks: A Systems Perspective*. Norwood, MA: Artech House.

9. Carle, J., and D. Simplot-Ryl. February 2004. "Energy-Efficient Area Monitoring for Sensor Networks." *Computer* 37 (2): 40–46. IEEE Computer Society.

10. Chaczko, Z., R. Klempous, J. Nikodem, and M. Nikodem. March 2007. "Methods of Sensors Localization in Wireless Sensor Networks." In *Proceedings of 14th Annual IEEE International Conference and Workshops on the Engineering of Computer-Based Systems*, Tucson, AZ, 145–52.

11. Chang, C., and A. Sahai. October 2004. "Estimation Bounds for Localization." In *Proceedings of First Annual IEEE Communications Society Conference on Sensor and Ad Hoc Communications and Networks 2004, IEEE SECON 2004*, 415–24.

12. Chen, Y., and H. Kobayashi. April 2002. "Signal Strength Based Indoor Geolocation." In *Proceedings of IEEE International Conference on Communications 2002, ICC 2002*, 436–39.

13. Chintalapudi, K. K., A. Dhariwal, R. Govindan, and G. Sukhatme. March 2004. "Ad-Hoc Localization Using Ranging and Sectoring." In *Proceedings of Twenty-Third Annual Joint Conference of the IEEE Computer and Communications Societies, INFOCOM 2004*, Hong Kong, vol. 4, 2662–72.

14. Dargie, W., and C. Poellabauer. 2010. *Fundamentals of Wireless Sensor Networks: Theory and Practice*. West Sussex, UK: John Wiley & Sons, Ltd.

15. Effen, M. C., W. A. Moreno, M. A. Labrador, and K. P. Valavanis. October 2006. "Adapting Sequential Monte-Carlo Estimation to Cooperative Localization in Wireless Sensor Networks." In *Proceedings of 2006 IEEE International Conference on Mobile Adhoc and Sensor Systems*, MASS, 656–61. Vancouver, BC.

16. Gezici, S., Z. Tian, G. B. Giannakis, H. Kobayashi, A. F. Molisch, H. V. Poor, and Z. Sahinoglu. July 2005. "Localization Via Ultra-Wideband Radios: A look at positioning aspects for future sensor networks." *IEEE Signal Processing Magazine* 22 (4): 70–84. IEEE Signal Processing Society.

17. Heidari, M., F. O. Akgul, N. A. Alsindi, and K. Pahlavan. November 2007. "Neural Network Assisted Identification of the Absence of Direct Path in Indoor Localization." In *Proceedings of IEEE Global Telecommunications Conference 2007, GLOBECOM '07*, Washington, DC, 387–92.

18. Ji, X., and H. Zha. 2003. "Multidimensional Scaling Based Sensor Positioning Algorithms in Wireless Sensor Networks." In *Proceedings of the 1st Annual ACM Conference on Embedded Networked Sensor Systems, SenSys '03*, Los Angeles, CA, 328–29. Association for Computing Machinery.

19. Kannan, A. A., G. Mao, and B. Vucetic. May 2006. "Simulated Annealing Based Wireless Sensor Network Localization with Flip Ambiguity Mitigation." In *Proceedings of IEEE 63rd Vehicular Technology Conference*, 1022–26. Melbourne.

20. Karl H., and A. Willig. 2005. *Protocols and Architectures for Wireless Sensor Networks.* West Sussex, England: John Wiley & Sons.
21. Korel, B. T., and S. G. M. Koo. May 2007. "Addressing Context Awareness Techniques in Body Sensor Networks." In *Proceedings of 21st International Conference on Advanced Information Networking and Applications Workshops, AINAW '07,* Niagara Falls, ON, 798–803. IEEE Computer Society.
22. Kusy, B., A. Ledeczi, M. Maroti, and L. Meertens. 2006. "Node-Density Independent Localization." In *Proceedings of the Fifth International Conference on Information Processing in Sensor Networks, IPSN 2006,* Nashville, TN, 441–48.
23. Ladd, A. M., K. E. Bekris, A. P. Rudys, D. S. Wallach, and L. E. Kavraki. June 2004. "On the Feasibility of Using Wireless Ethernet for Indoor Localization." *IEEE Transactions on Robotics and Automations* 20 (3): 555–59. IEEE Robotics and Automation Society.
24. Langendoen, K. and N. Reijers. November 2003. "Distributed Localization in Wireless Sensor Networks: A Quantitative Comparison." *The International Journal of Computer and Telecommunications Networking* 43 (4): 499–518. North-Holland, Elsevier.
25. Lanzisera, S., D. T. Lin, and K. S. J. Pister. June 2006. "RF Time of Flight Ranging for Wireless Sensor Network Localization." In *Proceedings of IEEE International Workshop on Intelligent Solutions in Embedded Systems,* 1–12. Vienna.
26. Larsson, E. G. March 2004. "Cramer-Rao Bound Analysis of Distributed Positioning in Sensor Networks." *IEEE Signal Processing Letters* 11 (3): 334–37.
27. Li, J., J. Jannotti, D. S. J. De Couto, D. R. Karger, and R. Morris. August 2000. "A Scalable Location Service for Geographic Adhoc Routing." In *Proceedings of the 6th Annual International Conference on Mobile Computing and Networking, MOBICOM '00,* 120–30. Boston.
28. Mao, G., B. Fidan, and B. D. O. Anderson. July 2007. "Wireless Sensor Network Localization Techniques." *Computer Networks: The International Journal of Computer and Telecommunications Networking* 51 (10): 2529–53. Elsevier.
29. Mauve, M., A. Widmer, and H. Hartenstein. 2001. "A Survey on Position-Based Routing in Mobile Ad Hoc Networks." *IEEE Network Magazine* 15 (6): 30–39.
30. Munir, S. A., B. Ren, W. Jiao, B. Wang, D. Xie, and M. Ma. May 2007. "Mobile Wireless Sensor Network: Architecture and Enabling Technologies for Ubiquitous Computing." In *Proceedings of 21st International Conference on Advanced Information Networking and Applications Workshop, AINAW '07,* 113–20. Niagra Falls, ON.
31. Niculescu, D., and B. Nath. 2003. "DV Based Positioning in Ad Hoc Networks." *Telecommunication Systems* 22 (1): 267–80. Kluwer Academic Publishers.
32. Pahlavan, K., X. Li, and J. P. Makela. February 2002. "Indoor Geolocation Science and Technology." *IEEE Communications Magazine,* 40 (2): 112–18. IEEE Communications Society.
33. Patwari, N., J. N. Ash, S. Kyperountas, A. O. Hero III, R. L. Moses, and N. S. Correal. July 2005. "Locating the Nodes: Cooperative Localization in Wireless Sensor Networks." *IEEE Signal Processing Magazine,* 22 (4): 54–69. IEEE Signal Processing Society.
34. Patwari, N., A. O. Hero III, M. Perkins, N. S. Correal, and R. J. O'Dea. July 2003. "Relative Location Estimation in Wireless Sensor Networks." *IEEE Transactions on Signal Processing* 51 (8): 2137–48. IEEE Signal Processing Society.
35. Qi, H., P. T. Kuruganti, and Y. Xu. July 2002. "The Development of Localized Algorithms in Wireless Sensor Networks." *Sensors* 2 (7): 286–93.
36. Qi, Y., and H. Kobayashi. December 2003. "On Relation Among Time Delay and Signal Strength Based Geolocation Methods." In *Proceedings of IEEE Global Telecommunications Conference 2003, GLOBECOM '03,* San Francisco, CA, 4079–83.
37. Qi, Y., H. Kobayashi, and H. Suda. March 2006. "Analysis of Wireless Geolocation in a Non-Line-of-Sight Environment." *IEEE Transactions on Wireless Communications* 5 (3): 672–81.
38. Qi, Y., H. Suda, and H. Kobayashi. September 2004. "On Time-of-Arrival Positioning in a Multipath Environment." In *Proceedings of IEEE 60th Vehicular Technology Conference,* Los Angeles, CA, 3540–44.
39. Qin, F., C. Wei, and L. Kezhong. April 2010. "Node Localization with a Mobile Beacon based on Ant Colony Algorithm in Wireless Sensor Networks." In *Proceedings of 2010 International Conference on Communications and Mobile Computing, CMC,* 303–07. Shenzhen.

40. Romer, K. 2003. "The Lighthouse Location System for Smart Dust." In *Proceedings of 1st International Conference on Mobile Systems, Applications and Services, MobiSys '03,* San Francisco, CA, 15–30.

41. Rudafshani, M., and S. Datta. April 2007. "Localization in Wireless Sensor Networks." In *Proceedings of the 6th International Conference on Information Processing in Sensor Networks, IPSN '07,* Cambridge, MA, 51–60.

42. Savvides, A., W. L. Garber, R. L. Moses, and M. B. Srivastava. 2005. "An Analysis of Error Inducing Parameters in Multihop Sensor Node Localization." *IEEE Transactions on Mobile Computing* 4 (6): 567–77. IEEE Computer Society.

43. Sayed, A. H., A. Tarighat, and N. Khajehnouri. June 2005. "Network-Based Wireless Location: Challenges Faced in Developing Techniques for Accurate Wireless Location Information." *IEEE Signal Processing Magazine* 22 (4): 24–40. IEEE Signal Processing Society.

44. Sohraby, K., D. Minoli, and T. Znati. May 2007. *Wireless Sensor Networks: Technology, Protocols and Applications.* New Jersey: John Wiley & Sons.

45. Stoleru, R., and J. A. Stankovic. October 2004. "Probability Grid: A Location Estimation Scheme for Wireless Sensor Networks." In *Proceedings of First Annual IEEE Communications Society Conference on Sensor and Ad Hoc Communications and Networks 2004, IEEE SECON 2004,* Santa Clara, CA, 430–38.

46. Sun, G., J. Chen, W. Guo, and K. J. R. Liu. July 2005. "Signal Processing Techniques in Network-Aided Positioning: A Survey." *IEEE Signal Processing Magazine* 22 (4): 12–23.

47. Tilak, S., V. Kolar, N. B. Abu-Ghazaleh, and K. D. Kang. April 2005. "Dynamic Localization Control for Mobile Sensor Networks." In *Proceedings of 24th IEEE International Performance, Computing, and Communications Conference, IPCCC 2005,* Limerick, Ireland, 587–92.

48. Torrieri, D. J. March 1984. "Statistical Theory of Passive Location Systems." *IEEE Transactions on Aerospace and Electronic Systems* AES-20 (2): 183–98.

49. Wang, G., G. Cao, T. La Porta, and W. Zhang. March 2005. "Sensor Relocation in Mobile Sensor Networks." In *Proceedings of 24th Annual Joint Conference on of the IEEE Computer and Communications Societies, INFOCOM 2005,* vol. 4, 2302–12. Miami, FL.

50. Wu, J., and I. Stojmenovic. February 2004. "Ad Hoc Networks." *Computer* 37 (2): 29–31. IEEE Computer Society.

51. Wylie, M. P., and J. Holtzman. 1996. "The Non-Line of Sight Problem in Mobile Location Estimation." In *Proceedings of 5th IEEE International Conference on Universal Personal Communications,* Cambridge, MA, vol. 2, 827–31.

52. Zaidi, A. S., and M. R. Suddle. September 2006. "Global Navigation Satellite Systems: A Survey." In *Proceedings of 2006 International Conference on Advances in Space Technologies,* 84–87. Islamabad, Pakistan.

53. Zhao, F., and L. Guibas. 2004. *Wireless Sensor Networks: An Information Processing Approach.* California: Morgan Kaufmann Publishers, Elsevier.

Chapter 9

Energy-Centric Simulation and Design Space Exploration for Wireless Sensor Networks

Fabien Mieyeville, David Navarro, Wan Du, and Mihai Galos

Contents

9.1 Introduction

Wireless sensor networks (WSNs) are highly distributed and self-organized systems. They are composed of many low-cost, highly dispersed tiny devices (called node or mote), presenting very strong constraints in terms of computation power, memory size, and energy minimization. The increasing integration process in electronic circuits paved the way for thousands of applications [1] based on spatially distributed communicating systems. Conventional applications of WSN are physical environment data collection on a given space and transmission of these data (after an eventual local processing) to a gathering device called collector node. This particular device is similar to the others with the exception of presenting an extended connectivity that it can be used to transmit data collected to a more powerful computation unit. This typical application of a WSN is illustrated in Figure 9.1.

Longevity of the network, whose elements are dispersed and often barely accessible, is the main constraint that drives most design approaches for WSNs. Robustness of the individual mote is important but not crucial since WSNs must be able to perform their task even in case of failure of some devices of the global system. Being deployed, every mote has embedded energy storage that will sustain its functionality. The longevity of the network is then related to longevity of its mote. Therefore, to optimize this longevity, WSNs are based on pulsed activity. Motes are put in reduced consumption mode (sleeping mode) for most of the time and are periodically waked up so as to acquire and eventually process data before sending them and going back to sleep. The time period attached to this behavior is typically seconds or more.

When developing and designing WSNS for a specific application, designers should be able to estimate the longevity of the network. Being a networked system, formalism of Open Systems Interconnection (OSI) can be used for design purpose. The commonly adopted OSI stack layer shown in Figure 9.2 and established by Akyildiz and Vuran [2] contributes to the fact that contrary to conventional systems where every stack can be isolated and treated specifically at the

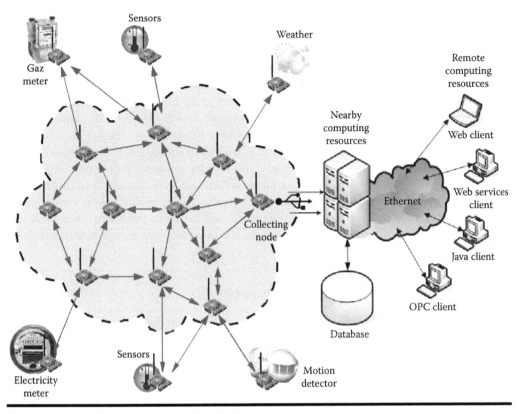

Figure 9.1   Wireless sensor network typical scenario.

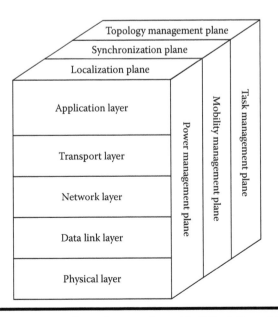

Figure 9.2   Wireless sensor network stack. (Data from Akyildiz, I. F., and M. C. Vuran, *Wireless Sensor Networks*, John Wiley & Sons Inc., New York, 2010.)

adequate level of abstraction, the high constraints and the limited hardware capacity of every mote make this layered approach inadequate for WSNs.

Indeed, the hardware is so highly constrained that every choice at every superior layer can have a dramatic consequence on the node and then on the network; global performances of the application and then the network and hardware/software performances of the node are highly intricate. That is why in WSNS, a vertical approach such as the one preconized by Akyildiz and Vuran [2] must be adopted preferably at the conventional layered approach. The three vertical planes recommended by Akyildiz and Vuran [2] are power management, mobility management, and task management. In this chapter, we will focus particularly on the power management plane since it is the most constraining in WSN design. This chapter will develop design for WSNS in several parts. First, WSN structure and mote software/hardware architecture will be introduced; energy constraints will conclude this introduction. Then, after having established the key elements in WSN design, a succinct overview of WSN simulator will be given. The last part will focus on WSN framework based on hierarchical approach and hardware/software-defined models enabling architectural exploration of WSN design space.

9.2 Wireless Sensor Network: Overview

The objective of this part is not to proceed to an extensive description of WSNs but rather to introduce key elements necessary for the apprehension of the bottlenecks in WSN design. Applications are deployed on WSNs through the use of middleware that physically deploy programming algorithms on every node to fulfill the global application; since it out of the scope of this chapter, this aspect will not be developed.* This chapter will focus more particularly on the three bottom layers of the OSI stack where constraints are the most pressing: physical layer, data link layer, and network layer are the main key elements of global performances of the network and the applications.

9.2.1 Mote: Hardware and Software Architecture

A mote, the elementary part of any WSN, is composed of one or several sensors associated with microcontroller driving radio frequency (RF) transceiver, which is powered by a battery (Figure 9.3). Size of a mote can vary from a matchbox to a 1-mm^3-sized box [3].

9.2.1.1 Hardware Architecture

Energy availability weakness takes its roots in this small size and results in the use of low-power-consuming devices: 8–16-bit microcontroller, small memory size, and pulsed RF devices. Eight to sixteen–bit architecture currently represents roughly 75% of hardware mote solutions [4].

Among this microcontroller architecture, two types of products are dominant: MSP430 microcontroller architecture from Texas Instrument and ATmega128 architecture from Atmel [4]; details about architecture repartitions are summarized in Figure 9.4. Most designers of WSNs and simulators (discussed in Section 9.5) focus on one or more architectures, but a global approach necessitates a good knowledge of these two main microcontroller families.

For the RF transceiver architecture, Chipcon-based products from Texas Instrument are dominant [4] with half of the RF solution emitting in the Industrial Scientific and Medical (ISM)

---

* For more details on middleware, the authors recommend the reading of S. Hadim [70].

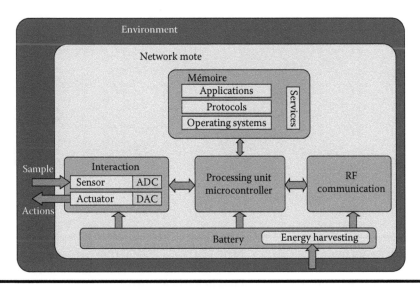

Figure 9.3 Mote hardware architecture. (Data from Carolina Fortuna, "Why is Sensor Data Hard to Get?" COIN-ACTIVE Summer School on Advanced Technologies for Knowledge Intensive Networked Organizations in Aachen, 2010.)

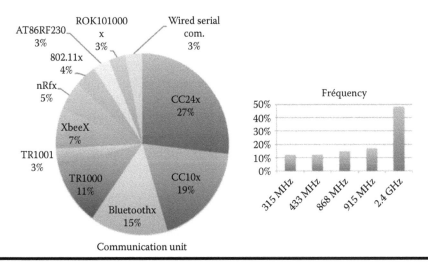

Figure 9.4 RF architecture repartition in mote platform. (Data from Carolina Fortuna, "Why is Sensor Data Hard to Get?" COIN-ACTIVE Summer School on Advanced Technologies for Knowledge Intensive Networked Organizations in Aachen, 2010.)

band of 2.4 GHz (Figure 9.5). If existing, wired-based connection and infrared communication are scarcely used in WSNs.

Available commercial mote platforms commonly used are MICA2 [5] and MICAz [6] from Crossbow based on ATmega128 architecture and CC1000 and CC2420 transceiver architecture respectively. A more recent architecture, which resulted in several other commercial emanations, is the TelosB architecture [7] based on MSP430 and CC2420 from Crossbow.

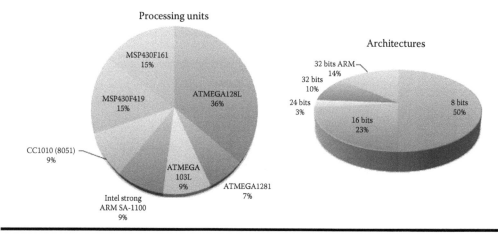

Figure 9.5 Hardware mote repartition in mote platforms. (Data from Carolina Fortuna, "Why is Sensor Data Hard to Get?" COIN-ACTIVE Summer School on Advanced Technologies for Knowledge Intensive Networked Organizations in Aachen, 2010.)

9.2.1.2 Software Architecture

Very few computation tasks are executed inside a mote: nevertheless an operating system (OS) is often implemented even if a simple task scheduler could be sufficient. OSs in WSNs are used for providing abstraction layer in hardware devices and for enabling relatively easy programming of mote without using microcontroller development environment tools. Energy constraints, limited computation capacity, and mainly small memory size have necessitated the development of specific OS with small footprint. If about 37 OSs are currently available for WSNs, in which only 4 are mainly used [8]: TinyOS [9], SOS [10], MantisOS [11], and Contiki [12]. Current literature exploration [13] shows that 81% of scientific publication in the domain are using TinyOS, 9% Contiki, 8% MantisOS, and the others share the remaining 2%. Each of the three OSs has its particularity and its hardware platform support. Table 9.1 summarizes these main elements.

Knowledge of the embedded OS is important for WSN design since OS has a dominant impact [14] on the node consumption; see Figure 9.6 for a typical application of data acquisition and multi-hop RF emission for a MICA2 node.

Furthermore, the choice of the OS will also determine the reconfiguration possibilities and their cost according to whether the OS architecture is monolithic (e.g., TinyOS), which results in high reconfiguration energy cost, or modular (MantisOS or Contiki), which enables partial reconfiguration with reduced energy consumption. Since software implementation is as important as hardware architecture, it will also determine which simulator can or cannot be used (discussed in Sections 9.4 and 9.5).

9.2.1.3 Energy Approach

To determine the aspects that should be optimized to minimize power consumption, a thorough analysis of hardware consumption at the mote level must be done. Based on the works of Landsiedel et al. (previously mentioned) [14] performed on MICA2 mote, the consumption repartition (Figure 9.6) is commonly accepted as global behavior.

Since the mote is in sleeping mode most of the time, the CPU idle state represents a significant part of the energy budget (20%). But the most consuming element is undoubtedly the RF transceiver

Table 9.1   Overview of Operating Systems for WSNs

OS	Structure	ROM Size (kB)	RAM Size (bits)	Process	Hardware Platform Support
TinyOS	Events	3.4	336	Tasks, commands, and event management	BTnode, EyesIF X v1, EyesIF Xv2, IMote, IMote 1.0, IMote 2.0, Iris, KMote, MICA, MICA2, MICAz, Rene, SenseNode, TelosB, T-Mote Sky, Shimmer
Contiki	Events	3.8	230	Protothread	T-Mote Sky, TelosB, avr MCU, MSP430 MCU, x86, 6502
Mantis OS	Threads	4	500	Threads	MICA2, MICAz, Nymph, TelosB
SOS	Events	20	1163	Tasks defined as modules	XYZ, T-Mote Sky, KMote, MICA2, MICAz, TelosB, Avrora, Protosb, Cricket, Cyclops, emu

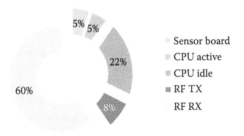

Figure 9.6   CPU/RF transceiver consumption in node. (Data from Landsiedel, O., K. Wehrle, and S. Gotz, In *Proceedings of the 2nd IEEE Workshop on Embedded Networked Sensors*, 37–44, IEEE Computer Society, Washington, DC, 2005.)

[15]. That is the reason why most research works are based on reducing the RF emission part. But, as it will be developed later, design approaches based exclusively on RF consumption cannot be optimal.

9.2.2   Communication between Mote

RF link is the backbone of any WSN for more than one reason. First, as communication element, the link sets network performances and then application performances. Furthermore, as mentioned in Section 9.2.13, RF communication is the most consuming part.

Because of this consumption aspect, classical RF protocols, not specified for low power consumption, are not suited for WSNs. A proprietary standard low-cost and low-consumption

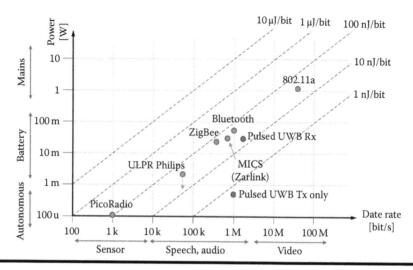

Figure 9.7 RF protocols adequation for wireless sensor networks. (Data from Bird, N. C., "RF challenges for WSN and Beyond," *2nd Workshop on Wireless Sensor Networks Research (WISEN '07)*, Dublin, Ireland, December 2007.)

protocol is currently emerging, based on the IEEE 802.15.4 norm [16]: ZigBee [17]. In the IEEE 802.15.4 standard, three different bands selected in the ISM band are defined for communication, that is, 2.4 GHz (global), 915 MHz (the Americas), and 868 MHz (Europe). If in the lower bands, binary phase-shift keying (BPSK) is used; offset quadrature phase-shift keying (O-QPSK) is preferred in the 2.4-GHz band; the medium access control (MAC) layer provides communication for star, mesh, and cluster tree-based topologies with controllers. The transmission range of the nodes is assumed to be 10–100 m with data rates of 20–250 kbps [16].

This norm has many common points with the Bluetooth technology while targeted on low-power-consuming applications. This protocol can be found in many civilian applications [2]. Many alternative solutions exist, but ZigBee is still the preferred solution [18,19] when deploying WSNs in spite of its real-time inadequacy and its optimization for low duty-cycle applications.* Figure 9.7 [20] gives an overview of protocols that can be used in WSNs according to energy per bit consumption, data rate, and emitting power (impacting range).

Currently, numerous RF transceivers offer a partial or full hardware support for IEEE 802.15.4 protocol or even ZigBee stack, making it easier to use since communication part is managed by the RF transceiver without the need of supplementary coding at the microcontroller level. Since IEEE 802.15.4 and ZigBee stack are an important part of WSNs and the preferred solution for many designers, more discussion is given in Section 9.2.3.

9.2.3 IEEE 802.15.4

In the traditional OSI layer representation, the norm IEEE 802.15.4 covers both physical layer and data link layer; it covers the physical transformation of data for transmission through the aerial way and the rules for accessing the communicating layer—MAC layer. Figure 9.8 represents the

---

* ZigBee is optimized for applications whose duty cycle (representing activity time compared to sleeping mode) is less than 0.1%.

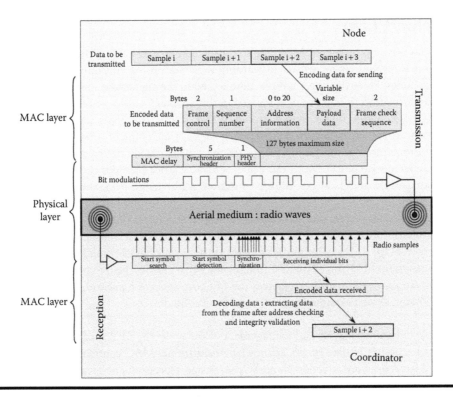

Figure 9.8 Anatomy of a RF communication.

unfolding of a point-to-point RF communication from data acquisition to gathering by the coordinator. In the 2.4-GHz band, IEEE 802.15.4 offers a maximal communication rate of 250 kbps.

Unfortunately, due to encapsulation of data necessary for routing of the information and for robustness of the RF communication, the effective data rate (representing the number of samples that can be emitted by second) falls to 127 kbps if the carrier sense, multiple access/collision avoidance (CSMA/CA) algorithm with a maximum payload size of 114 bytes is used and if the hypothesis of a RF channel is always free. If considering the probability of 25% of finding a busy channel necessitating the node to wait in active mode until the channel becomes free before transmission to avoid collision of data, this data rate falls to 101 kbps [21].

In IEEE 802.15.4, two ways can be used to access the RF channel: non-beacon and beacon mode. In non-beacon mode, every node can try and access the channel for emitting its data; so as to avoid collision of data in this competition to access the channel, the CSMA/CA algorithm is used. According to this algorithm, every node will test the channel to determine whether it is free or not. In the latter case, the node will wait for a randomly determined time before trying to reemit; both maximal waiting time and number of retries can be determined by parameters of the norm. This mode maximizes the usage of bandwidth, but offers no guarantee in terms of latency since every node is competing to access the channel. The beacon mode puts the coordinator node in charge of synchronizing the network; it emits periodically a beacon on which every node will synchronize itself and then wait for its turn to emit. This mode guarantees a minimum latency of communication but is not optimal since some slots not used by nodes cannot be allocated to other necessitating higher bandwidth. A superframe structure is used in this beacon mode with parameters such as setting active period, inactive period, number of slots, allocation process for the slot, and duration between two beacons.

9.2.4 Network Layer

If data link layer is responsible for MAC, flow control, and error checking, the upper layer called network layer is responsible for packet forwarding (host addressing) including routing through intermediate nodes. Since the upper layer is more linked to the application and has less impact on the global performances, it will not be discussed in this chapter.*

ZigBee, the emerging de facto standard in WSNs, covers the upper layers over the IEEE 802.15.4 norm as shown in Figure 9.9. From an application point of view, ZigBee gives the Application Programming Interface (API) to designers, which facilitates creation of applications based on ZigBee device object. From our point of view, ZigBee provides the necessary tools for network management and data traveling structuration. Main design parameters are as follows:

■ Network topology (tree, star, or mesh) as represented in Figure 9.10.
■ Node classification between full functionality devices (FFDs) and reduced functionality devices (RFDs). RFDs are minimalist hardware node that can just emit and receive data and join a network. FFDs have extended functionality that enables them to manage the network. RFDs can only talk to a FFD, whereas the FFD can serve as a network coordinator, a link coordinator, as well as simply another communication device.

To enable dynamic behavior and management of the network, ZigBee can use four different frames of IEEE 802.15.4: data frame, acknowledgment frame, MAC command frame, and beacon frame. After a brief introduction about WSNs, design aspects will be presented.

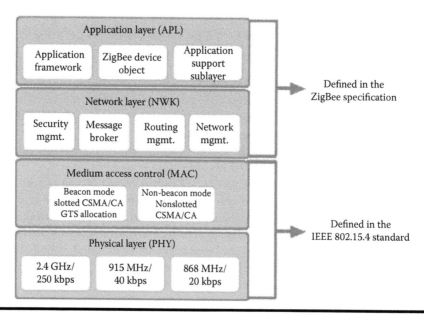

Figure 9.9 Overview of ZigBee layers.

---

* Since addressing hardware and software node-level abstraction up to network layer is challenging, upper layers are currently not concerned in the WSN design approach and are more of middleware domain.

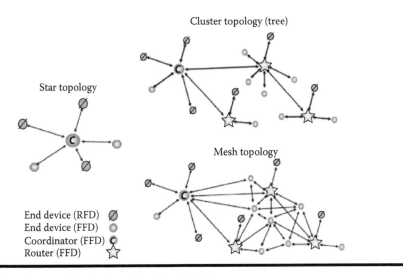

Figure 9.10   Topology and node functionality in ZigBee.

9.3 Design of Wireless Sensor Networks: Challenges and Constraints

9.3.1 Introduction

The main challenge in designing WSNs is that the target application is realized by an autonomous distributed network of communicating devices. It is therefore necessary to design an application that will be conducted by individual programming of dozens (hundreds and thousands) of nodes working through a RF network. Physical deployment of such systems as a method of design is obviously inadequate, and the use of pure analytical methods is to be avoided because of the complexity of the problem and the numerous interactions [22]. The main difficulty in this area is to combine methods for designing systems capable of managing the interaction of many systems with simulation methods specific to each separate physical system, while keeping the simulation time consistent. The difficulty lies in defining the level of abstraction for the study design required.

9.3.2 WSN Design Stakes

As introduced in Section 9.1, constraints on WSNs make a hierarchical layered approach based on OSI model inappropriate. A vertical translayer design approach is necessary as shown in Figure 9.2. Recommendations from Akyildiz and Vuran [23] are in favor of a three-plane approach: power management, mobility management, and task management. The fact that constraints inherent to WSNs result in a very strong impact of the material layer on the application layer implies that any WSN design tool must be able to consider the full hierarchy of the OSI layer since every layer is interpenetrated. Taking into account the physical layer of nodes (including complete hardware architecture transceiver, microcontroller, battery, and interfaces with the physical world) at the application layer is therefore essential.* We must be able to propagate through the layers the impact of any material changes at the node level in a bottom-up approach as well as be able to pass

---

* This is the reason why the classical design approach in WSN based on reducing power consumption by reducing RF consumption in WSN, while effective, is incomplete.

the specifications generated by an application down to the node's hardware so as to establish the best-suited architectures in a top-down approach. This design process must integrate every parameter: from physical node (hardware architecture and software implementation) up to transport layer, passing by data layer (MAC parameters, etc.) and network layer (topology, etc.).

This hierarchical approach can be easily linked to design stakes in WSNs such as established by Yick et al. [19], represented in Figure 9.11.

This classification defines three areas for action in the field of WSNs:

1. A low abstraction level centered on the node focusing on hardware architecture and node software implementation.
2. An area focused on communication within the network, from point-to-point communication, fundamental unit of the communication network, to the transport layer.
3. A field grouping service, which includes all the techniques applied to the global network as a whole to achieve and optimize the implementation of the application.

The two extreme domains of this classification, that is, sensors and actuators and their implementation technology on the one hand and the application itself on the other hand, can be considered independent of the WSN design optimization process even though they are the basis and the end-user finality.

WSN design tools must be capable of apprehending these three areas so as to meet any application requirement. If many simulators provide full support for one (sometimes two) of these areas,

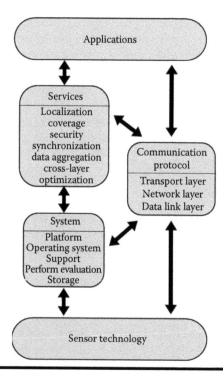

Figure 9.11   Challenges in designing wireless sensor networks. (Reprinted from J. Yick, B. Mukherjee, and D. Ghosal, "Wireless sensor network survey," *Computer Networks,* **52: 2292–330, Copyright 2008, with permission from Elsevier.)**

few of them implement the totality. Moreover, this classification makes easier energy-minimal-based design process, explained in Section 9.3.3.

9.3.3 Energy Constraint in WSNs

While in the field of WSNs, physical death of a node must not compromise the application; there is nevertheless a loss rate above which the application functionality cannot be ensured. If constraints on the design of a WSN are numerous (limited power supply, limited computing power, small storage capacity, etc.), they all result from an essential one, autonomy. Difficult accessibility of nodes reinforces the stress of autonomy. Furthermore, nodes of WSNs aim at being as small as possible; then, the size of battery is limited, which reinforces the constraint on energy consumption. To extend the life of the network, the nodes must have computing capabilities (power and memory), limited to minimize consumption. As we can see, the energy embedded in the node is in the heart of the design of a WSN.

As expected, energy minimization can be addressed from any of the communication/system/service area as well as at any abstraction level from the OSI sensor stack. Numerous energy conservation strategies can be implemented. Anastasi et al. [24] established, in 2008, a systematic and comprehensive taxonomy of the energy conservation schemes, which is presented in Figure 9.12.

Most WSN design approaches focus only on one of these domains, but an optimal design process should address this problem at each level of abstraction by combining different energy minimization techniques. Section 9.3.4 will develop both the specifications and the performance metrics for WSN simulation tools.

9.3.4 WSN Design Framework: Specifications

Given the challenges and constraints as discussed in Sections 9.3.2 and 9.3.3, we have established a number of criteria to be observed for WSN design tool.

Design under uncertainties: Designing WSN means predicting the behavior of a distributed real-time system. This requires accurate models (or models presenting reduced uncertainties) for RF propagation channels, physical environment, and the hardware architecture of the nodes as well as the battery. According to the level of abstraction, models are more or less precise. The knowledge of models' uncertainties is fundamental so as to evaluate viability of simulation results. Ignorance of uncertainties can severely impact the analysis of simulation results and then misconduct the designer in its design process. For example, strong inaccuracy on a battery model can significantly affect the routing of a packet through the network [25] and therefore the accuracy of the simulation. It is therefore important to know the uncertainties associated with the model to control the simulation error.

Scalability: As a distributed system, the scaling of the WSN and the support of a large number of nodes without negative impact on the simulation time and memory needed are important parts of developing any design tool.

Extensibility: Because of the long-term deployment of WSNs, it is vital to provide integration into an existing network of new node and/or modification of software programs loaded on each node. A modular structure, in which special care must be taken with respect to interfaces, allows the user of such changes on the network.

Heterogeneity: If the initial design of a WSN generally involves only a single hardware architecture, support for different hardware architecture is critical at two levels. During the design phase, the management of heterogeneity allows for architectural exploration and therefore a physical

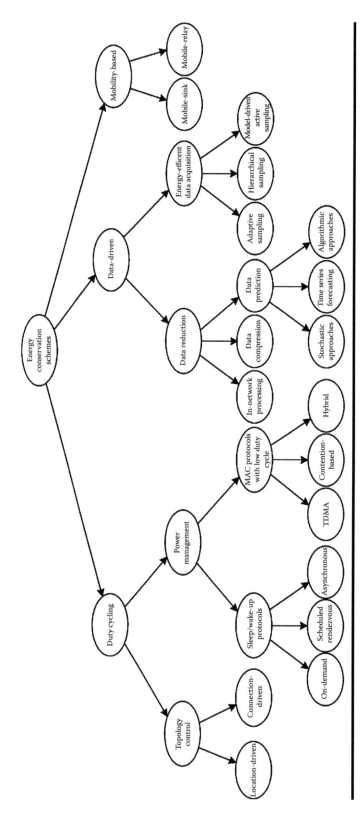

Figure 9.12 Taxonomy of energy conservation schemes. (Reprinted from G. Anastasi, M. Conti, M. Di. Francsesco, and A. Passarella, "Energy Conservation in Wireless Sensor Networks," *Ad Hoc Networks*, 7(3), 537–68. Copyright 2009, with permission from Elsevier.)

optimization of the network, and during maintenance of the network, it may be necessary (for performance constraints or simply obsolete systems deployed) to integrate, into an existing network, nodes materially different from initially deployed ones. Support for various hardware architectures is then necessary for any design [26].

A graphical user interface (GUI): Users involved in the development of WSNs are not necessarily proficient in network systems or in hardware architecture. Many WSN end-user designers are either application developers aiming at using WSN for final deployment, physical sensor designers willing to integrate their product in distributed smart systems, or embedded system designers needing tools to check the validation of a protocol or to estimate the contribution of a sensor. Thus, a GUI is necessary for easy handling of the tool and therefore its use in a sustainable manner. The GUI must offer interface for setting and launching of the simulation as well as support for graphical display of results and console for log files.

Energy consumption estimation: Being the main constraint in WSNs, the simulator must at least deliver an estimation of the lifetime of the designed WSN and in ideal cases should provide support and tools for optimization of the energy consumption at different levels of abstraction.

9.3.5 Performances Metrics

Metrics to consider in the design of WSNs depend on the level of abstraction considered. However, two general concepts can be identified:

1. The achievement of the overall implementation depends on the quality of communication within the network; performance factors to be monitored are the success rate of packet transmission or latency introduced by the network.
2. Energy efficiency at the node is critical since the length of that life depends on the sustainability of the application.

We can therefore establish two families of metrics: metrics related to the network and the communication aspect and those related to the node hardware itself and thus to its energy consumption.

The reference metric commonly adopted in the WSN world is the energy per bit successfully transmitted (Energy per Useful Bit [27]), which relates the energy consumption of any node to a unit data transmission. If such a criterion can be used to evaluate the global effectiveness of a WSN, it is insufficient when optimizing any WSN because of its aggregative nature. Indeed, accurate identification of most consuming part (hardware/software and node level/network level) is impossible. We propose a greater number of metrics to evaluate network performance and also to map precisely consumption within the node. These metrics are summarized in Table 9.2.

This set of parameters is the minimal performance metric that any simulator should evaluate since it allows a network-level view, reflective of communication performances, and offers a hardware/software analysis at the node level, indispensable for any accurate energy analysis and minimization.

9.3.6 MANET Simulation Studies: The Incredibles

If commonly accepted, use of simulation tools in the WSN is a subject of controversy. It is now established that these tools are effective in the comparison of solutions but not for validation before physical deployment. In fact, the error made by these tools is at best uncertain,

Table 9.2   Metrics of Performance in WSN

	Parameters	*Unit*	*Description*
Network level	Packet delivery throughput	Bit per second (bps)	Average throughput of the network
	Packet delivery rate	Percentage (%)	Total number of packets successfully transmitted divided by total number of packets emitted
	Energy consumption per node	μJ	Average energy consumption per node
	Power consumption per node	μJ	Average power consumption per node
	Network lifetime	Hours	Lifetime of the network from the beginning to a predetermined number of nodes' death
	Latency	ms	Average latency of the packets between initial sender and coordinator
Node level	Collisions		Number of collisions happening to a node. Essential when identifying bottlenecks in communication and to prevent premature exhaustion of node
	Power consumption per packet	μW	Average power consumption for transmitting successfully a packet
	Power consumption of microcontroller	μW	Average power consumption of the microcontroller
	Power consumption of transceiver	μW	Average power consumption of the radio-frequency transceiver
	Power consumption of SPI	μW	Average power consumption of the communication between microcontroller and RF transceiver. Depending on the hardware support of communication by RF transceiver, this power consumption can be highly impacted
	Power consumption of ADC	μW	Average power consumption of analog-to-digital conversion of physical sampling extracted from the environment by sensors (amplifier consumption included)

often unknown; very little work on validation experiments are conducted on the WSN design tools, and when made, they present significant differences compared with simulation results [22,28]. If this uncertainty with respect to reality has little impact on the comparison of protocols (the same mistake being made for all simulated solutions), it makes these tools inadequate for integration into a standard design flow as validation method before physical deployment. The study "MANET simulation studies: The incredibles" conducted by Stuart Kurkowski [29], in *Proceedings of MobiHoc* during the years 2000–2005, shows that less than 15% of publications offer reproducible results. Main causes are the lack of mention of the version of the simulator,* incomplete parameter set values, and the inability to access the simulator codes when it comes to solutions developed in-house without mentioning problems of simulation initialization or random number generation that is not mentioned (less than 7% of published works). Worse, most of the time, the number of iterations of the simulations is often omitted, which is an aberration in a domain where statistical is crucial. Errors or uncertainties are scarcely addressed since only 12.5% of publications use confidence interval on the plot; hence, only approximately 12% of the MobiHoc simulation results appear to be based on sound statistical techniques. These criticisms are also emphasized in Andel and Yasinsac's works [30]:

- No repeatability by third parties.
- No reliable statistics: Number of simulations and methods for generating random seeds required for the simulation.
- No sensitivity analysis of parameters.
- Approximation as to the models used: inadequate radio models, application unrealistic, no actual implementation.

He combines this analysis of a number of proposals to bring more rigor in the field of WSN simulation tools; according to Andel and Yasinsac, WSN simulation tools can only be used as proof of concept, and their use for protocols comparison should be avoided.

This survey by Kurkowski et al. [29], in 2005, was extended by the works of Ivan Stojmenovic [31] in 2008, who pointed out the same shortcomings as those evocated by S. Kurkowski [29] in 2005. Therefore, to alleviate these recurring problems, he also provides concepts for reliable tools and their uses. His proposals are all the more interesting other than that they go against the classical recommendations as established by Andel and Yasinsac (Table 9.3).

Where the general trend is pushed to the refinement of models and exhaustiveness of the parameters and variables, Stojmenovic recommends simplifying the simulations to improve visibility, tractability of the phenomena based on the parameters, and reproducibility of the simulations by third parties. If thorough evaluation and complex models are indeed needed for accurate simulations, analysis of the results and identification of the key parameters are all the more difficult because interactions between models and parameters are numerous. Such simulations provide some useful insights about performances but remain globally inconclusive, providing very few explanations and very few hints for future progress. A more progressive approach, such as recognized by Stojmenovic, allows a gradual understanding of phenomena at stake in intricate distributed systems such as WSNs and is totally compatible with a top-down design approach.

---

* When the name of the simulator is mentioned since 30% of the papers omit to mention the simulator used.

Table 9.3   Recommendations for WSN Simulation

Problem	Current Advice	New Advice
Proof of concept	Exhaustive evaluation, test bed	Basic simulation using assumptions in designed protocol
Comparison	Do not directly make the comparison of multiple protocols against one another	Comparison with truly competing existing protocols
Literature review	Partial is sufficient	Thorough and critical
Number of variables considered	Multiples variables of a realistic scenario	Incremental approach by solving one problem at a time, study one variable before adding another
Complexity of the models	Complex and realistic modeling	Simple modeling to preserve tractability
Simulation process	Immediate simulation with complex modeling	Parallel advance of useful modeling and protocol design

9.4 Simulators for WSNs

WSN has initially risen interest to communities from networks and sensor. The network community has seen in WSNs a new field where applying their expertise on networks with a strong emphasis on lowering power consumption so to reduce energy expense. The sensor community has identified in WSNs the extension of sensors they developed, bringing them the opportunity of behaving as communicating smart sensors. Simulators have been initially developed by these two communities, which have extended their own simulation tools to WSNs. Then, we can find network-based simulators and hardware-based simulators. Section 9.4.1 will briefly discuss these two aspects. For an in-depth analysis of taxonomy of simulators for WSNs, the reader is referred to the works of W. Du et al. [32] or Dwivedi and Vyas. [33].

9.4.1 Network Simulators Extended to WSNs

Most of the network-oriented simulators for WSNs are based on the legacy computer network simulators. Classical network simulators are based on the communication between Intellectual Property (IP) blocks representing hardware modules. In WSNs, nodes replace the IP blocks and are instantiated differently giving two classes of simulators: the network simulator with node models (NSNMs) and the network simulator with node emulators (NSNEs). NSNMs are event-driven simulators putting strong emphasis on the radio link and the network modeling. Hence, node models are often simple, commonly reduced to basic energetic models based on temporal profile of energy consumption.

NSNEs are using network simulators coupled with node emulators (more precisely microcontroller-based emulators). These simulators inherit their temporal precision from the hardware emulators' accuracy (cycle-accurate). Yet, these emulators often restrict themselves to the hardware architecture of the node's microcontroller, scarcely integrating RF transceiver structure. Moreover, coherence constraints and synchronization between the network simulator and the node emulator rapidly result in huge computation time consumption.

9.4.2 Node Simulators Extended to WSNs

Simulators of this category are node simulators whose capability has been extended by the adjunction of network simulation. Taking their roots in hardware community, these simulators fall into two categories: node simulators based on models using hardware description language such as SystemC [34] (node system simulator with network models [NSSNMs]) and hardware emulators focusing either on hardware instruction set or on embedded OS (node emulators with network models [NENMs]).

NSSNMs include a simulation kernel that offers support for concurrency and synchronization between the different hardware devices constituting the node. The modeling language used enables the codesign (hardware/software partitioning) and cosimulation. The node's hardware can be represented from a hierarchical point of view (at the system level, the transactional level, or the register transfer level [RTL]). Abstraction level determines simulation and precision error and must be associated with methodology for managing the error between hierarchical-level management. Coming from the System-On-Chip (SoC) community, these simulators present strong similarities with electronic design automation (EDA) design flow. Fast simulations at high abstraction can be performed to establish feasibility and preliminary performances as well as time-consuming precise simulations at node's hardware architecture level that takes into account software implementation.

NENMs use instruction set simulators (ISSs) or code-specific emulators for the nodes' behavioral description. ISSs are specific to microcontroller architecture; they implement instruction reading and update variables that represent register behavior. Emulators are specific to code implementation often linked to an OS such as TinyOS, Contiki, or SOS. They can directly execute programming codes to be inserted on the node.

The main drawbacks of these tools are the weakness of network models.

9.5 Overview of Commonly Used Simulators

In this section, a brief overview of commonly used simulators is given; for an exhaustive list, the reader is referred to the recent works of Dwivedi and Vyas [33] referencing 63 simulators/simulation frameworks, 14 emulators, 19 data visualization tools, 46 test beds, 26 debugging tools/services/concepts, 10 code-updation/reprogramming tools, and 8 network monitors used worldwide for WSN researches.

9.5.1 NS2

NS2 [35], an event-driven object-oriented network simulator belonging to NSNM, is the most used simulator [29] in mobile ad hoc networks (MANETs) domain. Simulations are realized in C++ language and object-oriented Tcl (OTcl). Protocols and extension libraries are implemented in C++. Creation, control, and management of simulations are realized in OTcl. Extension policy of NS2 library has greatly contributed to its popularity, many protocols being implemented by scientific community. WSN-specific protocols have been implemented in NS2, among which is a version of the IEEE 802.15.4 norm. Large-sized networks are difficult to implement because of their memory usage and simulation time [36]. Furthermore, detailed energy models for the different hardware and software elements of the node are lacking, resulting in poor precision at low level of abstraction.

Among the extensions of NS2 dedicated to WSNs, SensorSim [37], based on an IEEE 802.15.4 network model of NS2, is to be mentioned. It offers node modeling divided into two parts: a software model (called function model) and a hardware model. Several power-consumption models of several hardware architectures have been implemented. Hardware models are based on state representation driven by the software model; it then allows an accurate estimation of power consumption. By using SensorWare, a middleware for efficient programming of the network, SensorSim can be coupled to real physical node through the use of a gateway node. Hence, SensorWare scripts can be executed both on virtual nodes represented by models and on physical nodes. The main weakness of SensorSim lies in its generic hardware units that are not accurate when considering a specific target. Unfortunately, SensorSim is not supported any more.

Criticisms often made about NS2 are about the interdependences between modules resulting from its object-oriented structure. Hence, developing protocols for NS2 library is complex and requires a thorough knowledge of source code of NS2 from developers. In the network community where standard protocols are clearly identified, such a limitation can be tolerated, but in WSNs where no real standard has been clearly adopted* and where research in protocol domain remains dominant, these mixing up of modules become a hindrance to WSN-specific library development.

9.5.2 NS3

Third generation of NS simulator debuted in July 2006. If NS3 is, as its predecessor, based on C++, OTcl is neglected in favor of C++ (network models) and the Python language (optional). In addition, it incorporates GTNet [38], a simulator known for its support of scalability. These choices are made at the expense of backward compatibility that requires the manual complete rewriting of any model developed under NS2. This incompatibility explains the sustained use of NS2 for which many protocols exist. For more details on the differences between these two generations, the reader is referred to the comparative study of Font et al. [39].

9.5.3 OMNeT++

Second simulator in the category of NSNM,† OMNeT++ [40], is a simulator adopting a modular approach developed in an integrated development environment (IDE) based on Eclipse environment with a graphical execution. The IDE supports all stages of the simulation: development, creation, configuration, and execution of simulation models and analysis results. OMNeT++ is composed of modules that communicate through messages. The elementary modules implement the simplest atomic behavior of a model (typically a particular protocol) and can be combined to create more complex composite modules. OMNeT++ provides the infrastructure to assemble the simulations of these models and manage their configuration through a specific language named NED (Network Description). The functionality of OMNeT++ can easily be extended with interfaces for real-time simulation, emulation, distributed computing, or integration of SystemC models.

OMNeT++ was designed to overcome the development problems in NS2 [40] and is becoming even more popular. Often compared, they are the two most widely used simulators, according to Weingartner et al. [41]. Many WSN simulators are based on OMNeT++. MiXiM [42] (formerly Mobility Framework [43]) is a design platform based on OMNeT++ and dedicated to the simulation of wireless network and mobile.

PAWiS [44] is another design environment dedicated to WSNs. Like SensorSim, it takes into account the internal architecture of the node (modules) as well as the surrounding environment (network, disturbances, and physical environment) and allows the inclusion of specific WSN phenomena

---

* Even if ZigBee is, from a commercial point of view, commonly accepted as de facto standard, its inadequacy for real-time support and the increasing need for always lower consumption keep the protocols domain in the most active research fields of WSN.

† Technically, OMNeT++ is an all-around simulation environment based on discrete events.

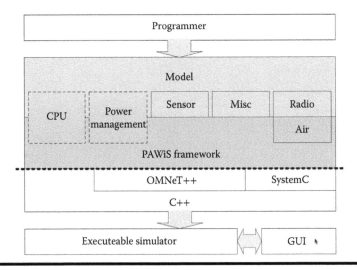

Figure 9.13   Structure of PAWiS simulator.

such as the analysis of consumption at different levels of granularity, mobility support, dynamic phenomena (environment and channel), and temporal aspects such as latency. Standard interfaces and a power analysis tool complete the tool (Figure 9.13). It differs from Mobility Framework by a different treatment of the communication channel and a power consumption analysis tool.

9.5.4 TOSSIM and Its Derivatives

TOSSIM [45] and PowerTOSSIM [46] are two NENMs dedicated to the execution of TinyOS on MICA platform. Software development of WSNs can be simplified by the use of these tools; they allow the development of algorithms and the observation of the behavior of systems and the interactions between the nodes. Every TinyOS application can be compiled into the simulation environment by replacing TinyOS software modules that interface hardware device by emulation libraries (counter, communication with the channel, sensors). TOSSIM can reproduce the behavior of networks of thousands of nodes with a fine-grained accuracy (bitwise). TOSSIM facilitates the porting of applications between the simulation environment and the physical nodes. PowerTOSSIM is the extension of TOSSIM dedicated to the evaluation of consumption. The major drawback of these platforms is the hardware limitation (restricted to MICA nodes—MICAz with PowerTOSSIM [47]—powered by ATmega architecture) and the software restriction (exclusively TinyOS/nesC).

9.5.5 Avrora

Avrora [48], in the family of NENM, is a sensor network simulator instruction written in Java. It combines the precision of ATEMU [49] (precise level cycle–cycle-accurate) with the scalability of TOSSIM (by displaying a half simulation time compared to TOSSIM [48]), up to 10,000 nodes. Furthermore, Avrora is independent of language and the embedded OSs. The disadvantage of such a tool is its hardware support limited to ATmega128 architecture from ATMEL (node MICA and MICAz). Since ATmega128L has been used in most of the first-generation nodes, the latest generation of nodes is implementing newer architecture such as the ultra low-power MSP430

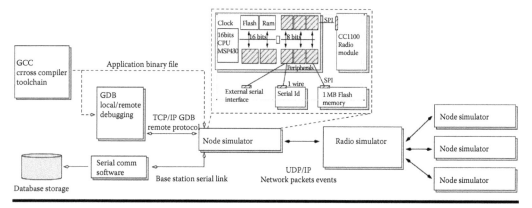

Figure 9.14 Structure of Worldsens framework. (Data from http://wsim.gforge.inria.ft/index.html.)

microcontroller from Texas Instruments, making the use of Avrora scarce. Moreover, using a high-level language, Avrora cannot be easily integrated into a conventional hardware design flow.

9.5.6 Worldsens

Worldsens [50] is a development environment of the NSNE family, which combines a network simulator WSNet and an ISS-based emulator node WSIM. This environment allows designers, by cosimulation, to take into account both the physical aspects and the network aspects of nodes. The overall architecture of the platform is shown in Figure 9.14.

The network simulator splits the communicating sensor into several modules (radio, antenna, battery, MAC, routing, and application mobility), without restriction on the internal architecture of the node and the environment (temperature, pressure, etc.), and splits the radio channel into three modules (propagation, interference, and modulation). This architecture allows WSNet to consider, with accuracy, the physical channel as well as the effects of MAC layer and above. WSIM integrates a hardware model that simulates cycle-precise behavior of each hardware device composing the physical node, including running embedded software. Worldsens is based on the cosimulation of WSIM and WSNet; each node of the network is simulated by an instance of WSIM (application binary code input). This native code is executed and emulated in each hardware block (microcontroller, transceiver, memory, etc.). When the transceiver transmits a message through RF channel, the information is transferred to the simulator WSNet, allowing granularity at byte-level simulation on a network (Figure 9.14).

This design platform has two major drawbacks: firstly, in its current version, node architecture is limited to systems based on MSP430 microcontroller from Texas Instruments and on RF transceivers from the same manufacturer (Chipcon CC1000, CC1001, CC2500, and CC2420), which limits the use of this tool and does not allow its use for design space architectural exploration. Secondly, cosimulation generates significant simulation time that can be prohibitive in early design–stage steps.

9.5.7 WISENES

WIreless SEnsor NEtwork Simulator (WISENES) [51] was developed by the University of Tampere (Finland) for the design, simulation, and the performance evaluation of WSNs. The WSN structure is described in Specification and Description Language (SDL), then simulated in WISENES, and finally implemented on the physical platform manually or through the use of

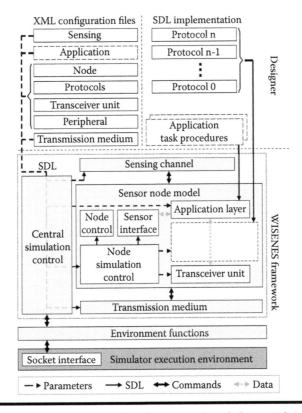

Figure 9.15 Instantiation of WISENES simulator. (Reprinted from *Ad Hoc Networks*, 6(6), Kuorilehto, M., M. Hännikäinen, and T. D. Hämäläinen, *Rapid Design and Evaluation Framework for Wireless Sensor Networks*, 909–35, Copyright 2008, with permission from Elsevier.)

an automatic code generator. The specificity of WISENES is its ability to back-annotate the top-level model with results from measurements realized on the physical hardware platform, which enables accurate simulations without sacrificing simulation time (Figure 9.15). If it can handle up to 10,000 nodes, WISENES does not take into account the precise hardware architecture and remains succinct in the RF channel [51].

9.5.8 IDEA1

IDEA1 [52] (hIerarchical DEsign plAtform for sensor networks exploration) is a WSN design framework developed on SystemC and C++. Supporting the transaction-level modeling (TLM), a hierarchical approach can be adopted so as to "hide" parameters from lower abstraction layer if necessary. IDEA1 enables the design space exploration at an early stage for WSN system designers. Users can use it to evaluate the performances (e.g., packet delivery rate, transmission latency, and power consumption) of the network composed of the COTS (Component Off-The-Shelf) node system platforms or their own SoC designs.

IDEA1 is a component-based simulation framework. Every component is modeled as an individual SystemC module communicating with each other through channels. The architecture of IDEA1 is illustrated in Figure 9.16. The SystemC kernel acts as the simulation engine. It schedules the execution of processes and updates the state of all modules at every simulation

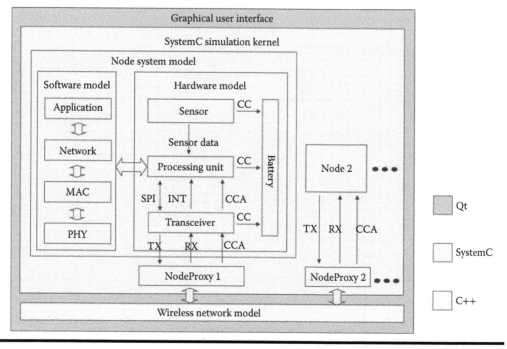

Figure 9.16 Architecture of IDEA1. (Data from http://www.idea1.fr.)

Table 9.4 Heterogeneity Support by IDEA1

Microcontroller Unit	RF Transceiver
ATMEL ATMega128	TI CC2420
Microchip PIC16LF88	TI CC1000
TI MSP430	Microchip MRF24J40 Nordic nRF24L01+

cycle. All active processes are invoked in order at the same simulator time, which creates an illusion of concurrency. Network modeling and RF communication are implemented in C++ so as to reduce simulation time. Each SystemC module communicates with the RF channel through proxy modules. Structuration of IDEA1 enables evaluation of performances in WSN and offers access to fine-grained analysis of energy consumption at hardware level as well as at network level. IDEA1 has been validated with experimental results presenting an error less than 4% (for performances at network levels as well as for detailed energy consumption at node level) and presents faster simulation time than NS2 while taking into account low-level parameters contrary to NS2.

Heterogeneity support is a key element of IDEA1 and distinguishes it from other simulators. Microcontroller and RF transceiver support are summarized in Table 9.4. This library enables the simulation of most of commercial and homemade platform available nowadays and then offers support for design space exploration for WSNs.

Most of the design consideration and modeling approaches developed in this chapter result from experience gained in the development of this framework and the study of its equivalents.

9.6 Modeling Methods for Wireless Sensor Networks

This section will deal with modeling methods for WSNs; we will treat WSN node modeling with emphasis on hardware architecture, software implementation, energy modeling, and RF communication modeling.

9.6.1 Hardware Modeling

Commonly used approach for modeling node in WSNs is to resort to finite state machine (FSM). Each hardware module (composite or not) can be described as a succession of states that are interconnected and triggered by external or internal events. Modeling of the network necessitates often analytical or behavioral formulation when finite element description is not used.

The main advantage of a FSM-based description is the modularity of this approach and the possibility to strongly couple hardware and software aspects without mentioning the fact that many hardware modules used in nodes of WSNs are described in datasheet under this form, hence facilitating the modeling (Figure 9.17). We will present here a conventional modeling approach that we used to develop models in our WSN framework.

9.6.1.1 Microcontroller Modeling

For microcontroller, a generic FSM description corresponding to classical WSN application is implemented. The generic classical embedded application is composed of periodic sampling from sensors, data storage, eventual aggregation of data, optional local computation, and sending data. We then obtain the representation (Figure 9.18; called hardware defined model of microcontroller) that is very similar to most models used in the WSN simulation community.

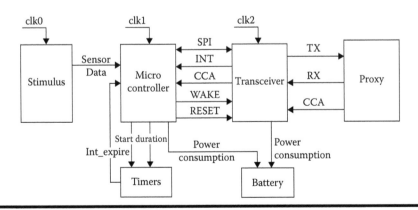

Figure 9.17 Classical modeling representation of WSN node.

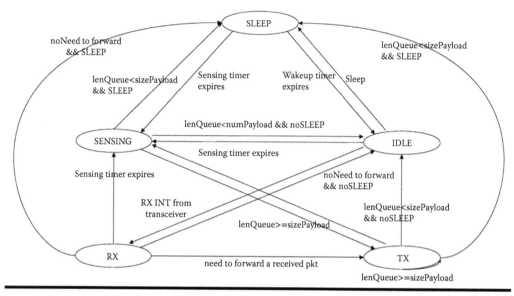

Figure 9.18 Generic model finite state machine for microcontroller.

This generic representation can be tuned so as to match any specific hardware architecture as well as embedded software implementation. Principal parameters of each state are transition time, duration (explicit or implicit, resulting from simulation), and current/voltage consumption.

The FSM is controlled by interrupts generated by the conversion or the RF transceiver. The microcontroller is then periodically awakened by one of its internal counter to acquire sensor data and to store it in one of its internal registers. Depending on the settings of the application, the node processes the data and tries to proceed to immediate transmission or accumulate it with older samples and waits until having reached a certain number of samples before transmission. The sending (TX) state and the receiving (RX) state are controlled through the Serial Peripheral Interface (SPI) link. Depending on the configuration and application, the RX and TX modes can be asked to interact directly (e.g., in the case of multi-hop transmission).

According to the transceiver support of the standard chosen for RF communication, the selected TX and RX states can be more or less complex. Indeed, in the case of a hardware support of a standard by the transceiver, RX and TX states, at microcontroller level, will be limited to simple sending of data through the SPI communication. On the contrary, in the case of software support of RF standard, these states will be more complex as they will support the smooth running of communication. Hence, the microcontroller must send to the transceiver all the information necessary for communication in addition to data (channel access checking, generation of the preamble, address management, etc.). This specialization of TX and RX states is made during the implementation of the software aspect in the hardware-defined model of the microcontroller. These two states can be implemented as FSM since this approach is hierarchical and authorizes the nesting of descriptions.

Figure 9.19 shows, as example, the implementation of the TX state of a software implementation of CSMA/CA algorithm in IEEE 802.15.4. The TX state of the generic model in Figure 9.18 is replaced by the submodel FSM corresponding to the software management of this algorithm.

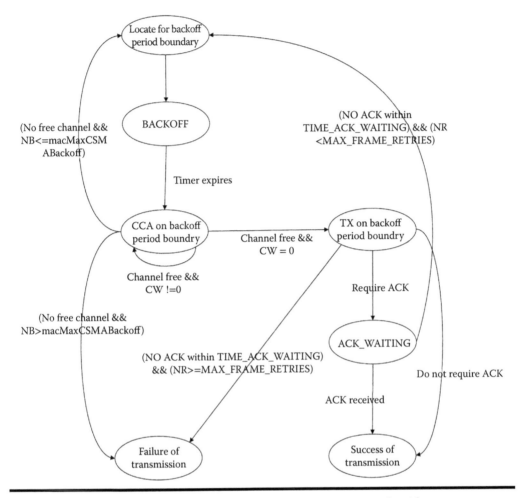

Figure 9.19 **Substate transmission model of IEEE 802.15.4 CSMA/CA algorithm.**

9.6.1.2 Radio Frequency Transceiver Model

The RF transceiver is also modeled as one or more FSMs (depending on supported protocols, several models can be developed), generally extracted from the technical datasheet (Figure 9.20).

Generic model makes appear a power-down mode in which RF transceiver has low consumption. When awoken RF transceiver switches to idle mode, ready to receive or transmit data after a calibration state. Each RX or TX state can be as complex as necessary depending on the selected protocol.

9.6.2 Software Modeling

Application through the use of WSNs is achieved by implementation of software in hardware mote. Section 9.6.1 described the FSM approach widely used in WSN simulation community. This description of hardware implementation can be altered to take into account software implementation. Three mains methods can be used: extraction of number of cycle of code from compiler, physical measurement, or use of ISS.

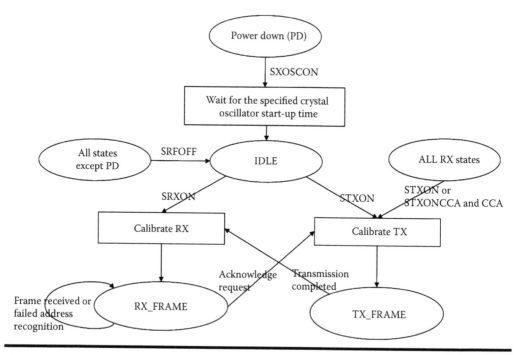

Figure 9.20   Generic model of RF transceiver.

Physical measurement needs people to implement effectively part of the code on the hardware mote and makes measurement for duration of code to be evaluated (it can be very difficult when code targeted necessitates data provided by other nodes of the network). This kind of experience has been realized for TelosB node [53]. This type of approach can be used when physical node can be integrated in the software design flow such as Worldsens or WISENES. The best example of this type of approach is AEON [14], implemented on top of Avrora.

Other approaches (ISS and extraction from compiler) present the drawback of considering only one node and not the entire network. The solutions can present problems for accurate evaluation of codes based on data provided by the network. Real code implementation implies that code must be compiled and then deployed on hardware (single node) for physical measurement or emulator so as to extract duration and transition time of the code in correlation with the state of microcontroller or RF transceiver. The use of ISS enables faster processing since it enables to establish the number of cycles of code implemented. This type of approach is found in PowerTOSSIM, IDEA1, Worldsens, and so on.

For global consideration of hardware architecture and software implementation at node level, we recommend the approach described in Figure 9.21.

9.6.3 Hardware-Defined Software-Implemented Model

For global consideration of hardware architecture and software implementation at node level, we recommend the approach described in Figure 9.21 that is very similar to those adopted in Worldsens or in WISENES. From library, the user must be able to select a microcontroller architecture and a RF transceiver structure. A preliminary FSM model is then established for microcontroller and RF transceiver. Being based on hardware performances only, they are called hardware-defined

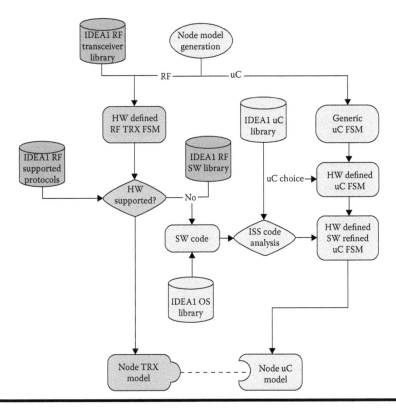

Figure 9.21 Global design methodology for FSM-based modeling in WSN.

FSM. Influence of software must be added to this model. Depending on the eventual hardware support of protocols by the RF transceiver, specific code to RF transmission could be included in the microcontroller. This part is added to the software to be implemented in the microcontroller unit. By using an ISS, source code can be analyzed and metrics are extracted from the code; the hardware-defined microcontroller FSM model is modified to implement software influence in the model to obtain a hardware-defined software-refined FSM model from microcontroller. This model is linked to RF transceiver definite FSM model (including eventual hardware support of protocol) to obtain the final hardware-defined software-refined node model.

9.6.4 Energy Model

There are a number of energy models relying on the establishment of analytical equations based on various physical parameters or network protocols [54]. These models generally determine the average power consumption per node in the network for a given application using a defined protocol.

Analytical approach use formal description to establish energy consumption. Besides the fact that these formulas are specific to a protocol [55,56] and then must be established for each protocol variations, they focus on the RF aspect (since most consuming) to the detriment of other considerations such as those summarized in Table 9.5.

Establishing a comprehensive model taking into account all these parameters is impossible. If the analytical approach can be used in the preparation of a specification of WSNs in an early design phase, it must be enhanced (or replaced) by an approach that allows the consideration of all these aspects and the inherent probabilistic events associated.

Table 9.5 Main Parameters Affecting Power Consumption

Power supply	Discharge rate
	Relaxation effect
	Battery dimensions
	Supply voltage
	Type of electrode material used and diffusion rate of the active materials in electrolyte for conventional cell
	Ratio and duty cycle of peak to average current
Sensors	Physical to electrical signal conversion
	Complexity of supporting components
	Signal sampling
	Signal conditioning
	Dynamic range (ratio between the smallest and largest possible measurements)
ADC	Sampling rate
	Aliasing
	Dither
Microcontroller	Core operating frequencies
	Power proportional to process and computational load
	Ambient temperature
	Application code
	Peripheral utilization
Radio	Modulation scheme
	Data rate
	Transmission range (determined by the transmission power)
	Operational duty cycle

Many simulators (from NSSNM and NENM families among which IDEA1) use an energy model* rather than based on computation results delivered by the simulation kernel. When representing node with FSM, the simulation kernel can estimate the actual consumption of each state (as long as parameters such as voltage and current are provided) from its calculation of state duration. This approach has the merit of being very close to the actual measurement method for consumption of any hardware node.

---

* One can hardly talk about energy model but rather a method of evaluation of the energy consumption. The name of the energy model being commonly used in literature for this approach will be used despite its inaccuracy.

The energy consumed by a node i of a WSN can be calculated as follows:

$$E_i = \sum_{j=0}^{N} \left(\sum_{k=0}^{M} E_{ijk} + \sum_{l=O}^{P} E_{ijl} \right) = \sum_{j=0}^{N} \left(\sum_{k=0}^{M} V.I_{ijk}.t_{ijk} + \sum_{l=O}^{P} V.I_{ijl}.t_{ijl} \right)$$

E_{ijk} represents the energy consumption of the kth state of the jth component of node i, and E_{ijl} stands for the energy consumption of the lth state transition of the jth component of node i. The node has N components consuming energy. Each component has M states and O transitions.* During the simulation, the time spent on different states and transitions, t_{ijk} and t_{ijl}, is known. Based on this information, the battery module (or directly the kernel simulation) calculates the energy consumptions of each component as well as the network lifetime during runtime.

This approach can be found in WISENES, PAWiS, IDEA1, PowerTOSSIM, and so on. Depending on the simulator, transition state can be neglected (WISENES). Hence, energy analysis of node relies exclusively on the accuracy of the FSM description, spanning from an average over the states to a detailed classification of each important consuming states.

9.6.5 Radio Frequency Communication Modeling

Modeling of RF channel is a research topic in itself. The simple question of choosing the right level of abstraction for a sufficient accuracy of physical phenomena representation is itself a major debate among experts [57]. If many works do exist in the field of the IEEE 802.11 standard [58–60], there are far fewer in the smaller community of WSN protocols [28,61,62], despite the established fact that first-order models are inadequate [63]. Basic models commonly used are the classical free space [64] (for which improvements can be implemented as proposed by Giacomin et al. [65]) and the indoor propagation model of ITU [66].

Establishing realistic radio propagation models compatible with time simulation of networks is one of the current bottlenecks in WSN simulation, and the adjunction of mobility with dynamic RF channel is adding to the complexity of that key problem.

9.7 Design Space Exploration in WSNs

Conventional use of simulators in WSNs restricts them to predeployment feasibility and rough estimation of performances and network lifetime. Despite the increasing or available platforms, common WSN design approach consists of selecting a hardware platform (based on various criteria such as microcontroller and transceiver architecture, supported OS, and personal knowledge of hardware) and then establishing its performances with some protocols' exploration to achieve better results. From our point of view, a WSN design tool should allow designers to explore hardware architectural solutions and software implementation. In this section, we will briefly present a design space exploration with IDEA1 for selecting the best-suited architecture for a mechatronic application (Figure 9.22).

---

* From an analytical point of view, this approach underpins that all components have the same number of states; in fact, M is the largest number of states met on the set of N components. Thus, a component having number of states less than N (e.g., H) has all states greater than H null ($E_{ij(k > H)} = 0$).

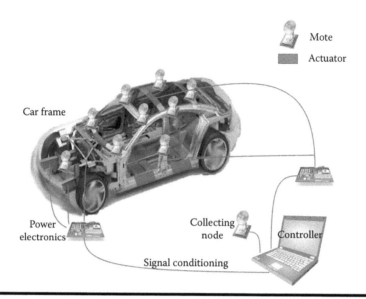

Figure 9.22 Active vibration control for an automobile.

9.7.1 Specifications

The targeted application is for exploration of WSN deployment for active control of vibrations for acoustic comfort in automobiles. Mechanical specifications define a sampling rate of 1 kHz and a sample size of 8 bits. While not totally appropriate for high data rate application, ZigBee can be considered as a viable alternative for automotive applications [67]. According to Tamar ElBatt [68], only IEEE 802.15.4 Guaranteed Time Slot (GTS) can offer the appropriate support, yet presenting a worst-case latency of 16 ms. In consequence of mechanical specifications and Tamar ElBatt preconization, a star topology composed of eight nodes and a coordinator is chosen. Since conventional GTS only offers support to up to seven nodes, we instantiate the variant proposed by Cheng et al. [69] (named TDMA–Based GTS) recommended for industrial applications supporting multiple nodes with reduced latency.

9.7.2 Hardware Platform Presentation

Hardware platforms explored in this work are as follows:

- MICAz, a well-known platform often used as reference composed of an ATMEGA128 microcontroller and a CC2420 transceiver (partial hardware support of IEEE 802.15.4).
- N@L, a specific node developed for this application based on Microchip hardware: PIC16LF88 microcontroller selected for its nanoWatt technology and the MRF24J40 transceiver chosen for its full hardware support of IEEE 802.154.

9.7.3 Results

Both non-beacon and beacon mode have been explored in this work. As attended, only beacon mode offers a near 100% packet delivery rate. Both IEEE 802.15.4 GTS (with a network reduced to seven nodes and a coordinator) and TDMA–Based GTS have been studied. Best

Table 9.6   Network Level Results from Design Space Exploration

Results	*IEEE 802154 GTS*		*TDMA-Based GTS*	
Node	MICAz	N@L	MICAz	N@L
Payload	30	15	10	19
BO	1	0	n/a	n/a
BI (µs)	30,720	15,360	10,000	19,000
PDR: Packet delivery rate (%)	97.4	97.4	100	100
AL: Average latency (µs)	53,854	42,777	6,953	12,508
ECPkt: Energy consumption per packet (µJ/pkt)	1,283	1,001	425	408
NAPC: Node average power consumption (µW)	41,071	64,630	42,300	21,264
NAPC microcontroller (µW)	29,915	4,448	29,928	4,573
NAPC transceiver (µW)	11,157	60,182	12,371	16,691

results (considering specifications) are presented in Table 9.6. What is striking is the noticeable differences both in protocol parameters used for achievement of specifications and in performance results for each architecture. The first conclusion is the necessity to aggregate sampled data; it is impossible to send data as soon as they are collected. Depending on the protocol, they must be sent by packets of 10–30. This can have a strong impact on control law that should be implemented in active control. A payload of 10 with an average latency of 6.95 ms means that when data are delivered to coordinator nodes, the first sample of the packet is 16.95 ms old (sampling period of 1 ms). This has a huge impact of the establishment of the control law that can drive the designer to proceed to undersampling or to use retarded control law so as to work on larger number of samples. TDMA–Based GTS algorithm presents best overall performances. For this algorithm, if MICAz node has the lowest payload and better latency, its consumption is twice compared to the N@L platform. For more in-depth analysis, IDEA1 can extract results at node level as shown in Figure 9.23.

General tendencies specific to node hardware architecture can be highlighted. Microcontroller from Microchip, taking benefices of its newer architecture and its nanoWatt technology, is less consuming than the ATmega128, particularly in the process of integrated analog-to-digital conversion. The partial hardware support of IEEE 802.15.4 of MICAz's transceiver results in an increased workload on the microcontroller that must manage the missing part of support by software, resulting in an important consumption. Full hardware support of IEEE 802.15.4 by N@L's transceiver results in a weak consumption for PIC16LF88. From a transceiver point of view, the Chipcon circuit is less consuming than the Microchip one with a huge consumption to be noticed for the IEEE 802.15.4 despite its full hardware support of the norm. This unexpected result can be observed due to the high accuracy of low-level abstraction layer of IDEA1 that enables the identification of bottleneck through the different layers of abstraction. A thorough analysis shows that the hardware implementation of IEEE 802.15.4 norm in MRF24J40 allows the transceiver in active mode between the beacon and the slot, which is allocated to the node instead of putting

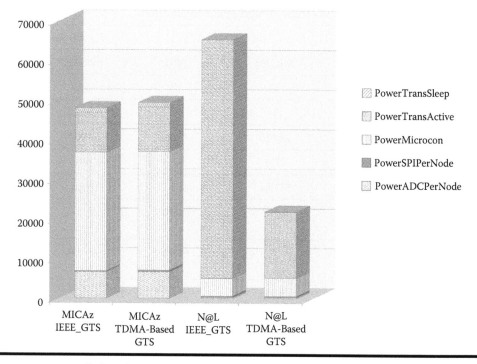

Figure 9.23 **Detailed analysis at node level.**

it in sleep mode. This poor hardware implementation presents a major drawback for this architecture. When implementing GTS by software as made on TDMA–Based GTS for MRF24J40, this absurd behavior disappears.

9.8 Conclusion

In this chapter, simulation of WSN stakes has been presented. From the stakes and constraints to be considered to the modeling methods commonly used for WSN simulation, simulation bottlenecks have been identified and highlighted. Since energy is at the heart of WSNs, a particular emphasis has been put on its treatment in WSN design process. Heterogeneity and necessity for considering low-level abstraction layer at network layer (and above) have been particularly treated. A design space exploration example under IDEA1 framework has been introduced to illustrate the problems and to show that the software architecture of the node can be ignored when designing WSNs. If few simulators present the capacity to proceed for this type of approach, IDEA1, due to its large model library and its hierarchical approach, can be used in a top-down and bottom-up design approach, similar to the design process commonly used in microelectronic industry, making it more a real WSN design framework than a "simple" simulator.

References

1. Feng, M. W. 2008. "Wireless Sensor Network Industrial View? What Will Be the Killer Apps for Wireless Sensor Network?" In *International Conference on Sensor Networks, Ubiquitous, and Trustworthy Computing*, vol. 0, 270. IEEE Computer Society.
2. Akyildiz, I. F., and M. C. Vuran. 2010. *Wireless Sensor Networks*. UK: John Wiley & Sons Inc.

3. Johnson, R. C. 2011. "Millimeter-Scale Computer Targets Environmental, Medical Monitoring." *EE Times* 1597: 26.
4. Carolina Fortuna. 2010. "Why is Sensor Data Hard to Get?" COIN-ACTIVE Summer School on Advanced Technologies for Knowledge Intensive Networked Organizations in Aachen.
5. Mica2 datasheet. Crossbow Technology, Inc.
6. Micaz datasheet. XBow, 2010.
7. Polastre, J., R. Szewczyk, and D. E. Culler. 2005. "Telos: Enabling Ultra-Low-Power Wireless Research." In *Information Processing in Sensor Networks*, 364–9.
8. Farooq, M. O., and T. Kunz. May 2011. "Operating Systems for Wireless Sensor Networks: A Survey." *Sensors* 11 (6): 5900–30.
9. Hill, J., R. Szewczyk, A. Woo, S. Hollar, D. E. Culler, and K. S. J. Pister. 2000. "System Architecture Directions for Networked Sensors." In *Architectural Support for Programming Languages and Operating systems*, 93–104.
10. Han, C. C., R. Kumar, R. Shea, E. Kohler, and M. Srivastava. 2005. "A Dynamic Operating System for Sensor Nodes." In *Mobisys '05: Proceedings of the 3rd International Conference on Mobile Systems, Applications, and Services*, 163–76. New York, NY: ACM.
11. Bhatti, S., J. Carlson, H. Dai, J. Deng, J. Rose, A. Sheth, B. Shucker, C. Gruenwald, A. Torgerson, and R. Han. 2005. "MANTIS OS: An Embedded Multithreaded Operating System for Wireless Micro Sensor Platforms." *Mobile Networks and Applications* 10 (4): 563–79.
12. Dunkels, A., B. Gronvall, and T. Voigt. 2004. "Contiki: A Lightweight and Flexible Operating System for Tiny Networked Sensors." In *Lcn '04: Proceedings of the 29th Annual IEEE International Conference on Local Computer Networks*, vol. 0, 455–62. Los Alamitos, CA: IEEE Computer Society.
13. Lajara R., J. Pelegrí-Sebastiá, and J. J. Perez Solano. 2010. "Power Consumption Analysis of Operating Systems for Wireless Sensor Networks." *Sensors* 10 (6): 5809–26.
14. Landsiedel, O., K. Wehrle, and S. Gotz. 2005. "Accurate Prediction of Power Consumption in Sensor Networks." In *Proceedings of the 2nd IEEE Workshop on Embedded Networked Sensors*, 37–44. Washington, DC: IEEE Computer Society.
15. Raghunathan, V., C. Schurgers, S. Park, and M. B. Srivastava. March. 2002. "Energy-Aware Wireless Microsensor Networks." *IEEE Signal Processing Magazine* 19 (2): 40–50.
16. *IEEE Standard for Information technology - Telecommunications and information exchange between systems- Local and metropolitan area networks- Specific requirements Part 15.4: Wireless Medium Access Control (MAC) and Physical Layer (PHY) Specifications for Low-Rate Wireless Personal Area Networks (WPANs)*. Institute of Electrical and Electronics Engineers, 2006.
17. ZigBee Alliance. 2012. Zigbee specifications. http://www.zigbee.org.
18. Mahalik, N. P. 2006. *Sensor Networks and Configuration: Fundamentals, Standards, Platforms, and Applications*. Secaucus, NJ: Springer-Verlag New York, Inc.
19. Yick, J., B. Mukherjee, and D. Ghosal. August 2008. "Wireless Sensor Network Survey." *Computer Networks* 52: 2292–330.
20. Bird, N. C. December 2007. "RF challenges for WSN and Beyond." 2nd Workshop on Wireless Sensor Networks Research, WiSen2007, Dublin, Ireland.
21. Jennic, August 2006. "Application Note JN-AN-1035 - Calculating 802.15.4 Data Rates." Technical Report.
22. Shakshuki, E. M., H. Malik, and T. R. Sheltami. 2011. "A Comparative Study on Simulation vs. Real Time Deployment in Wireless Sensor Networks." *Journal of Systems and Software* 84 (1): 45–54.
23. Akyildiz, I. F., W. Su, Y. Sankarasubramaniam, and E. Cayirci. 2002. "Wireless Sensor Networks: A Survey." *Computer Networks* 38: 393–422.
24. Anastasi, G., M. Conti, M. Di. Francesco, and A. Passarella. May 2009. "Energy Conservation in Wireless Sensor Networks: A Survey." *Ad Hoc Networks* 7 (3): 537–68.
25. Varshney, M., and R. Bagrodia. 2004. "Detailed Models for Sensor Network Simulations and their Impact on Network Performance." In *Proceedings of the 7th ACM International Symposium on Modeling, Analysis and Simulation of Wireless and Mobile Systems*, 70–7. Venice, Italy: ACM.
26. Girod, L., T. Stathopoulos, N. Ramanathan, J. Elson, T. Osterweil, E. Schoellhammer, and D. Estrin. 2004. "A System for Simulation, Emulation, and Deployment of Heterogeneous Sensor Networks."

In *Proceedings of the 2nd International Conference on Embedded Networked Sensor Systems,* 201–13. Baltimore, MD.

27. Ammer, J., and J. Rabaey. June 2006. "The Energy-Per-Useful-Bit Metric for Evaluating and Optimizing Sensor Network Physical Layers." In *Proceedings of the IEEE International Workshop on Wireless Ad Hoc & Sensor Network.* New York: IEEE.

28. Kotz, D., C. Newport, R. S. Gray, J. Liu, Y. Yuan, and C. Elliott. 2004. "Experimental Evaluation of Wireless Simulation Assumptions." In *Proceedings of the 7th ACM International Symposium on Modeling, Analysis and Simulation of Wireless and Mobile Systems, mswim'04,* 78–82. New York: ACM Press.

29. Kurkowski, S., T. Camp, and M. Colagrosso. October 2005. "MANET Simulation Studies: The Incredibles." *SIGMOBILE Mobile Computing and Communications Review* 9 (4): 50–61.

30. Andel, T. R., and A. Yasinsac. 2006. "On the Credibility of MANET Simulations." *IEEE Computer* 39 (7): 48–54. New York: IEEE.

31. Stojmenovic, I. 2008. "Simulations in Wireless Sensor and Ad Hoc Networks: Matching and Advancing Models, Metrics, and Solutions." *Communications Magazine, IEEE* 46 (12): 102–7. New York: IEEE.

32. Du, W., D. Navarro, F. Mieyeville, and F. Gaffiot. 2010. "Towards a Taxonomy of Simulation Tools for Wireless Sensor Networks." In *Proceedings of the 3rd International ICST Conference on Simulation Tools and Techniques, SIMUTools '10.* 52:1–52:7. Brussels, Belgium: ICST (Institute for Computer Sciences, Social-Informatics and Telecommunications Engineering).

33. Dwivedi, A. K., and O. P. Vyas. 2011. "An Exploratory Study of Experimental Tools for Wireless Sensor Networks." *Wireless Sensor Network* 3: 215–40.

34. Riccobene, E., P. Scandurra, S. Bocchio, A. Rosti, L. Lavazza, and L. Mantellini. 2009. "SystemC/C-Based Model-Driven Design for Embedded Systems." *ACM Transactions on Embedded Computing Systems* 8 (4): 1–37.

35. Fall, K., and K. Varadhan. January 2009. *The ns Manual* (formerly ns Notes and Documentation).

36. Naoumov, V., and T. Gross. 2003. "Simulation of Large Ad Hoc Networks." In *Proceedings of the 6th ACM International Workshop on Modeling Analysis and Simulation of Wireless and Mobile Systems, MSWIM '03,* 50–7. New York, NY: ACM.

37. Park, S., A. Savvides, and M. B. Srivastava. 2000. "SensorSim: A Simulation Framework for Sensor Networks." In *Mswim '00: Proceedings of the 3rd ACM International Workshop on Modeling, Analysis and Simulation of Wireless and Mobile Systems,* 104–11. New York, NY: ACM.

38. Riley, G. F. 2003. "Large-Scale Network Simulations with GTNetS." In *Proceedings of the 2003 Winter Simulation Conference,* vol. 1, 676–84.

39. Font, J. L., P. Iñigo, M. Domínguez, J. L. Sevillano, and C. Amaya. 2011. "Analysis of Source Code Metrics from ns-2 and ns-3 Network Simulators." *Simulation Modelling Practice and Theory* 19 (5): 1330–46.

40. Mallanda, C., A. Suri, V. Kunchakarra, S. S. Iyengar, R. Kannan, A. Durresi, and S. Sastry. Simulating wireless sensor networks with OMNeT++.

41. Weingartner, E., H. vom Lehn, and K. Wehrle. 2009. "A Performance Comparison of Recent Network Simulators." In *Proceedings of the IEEE International Conference on Communications 2009, ICC 20 09.* Dresden, Germany: IEEE.

42. Köpke, A., M. Swigulski, K. Wessel, D. Willkomm, P. T. Klein Haneveld, T. E. V. Parker, O. W. Visser, H. S. Lichte, and S. Valentin. 2008. "Simulating Wireless and Mobile Networks in OMNeT++: The MiXiM vision." In *Proceedings of the 1st International Conference on Simulation Tools and Techniques for Communications, Networks and Systems & Workshops,* SIMUTOOLS '08, 71:1–71:8. Brussels, Belgium: ICST (Institute for Computer Sciences, Social-Informatics and Telecommunications Engineering).

43. Drytkiewicz, W., S. Sroka, V. Handziski, A. Köpke, H. Karl, and Technische Universität Berlin. 2003. A Mobility Framework for OMNeT++.

44. Glaser, J., D. Weber, S. A. Madani, and S. Mahlknecht. 2008. "Power Aware Simulation Framework for Wireless Sensor Networks and Nodes." *EURASIP Journal on Embedded Systems* 2008: 1–16.

45. Levis, P., N. Lee, M. Welsh, and D. Culler. 2003. "TOSSIM: Accurate and Scalable Simulation of Entire TinyOS Applications." In *Sensys '03: Proceedings of the 1st International Conference on Embedded Networked Sensor Systems,* 126–37. New York, NY: ACM.

46. Shnayder, V., M. Hempstead, B.-R. Chen, G. W. Allen, and M. Welsh. 2004. "Simulating the Power Consumption of Large-Scale Sensor Network Applications." In *Proceedings of the 2nd International Conference on Embedded Networked Sensor Systems*, SenSys '04, 188–200. New York, NY: ACM.

47. Perla, E., A. O. Cathain, R. S. Carbajo, M. Huggard, and C. M. Goldrick. 2008. "PowerTOSSIM z: Realistic Energy Modelling for Wireless Sensor Network Environments." In *Proceedings of the 3rd ACM Workshop on Performance Monitoring and Measurement of Heterogeneous Wireless and Wired Networks*, PM2HW2N '08, 35–42. New York, NY: ACM.

48. Titzer, B. L., et al. 2005. "Avrora: Scalable Sensor Network Simulation With Precise Timing." *In Proc. of the 4th Int'l Conf. on Information Processing in Sensor Networks*, 477–482.

49. Polley, J., D. Blazakis, J. McGee, D. Rusk, and J. S. Baras. 2004. "Atemu: A Fine-Grained Sensor Network Simulator." In *First Annual IEEE Communications Society Conference on Sensor and Ad Hoc Communications and Networks, IEEE Secon 2004*, 145–52. New York: IEEE.

50. Fraboulet, A., G. Chelius, and E. Fleury. 2007. "Worldsens: Development and Prototyping Tools for Application Specific Wireless Sensors Networks." In *Proceedings of the 6th International Conference on Information Processing in Sensor Networks*, IPSN '07, 176–85. New York, NY: ACM.

51. Kuorilehto, M., M. Hännikäinen, and T. D. Hämäläinen. 2008. "Rapid Design and Evaluation Framework for Wireless Sensor Networks." *Ad Hoc Networks* 6 (6): 909–35.

52. Du, W., F. Mieyeville, D. Navarro, and I. O'Connor. 2011. "Idea1: A Validated Systemc-Based System-Level Design and Simulation Environment for Wireless Sensor Networks." *EURASIP Journal on Wireless Communications and Networking* 2011 (1): 143.

53. Prayati, A., C. Antonopoulos, T. Stoyanova, C. Koulamas, and G. Papadopoulos. 2010. "A Modeling Approach on the TelosB WSN Platform Power Consumption." *Journal of Systems and Software* 83 (8).

54. Holland, M., T. Wang, B. Tavli, A. Seyedi, and W. Heinzelman. February 2011. "Optimizing Physical-Layer Parameters for Wireless Sensor Networks." *ACM Transactions on Sensor Networks* 7: 28:1–28:20.

55. Buratti, C. 2009. "A Mathematical Model for Performance of IEEE 802.15.4 Beacon-Enabled Mode." In *IWCMC*, eds. M. Guizani, P. Müller, K.-P. Fähnrich, A. V. Vasilakos, Y. Zhang, and J. Zhang, 1184–90. New York: ACM.

56. Buratti, C., and R. Verdone. 2009. "Performance Analysis of IEEE 802.15.4 Non Beacon-Enabled Mode." *IEEE Transactions on Vehicular Technology* 58 (7): 3480–93.

57. Iyer, A., C. Rosenberg, and A. Karnik. May 2009. "What is the Right Model for Wireless Channel Interference?" *IEEE Transactions on Wireless Communications* 8 (5): 2662–71.

58. Aráuz, J., and P. Krishnamurthy. 2003. "Markov Modeling of 802.11 Channels." In *58th IEEE Vehicular Technology Conference, vtc 2003-fall*, vol. 2, 771–5. New York: IEEE.

59. Konrad, A., B. Y. Zhao, A. D. Joseph, and R. Ludwig. 2003. "A Markov-Based Channel Model Algorithm for Wireless Networks." *Wireless Networks* 9 (3): 189–99.

60. Khayam, S. A., and H. Radha. 2006. "Constant-Complexity Models for Wireless Channels." In *Proceedings of the 25th Annual Joint Conference on the IEEE Computer and Communications Societies, infocom '06*, 1–11.

61. Zhao, J., and R. Govindan. 2003. "Understanding Packet Delivery Performance in Dense Wireless Sensor Networks." In *Proceedings of the 1st International Conference on Embedded Networked Sensor Systems, SenSys '03*, 1–13. New York, NY: ACM.

62. Lee, H., A. Cerpa, and P. Levis. 2007. "Improving Wireless Simulation Through Noise Modeling." In *Proceedings of the 6th International Conference on Information Processing in Sensor Networks*, IPSN '07, 21–30. New York, NY: ACM.

63. Iqbal, A., and S. A. Khayam. May 2008. "Improving WSN Simulation and Analysis Accuracy Using Two-Tier Channel Models." In *2008 IEEE International Conference on Communications*, 349–53. New York: IEEE.

64. Linmartz, J.-P. M. G. 1996. *Wireless Communication*. Amsterdam: Baltzer Science Publishers.

65. Giacomin, J. C., L. H. A. Correia, T. Heimfarth, V. F. Silva, G. M. Pereira, and J. L. P. de Santana. 2010. "Radio Channel Model of Wireless Sensor Networks Operating in 2.4 ghz ism." *INFOCOMP Journal of Computer Science* 9 (1): 98–106.

66. Propagation Data and Prediction Methods for the Planning of Indoor Radiocomm. Systems and Radio Local Area Networks in the Frequency Range 900 mhz to 100 ghz. 1999. Recommendation ITU-R P.1238-1.

67. Nolte, T., H. Hansson, and L. Lo. Bello. July 2005. "Wireless Automotive Communications." In *Euromicro Conference on Real-Time Systems,* 35–8. Palma de Mayorca, Spain.

68. ElBatt, T., C. Saraydar, M. Ames, and T. Talty. 2006. "Potential for Intra-Vehicle Wireless Automotive Sensor Networks." In *2006 IEEE Sarnoff Symposium*, 27–8. New York: IEEE.

69. Chen, F., T. Talanis, R. German, and F. Dressler. July 2009. "Real-Time Enabled IEEE 802.15.4 Sensor Networks in Industrial Automation." In *IEEE Symposium on Industrial Embedded Systems, sies 2009*, 136–9. Lausanne, Switzerland: New York: IEEE.

70. Hadim, S., and N. Mohamed. 2006. "Middleware: Middleware Challenges and Approaches for Wireless Sensor Networks." In *IEEE Distributed Systems Online archive*, 7(3). Piscataway, NJ: IEEE Education Activities Department.

Chapter 10

Quality of Service MAC for Wireless Sensor Networks

Bilal Muhammad Khan and Rabia Bilal

Contents

10.1 Introduction

Since wireless sensor networks (WSNs) are being used in many emerging applications, the requirement of providing high quality of service (QoS) is becoming ever more necessary. This highlights major issues like collision, scalability, latency, throughput, and energy consumption. In addition, mobile sensor network faces further challenges like link failure, neighborhood information, association, scheduling, synchronization, and collision. Medium access control (MAC) protocols play vital role in solving these key issues. This chapter presents the fundamentals of MAC protocols and explains the specific requirements and problems these protocols have to withstand for WSN. The QoS is addressed for both static and mobile sensor networks with detailed case study of the Institute of Electrical Electronics Engineering (IEEE) 802.15.4 wireless personal area network (WPAN) standard. Research challenges with literature survey and further directions are also discussed. The chapter ends with conclusions and references.

10.2 Fundamentals of Wireless MAC Protocols

MAC layer is a part of the data link layer in the Open System Internet (OSI) reference model. The task of the MAC layer is clear and well defined; it determines the access mechanism and time for a node to try to transmit data, control or manage packets, to another node in the case of unicast scenario or to a set of nodes in the multicast scenario. The two important responsibilities of the remaining part of the data link layer are error control and flow control. Error control techniques are used to ensure that transmission incurs no errors and to take corrective measures if there are errors in transmission, whereas flow control regulates the rate of transmission, which is crucial in the case of slow receivers to prevent them from being overloaded with data.

10.2.1 Requirements and Design Constraints
for Wireless MAC Protocols

The most important performance measurements for MAC protocols are throughput, fairness, stability, low access delay, low transmission delay, low overhead, and less energy consumption. There are several causes for overhead in MAC protocols; it can be the result of length per packet, such as

MAC header and trailer, collisions, by exchange of extra control packets. Packet collision tends to occur if the MAC protocol allows two or more nodes to access the shared communication medium at the same time. Collision is the result of the inability of the receiver to decode the packets correctly; this triggers the mechanism of retransmission of data packets. For time-critical applications, this delay and retransmission process is unacceptable; in some applications, the transmission of important packets is preferred over unimportant ones, which generates another class of MAC protocols, which are based on the concept of priorities.

The operation and performance of MAC protocols is heavily influenced by the properties of the physical layer. As the signal is transmitted over the wireless link, its strength decreases with the increase in the distance between transmitting and receiving nodes. This loss in signal strength is a very important factor in terms of successfully demodulating the signal at the receiving node since transceivers need minimum signal strength to carry out this task. This leads to a maximum range that a sensor node can reach with a given transmit power. However, if the two nodes are out of reach from each other, they will be unable to receive each other's ongoing communication, which gives rise to one of the challenges that MAC protocols have to overcome in the wireless domain of hidden and exposed nodes [1].

10.2.1.1 Hidden Node Problem

The hidden node problem occurs in Carrier Sense Multiple Access (CSMA) protocols in which a node senses the medium before the start of transmission. If the medium is found to be busy, the node defers its transmission to avoid collision and retransmission. Figure 10.1 gives a typical scenario for hidden node problem. There are three nodes A, B, and C in the network. The arrangement of the nodes is such that node A and node B are in communication range as well as node C and node B; however, node A and node C are out of communication range from each other. Assuming that node A starts to communicate data packets toward node B, after some time, node C senses the channel and finds it idle as it is out of the communication range of node A, and it also starts to transmit packets toward node B resulting in collision due to hidden node. Using simple CSMA MAC protocol in such a scenario leads to collision and needless retransmission of data packets.

10.2.1.2 Exposed Node Problem

Consider Figure 10.2 in which there are four nodes A, B, C, and D. Node C and node D are out of the communication range from node A and node B respectively. If node B starts to transmit data to node A, node C attempts to transmit data toward node D after some time, and in normal

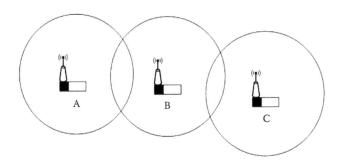

Figure 10.1   Hidden node problem in wireless networks.

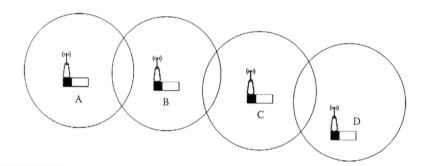

Figure 10.2 Exposed node problem in wireless networks.

circumstances, this communication is possible; however, in the case of wireless communication, especially using simple CSMA protocol, node C transmission will be suppressed, this is due to the fact that before transmission, node C, according to the CSMA procedure, tends to sense the channel and although the ongoing transmission is between node B and node A, node C considers that the medium is busy and defers its transmission, causing waste of bandwidth.

The solutions for hidden and exposed node terminals are proposed in the shape of busy tone [1] and request to send (RTS)/clear to send (CTS) handshaking signals used in IEEE 802.11 [2] and are implemented in MACA [3] and MACAW [4] MAC protocols.

10.3 Classification of QoS Wireless MAC Protocols

There are several wireless MAC protocols available in literature. These protocols can be broadly classified into fixed assignment protocols, demand assignment protocols, and random access protocols.

10.3.1 Fixed Assignment Protocols

The fixed assignment class of protocols is collision free, and all the available resources are divided between the nodes such that each node can use its resources exclusively. In case of topology changes due to mobility, joining of new nodes or malfunction of nodes, or changes in the load pattern, different signaling is used to renegotiate the resources to the remaining nodes in the network. Hence, such type of protocols may not be viable in terms of scalable networks.

Typical protocols that fall in this category are time division multiple access (TDMA), frequency division multiple access (FDMA), code division multiple access (CDMA), and space division multiple access (SDMA). TDMA [5] divides the time into several superframes depending on the size of network, and these superframes are further subdivided into time slots. These time slots are assigned to the nodes in the network, which transmit data in the specified slots periodically in every superframe. To avoid overlapping of different time slots, the TDMA system requires synchronization. In the case of FDMA, the frequency band is divided into a number of subchannels, and the nodes are individually assigned these subchannels to transmit the data. In such protocols, frequency synchronization is required, and the transceiver of FDMA is more complex than that of TDMA. In CDMA [6–8], the nodes spread the signal over a much larger bandwidth using different codes to separate their transmissions. The receiving node must know the codes used by each transmitter. However, in SDMA, spatial separations are used by nodes to separate their transmissions. The SDMA protocols require an array of antenna and complex signal processing algorithms [9]; hence, it is not a suitable candidate to be used in resource-constrained WSNs.

10.3.2 Demand Assignment Protocols

In this class of protocols, the resources are allocated to the nodes on a temporary short-term basis upon request and usually last long enough to support the data burst. This class of protocol is divided into centralized and distributed protocols. The examples of centralized protocols include HIPERLAN/2 [10–14], DQRUMA [15], and MASCARA protocols [16] as well as polling schemes [17–19]. In such protocols, nodes send out requests for the allocation of resources to the central coordinator, which accepts or rejects the request depending on the availability of resources. In the case of successful allocation, the node is sent acknowledgment with the information of the allocated resources, that is, the time slot in TDMA and the duration in which the node has to complete its transmission. The request from the nodes to the coordinator is usually contention based. Alternatively, the nodes combine the requests with the data packets transmitted in their specified time slot avoiding transmission of a separate request. Another approach is that the coordinator polls all the joining nodes for their request to reserve or allocate the resources. In such protocols, the central node or the coordinator needs to be awake all the time to receive any incoming request and to allocate the resources to the requesting node. The deallocation of resources is performed by the central coordinator if the node is not using its allocated slot. Thus, the node does not have to send an extra deallocation packet to the coordinator. This class of protocol requires the coordinator perform several tasks that are energy exhaustive. Thus, the protocols are suitable for the networks that have no constraints over energy or more than one node having no constraints of energy.

Token bus [20] is an example of a distributed demand assignment protocol. The node capturing the token has the right to initiate the transmission. Token frame is rotated among the nodes that are in logical ring topology. Special management procedures are devised to include or exclude a node from the ring as well as to handle failures such as the token being lost. Token passing protocols have also been considered for wireless applications [21–23], but they suffer greatly due to the dynamic changing channel conditions and struggle to maintain logical ring topology. Due to the ad hoc nature of the network, maintaining ring topology involves a significant amount of signaling on top of token passing. Moreover, the node transceiver has to be in the ON state as the timing of token passing is variable to avoid token loss and breaking of the ring.

10.3.3 Random Access Protocols

The first random access protocol was ALOHA or slotted ALOHA [24]. In the case of ALOHA, the node that wants to transmit the packet starts the transmission immediately without coordinating with other nodes, thus increasing the chances of collisions at the receiver. To find out about the successful transmission, the receiver sends an acknowledgment signal to the transmitter; in the absence of the acknowledgment, the transmitter assumes that a collision occurred at the receiver. In this condition, the transmitter backs off for a random amount of time and starts the process again. Thus, ALOHA provides short access under low traffic load; however, under high traffic and dense network conditions, the number of collisions increases and the network efficiency and throughput is degraded significantly. In the case of slotted ALOHA, the superframe is subdivided into time slots of equal lengths. The nodes have to synchronize to transmit the data in these time slots as only those nodes will be allowed to transmit, which begin transmission at the start of the time slot; during a specific time slot, any other node that wants to transmit data has to wait for the starting of a new time slot; thus, by synchronization, number of collisions is minimized and the throughput efficiency of the slotted ALOHA is better than the pure ALOHA.

To improve the efficiency of ALOHA protocols and maintain an acceptable QoS, CSMA protocols [25] are proposed. In these protocols, the transmitting node senses the channel; this procedure is known as carrier sensing; if the channel is available or in other terms there is no ongoing transmission over the channel, the node starts transmitting. If the node finds the channel busy during the sensing procedure, then it defers its transmission by using several possible algorithms. In case of nonpersistent CSMA, the node waits for a random time after it senses the channel. Before this random waiting time, the node does not care about the state of the channel. In persistent CSMA, the node waits for the ongoing transmission and then follows the backoff algorithm. In case of p-persistent, the CSMA transmitting node starts the transmission in a time slot with some probability p and waits for the next slot for transmission with probability $1 - p$. If some other node in the network starts to transmit, the node defers and repeats the entire procedure. Selection of smaller values of p makes collisions highly unlikely but at the cost of large access delays; conversely, larger values of p yield higher number of collisions.

In the case of the backoff algorithm used in distributed coordination function (DCF), the transmitting node chooses a random value and starts its timer. The timer is decremented after each time slot; if another node starts to transmit in these time slots, the node freezes its timer and resumes after the transmission is over. If the timer of the node reaches zero, the node starts transmitting its data packet over the channel. If the node transmission confronts an error, such as nonreception of acknowledgment frame from the intended receiver, then it doubles the value of the contention window (CW) and the whole procedure of contention continues again. The CSMA algorithm is susceptible to hidden terminal problems.

10.4 QoS Challenges in Wireless Sensor Networks

In the case of WSNs, the requirement is different from traditional wireless MAC protocols. The major concern is to conserve energy. The significance of energy efficiency in the design of MAC protocols is relatively new, and several of the traditional protocols such as ALOHA and CSMA have no provision to accommodate this feature. Energy conservation accompanied by typical performance factors such as fairness, throughput, and latency is discussed by Ye et al., Lin et al., and Schurgers et al. [26-28]. Moreover, scalability and robustness also emerge as prime and significant requirements in the design of MAC protocol for WSNs. These QoS challenges for WSNs are explained in this section.

- *Resource constraints*: WSNs typically lack in bandwidth, memory, energy, and processing capabilities. However, energy is by far the most crucial factor as in many scenarios it is almost impractical to replace or recharge batteries of sensor nodes. Therefore, it is of paramount importance that any proposed QoS MAC should be simple to operate on a highly resource-constrained sensor node.
- *Collisions*: Collision causes unnecessary waste of energy in the retransmission of the collided packet at the transmitting node as well as receiving the same packet again at the receiving node. Collision incurs high latencies and affects the throughput of the network. Tackling the problem of collision is the prime goal in designing any QoS MAC protocol.

- *Node deployment*: Node deployment in WSNs can be deterministic or random. In case of deterministic deployment, the nodes are deployed in fixed known topology and routing can be performed through prescheduled paths. Whereas in random deployment, the nodes organize them self-randomly.

- *Topology changes*: Due to node mobility, link failure, energy depletion, or security attacks, the topology is changed. One more factor that is inherent to WSN is that to save energy, most MAC protocols use sleep-listen mode and turn off the radio completely for energy-saving purpose; this results in topology change as well. Hence, the dynamic nature of topology change in WSN introduces an extra challenge for QoS support.

- *Data redundancy*: Most of the applications for WSNs comprise homogenous nodes; therefore, an event can be detected by several nodes. This redundancy is helpful in reliable data transmission; it also causes unnecessary data delivery that results in congestion. There are data aggregation protocols [29,30] that deal with such type of problems, but they may also introduce additional delays and complexity in the system. Hence, efficient and effective QoS mechanisms are required to deal with data redundancy issue.

- *Real-time traffic*: There are certain applications in which data is valid for a certain amount of time frame. For such kind of application, QoS requirements are very high, and a suitable protocol is required, which meets this criterion without compromising other attributes of MAC.

- *Overhearing*: Wireless medium is broadcast in nature, and all the nodes that are within the radio range of the source node overhear the transmitted packet. The node to which the packet is not destined drops the packet on receiving it [31,32]. In cases of dense networks, avoidance of overhearing can contribute to substantial savings of energy. However, completely avoiding overhearing is also not an efficient scheme as nodes also want to be aware of channel state as well as the state of the neighboring nodes.

- *Protocol overheads*: To tackle different problems such as collision, MAC introduces protocol overhead related to control frames. RTS/CTS control packets are overhead and occur on a per packet transmission basis. In one way, these overheads tend to improve the overall QoS of the network by addressing some of the major hurdles faced by MAC protocols, but on the contrary, excessive use of these overheads can cause loss of energy and increase in latency and make the network less efficient.

- *Idle listening*: A node in the state of readiness to receive a data packet but not receiving it is said to be in the state of idle listening. Idle listening also contributes to the energy loss of the network.

The preceding are most of the challenges related to WSN, especially to maintain QoS. These challenges make it difficult for providing deterministic QoS guarantees such as hard time bound for data arrival, packet loss, or guaranteed bandwidth. However, different applications have different QoS requirements and it is feasible to provide an acceptable QoS for these various applications as discussed in the rest of the chapter.

10.5 QoS-Aware MAC Protocols for WSNs

Most of the MAC protocols designed for WSN try to tackle one of the abovementioned problems to save energy and to improve QoS of the network. These protocols can be broadly classified into low duty cycle protocols, contention-based protocols, and schedule-based protocols. In Section 10.5.1, these classes and major protocols within them are discussed along with how much impact they made on achieving the challenge of maintaining QoS and energy conservation for WSNs.

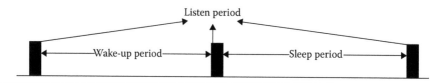

Figure 10.3   Periodic wake-up scheme.

10.5.1 Low Duty Cycle Protocols

To tackle the problem of idle listening, low duty cycle protocols are proposed. The main theme behind these protocols is to avoid spending valuable energy in idle state and reduce the communication activities of a sensor node. In such schemes, the node spends most of the time in sleep state and only wakes up when it is about to either transmit or receive data packets.

To implement such protocols, several different approaches are used; a cycled receiver approach is used by Dam and Langendoen [33]. In this approach, a node spends most of the time sleeping while it wakes up periodically to receive the packets from the neighboring nodes. The scheme is divided into listen, wake-up, and sleep periods as illustrated in Figure 10.3. In such schemes, the node listens to the channel during the listening period and goes back to sleep mode if there is no activity over the channel or in other words no node tries to direct packets toward the listening node. To communicate, the transmitting node must acquire knowledge of the listening period for its receiving node. This can be achieved by letting the node send a short beacon at the start of its listening period [33]; another approach is the transmitting node continuously sends request packets toward the destination until one hits the destination listen period. There are several problems in such schemes and algorithms; by choosing small duty cycle, the traffic from the neighboring nodes toward the destination node causes congestion; especially in high traffic load scenarios, this causes severe degradation in network performance; however, by selecting low duty cycle, a considerable amount of energy can be conserved. On the contrary, a long sleep period causes high latency; in the case of multihop scenarios, the per-hop latency significantly increases causing high end to end latency.

There are several variants of low duty cycle protocols available; some of the major protocols are discussed in Section 10.5.1.1.

10.5.1.1 Sparse Topology and Energy Management

The sparse topology and energy management (STEM) protocol [34] was developed to provide a solution for the idle listening problem. STEM is suitable for those networks that occasionally transmit data or that have no hard latency requirements; an example of such networks could be a habitat monitoring system in which the node has to wait a long time before it transmits any considerable change in the environment; thus, to enhance battery life and reduce the amount of energy consumed, STEM protocol can provide a potential solution for idle listening problems in such types of networks. Figure 10.4 illustrates the STEM protocol duty cycle for a single node. It can be seen from the figure that STEM uses two channels that require two transceivers in each node, one for the wake-up channel and the other for the data channel. The data channel is always in sleep mode, except when transmitting or receiving the data packets. The wake-up channel is subdivided into listen period, wake-up period, and sleep period. If the node does not receive any signal or request from the neighboring node, it goes directly to the sleep period, switching its transceivers to off state as well. However, upon detection of any request, the node starts packet transfer on the data

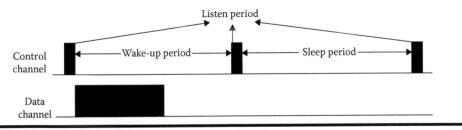

Figure 10.4   STEM duty cycle.

channel. STEM protocol has two different variants on the basis of methodology for the transmitter to acquire receiver acknowledgment; these variants are STEM-B and STEM-T.

10.5.1.1.1 STEM-B

The transmitter sends a beacon signal during the wake-up period periodically without the performance of carrier sensing. The beacon contains the addresses of the transmitter and the intended receiver. As soon as the receiver receives the beacon, it sends acknowledgment in its wake-up period causing the transmitter to stop transmitting any further beacons. Both the transmitting and the receiving nodes switch to the data channel and start data transmission. Any other node that receives the beacon, which is not intended toward it, will ignore it and go to sleep mode to conserve energy.

10.5.1.1.2 STEM-T

The transmitter in this case sends busy tone on the control channel for a long time, which is enough to hit the receiving node's listen period. Since the busy tone does not contain any address information, all the nodes that are in the radio range of the transmitter having received this busy tone switch to the data channel. It is on this channel that after the reception of the initial packets, the nodes will be able to find whether the packets are destined toward it or not. The nodes that are simply not the intended receiver on this stage will go to the sleep mode.

From the preceding discussion, it can be clearly established that for a scenario in which the number of transmitters are more than one, STEM-B may be likely to suffer with high collisions of beacons as the beacons are transmitted without any carrier sensing. In terms of latency analysis carried out by Callaway [34], it is shown that STEM-B provides better latency than STEM-T. The reason behind it is that in both the cases, the transmitter sends beacon or busy tone over the control channel; however, in STEM-B, as the receiver receives the appropriate beacon, it sends the acknowledgment and both the transmitter and the receiver switch to the data channel and start data communication, whereas in STEM-T, there is no such acknowledgment packet concept and hence the transmitter has to finish the entire duration of its busy tone before switching to the data channel. In terms of energy consumption, STEM-T outperforms STEM-B as the acknowledgment is not transmitted as well as that the listening time for the busy tone is smaller than that of beacon reception.

10.5.1.2 Sensor MAC

Sensor MAC (S-MAC) protocol proposed by Ye et al. [32] and Callaway [35] tends to give a possible solution for idle listening, collisions, and overhearing problems. Unlike STEM protocols, S-MAC does not have two channels. S-MAC adopts a periodic wake-up scheme subdivided into wake-up and sleep periods. However, unlike STEM protocol, during the listen period in S-MAC, a node can receive and transmit packets. Figure 10.5 illustrates the duty cycle adopted by S-MAC protocol. S-MAC tries to coordinate with other neighboring nodes such that their listen period starts at the same time. In S-MAC protocol, listen period of the node is further subdivided into three states.

The first state is also known as the synchronization state. In this state, a node accepts its neighboring nodes' synchronization packet that contains the neighboring node schedule; the receiving node stores this information in a table, also known as a schedule table. The synchronization phase is subdivided into time slots, and all the neighboring nodes of the intended receiver contend using CSMA protocol with backoff algorithm. In this case, each neighboring node of the intended receiver chooses a random time slot and starts to transmit its synchronization packet if no signal or activity is detected in that particular slot, otherwise the node will go to sleep mode and wait for the next wake-up cycle of the intended receiving node. On the contrary, to maintain synchronization in the network and to allow new nodes to be aware of their surrounding topology, the node sends a synchronization packet in the broadcast to all the neighboring nodes; in this way, all the nodes that are initially part of the system can synchronize again and the new node can be accommodated in the network. This period is known as synchronization period.

After the synchronization state, the next state is RTS state. In this state, the node listens for any RTS packet from the neighboring nodes. S-MAC uses RTS/CTS handshake to minimize collision due to hidden node problem. The interested nodes having data to transmit contend using CSMA with backoff to send RTS packet toward the intended receiver.

Finally, the node after receiving the RTS packet sends out a CTS packet to the transmitting node; after this, data packet exchange starts between these two nodes.

Hence, by using a combination of RTS/CTS handshake, virtual carrier sensing, and network allocation vector (NAV) tables, S-MAC protocol tries to tackle the problem of overhearing, idle listening, and collision. Hence, a virtual cluster is formed in S-MAC protocol if all the nodes within the network know about their neighboring node schedule and they all wake up at the same time to perform data or control packet exchange activity. In general, the working of S-MAC protocol uses the following steps. A new node switches on its transceiver during the listen period for a time equal to synchronization period. If the node receives any synchronization packet from any of the neighboring nodes, it adapts to this schedule and broadcasts it again in the next listen period of its neighboring node. If this is not to be the case, then the new node randomly picks up its own schedule and broadcasts it. However, if the node receives another schedule during the broadcast synchronization period, it drops its own schedule and adapts to the new one. Moreover, there can be a scenario in which the node receives a new schedule after it has chosen itself and its schedule has been adopted by some of the other neighboring nodes. In this particular condition,

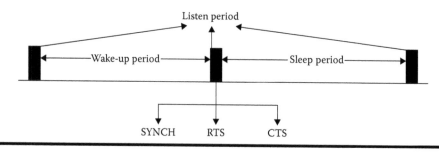

Figure 10.5 SMAC duty cycle.

the node will retain both schedules and transmit synchronization packets in the two different schedules that it receives. On the other-hand if the node knows that there is no other neighboring node sharing its previously adopted schedule, then the node drops its own schedule and adopts the new one.

In this case, S-MAC makes virtual clusters of different schedules; the border nodes of these clusters have to adopt both the schedules and are most likely to lose more energy to work like a bridging node between the two virtual clusters having different synchronization schedules. S-MAC in terms of latency pays a high price as most of the time the nodes are in sleep mode. A variant to S-MAC is proposed by Ye et al. [32] that introduces the concept of adaptive listening to reduce the latency of the S-MAC protocol itself. In such schemes, a node listens to the RTS/CTS handshake between the two nodes. These control packets also carry the duration of the transmission between the two nodes. The overhearing node for these packets knows the fact that it is also in the neighborhood of the intended receiving node and hence it increases its duration of listening just in case the packet is destined to some other node that is one hop away from the actual destination node and the overhearing node can provide this hop. In this way, the adaptive listening procedure actually reduces the per-hop latency of the packet.

S-MAC protocol has one major drawback that it is hard to adopt the wake-up and listen period according to the changing traffic load. Another variation of S-MAC is Timeout-MAC (T-MAC) [33]. This protocol enables nodes to shorten the time of listen period if the node does not receive any activity on the channel. If a node does not receive any signal during a set of defined duration within its listen period, it is allowed to go into sleep mode, whereas in the case of S-MAC, the node has to keep listening the full time of listen period before it makes this decision.

10.5.1.3 Mediation Device Protocol

Mediation device (MD) protocol [34] can work concurrently with WSN industrial standard IEEE 802.15.4, especially in peer-to-peer mode [35–37]. Like all other preceding protocols, the prime objective of this protocol is also to conserve energy, and hence, it allows nodes in WSNs to periodically go to sleep mode and wake up only for a short duration of time to exchange the data. The protocol does not involve any complex arrangements between the nodes; each node has its own individual sleep schedule and is not required to gain information about schedule for other nodes.

In general, when a node wakes up, it transmits a query beacon; this beacon contains the address of the node. After transmitting the query beacon, the node waits for some time according to the protocol to receive any packet or signal. If the node during this wait time does not receive any packet, it goes to sleep mode. For a transmitting node when it transmits the packet, it has to synchronize with the schedule of the neighboring receiving node; to achieve this, the transmitter would stay awake to receive a query beacon; this whole exercise consumes a lot of energy of the transmitter, which is undesirable especially in the case of WSNs. The problem is tackled by using dynamic synchronization in the protocol; to achieve such synchronization, MD is used. The protocol assumes that the MD is not energy constrained and has a full duty cycle. Because the device has a full duty cycle, it will remain active all the time and is able to receive packets at anytime from all the nodes within its radio range and will be aware of the schedules of all the nodes in the surrounding area. The scenario is depicted in Figure 10.6. In the figure, there are three nodes X, Y, and Z. The node Z acts as MD for nodes X and Y. Node X wakes up and sends its query beacon toward node Z; since the node X has data to transmit, it sends RTS signal toward node Z; after transmitting the RTS packet, the node X waits for some time, but as it does not receive

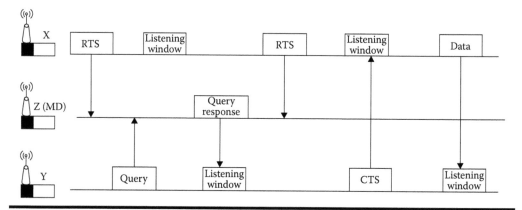

Figure 10.6   Illustration of mediation device protocol.

any signal from any other node, it goes to sleep mode. As node Y wakes up, which is the intended receiving node in this scenario, it sends a query beacon toward node Z. The MD Z sends query response packet toward node Y, which has the wake-up detail and address of node X. In this way, node Y knows about the wake-up cycle for node X, more specifically, the receiving window timing just after the query beacon from node X. Hence, in the next wake-up cycle, when node X sends its RTS packet, node Y responds with a CTS packet, and the data packet transmission occurs between the two nodes; after the successful transmission, the node Y sends acknowledgment for the data reception, dislocates itself from node X wake-up schedule, and follows its own previous wake-up cycle.

Under the main assumption that the MD is not energy constrained and can have a full duty cycle, the protocol has some advantages. Firstly, it does not require any global time synchronization between the nodes, all the synchronization is done through the MD. Secondly, the burden of energy is only shifted to MD and not the transmitting or receiving nodes. This enables them to have a low duty cycle and conserve as much energy as they possibly can. However, the idea of having a MD not affected by energy especially in the case of WSNs is not practical. Moreover, the nodes as they wake up send query messages; in the case of a dense network where nodes can have their duty cycle overlap, there can be a collision of query beacons at the start of the wake-up cycle. The collisions will increase as more and more active nodes enter the network. Finally, to cover the entire network, the number of MDs should be more than one, again making it difficult to support such devices in energy-constrained WSNs.

10.5.2 Schedule-Based Protocols

Many MAC protocol designs for WSNs fall in the category of schedule-based protocols. The advantage of schedule-based protocols is that, firstly, they tackle the problem of idle listening by allocating or assigning time slots for the transmission and reception nodes so that the nodes will stay in sleep mode after their allocated time. Secondly, a transmission schedule can be implemented in such a way that there is no collision at the receiving node and hence avoiding specific and complex algorithms to avoid collisions due to hidden nodes. However, in terms of drawback, such protocols use extensive overhead signaling for maintaining scheduling in cases of variable topology; such overheads are too costly in terms of resource-constrained WSNs. Moreover, the schedule is not adaptable to different load conditions, especially if the load varies frequently.

To maintain a schedule and the information regarding neighboring node schedules, memory is required, which is not that much in abundance in the case of tiny sensor networks. Following are some of the major schedule-based protocols designed for WSNs to handle such problems and difficulties.

10.5.2.1 Low Energy Adaptive Clustering Hierarchy Protocol

The low energy adaptive clustering hierarchy protocol (LEACH) protocol is presented by Heinzelman et al. [38]; it uses TDMA-based MAC protocol with clustering and routing technique. The protocol targets dense homogeneous WSNs with energy-constrained nodes. The protocol divides the network into several clusters, and each cluster has a cluster head that is responsible for creating and maintaining the TDMA schedule among the nodes, and the rest of the nodes are the member nodes. The member nodes use these assigned TDMA slots to exchange data among themselves and the cluster heads. The cluster heads then aggregate all the data and transmit it to the sink node or to another node for relaying purposes. Since most of the time the sink node is far away in large networks, the cluster head uses more power to transmit the data toward the sink, whereas the rest of the nodes use the same low power to reach the cluster head. The cluster heads are the nodes that consume a lot of energy since they are responsible for communicating data toward the sink and they have a full duty cycle as well. Hence, if a single node is chosen to be the cluster head for a long duration, it will deplete all its energy resources in a quick time, which results in a communication breakdown between the cluster and the rest of the network, including the network sink. To overcome this problem, the cluster head responsibility is rotated within the cluster so that no one node exhausts its resources. The process of becoming cluster head is independent and each node decides on its own that it has to act as a cluster head without any kind of signaling and election process. The decision to become cluster head depends on when the node became the cluster head for the last time, and if the time duration is longer, then the chances of a node opting to become cluster head become higher; the process is also described by Heinzelman et al. [39]. Moreover, the nodes that choose to remain as non–cluster head chose their cluster head on the basis of received signal strength from the potential cluster heads.

After the formation of a cluster, each cluster head randomly selects a CDMA code for its cluster, which it broadcasts to all its member nodes. The reason behind this is to stop the interference from the border nodes sitting at the edges of each cluster. The protocol comprises different phases. The initial phase is known as the setup phase in which the node elects itself as the cluster head. The next phase is termed as the advertisement phase in which the cluster head informs all neighbors with an advertisement packet. The cluster head for this phase uses the CSMA protocol. The non–cluster head member nodes pick up the advertisement packet of the strongest signal and send their information to the cluster head in the joining phase. The nodes use the CSMA protocol in the join phase as well. After the join phase, the cluster head knows the number of members in its cluster and constructs a TDMA schedule for the neighbor node alongside picking up a random CDMA code for the cluster communication. The cluster heads then broadcast the schedule along with the code to the respective nodes. After this phase, TDMA steady-state phase starts in which nodes exchange data packets with the cluster heads.

LEACH protocol suffers from collision as well; due to the collisions between advertisement or join packets, the protocol does not guarantee that all the non–cluster nodes will be part of a cluster. The cluster head has to be switched on all the time, whereas the non–cluster head should be awake during setup phase. This creates an extra burden in terms of loss of energy, and the entire setup time causes an increase in the latency of data transmission. Moreover, for large areas, LEACH protocol is not suitable as the cluster head has a finite amount of range beyond which it cannot communicate with the sink directly; this in terms of scalability is the major disadvantage of the protocol as no two cluster heads communicate with each other.

10.5.2.2 Self-Organizing Medium Access Control for Sensor Networks Protocol

The sensor medium access control (SMAC) presented by Sohrabi et al. [40] and Sohrabi and Pottie [41] is a part of WSN protocol suite that covers MAC, mobile nodes, multihop routing protocol, neighbor discovery, and local routing protocol. SMACS assume that the available spectrum is divided into several channels and each node can tune its transceiver to any one of them as well as the nodes have several CDMA codes at their disposal. As far as topology is concerned, the network is static for a long time. Each node has fixed time slots and these are divided into superframes. All the nodes within the network have the same superframe length; the superframes are also divided into time slots. The general working principle of this protocol is illustrated in Figure 10.7. Assume that there are two nodes in the networks X and Y as shown in Figure 10.7. In the start of SMACS protocol, the nodes perform neighborhood discovery. Consider node X wakes up and listens to the frequency channels; if the node does not find any activity or any message over the link, it transmits an invitation message with its own address and number of neighbors attached to it; assume that the total number of nodes attached to node X is zero. When node Y receives a message from node X, it waits for some duration of time and then replies with a message containing nodes X and Y addresses and the amount of neighbors attached with it, so far assuming that the total number of neighbors with node Y is also zero. As node X receives this message, it invites Y to construct a link by sending another message to Y containing Y's address only. This message indicates to node Y that it has been selected and that it can select any time slot for the transmission and reception of data as no other nodes are attached with node X. In response, node Y establishes a link with node X by transmitting a link specification in which it gives the information of the time slots and code over which these two nodes will communicate with each other.

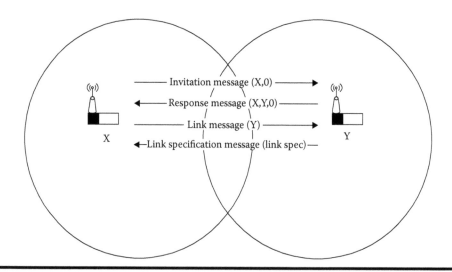

Figure 10.7   Neighborhood discovery mechanism in SMAC.

In a scenario when both the nodes have neighbor nodes attached to them, the protocol runs by starting with the transmission of an invitation message from node X with the address of itself and information regarding the attached nodes. Node Y also responds by sending the address of itself and node X with all the information regarding neighboring nodes. In response to that, node X sends its entire schedule to node Y, which has all the timing details and slots that are reserved for communication with other nodes. Node Y after carefully selecting mutually common time slots, which are not overlapping with the existing neighbor nodes of either of nodes, sends back the link specification to node X. In this manner, the entire protocol works in each cycle.

The major drawback of this protocol is the length of superframe. If the superframe is too short, then all the neighbor nodes will not be visible to the node. Another drawback will be in terms of network load and number of nodes. If the network load is low but the number of nodes in the neighborhood is high, the node will be awake on every set schedule just to find no data transfer. The protocol incurs high overheads and can consume energy.

10.5.3 Contention-Based Protocols

In contention-based protocol, the nodes in a network have to compete to transmit their data from one node to another and eventually to the sink node. The quality of the network and the performance of the MAC protocols significantly depend on the number of active nodes, traffic load, and topology of network. If in a given network the number of active nodes are increased, this would lead to an increase in the chances of collision due to nodes that select the same slot to transmit but are also in the communication range of each other as well those nodes that are hidden to the transmitting node. This causes a significant loss of transmitter and receiving node's energy as well as adding to network latency and overall throughput degradation of the entire network. The two most important and common contention-based protocols that are already described in Section 10.4 are ALOHA and CSMA. In Section 10.5.3.1, variants of these protocols are discussed, which are especially designed to fulfill the needs of WSNs and tackle the problems that are unique to such networks.

10.5.3.1 Carrier Sense Multiple Access Protocols

In Ref. [42], several CSMA variants are discussed and their performance measures are presented. The study includes fairness and energy consumption in these protocols. The author considers multihop networks with single and multiple sinks. The network is considered as being in the idle state for a long duration and becomes active after sensing something substantial. Upon triggering, all the nodes within the network become active and try to communicate the data; this results in collisions and degradation of QoS of the network. The state diagram used to represent CSMA by Woo and Culler [42] is presented in Figure 10.8 for clarity of the protocol.

It is shown in Figure 10.8 that a node undergoes several stages before transmitting data using the CSMA protocol. As the data packet reaches the buffer of the node, it starts a random delay and initializes its trial counter. During this random delay period, the node transceivers are in sleep mode, and when the delay period is over, the node goes to the listening state; if there is an ongoing activity over the channel, the node goes to the backoff stage once again, provided its number of trials is smaller than the maximum number of allowable trials. During this backoff, the node once again can go into sleep mode. After the backoff delay period is over, the node once again enters the listening state; if the medium is found busy, the node either drops the packet if the number of the retrial counter exceeds its maximum limit or it goes to backoff delay once again. However, if

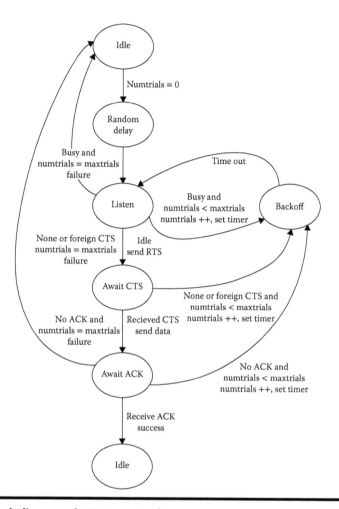

Figure 10.8 Block diagram of CSMA protocol.

the medium is found to be idle, the node sends an RTS signal toward the intended receiver; this handshaking signaling can be dropped if the load over the network is small. After a RTS, the node waits for a CTS signal for a bounded duration of time; if it receives a CTS, then it starts the data transmission and waits for acknowledgment of data reception; otherwise, the node goes to the backoff state once again if the number of retrials is still smaller than the maximum limit or drops the packet. Several variants of this protocol with random delay, nonrandom delay, fixed window backoff, exponentially increasing backoff, exponentially decreasing backoff, and no backoffs have been proposed and investigated in; Woo and Culler suggested that a protocol with random delay, backoff algorithm, and fixed listening time perform better in single-hop scenarios [42].

10.5.3.2 Power-Aware Multiaccess with Signaling

Power-aware multiaccess with signaling (PAMAS) protocol is presented by Raghavendra and Singh [43]; the protocol provides a mechanism against overhearing by combining RTS/CTS handshaking and the busy tone used by the MACA protocol presented by Karn [3]. PAMAS uses two channels, one for control signaling and the other for data packet transmission. If a node becomes activated and has some data packets in its buffer, it starts by sending a RTS packet over the control channel toward the intended receiving node without performing any carrier sensing. The RTS packet carries the addresses of both transmitting and receiving nodes including the length of transmission. As the intended receiver recycles the RTS, it responds by sending a CTS packet toward the transmitter, provided there is no ongoing transmission near the receiving node. As the transmitting nodes receive CTS packets, they start sending data packets over the data channel toward the receiver; however, the receiver receiving the data packet also sends a busy tone on the control channel. If the transmitting node does not receive any CTS packet from the intended receiver, it waits for a bounded time interval and then goes into an exponential backoff state.

As it is clear from the discussion that protocol requires some sort of synchronization or awareness to the node so that when they wake up from the sleep mode they should realize that the channel is available for any ongoing transmission taking place in its neighborhood; moreover, a mechanism is needed if CTS and RTS packets are corrupted and the node does not know exactly how long the transmission that is taking place in the neighborhood will last for. For this purpose, the protocol supports a probing procedure. The node runs the probing protocol to find out how long the data transmission, which is ongoing, will last so that it can start its data transmission setup, and most importantly, during the time of ongoing data transmission, it can switch its transceiver off and goes into sleep mode. In probing protocol, the node using a query sends out probing packets on the control channel, and any transmitter that finishes its transmission in the time interval sends to the query node the query response signal on the control channel; in this way, the node knows how long the ongoing transmission will take. Most importantly, in dense WSNs, this process requires more computation and retransmission of the probing packet until the node finds out the actual duration of the ongoing transmission.

The major drawback of the protocol is that it requires heavy computation in terms of probing, and in the case of WSNs, the resources are limited; moreover, the control packets are transmitted without any carrier sensing, this could also lead to high latency before the transmission of data packets and degrade the QoS of the network.

10.6 IEEE 802.15.4 MAC Protocol

The industrial standard for WSNs is the IEEE 802.15.4 developed in October 2003 [44,45]. The standard covers the physical layer and the MAC layer for low-rate WPANs. The targeted application for the IEEE 802.15.4 standard requires low bit rate and energy conservation. The physical layer provides a bit rate of 20 Kbps for a single channel in the frequency range of 868–868.6 MHz, 40 Kbps for 10 channels in the range of 905–928 MHz, and 250 Kbps for 16 channels in 2.4-GHz Industrial Scientific and Medical (ISM) band. There are a total of 27 channels available, but the MAC protocol used in this standard uses only one channel at a time as it is not multichannel. The MAC protocol uses both schedule-based and contention-based periods, and the nodes can be homogenous as well as heterogeneous within the same network.

10.6.1 Network Architecture

The MAC protocol uses two types of nodes: full function devices (FFDs) and reduced function devices (RFDs). The FFD nodes are divided into three operational categories: personal area network (PAN) coordinator, a simple coordinator, or just a simple device. The reduced function node works only as a simple device for transmission and reception. A device is associated with a coordinator node, which must be a FFD and only communicate with it, thus in principle forming a star network with the coordinator. However, the FFD nodes can communicate in a peer-to-peer fashion, and multiple coordinator nodes form a PAN and have a central coordinator known as a PAN coordinator, which also serves as the central command of the network as well as the sink. The role of coordinator node in IEEE 802.15.4 standard is defined as a node that manages a list of all the associated devices. Devices can associate, disassociate, and reassociate with the coordinator. The coordinator is responsible for the allocation of short addresses to all the devices associated with it. To maintain synchronization, the coordinator node in beacon-enabled mode sends a continuous beacon that has information about the start of the next superframe, outstanding requests, and other parameters; moreover, the coordinator upon the request of the node can also allocate a fixed slot for the transmission of data.

10.6.2 Superframe

In beacon-enabled mode, the coordinator organizes channel access and data transmission using the superframes presented in Figure 10.9. All the superframes are equal in length; the coordinator starts the superframe by sending a beacon packet that marks the starting of the frame and also contains information regarding the length of various components of the superframe. The superframe of IEEE 802.15.4 beacon-enabled mode is subdivided into an active period and an inactive period. During the inactive period, all the nodes including the coordinator switch off their radios and go to sleep mode. All nodes become active just before the inactive period of superframe ends to receive the beacon. The active period is divided into 16 time slots; the first time slot is occupied by the beacon frame whereas the remaining is portioned as contention access period and guaranteed time slot period, which comprises a maximum of seven time slots and it is solely at the discretion of the coordinator to allocate the requesting node. Only the node that has been allocated the Guaranteed Time Slot (GTS) is activated during its specific time slot whereas the rest of the nodes in the network go to sleep mode.

10.6.3 Slotted CSMA/CA Protocol

The node in Contention Access Period (CAP) of the superframe uses the slotted CSMA/CA protocol for data transmission. There is a difference in CSMA/CA protocol use in wireless local area network (LAN) IEEE 802.11 and WSN IEEE 802.15.4; in the latter, the protocol has no defense mechanism against the hidden nodes, that is, the protocol does not support the RTS/CTS handshake algorithm. The operation of this protocol is described in Figure 10.10. The 16 time slots in the superframe are further subdivided into smaller slots known as the backoff period (BP).

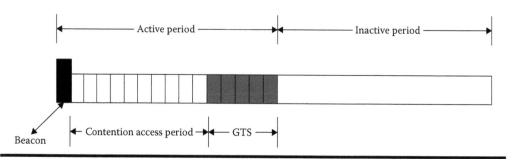

Figure 10.9 Superframe structure of IEEE 802.15.4.

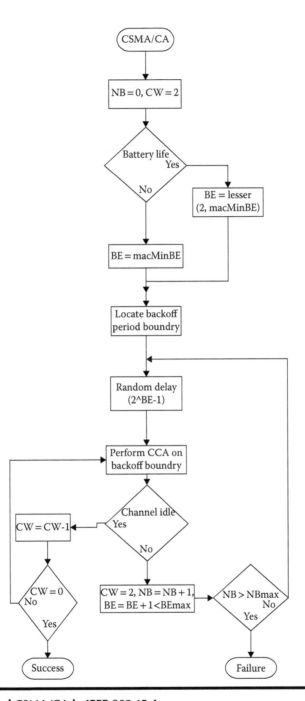

Figure 10.10   Slotted CSMA/CA in IEEE 802.15.4.

One BP has a length of 20 channel symbols; the CSMA/CA considers each BP as a single slot. The protocol contains three variables: NB (number of backoffs), CW (contention window), and BE (backoff exponent). When a node has packets in the buffer, it initializes these parameters to the values of NB = 0, CW = 2, and BE = macMinBE. The node waits for the next BP and computes a random number in the range of [0, $2^{BE} - 1$]. This random number is the backoff delay and the device waits for this period before attempting any further steps toward the transmission of data. After the delay period is over, the node with the next BP performs clear channel access (CCA); if the node finds the channel to be idle, then it decrements the value of CW and with the next BP boundary performs CCA again; if the channel is found to be idle again, the node assumes that it has won the contention and starts data transmission. If during any one of the CCA processes the channel is found to be busy, the node increments the value of NB as well as BE; if the number of retrials exceeds the allowable NB limits, the packet is dropped; otherwise, the node selects once again a random backoff delay in the range of $[0, 2^{BE} - 1]$; moreover, for each retrial, the value of CW is also initialized to its original value of 2.

10.6.4 Unslotted CSMA/CA Protocol

As discussed in Section 10.4, IEEE 802.15.4 works in beacon-enabled and non-beacon-enabled modes. In non-beacon-enabled mode, the protocol uses unslotted CSMA/CA MAC. Figure 10.11 shows the operation of unslotted CSMA/CA. In this mode, the coordinator node does not send

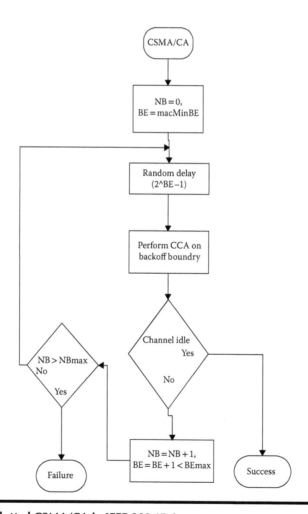

Figure 10.11   Unslotted CSMA/CA in IEEE 802.15.4.

any beacon and there are no GTS slots; hence, in this mode, the nodes are not synchronized. The node performs CCA once and then transmits the data packet. In this protocol, the coordinator is always in active state whereas all other nodes in the network can go to sleep mode. The major drawback of this protocol is that as the network is scaled up more and more, nodes will face the problem of collision as there is no set universally defined superframe available, which all the nodes follow. Moreover, the latency of the network will also increase as the sleep and wake-up schedule of every single node in the network will differ from the other nodes because of lack of synchronization.

10.7 Mobile WSN Key Problems

Mobility in WSNs raises new challenges and problems that occur at MAC level due to the existing link failure and new link formation. This in turn introduces delays that are associated to establish connectivity. Moreover, due to mobility of the neighborhood, information is variable, which causes problems in scheduling. The mobile nodes may experience loss of connectivity in a network using schedule-based protocols. The reason is that as the node enters into the radio range of a new cluster, it has to wait for the next schedule synchronization point; this leads to temporary loss of connectivity and an increased latency for the packet which the mobile node wants to transmit. Moreover, mobility may also increase the chances of number of collisions in the network, as the mobile node has no information about the neighboring nodes in the new cluster, thus can interfere in the ongoing transmission between the two nodes and cause collision in the network. This collision will lead to increase in number of retransmissions, resulting in increase in packet latency, high energy consumption, and often resulting in packet drop.

10.7.1 Effect of Mobility on the Quality of Service

Node mobility in WSN under certain scenarios can be beneficial, as in case of increasing network lifetime and network coverage [46–48]. However, as discussed in Section 10.7, mobility also imposes new challenges and problems on the design of MAC protocols. As the node moves away from the communicating node, frame loss and packet drop can occur due to the significant signal strength variations [49]. Moreover, due to hidden node and nonsynchronization with the new cluster, the mobile nodes contribute to the increase in number of collisions, which plays a significant role in the degradation of QoS in wireless network. Especially, in the case of WSN, where the resources are limited and energy conservation is significant, collision causes excessive loss of energy. Due to the loss of packets, the number of retransmission increases, resulting in severe degradation in throughput, loss of energy, inefficient bandwidth utilization, and higher latency for the network.

10.7.2 Challenges and Motivation for Mobile WSN

Designing MAC protocol for mobile scenario is a challenging task, especially in case of WSNs that are inherent resource constrained. The prime objective of such type of protocols in mobile scenario is to maintain connectivity and acceptable QoS while incurring less collision in the network. The protocols should be less complex and conserve energy. Moreover, the time associated for connectivity, neighborhood discovery, and synchronization should be minimized as these factors contribute significantly to the increase in latency.

10.7.3 Classification of Mobility in Wireless Sensor Networks

The breaking of existing link and formation of new link due to one-hop neighbors moving out of radio range due to mobility causes packet loss, increase in latency, increase in number of collisions, and high energy consumption [50–52]. Mobility causes rapid changes in topology of the network [53] and, hence, becomes increasingly challenging especially for MAC protocols to adapt to such fast changes.

Mobility in WSNs can be broadly classified into the following three main categories:

1. Nodes are mobile while the sinks are stationary.
2. Sinks are mobile and nodes are stationary.
3. Both nodes and sinks are mobile.

In a typical WSN, the sink is located centrally and stationary; however, the area around the sink experiences high level of communication activities, and due to this reason, the energy supplies of the node in the vicinity of sink deplete considerably fast. This may result in the disconnection of the sink with the entire network. To mitigate this problem, mobile sink concept is introduced. Gandham et al. [54] propose that sink relocation methods on the basis of energy consumption of individual sensors are evenly distributed and overall network energy is minimized. Load balancing and data transmission is investigated in Ref. [55]; the authors propose joint mobility and routing scheme to maximize the network lifetime. In this scheme, the mobile sink moves on a circular trajectory inside the area where the nodes are deployed. The sensing nodes that are inside this trajectory send the data to the mobile sink using the shortest possible path to conserve energy.

To address the problems of connectivity in sparse network where the number of sensing nodes are low, mobile data collector (MDC) scheme is introduced [56]. In this scheme, the mobile node gathers all the buffer information stored in an individual node. Investigation over different mobility patterns is presented by Kansal et al. [57]. A random mobility pattern is introduced for MDC by Shah et al. [58]. In this scheme, a MDC randomly moves to collect the data buffered in the nodes of a sparse network. Since the movement of the data collector is random, the network experiences latency. Knowledge-based mobility or predictable mobility for MDC is introduced to meet some latency bounds as well as save the energy of the individual nodes [59]. The nodes in such scenario are aware of data transfer time and based on this go into sleep mode to save energy, hence increasing network lifetime [60]. In majority of WSNs, sensor nodes generate data at different rates. Data losses in sensor node occur if the buffer or if the node is full and the contents of the buffer are not being transferred to the MDC. To resolve such issues, a scheme referred to as controlled mobility [61,62] is proposed. To achieve controlled mobility, mobile element (ME) scheduling is proposed [63]. The ME is scheduled in real time to visit the nodes before the buffers are full. To improve the protocol, two more proposals Earliest Deadline First (EDF) and Minimum Weight Sum First (MWSF) are proposed [63].

In EDF, the decision of the next node to be visited by ME depends on the node having its buffer overflow deadline first. EDF scheme leads to significant amount of data loss due to the fact that nodes are far away as the network is sparse and two consecutive nodes can have the same deadline for buffer overflow. In MWSF, this problem is tackled by introducing distance factor. The ME scheduling not only depends on the earliest deadline of the buffer overflow but it also takes into account the fact of how far the nodes are within the network. Even MWSF introduces the distance factor before deciding which node to visit, but the back and forth movement to reach faraway nodes is not completely avoided. Multiple MDC with relay data collection algorithm is proposed to increase the scalability of the sparse network and to curb the problem of distance between the nodes [64]. In this algorithm, nodes that are out of range from MDC find the nearest sensor nodes to the MDC and relay their data to them using the shortest possible route.

10.7.4 Case Study for IEEE 802.15.4 MAC Protocol for Mobile Sensor Network

In Ref. [65], IEEE 802.15.4 industrial standard CSMA/CA MAC protocol is presented. CSMA/CA uses random backoff values as collision resolution algorithm (CRA). Since WSN is resource constrained, the entire backoff phenomenon is performed blindly without the knowledge of the channel condition; this factor contributes to high number of collisions, especially when the number of active nodes are high. The IEEE 802.15.4 MAC protocol was mainly used in static scenarios, and initial research work and performance studies are done while considering static applications [37,66–68]. Recently, the use of mobile nodes in WSN becomes a very hot topic [62,69,70]. Introducing node mobility in WSN raises many challenges and problems, especially at MAC level. IEEE 802.15.4 MAC CSMA/CA is also used in mobile applications; however, the QoS is very poor [71–73]. The protocol incurs high latency over the mobile node as it requires long association process for the node to leave from one cluster to another cluster. If the node misses the beacon from the new cluster head consecutively for four times, then the node will go into orphan realignment process, hence inducing more delays. For the node to join the new cluster and not to miss the beacon from the new cluster head, it tends to wake up for long duration, which in turn significantly increases the energy consumption of the individual node. Moreover, as the mobile node succeeds in migrating from one cluster to another, it has to compete with the already available active nodes in the cluster, which contributes to significant number of collisions and packet drops, resulting in overall degradation of network QoS.

10.7.4.1 Association, Synchronization, and Orphan Scan in IEEE 802.15.4

The process of association and synchronization in IEEE 802.15.4 CSMA/CA MAC protocol is presented in Figure 10.12. The node association starts with an active scan procedure that scans all listed channels by sending beacon requests to all nearby coordinators. All the information received in a beacon frame will be recorded in a PAN descriptor. The results of the channel scan will be used to choose a suitable PAN. The node then sends a request to associate with the chosen coordinator. The node updates its current channel and PAN ID while waiting for an acknowledgment from the coordinator. Upon receiving an acknowledgment, the node then waits for the association results. The coordinator will take *aResponseWaitTime* symbols (*32*aBaseSuperframeDuration, about 0.49 seconds*) to determine whether the current resources are available on the PAN to allow the node to associate. If sufficient resources are available, the coordinator then allocates a short address to the node and sends an association response command containing a new address and a status, indicating a successful association. If there are no sufficient resources, the node will receive an association response command with a failure status.

After the node associates with its coordinator, it will send a request to synchronize and start tracking the beacons regularly. If the node fails to receive a beacon *aMaxLostBeacons* times (equal to four times), it may conclude that it has been orphaned. The node then has the option either to

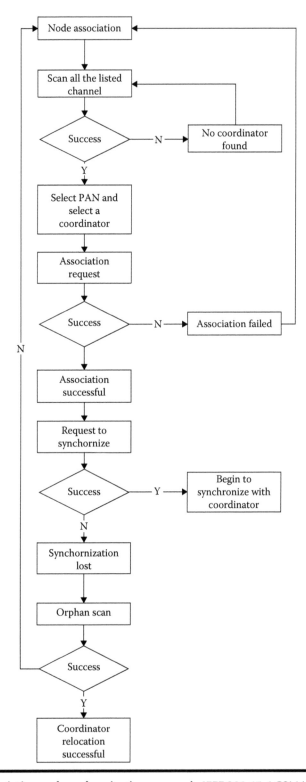

Figure 10.12   Association and synchronization process in IEEE 802.15.4 CSMA/CA MAC protocol.

perform the orphan device realignment procedure or to perform the association procedure. If the node chooses to perform an orphan device realignment, it will do the orphan scanning by sending an orphan notification command to relocate its coordinator. The node waits for *aResponseWait-Time* symbols to receive a coordinator realignment command. The coordinator that receives the orphan notification command will search its list looking for the record of that node. If the coordinator finds the record, it will send a coordinator realignment command to the orphaned node together with its current PAN ID, MAC PAN ID, logical channel, and the orphaned node's short address. The process of searching the record and sending the coordinator realignment command takes within *aResponseWaitTime* symbols.

10.7.5 Other Mobile MAC Protocol for Wireless Sensor Networks

Mobile MAC (MOB-MAC) is presented for mobile WSN scenario [49]. MOB-MAC uses an adaptive frame size predictor to significantly reduce the energy consumption. A smaller frame size is predicted when the signal characteristics become poor. However, the protocol incurs heavy delays due to the variable size in the frame and is not suitable for mobile real-time applications. Adaptive mobility MAC (AM-MAC) is presented by Choi et al. [74]; it is the modification of S-MAC to make it more useful in mobile applications. In AM-MAC, as the mobile node reaches the border node of the second cluster, it copies and holds the schedule of the approaching virtual cluster as well as the current virtual cluster. By adopting this phenomenon, the protocol provides fast connection for the mobile node moving from one cluster and entering the other. The main drawback is the node has to wake up according to both the schedules but cannot transmit the received data packet during the wake-up schedule other than the current cluster. This contributes to significant delay and loss of energy due to idle wake-up. Another variation of S-MAC is presented by the name of mobility sensor MAC (MS-MAC) [75]. It uses signal strength mechanism to facilitate fast connection between the mobile node and the new cluster. If the node experiences change in received signal strength, then it assumes that the transmitting node is mobile. In this case, the sender node sends not only the schedule but also the mobility information in the synchronous message. This information is used by the neighboring node to form an active zone around the mobile node so that whenever the node reaches this active zone it may be able to update the schedule according to the new cluster. The drawback of this protocol is idle listening and loss of energy. Moreover, nodes in the so-called active zone spend most of the time receiving synchronous messages rather than actual data packets, thus resulting in low throughput and increase latency. Another real-time mobile sensor protocol, dynamic sensor MAC (DS-MAC), is presented by Lin et al. [76]. In this protocol, according to the energy consumption and latency requirement, the duty cycle is doubled or halved. If the value of energy consumption is lower, the threshold value (T_e) then the protocol doubles its wake-up duty cycle to transfer more data, thus increasing the throughput and decreasing the latency of the network. However, the protocol requires overheads, and during the process of doubling of wake-up duty cycle if the value crosses T_e, the protocol continues to transmit data using double duty cycle resulting in loss of energy. Mobility-aware delay sensitive MAC (MD-MAC) is proposed by Hameed et al. [77], which is the extension of DS-MAC and MS-MAC. The protocol enforces T_e value, and during any time if the value of energy consumption is doubled, then it halves the duty cycle. Moreover, the mobile node undergoes only neighborhood discovery rather than other neighboring node that forms an active zone as in the case of MS-MAC. MD-MAC is complex and requires high overheads.

10.8 Open Issues and Future Research Directions

WSNs are being rapidly used in various applications; these applications have different QoS requirements. This expansion in the utilization of WSN creates new future directions and generates various challenges to overcome to meet and satisfy the demands of various applications. Following are some of the major issues and research directions that are open for investigation due to the increase in applications for WSNs.

Heterogeneous networks: Traditionally, WSNs comprise homogenous nodes, but as the WSNs are being used in diverse applications, the networks now comprise of mostly heterogeneous nodes. Hence, one research direction can be to develop novel MAC protocols, which have the ability to fulfill the diverse QoS requirements of heterogeneous sensor networks.

Multimedia applications: Due to the size and ease of deployment, WSNs are very strong candidates for multimedia applications. As it is known, the delivery requirements of multimedia data are different than that of scalar data. Such type of data requires higher throughput, bounded delay, low congestion, and reliability. Therefore, another potential area for research is developing novel MAC protocols that meet the requirements of multimedia WSNs.

Multiple applications over single platform: With the advent of new powerful operating system for WSN, it is now possible to run multiple applications over a single node. This, on the contrary, requires large amount of data flow over the network and becomes even more challenging if this data has different priority levels. Development of protocols that support QoS for multiple applications having diverse priority levels is another research direction that should be further investigated.

- *Coexisting networks*: As WSN is being used in multiple application scenarios, more than one WSN can be in the neighborhood of other sensor networks. This causes sharing of resources that are already limited. Moreover, different networks in the neighborhood need to collaborate to achieve fairness in sharing the wireless resources. Hence, collaborative QoS providing MAC protocols can be another topic of research.
- *Mobile sensor networks*: Traditionally, majority of the applications in which WSN is used were static; as new applications emerge, WSNs are also being used in mobile applications. This brings up new challenges and problems to tackle. Mobile WSN can be one of the existing and challenging research directions, which need further investigation to meet QoS requirements.
- *Energy efficiency*: Mostly, QoS and energy conservation are reciprocal to each other. As the applications are varied, the requirement of QoS along with energy conservation is increased. This provides a unique challenge and needs further research and analysis to develop such a novel MAC protocol that can satisfy both energy and acceptable QoS for modern applications.
- *Hardware implementation*: Majority of the protocol available in the literature show the performance on the basis of simulation results. However, there is a strong need to implement these protocols on hardware platforms and verify their performance in practical scenario. Implementation of new and old protocols over test bed and relaxing the assumptions over the protocols that are set due to simulation environment is another open research area that has a lot of potential.
- *Cross-layer approach*: Providing QoS solution for WSNs is not an easy task since these networks are resource constrained. Therefore, it is worth investigating cross-layer approach to relax the limitation of MAC layer alone to achieve the required QoS for emerging applications.

10.9 Conclusions

At present, WSNs are used not only for low data rate applications but also for more complex applications that require reliability and time-bounded and efficient transmission and reception of data. Moreover, the WSNs comprise heterogeneous nodes that range from simple data collecting sensor nodes to video and microphones. Such increase in utilization of sensor nodes in various applications requires high QoS, providing protocols for WSNs. Keeping in view of such requirements, this chapter focuses on challenges, metrics, parameters, and requirements of QoS-aware MAC protocols for WSNs. The chapter provides in-depth and detail analysis and comparisons between the developed protocols and their achievements. Moreover, mobile sensor networks are also presented and a case study of IEEE 802.15.4 in mobile scenario is discussed. The chapter also provides designing trade-offs and open research issues in the field of QoS, providing MAC protocols for WSNs to contribute to further research efforts in the field of WSNs.

References

1. Tobagi, F. A., and L. Kleinrock. 1975. "Packet Switching in Radio Terminals. Part II: The Hidden Terminal Problem in CSMA and Busy Tone Solution." *IEEE Transaction on Communications* 23 (12): 1417–33.
2. The Editors of IEEE 802.11; IEEE standard for Wireless LAN Medium Access Control (MAC) and Physical Layer (PHY) Specifications. November 1997.
3. Karn, P. September 1990. "A New Channel Access Method for Packet Radio." In *Proceedings of ARRL/CRRL Amateur Radio, 9th Computer Network Conference*, 134–40.
4. Bharghavan, V. 1994. "MACAW: A Medium Access for Wireless LANS." In *Proceedings of ACM SIGCOMM 94*, London, UK.
5. Rubin, I. 1979. "Access Control Disciplines for Multi-Access Communication Channels reservation and TDMA Schemes." *IEEE Transaction on Information Theory* 25 (5): 516–36.
6. Glisic, S., and B. Vucetic. 1997. *Spread Spectrums CDMA Systems for Wireless Communications*. Boston, MA: Artech House.
7. Goiser, A. M. J. 1998. *Spread Spectrum Techniques*. New York: Springer Verlag.
8. Ross, A. H. M., and K. S. Gilhausen. 1996. "CDMA Technology and IS-95 North American Standard." In *the Communications Handbook*, 430–48. New York: CRC Press in cooperation with IEEE Press.
9. Lehne, P. H., and M. Petersen. 1999. "An overview of Smart Antenna Technology for Mobile Communication Systems." *IEEE Communication Surveys and Tutorials* 2 (4).
10. Doufexi, A., S. Armour, M. Butler, A. Nix, D. Bull, and J. McGehan. 2002. "A Comparison of Hyper LAN/2 and IEEE 802.11a Wireless LAN Standard." *IEEE Communication Magazine* 40 (5): 172–80.
11. ETSI. TR 101 683, HYPERLAN Type 2: System Overview. February 2000.
12. ETSI. TR 101 475, BRAN HYPERLAN Type 2: Physical Layer. March 2000.
13. ETSI. TR 101 761-1, BRAN HYPERLAN Type 2: Data Link Control Layer Part 1 Basic Data Transport Function. March 2000.
14. ETSI. TR 101 761-2, BRAN HYPERLAN Type 2: Data Link Control Layer Part 2, Radio Link Control Protocol Basic Functions. March 2000.
15. Karol, M. J., Z. Liu, and K. Y. Eng. 1995. "An Efficient Demand-Assignment Multiple Access Protocol for Wireless (ATM) Networks." *Wireless Networks* 1 (3): 269–79.
16. Passas, N., S. Paskalis, D. Vali, and L. Merakos. 1997. "Quality of Service Oriented Medium Access Control for Wireless ATM Networks." *IEEE Communication Magazine* 35 (11): 42–50.
17. Sharon, O., and E. Altman. 2001. "An Efficient Polling MAC for Wireless LAN." *IEEE/ACM Transaction on Networking* 9 (4): 439–51.
18. Takagi, H. 1986. *Analysis of Polling Systems*. Cambridge, MA: MIT Press.
19. Tobagi, F. A., and L. Kleinrock. 1976. "Packet Switching in Radio Channels, Part III Polling and Dynamic Split Channels Reservation Multiple Access." *IEEE Transaction on Communications* 24 (8): 832–45.

20. IEEE 802.4 Token Passing Buss Access Method. 1985.
21. Moon, H. J., H. S. Park, S. C. Ahn, and W. H. Kwon. 1998. "Performance Degradation of IEEE 802.4 Token Bus Network in Noisy Environment." *Computer Communications* 21: 547–57.
22. Malpani, N., Y. Chen, N. Vadiya, and J. Welch. 2004. "Distributed Token Circulation on Mobile Adhoc Networks." *IEEE Transaction on Mobile Computing* 4 (2): 154–65.
23. Willig, A., and A. Wolisz. 2001. "Ring Stability of the PROFIBUS Token Passing Protocol over Error Prone Links." *IEEE Transactions on Industrial Electronics* 48 (5): 1025–33.
24. Abramson, N. 1985. "Development of the ALOHANT." *IEEE Transaction on Information Theory* 31 (2): 119–23.
25. Kelinrok, L., and F. A. Tobagi. 1975. "Packet Switching in Radio Channels Part 1, Carrier Sense Multiple Access Models and their Throughput/Delay Characteristics." *IEEE Transaction on Communications* 23 (12): 1400–16.
26. Ye, W., J. Heidemann, and D. Estrin. 2004. "Medium Access Protocol with Coordinated Adaptive Sleeping for Wireless Sensor Networks." *IEEE/ACM Transaction on Networking.*
27. Lin, E. Y. A., J. M. Rabaey, and A. Wolisz. June 2004. "Power Efficient Rendez-vous Schemes for Dense Wireless Sensor Networks." In *Proceedings of ICC 2004*, Paris, France.
28. Schurgers, C., V. Tsiatsis, S. Ganeriwal, and M. Srivastava. 2002. "Optimizing Sensor Network in the Energy Latency Density Design Space." *IEEE Transaction on Mobile Computing* 1 (1): 70–80.
29. Taneja, J., J. Jeong, and D. Culler. 2008. "Design, Modeling and Capacity Planning for Micro-Solar Power Wireless Sensor Networks." In *7th International Conference on Information Processing in Sensor Networks, IPSN08*, 407–18, Washington DC, IEEE Computer Society.
30. Intanagonwiwat, C., R. Govindan, D. Estrin, J. Heidemann, and F. Silva. 2003. "Directed Diffusion for Wireless Sensor Networking." *IEEE/ACM Transaction on Networks* 11 (1): 2–16.
31. Goldsmith, A. J., and S. B. Wicker. 2002. "Design Challenges for Energy Constrained Ad-Hoc Wireless Networks." *IEEE Wireless Communications* 9 (4): 8–27.
32. Ye, W., J. Heidemann, and D. Estrin. 2002. "An Energy-Efficient MAC Protocol for Wireless Sensor Networks." In *IEEE INFOCOM 2002*, 1567–76.
33. Dam, T. V., and K. Langendoen. November 2003. "An Adaptive Energy Efficient MAC Protocol for Wireless Sensor Networks." In *1st International Conference on Embedded Networked Sensor Systems*, 171–80.
34. Callaway, E. H. 2003. *Wireless Sensor Networks Architecture and Protocols*. Boca Raton, FL: CRC Press.
35. Callaway, E., P. Gorday, L. Hester, J. A. Gutierrez, M. Naeve, B. Heile, and V. Bahl. 2002. "Home Network with IEEE 802.15.4: A Developing Standard for Low Rate Wireless Personal Area Network." *IEEE Communication Magazine* 40 (8): 70–7.
36. LAN/MAN standard committee of the IEEE Computer Society. October 2003. IEEE Standard for Information Technology, Telecommunications and Information Exchange between Systems, Local and Metropolitan Area Network Specific Requirements Part 15.4, Wireless Medium Access Control and Physical Layer Specifications for Low Rate Wireless Personal Area Network.
37. Lu, G., B. Krishnamachari, and C. S. Raghavendra. April 2004. "Performance Evaluation of the IEEE 802.15.4 MAC for Low Rate Low Power Wireless Networks." In *IEEE International Conference on Performance Computing and Communications*, 701–6, Phoenix, AZ.
38. Heinzelman, W. B., A. P. Chandrakasan, and H. Balakrishnan. 2002. "Adaptive Protocol for Information Dissemination in Wireless Sensor Networks." *IEEE Transaction on Wireless Networking* 1 (4): 660–70.
39. Heinzelman, W. B., A. P. Chandrakasan, and H. Balakrishnan. January 2000. "Energy Efficient Communication Protocol for Wireless Microsensor Networks." In *3rd Hawaii International Conference on System Services*, 174–85, Hawaii.
40. Sohrabi, K., J. Gao, V. Ailawadhi, and G. J. Pottie. 2000. "Protocol for Self-Organization of a Wireless Sensor Network." *IEEE Personal Communications* 7 (5): 16–27.
41. Sohrabi, K., and G. J. Pottie. 1999. "Performance of a Novel Self-Organize Protocol for Wireless Ad-Hoc Sensor Networks." In *IEEE 5th Vehicular Technology Conference*, 1222–6.
42. Woo, A., and D. Culler. 2001. "A Transmission Control Scheme for Media Access in Sensor Networks." In *ACM/IEEE International Conference on Mobile Computing and Networking, Mobicom 2001*, 221–35.

43. Raghavendra, C. S., and S. Singh. 1998. "PAMAS Power Aware Multi Access Protocol with Signaling for Ad-hoc Networks." *ACM Computer Communications*, 27: 5–26.

44. Zheng, J., and M. J. Lee. 2004. "Will IEEE 802.15.4 Make Ubiquitous Networking a Reality?" *IEEE Communication Magazine* 42 (6): 140–6.

45. Gutierrez, J. A., M. Naeve, E. Callaway, V. Mitter, and B. Heile. 2001. "IEEE 802.15.4 A Developing Standard for Low Power Low Cost Wireless Personal Area Network." *IEEE Network Magazine* 15 (2): 12–19.

46. Grossglauser, M., and D. Tse. 2001. "Mobility Increases the Capacity of Adhoc Wireless Networks." *IEEE Infocom 2001: The Conference on Computer Communications* 1 (3): 1360–9.

47. Luo, J., J. Panchard, M. Piorkowski, M. Grossglauser, and J. P. Hubaux, June 2006. "MobiRoute: Routing towards a Mobile Sink for Improving Lifetime in Sensor Networks." In *2nd IEEE/ACM International Conference on Distributed Computing in Sensor Systems*, 480–97, San Francisco, CA.

48. Vincze, Z., and R. Vida. October 2005. "Multi-Hop Wireless Sensor Networks with Mobile Sink." In *ACM Conference on Emerging Network Experiment and Technology*, 302–3, Toulouse, France.

49. Raviraj, P., H. Sharif, M. Hempel, and S. Ci. 14–17 August 2005. "MOBMAC: An Energy Efficient and Low Latency MAC for Mobile Wireless Sensor Networks." *IEEE Systems Communications* 370–5.

50. Munir, S. A., B. Ren, W. Jiao, B. Wang, D. Xie, and J. Ma. May 2007. "Mobile Wireless Sensor Network Architecture and Enabling Technologies for Ubiquitous." In *Conference on Advanced Information Networking and Applications Workshops, AINAW '07*, 113–20.

51. Rahimi, M., H. Shah, G. S. Sukhatme, J. Heideman, and D. Estrin. May 2003. "Studying the Feasibility of Energy Harvesting in a Mobile Sensor Networks." In *Proceedings of the IEEE International Conference on Robotics and Automation*, vol. 1, 19–24, Taipei, China.

52. Chakrabarti, A., A. Sabharwal, and B. Aazhang. April 2003. "Using Predictable Observer Mobility for Power-Efficient Design of Sensor Networks." In *2nd International Workshop on Information Processing in Sensor Networks*, vol. 2634, 129–45.

53. Ali, M., T. Suleman, and Z. A. Uzmi. 2005. "MMAC: A Mobility-Adaptive, Collision-Free Mac Protocol for Wireless Sensor Networks." In *Proceedings of the 24th IEEE IPCCC'05*, 401–7, Phoenix, AZ.

54. Gandham S. R. et al. 2003. "Energy Efficient Schemes for Wireless Sensor Networks with Multiple Mobile Base Stations." In *Proceedings of IEEE GLOBECOM*.

55. Luo, J., and Hubaux, J.-P. 2005. "Joint Mobility and Routing for Lifetime Elongation in Wireless Sensor Networks." In *Proceedings of IEEE INFOCOM*.

56. Ekici, E., Y. Gu, and D. Bozdag. July 2006. "Mobility-Based Communication in Wireless Sensor Networks." *IEEE Communications Magazine* 44 (7): 56–62.

57. Kansal, A. et al. 2004. "Intelligent Fluid Infrastructure for Embedded Networks." In *Proceedings of 2nd International. Conference Mobile Systems Applications and Services*.

58. Shah, R. et al. 2003. "Data Mules: Modelling a Three-Tier Architecture for Sparse Sensor Networks." In *Proceedings of IEEE Workshop Sensor Network Protocols and Applications*.

59. Ghassemian, M., and H. Aghvami. June 2008. "An Investigation of the Impact of Mobility on the Protocol Performance in Wireless Sensor Networks." In *24th Biennial Symposium on Communications*, 310–5.

60. Narwaz, S., M. Hussain, S. Watson, N. Trigoni, and P. N. Green. 2009. "An Underwater Robotic Network for Monitoring Nuclear Waste Storage Pools." In *Sensors and Software Systems*. Springer.

61. Pandya, A., A. Kansal, and G. Pottie. March 2008. "Goodput and Delay in Networks with Controlled Mobility." In *2008 IEEE Aerospace Conference*, 1323–30.

62. Dantu, K., M. Rahimi, H. Shah, S. Babel, A. Dhariwal, and G. S. Sukhatme. April 2005. "Robomote: Enabling Mobility in Sensor Networks." In *IEEE/ACM, 4th International Conference on Information Processing in Sensor Networks, (IPSN/SPOTS)*, 404–9.

63. Somasundara, A. A., A. Ramamoorthy, and M. B. Srivastava. 2004. "Mobile Element Scheduling for Efficient Data Collection in Wireless Sensor Networks with Dynamic Deadlines." In *Proceedings of the 25th IEEE International Real-Time Systems Symposium*.

64. Jea, D., A. A. Somasundara, and M. B. Srivastava. 2005. "Multiple Controlled Mobile Elements (Data Mules) for Data Collection in Sensor Networks." In *Proceedings of IEEE/ACM International. Conference on Distributed Computing. in Sensor Systems*.

65. Lam, S. 1980. "A Carrier Sense Multiple Access Protocol for Local Networks." *Computer Networks* 4: 21–32.

66. Woon, W. T. H., and T. C. Wan. 2008. "Performance Evaluation of IEEE 802.15.4 Wireless Multi-Hop Networks: Simulation and Testbed Approach." *International Journal of Ad-Hoc and Ubiquitous Computing* 3 (1): 57–66.

67. Zheng, J., and M. J. Lee. 2006. "A Comprehensive Performance Study of IEEE 802.15.4." In *Sensor Network Operations*, Chapter 4, 218–37. IEEE Press, Wiley Interscience.

68. Koubaa, A., M. Alves, E. Tovar, and Y. Q. Song. April 2006. "On The Performance Limits of Slotted CSMA/CA in IEEE 802.15.4 for Broadcast Transmissions in Wireless Sensor Networks." IPP-HURRAY Technical Report, TR-060401.

69. Laibowitz, M., and J. A. Paradiso. May 2005. "Parasitic Mobility for Pervasive Networks." In *3rd International Conference on Pervasive Computing, PERVASIVE 2005*, Munich, Germany.

70. Hu, L., and D. Evans. September 2004. "Localization for Mobile Sensor Networks." In ACM Mobi-Com 2004.

71. Chen, C., and J. Ma. 2007. "Simulation Study of AODV Performance over IEEE 802.15.4 MAC in WSN with Mobile Sinks." In *Proceedings of Advanced Information Networking and Applications Workshop 2007, AINAW'07*, 159–63.

72. Attia, S. B., A. Cunha, A. Koubaa, and M. Alves. July 2007. "Fault Tolerance Mechanism for Zigbee Wireless Sensor Networks." In *19th Euromicro Conference on Real Time Systems, ECRTS'07*, Pisa, Italy.

73. Zen, K., D. Habibi, A. Rassau, and I. Ahmed. 2008. "Performance Evaluation of IEEE 802.15.4 for Mobile Sensor Networks." In *5th International Conference on Wireless and Optical Communications Networks*, Surabaya, Indonesia.

74. Choi, S.-C., J.-W. Lee, and Y. Kim. 2008. "An Adaptive Mobility-Supporting MAC Protocol for Mobile Sensor Networks." In *IEEE Vehicular Technology Conference*, 168–72.

75. Pham, H., and S. Jha. 2004. "An Adaptive Mobility-Aware MAC Protocol for Sensor Networks (MS-MAC)." *In Proceedings of the IEEE International Conference on Mobile Ad-Hoc and Sensor Systems (MASS)*, 214–26.

76. Lin, P., C. Qiao, and X. Wang. 2004. "Medium Access Control with a Dynamic Duty Cycle for Sensor Networks." In *Proceedings of the IEEE Wireless Communications and Networking Conference (WCNC)*, vol. 3, 1534–39.

77. Hameed, S. A., E. M. Shaaban, H. M. Faheem, and M. S. Ghoniemy. October 2009. "Mobility-Aware MAC protocol for Delay Sensitive Wireless Sensor Networks." In *IEEE Ultra Modern Telecommunications & Workshops*, 1–8.

PROTOCOLS AND DATA GATHERING ISSUES

Chapter 11

Investigation on Protocols for Wireless Sensor Networks

A. K. Dwivedi and O. P. Vyas

Contents

11.1 Introduction

Advances in wireless sensor networks (WSNs) have led to many new protocols. A careful approach is needed while proposing, designing, selecting, and classifying protocol for WSNs, because WSNs are challenging due to their inherent characteristics such as energy efficiency and awareness, connection maintenance, minimum resource usage limitation, low latency, and load balancing in terms of energy used by sensor nodes, where energy awareness is an essential consideration.

A three-dimensional sensor network protocol stack for WSNs is presented by Akyildiz et al. [1], which comprises five layers with three planes. Marco [2] presented a simplified protocol stack for WSNs as well. The generalized protocol stack for WSNs is shown in Figure 11.1. Whereas in practical WSNs in which sensor nodes use TinyOS operating systems, there does not exist a

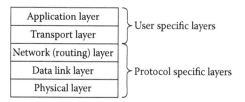

Figure 11.1 A generalized protocol stack for WSNs.

clear notion for demarcating between the various layers [3]. According to the author of the same technical report, TinyOS specifications divide the entire communication stack into three major layers. The upper layer is the Application Layer, which is quite specific to the usage and deployment environment of the WSN. The middle Layer is the Data Link (MAC) Layer, which deals with issues such as power control, time scheduling, and synchronization among the nodes. Third is the Physical Layer, which deals with all radio-related tasks.

The objective of this contribution is to present all fundamental theoretical aspects in analytical way with deep focus on each layer of the generalized WSN protocol stack, cross-layer design protocols, and highlighting the importance of radio irregularity.

Initially, in research contributions most of the consideration has been given to the network layer (routing) protocols because they might differ depending on the network architectures and their applications. Akkaya and Younis [4] survey some routing protocols for WSNs and present a classification for the various pursued approaches. Three main categories are explored in their paper: data centric, hierarchical, and location based. In another research paper [5], Chalak et al. present a comparative study of routing protocols in WSNs.

Becker et al. [6] present a study on the usability of several routing protocols in a real-world environmental monitoring task and show how the performance of wireless routing protocols can be improved significantly if adapted carefully for the use in WSN. The performance of the different available routing protocols is then measured and compared through actual deployment of the WSN using Cricket motes, which have been designed by University of California, Berkeley.

Boukerche et al. [7] present applications that require fine-grain monitoring of physical environments subjected to critical conditions such as fire, leaking of toxic gases, and explosions, which pose a great challenge to sensor network protocols. These networks have to provide a fast, reliable, fault-tolerant, and energy-aware channel for events diffusion, which meets the requirements of query-based, event-driven, and periodic sensor networks application scenarios. These requirements have to be met even in the presence of emergency conditions that can lead to node failures and path disruption to the sink. This paper proposes two routing protocols: (1) periodic, event-driven, and query-based protocol (PEQ) and (2) its variation CPEQ, two fault-tolerant and low-latency algorithms that meet sensor network requirements for critical conditions supervision in context-aware physical environments.

In another research contribution [8], Li and Newe focus on the Selection of Routing Protocol for Applications. Due to the differences between WSNs and other wireless networks, new network architectures have been developed and many new routing protocols have been proposed for these architectures. Moreover, many applications and systems with enormously varying requirements and characteristics have been designed and implemented. Since the selection of routing protocols for WSNs is closely related with the application requirements and routing protocol characteristics, there is no common architecture or routing protocol solution. The same paper [8] presents a table that enables deployment engineers to select appropriate routing protocols that are suitable for a particular application and that will maximize the return from the deployed network.

With time, researchers explored the WSNs in terms of new applications, its architectures, associated technologies, and new research challenges. Initially, researchers paid attention to the network layer, but now they think and work beyond this, which means that now researchers focus on other layers of protocol stack, with the exception of the network (routing) layer. Furthermore, they work on cross-layer approaches to achieve energy efficiency, reliable connection maintenance, minimum resource usage, low latency, and load balancing in terms of energy used by sensor nodes.

11.2 Prior Work

According to Raman and Kameswari [9], WSN-specific protocols are extremely dependent on application scenarios, but most WSN-specific protocols do not mention or have any prior information about any specific application in their design. After a deep investigation on WSN-specific protocols, various evaluation factors and design challenges are identified that are mostly applicable to all layers of the generalized protocol stack for WSNs. A statistic is presented here in this subsection. Security concerns in layered and cross-layer approaches are also discussed here in short.

11.2.1 Protocol Evaluation Factors for WSN-Specific Protocols

Table 11.1 shows some of the common parameters on which WSN-specific protocols of almost all layers (except some evaluation parameters, which are specific to an individual layer) must be evaluated during the designing of a new one.

Table 11.1 Evaluation Factors for WSN-Specific Protocols

Evaluation Parameter	Description
Power Usage	Sensor node's lifetime is clearly dependent on its power source; thus, useful power usage must involve transmitting/receiving data, processing query requests, and forwarding queries/data to neighboring nodes.
Data Aggregation	Substantial energy savings and traffic optimization can be obtained through data aggregation.
Scalability	The possibility to enlarge and reduce the network.
Reliability or Fault Tolerance	Fault tolerance is the ability to sustain WSN functionalities without any interruption due to node failures.
Latency (delay) and Overhead	Multihop relays and data aggregation cause data latency, and these important factors influence routing protocol design.
Data Delivery Model	Data delivery model (Continuous, Event-driven, Query-driven, and Hybrid) [10] determines when the data collected by the sensor node have to be delivered.
Quality of Service (QoS)	Quality service required by the application involves length of life time, data reliability, energy efficiency, location awareness, collaborative processing, etc. QoS factors will affect the selection of routing protocols for a particular application.

(Continued)

Table 11.1   Evaluation Factors for WSN-Specific Protocols (*Continued*)

Evaluation Parameter	Description
Security	Security concerns need special attention in current era, where data stealing and data diddling become major issues.
Node Deployment Options	Node deployment options affect the performance of routing protocol, basically in terms of energy consumptions.
Topology	Topology of a WSN affects many of its characteristics like latency, capacity, and robustness. Further, the complexity of data routing and processing depends on the network topology.
Sensor Density and Network Size	Sensor density of nodes affects the degree of coverage area of interest, whereas networks size affects reliability, accuracy, and data-processing algorithms.
Environment or Scenario	A critical parameter because node and network lifetime is directly dependent on it.
Byte Overhead [11]	Byte overhead means the total number of bytes in the routing control messages needed to find a route to the sink. For flooding, byte overhead means the total number of bytes in the extra messages flooded throughout the network. In both cases, the bytes in the data packets transmitted by nodes along the route from the originating node to the sink node are not counted as overhead.

11.2.2 Protocol Design Challenges

In various literatures or research contributions [1,4,12–17] related to WSNs, the following design challenges are identified for protocols:

- No global addressing scheme is possible because of the large numbers of wireless sensor nodes. Traditional TCP/IP-based protocols may not be directly applied to WSNs because overhead of ID maintenance becomes high for the deployment of large numbers of wireless sensor nodes.
- Unlike typical communication networks, in most WSN applications, multiple sources sense environmental data and transfer them to a particular base station or sink node.
- Careful resource management is required due to the highly constrained nature of sensor nodes in terms of energy, processing, transmission, and storage capabilities.
- Normally, node deployments in WSNs are static, except in some special cases where node deployments in WSNs can be dynamic (deploying nodes on some mobile devices/vehicles/robots). Some stationary WSNs may have few mobile nodes.
- Design requirements of a sensor network change with application because most of the WSNs are application specific.
- Since data collection is normally based on the location, position awareness of sensor nodes is important.
- Data collected by various sensors in WSNs are typically based on environmental parameters; hence, there is a chance of some redundancy in these data.
- Visibility [16] is a new metric for the WSN protocol design, whose objective is, "Minimize the energy cost of diagnosing the cause of a failure or behavior."
- "Spatial-Temporal Distribution" is also a new metric for WSN protocols presented by Liu and Cao [17].

Table 11.2 Protocol Design Challenges and Recommended Solutions

Design Factors or Challenges	Recommended Solutions
Energy Depletion	1. Maximize network lifetime. 2. Minimize data communications over the air. 3. Reduce the active duty cycle for each sensor node.
Scalability in term of wireless sensor nodes	1. Organize network in a hierarchical (clustered) manner. 2. Utilize localized algorithms with localized interactions among sensor nodes.
Robustness to dynamic environment	1. Self-organize networks. 2. Self-healing networks. 3. Self-configuring networks. 4. Self-adaptive networks.

According to Al Obaisat and Braun [18], these factors are identified, which creates a roadmap for designing a new WSN-specific protocol or algorithm: reliability or fault tolerance, density and network size/scalability, sensor network topology, energy consumption, hardware constraints, data aggregation/data fusion, transmission media, security, self-configuration, network dynamics, quality of service, coverage, and connectivity. In the same research contribution [18], some common design factors and challenges for WSN-specific protocols are suggested, and the authors also recommended solutions for these design factors and challenges, which are listed in Table 11.2.

11.2.3 Security Concerns at WSN-Specific Protocols

Currently much of the work is being done on providing layered security; in this approach, security solutions try to either provide security to all the layers of the protocol stack or just focus on a particular layer. Sharma and Ghose [19] focus on the limitations of this layered approach, which states that security services of layered approaches are layer-centric; some of these solutions address only a particular security treat or attack, and this approach has pitfalls such as redundancy and inflexibility in security.

A technical report [3] discusses about the security in WSN, which is a layer-based classification, and states that considering security threats on layer basis is a better option than considering security threats on attack basis.

An individual protocol layer of WSN protocol stack particularly focuses on different aspects for the security provisions in WSNs. Gawdan et al. [20] suggests that the upper layer of the protocol stack (application layer) focuses on key management mechanisms, which in turn supports encryption and decryption of the lower layers; that data link and network layers deal with the encryption of sensitive data and routing information; and that the lower layer (physical layer) provides information privacy using encoding.

Various types of security threats and respective countermeasures are proposed for WSN-specific protocol layers in different research contributions, such as [3,21,22]. With the help of these research contributions, we have described the identified security threats and respective countermeasures in Table 11.3.

Table 11.3 Security Threats and Respective Countermeasures in Case of WSNs

Protocol Layer	Identified Threats	Respective Countermeasures
Application Layer	Malicious Nodes	1. With the help of aggregation schemes/protocols
	Data's Confidentiality, Authentication, Integrity, Availability, Freshness, Time Synchronization, and Secure Localization	1. Tackle by using cryptography and key management, secure time synchronization, secure location discovery, secure routing, trust management system, secure data aggregation, and intrusion detection systems/protocols
Transport Layer	Flooding Attacks	1. Limiting the number of connections that a node can make prevents complete resource exhaustion 2. Client puzzles
	De-synchronization Attacks	1. Authentication mechanisms for all exchange packets
Network Layer	Wormhole	1. Physical monitoring of sensor devices and regular monitoring of network using source routing. As well as monitoring system may use Packet Leach techniques
	Selective Forwarding	1. Regular network monitoring using source routing
	D.o.S.	1. Protection of network specific data like network ID, etc. 2. Physical protection and inspection of network
	Sybil	1. Resetting of devices and changing of session keys
	Traffic Analysis	1. Sending of dummy packet in quite hours with regular monitoring of network
	Eavesdropping	1. Session keys protect NPDU from eavesdroppers
	Neglect and Greed	1. Redundancy and probing
	Homing	1. Encryption
	Misdirection	1. Egress filtering, authorization, and monitoring
	Black Holes	1. Authorization, monitoring, and redundancy
	Node Capture attacks	1. State replication across the network with majority voting to detect inconsistencies 2. Model node-capture resilient networks

Table 11.3 Security Threats and Respective Countermeasures in Case of WSNs (*Continued*)

Protocol Layer	Identified Threats	Respective Countermeasures
Data-Link (MAC) Layer	Collision	1. Use of error correcting codes (CRC) and time diversity
	Exhaustion (Continuous channel access)	1. Rate limiting 2. Protection of network ID and other information that is required for joining device
	Spoofing	1. For re-sending the messages use different paths
	Sybil	1. Changing of keys regularly
	De-synchronization	1. Using different neighbors for time synchronization
	Traffic Analysis	1. Sending of dummy packet in quite hours with regular monitoring of networks
	Eavesdropping	1. Session keys protect NPDU from eavesdroppers
	Unfairness	1. Use of small frames
Physical Layer	Interference and Jamming	1. Use of spread-spectrum/channel hoping techniques for radio communications and blacklisting
	Sybil	1. Physical protection of devices
	Tampering	1. Protection and changing of key

Attacks and countermeasures at the network layer (routing) protocols are also presented in a highly cited research contribution [23]. Karlof and Wagner [23] also focus on the physical and MAC layer; according to the authors, these two layers are susceptible to direct attack. Security threats on the physical layer can be countered by frequency hopping or spread spectrum communication, whereas attacks on the MAC layer can be alleviated by using techniques such as use of less susceptible protocol, good entropy management, and a cryptographically secure pseudo-random number generator.

The holistic security approach [24], is a good example of the aforementioned layered approach, which provides network security for all the layers of the protocol stack.

Since the security mechanism is required for each layer in the traditional layered approach, in view of limited computational resources available with wireless sensor nodes and their wastage, other options for network security must be considered, except layered approach. Cross-layer security solutions can be the best option. Cross-layer design for the security of WSNs is presented by Xiao et al. [25]. Design guidelines for the cross-layer security solution are available in a research contribution [19]. Some proposed security schemes based on the concept of cross-layer design methodology include the Cross-Layer Integrated Framework for Security for WSN (CLIFFs WSN or simply CLIFFs) [19] and Cross-Layer-Based Comprehensive Security Framework (CLBCSF) [20].

11.3 Theoretical Aspects of Application Layer Protocols for WSNs

There are a wide range of areas where WSNs can be applied such as disaster relief/emergency rescue operation, detecting chemical/biological/radiological/nuclear/explosive materials, habitat monitoring and tracking/monitoring patients and elderly people/healthcare applications, military battlefield awareness and surveillance/fleet monitoring/security and surveillance/homeland security, environmental monitoring, pipeline corrosion monitoring, monitoring conditions of buildings and bridges, industrial process monitoring and control, machine health monitoring, asset and warehouse management, home automation/home networks, building monitoring and control, and traffic control. Based on the environment's sensing parameters and applications, various types of application software can be built and used on the application layer [1].

The application layer includes all aspects related to the WSN protocol stack that are not covered by lower layers. According to Haenselmann [26], in WSNs the application will often implement almost all layers. The issues on the application layer design represent user application-related services. Application layer protocols are responsible for the user and application program support such as gathering and collecting data, signal processing, data fusion, and notification. The main objectives of the processes at the application layer are to create effective new capabilities for efficient extraction, manipulation, transport, and representation of information derived from sensor data [27].

The application layer allows the end application to communicate with the communication architecture by providing the appropriate communication services to the application. Three application layer protocols are examined by Akyildiz et al. [1]: Sensor Management Protocol (SMP), Task Assignment and Data Advertisement Protocol (TADAP), and Sensor Query and Data Dissemination Protocol (SQDDP).

SMP [1] provides the software operations needed to perform the following administrative tasks: introducing the rules related to data aggregation, attribute-based naming, clustering to the wireless sensor nodes, exchanging data related to the location-finding algorithms, time synchronization of the sensor nodes, moving sensor nodes, turning sensor nodes on and off, querying the sensor network configuration and the status of nodes, re-configuring the sensor network, and authentication, key distribution, and security in data communications. TADAP [1] is responsible to provide user software with efficient interfaces for interest dissemination, which is useful for lower layer operations. SQDDP [1] provides end-user applications with interfaces to issue queries, respond to queries, and collect incoming replies.

Another example that provides services required at application layer is mentioned by Marco [2], which is Sensor Network Services Platform (SNSP) [28]. According to Marco [2], the application layer defines a standard set of services and interface primitives available to a programmer independently on their implementation on every kind of platform.

11.4 Theoretical Aspects of Transport Layer Protocols for WSNs

Most WSN applications require that tiny wireless sensor nodes sense some physical or environmental data and transmit to the base station or sink node. This approach requires data delivery guarantee and tolerable end-to-end delay. An exhaustive list and analysis of WSN transport protocols are available in two research contributions [29,30]. A book chapter dedicated to WSN transport layer [31] focuses on requirements needed for WSN transport protocols; according to this research contribution, WSN transport layer should be able to

- Provide consistency with the application's data flow model like event-driven, continuous, real-time and/or hybrid,
- Provide upstream and downstream reliability, end-to-end reliability, and congestion control,
- Handle application's variable reliability model,
- Be fairly scalable with volume/density of sensor nodes in WSN,
- Cooperate with upper and lower layers of WSN protocol stack,
- Consume low energy as possible, and
- Minimize the usage of control messages without compromising the required level of data throughput.

Another research paper [32] highlights the requirements needed for WSN transport protocols; they include generic, heterogeneous data flow support, controlled variable reliability, congestion detection and avoidance, base station controlled network, scalability, and future enhancements and optimizations.

Reliable data transfer on WSNs is also very crucial. End-to-end retransmission and end-to-end congestion control are two basic techniques used in the Internet for reliable data transport, but these two are inappropriate with respect to resource constraint nature of wireless sensor nodes and with often lossy and unreliable communication links of WSNs. Based on the facts of some research papers, Tezcan and Wang [33] presented that existing WSN-specific transport layer protocols designed for upstream or downstream reliability in WSNs are either sink-to-sensor or sensor-to-sink reliable delivery. A comparison table of some reliable WSN transport protocols is presented by Pereira et al. [34]. Some techniques suitable for WSNs are evaluated by Kim et al. [35], like route fix, erasure code, and link-level retransmission. This research paper shows that the right and efficient combinations of primitives can yield more than 99% reliability with low overhead, providing a viable alternative to end-to-end retransmission over multiple hops.

Congestion control is a vital issue in large-scale WSNs. Congestion may bring degradation of overall channel quality and increased loss rates, leads to buffer drops and enlarged delays, and tends to be disgustingly unfair toward nodes whose data have to traverse a larger number of hops [36]. Monteiro et al. [36] present a performance comparison study with the focus on the most relevant congestion control transport protocols for WSNs. The main objective of this study was to identify some characteristics of these types of protocols, focusing on energy saving and penalty fidelity with the help of experimental results.

According to Pereira et al. [34], transport protocols for WSNs can be classified into two main functions:

1. Classification based on reliability
2. Classification based on congestion control

Based on the information available from research papers at that time, Rahman et al. [31] presented three different classifications for WSN-specific transport layer protocols:

- Classification based on Congestion Control and Reliability:
    - Protocols with a Congestion Control Mechanism
        - Protocols under this classification include ART, CCF, CODA, ESRT, FUSION, PCCP, RAP, SenTCP, Siphon, and STCP.
    - Protocols with Reliability Features
        - Protocols under this classification include ART, ESRT, GARUDA, PSFQ, RBC, RMST, STCP, Tiny TCP/IP, and Trickle.

- Classification based on evaluation type:
  - Protocols evaluated on Simulation
    - Protocols under this classification include ART, CCF, CODA, ESRT, GARUDA, PCCP, PSFQ, RAP, RMST, SenTCP, Siphon, STCP, Tiny TCP/IP, and Trickle.
  - Protocols evaluated on Experimental testbed
    - Protocols under this classification include CCF, CODA, FUSION, PSFQ, RBC, Siphon, Tiny TCP/IP, and Trickle.
- Some miscellaneous classifications:
  - Classification based on Fairness
    - Protocols under this classification include CCF, FUSION, PCCP, RAP, and SenTCP.
  - Classification based on Energy
    - Protocols under this classification include ART, CODA, ESRT, FUSION, GARUDA, PCCP, PSFQ, RMST, SenTCP, Siphon, STCP, Tiny TCP/IP, and Trickle.
  - Classification based on Traffic Type
    - Periodic
      - Protocols under this classification include FUSION, RAP, and STCP.
    - Event based
      - Protocols under this classification include ART, ESRT, FUSION, RAP, STCP, and RBC.
  - Classification based on Cross-Layer Issues
    - Protocols under this classification include FUSION, RAP, and RMST.

Following sections deal with some WSN transport layer protocols.

11.4.1 Wireless Modular Transport Protocol (WMTP) [37]

WMTP is a modular transport layer protocol for WSNs; its main objective is to use simultaneously all the main features commonly found in WSN transport protocols, like congestion control, reliability, and fairness. Through this way, the upper application layer is free to choose and use exact features that required. WMTP provides a set of uncommon features like throttling, flow-control, transport layer QoS, and optional integration with service discovery.

11.4.2 Pump Slowly, Fetch Quickly (PSFQ) [38]

PSFQ is an efficient reliable transport protocol for WSNs. PSFQ represents a simple approach with minimum requirements on the routing infrastructure; minimum signaling, thereby reducing the communication cost for data reliability; and, finally, responsive to high error rates, allowing successful operation even under highly error-prone conditions. PSFQ comprises three functions: message relaying (i.e., pump operation), relay-initiated error recovery (i.e., fetch operation), and selective status reporting (i.e., report operation). The basic idea behind PSFQ is to distribute data from a source node by pacing data at a relatively slow speed (pump slowly), but allowing nodes that experience data loss to fetch (i.e., recover) any missing segments from immediate neighbors very aggressively (local recovery, fetch quickly). It operates correctly in high error-prone environments.

11.4.3 Event-to-Sink Reliable Transport (ESRT) [39]

ESRT is a novel reliable transport solution for data-centric WSNs. The primary goals of this protocol are to achieve event-to-sink reliability and to provide reliable event detection without any intermediate caching requirements. Based on local buffer-level monitoring at a node level, ESRT uses a congestion detection mechanism. It also seeks to achieve the required event detection accuracy using minimum energy expenditure and has a congestion control component. It has no mechanism to support end-to-end reliability.

11.4.4 Sensor Transmission Control Protocol (STCP) [32]

A generic, scalable, and reliable transport layer solution for energy-constrained WSNs. Most of the STCP functionalities are implemented at the base station or sink node. It offers controlled variable reliability, congestion detection, and avoidance, and it supports multiple applications in the same network. Like UDP, for data-centric WSN applications, it does not provide any acknowledgment scheme.

11.4.5 Asymmetric and Reliable Transport (ART) [33]

It is a transport protocol designed to achieve event and query reliability. It provides end-to-end reliability in two directions based on energy-aware node classification (essential and nonessential sensor nodes) and a distributed energy aware congestion control mechanism. ART protocol operation includes three main functions: reliable query transfer, reliable event transfer, and distributed congestion control. After node classification, end-to-end reliable communication is achieved using asymmetric acknowledgment (ACK) and negative acknowledgment (NACK) signaling between essential nodes to sink node and sink node to sensor nodes, respectively. ACK handles event transfer, whereas NACK solves the reliable query delivery. Congestion is relieved by regulating traffic from nonessential sensor nodes.

11.4.6 Distributed Transport for Sensor Networks (DTSN) [40]

It is a novel, reliable transport protocol for converge-cast and uni-cast communications in WSNs. The main goals of DTSN are energy efficiency, reliability differentiation, and distributed functionality. DTSN specification can be divided into total reliability service and differentiated reliability service. Total reliability is achieved by a Selective Repeat Automatic Repeat Request (ARQ) mechanism that uses both negative acknowledgment (NACK) and positive acknowledgment (ACK) control packets. The differentiated reliability service is based on the total reliability service. Actually, basic DTSN ARQ policy is eventually combined with Forward Error Correction (FEC) strategy and is able to support several reliability grades based on the integration of partial buffering at the source, intermediate node caching, and erasure coding.

11.5 Theoretical Aspects of Network Layer (Routing) Protocols for WSNs

The network layer handles end-to-end communications, and this means that two systems can communicate through an arbitrary number of intermediate nodes called hops. Whereas transmission on the data link layer takes place only between two neighbors, the multihop communication on the network layer involves routing decisions. Since WSNs are a special class of infrastructure-less networks, they have to provide all the services they need themselves. As a result, each node has dual responsibility: it can be the source of information, or it may work as a router. Not all WSN applications require to route information between arbitrary pairs of nodes, but all nodes are sources that send data only to one or a small number of dedicated sinks/root. This simplifies routing decisions significantly.

There are various well-researched and established protocols for Mobile Ad hoc Network (MANET) under various routing categories, such as Pro-active or Table-driven routing, Reactive or On-demand routing, Flow-oriented routing, Adaptive or Situation-aware routing, Hybrid (both pro-active and reactive) routing, Hierarchical routing protocols, Host Specific Routing protocols, Geographical routing protocols, Power-aware routing protocols, Multicast routing, and Geographical multicast protocols or Geocasting. Due to the finite power available to each sensor node and with the limited processing power, regular ad hoc routing techniques cannot be directly applied to the sensor network domain.

Initially, existing ad hoc routing protocols had been employed in WSNs, but due to the characteristics of the WSN technology and the application, these protocols often perform badly. Performance of WSN-specific routing protocols are closely related to the architectural model.

Various classifications for WSNs are presented in a different literature for routing protocols (see Table 11.4).

Table 11.4 Main Categories and Subcategories of Protocol Classifications

S.N.	Main Category	Sub Categories
1.	Classification based on Network Structure [14]	1. Flat-based or Data-Centric routing 2. Hierarchical-based or Cluster-based routing 3. Location-based routing
2.	Classification based on Protocol Operation [14]	1. Multipath-based routing 2. Query-based routing 3. Negotiation-based routing 4. QoS-based routing 5. Noncoherent & Coherent data-processing-based routing
3.	Classification based on Sensor Node Architecture [14]	1. Protocols operating on flat topology (WSN consisting homogeneous nodes) 2. Protocols operating on hierarchical topology (WSN consisting heterogeneous nodes)
4.	Classification based on Packet Destination [15]	1. Gossiping and agent-based unicast forwarding 2. Energy-efficient unicast 3. Broadcast and multicast 4. Geographic routing 5. Mobile nodes
5.	Crossbow (Xbow) classification [41,42]	1. Basic routing (with normal or improved variants) 2. Reliable routing 3. Low-power routing 4. XMesh routing

Table 11.4   Main Categories and Subcategories of Protocol Classifications (*Continued*)

S.N.	Main Category	Sub Categories
6.	Cooperative routing [42]	1. An approach in which sensor nodes send data to a central node that join the data to reduce the cost in terms of energy consumption
7.	Classification based on State [13]	1. Stateful ad hoc routing 2. Stateless Geometric ad hoc routing
8.	Classification based on epidemic behavior [43]	1. Pull-based epidemic algorithm 2. Push-based epidemic algorithm 3. Pull–push-based epidemic algorithm
9.	Classification based on protocol's initialization point [44]	1. Source-initiated (Src-initiated) 2. Destination-initiated (Dst-initiated)
10.	Classification based on how the source finds the destination [44]	1. Proactive 2. Reactive 3. Hybrid
11.	Classification based on the basis of how to reduce useful energy consumption [45]	1. Protocols that control the transmission power level at each node by increasing network capacity while keeping the network connected 2. Protocols that make routing decisions based on power optimization goals 3. Protocols that control the network topology by determining which nodes should participate in the network operation (be awake) and which should not (remain asleep)
12.	Classification based on application [46]	1. Routing protocols suitable for Query-based WSN Applications 2. Routing protocols suitable for Event-driven WSN Applications 3. Routing protocols suitable for Periodic WSN Applications
13.	Classification by K. Sha et al. [47]	1. Greedy geographic routing 2. Load balanced routing 3. Energy aware routing 4. Fault tolerant routing 5. Information exploiting routing
14.	Classification by M. Perillo et al. [48]	1. Resource-Aware Routing 2. Energy-Aware Routing 3. Fidelity-Aware Routing 4. Data-Centric Routing Protocols 5. Geographic Routing 6. Clustering

Apart from these classifications, routing protocols can be classified based on whether they have security mechanism, whether they are energy efficient, whether they are suitable for real-time applications, and so on.

11.5.1 WSN Protocols Based on Network Structure

Based on the network structure, WSN routing protocols can be classified into [14]

■ Flat-based or Data-centric routing
■ Hierarchical-based or Cluster based routing
■ Location-based routing

11.5.1.1 Flat-Based or Data-Centric Routing

In flat-based routing algorithm, all nodes play the same role and mainly apply flood-based data transferring. The drawbacks of flooding include implosion, which is caused by duplicate messages being sent to the same node, overlap when two nodes sensing the same region send similar packets to the same neighbor, and resource blindness in consuming large amounts of energy without consideration for energy constraints. Examples of Flat-based routing algorithms or protocols are Directed Diffusion, Minimum Cost Forwarding, Coherent/No coherent Processing, Sensor Protocols for Information via Negotiation (SPIN) family of protocols (SPIN-PP, SPIN-EC, SPIN-BC, SPIN-RL, etc.), Rumor, Stream Enable Routing (SER), Gradient-Based Routing (GBR), Constrained Anisotropic Diffusion Routing (CADR), COUGAR, Active Query Forwarding in Sensor Networks (ACQUIRE), Threshold-sensitive Energy Efficient sensor Network protocol (TEEN), and Adaptive Periodic TEEN (APTEEN).

11.5.1.2 Hierarchical-Based or Cluster-Based Routing

Hierarchical routing is mainly two-layer routing, where one layer is used to select cluster heads and the other for routing. The main objective of this approach is to cluster the sensor nodes so that the cluster heads can do some aggregation and reduction of data to save energy. Nodes in hierarchical networks play different roles. Higher-energy nodes can be used to process and send the information; low-energy nodes can be used to perform the sensing in the proximity of the target. Hierarchical routing is utilized to perform energy-efficient routing in WSNs. Hierarchical routing is an efficient way to lower energy consumption within a cluster. Examples of Hierarchical-based routing algorithms or protocols are Simple Hierarchical Routing Protocol (SHRP), Low-Energy Adaptive Cluster Hierarchy (LEACH), LEACH-Centralized (LEACH-C), Power Efficient Gathering in Sensor Information System (PEGASIS), Hierarchical PEGASIS, Hierarchical Energy Aware Protocol for routing and Aggregation in Sensor networks (HEAP), Hierarchical Periodic, Event-driven and Query-based (HPEQ), Threshold Sensitive Energy Efficient Sensor Network Protocol (TEEN), Adaptive TEEN (APTEEN), Small MECN (SMECN), Self-Organizing Protocol (SOP), Geographic Adaptive Fidelity (GAF), and SPAN.

11.5.1.3 Location-Based Routing

This class of routing protocols utilizes the position information to pass on the data to the desired regions rather than the entire network. Most of the routing protocols for sensor networks require location information for sensor nodes. In most cases, location information is needed to calculate the distance between two particular nodes so that energy consumption can be estimated. Since there is no addressing scheme for sensor networks like IP addresses, they are spatially deployed on a region. Examples of Location-based routing algorithm/protocols are Minimum Energy Communication Network (MECN), Small MECN (SMECN), Geographic Adaptive Fidelity (GAF), and Geographic and Energy Aware Routing (GEAR).

11.5.2 WSN Protocols Based on Protocol Operation

Based on the protocol operation, WSN routing protocols can be categorized into [14]

- Multipath-based routing
- Query-based routing
- Negotiation-based routing
- QoS-based routing
- Noncoherent & Coherent data-processing-based routing

11.5.2.1 Multipath-Based Routing

This type of routing protocols uses multiple paths instead of a single path to enhance network performance. Directed diffusion is a good example of this type of routing.

11.5.2.2 Query-Based Routing

In this type of routing protocol, destination nodes propagate a query for data (sensing task) from a node through the network, and a node with these data send the data that matched the query back to the node that initiated the query. Directed diffusion, Rumor, ACQUIRE, and COUGAR are good examples of this type of routing protocol.

11.5.2.3 Negotiation-Based Routing Protocols

These protocols use high-level data descriptors to eliminate redundant data transmissions through negotiation. Communication decisions are also made based on the resources available to them. SPIN family protocols are good examples of negotiation-based routing protocols.

11.5.2.4 QoS-Based Routing

In this class of routing protocols, the network creates a balance between data quality and energy consumption. Particularly, the network has to assure certain QoS metrics (delay, energy, bandwidth, etc.) when delivering data to the base station. Stateless Protocol for Real-Time Communication in Sensor Networks (SPEED) is good example of this type of protocols.

11.5.2.5 Noncoherent and Coherent Data-Processing-Based Routing

In noncoherent data processing routing, nodes will locally process the raw data before they are sent to other nodes for further processing. The nodes that perform further processing are called aggregators. Noncoherent functions have fairly low data traffic loading. In coherent routing, the data are forwarded to aggregators after minimum processing. The minimum processing typically includes tasks like time stamping and duplicate suppression. To perform energy-efficient routing, coherent processing is normally selected. Since in coherent processing long data streams are generated, but with the help optimal paths, energy efficiency is achieved. Single Winner Algorithm (SWE) is a good example of noncoherent, while Multiple Winner Algorithm (MWE), a little variant of SWE, is a coherent processing algorithm.

11.5.3 WSN Protocols Based on Sensor Node Architecture

Based on sensor node architecture, WSN routing protocols can be classified as [14]

- Protocols operating on flat topology (WSN consisting homogeneous nodes)
- Protocols operating on hierarchical topology (WSN consisting heterogeneous nodes)

11.5.4 WSN Protocols Based on Packet Destination

According to packet destination (a single node, a set of nodes, or every node in network), WSN routing protocols can be classified as [15]

- Gossiping and agent-based uni-cast forwarding
- Energy-efficient uni-cast
- Broadcast and multicast
- Geographic routing
- Mobile nodes

11.5.4.1 Gossiping and Agent-Based Uni-Cast Forwarding

These schemas are an attempt of working without routing tables to minimize the overflow needed to build the tables, as much as results of the initial stages in which the tables were not built yet. The simplest choice is flooding (forwarding each message received), but it is not very efficient. Gossiping avoids the problem of implosion by selecting a random node to which to send the packet rather than broadcasting the packet blindly. However, this causes delays in propagation of data through the nodes.

11.5.4.2 Energy Efficient Uni-Cast

These techniques analyze the network nodes distribution to set the cost of transmitting over the link between two nodes and select an algorithm to calculate the minimum cost. There are many aspects to consider about the energy awareness:

- Minimize energy per packet
- Maximize network's lifetime
- Set routes according to the remaining energy
- Minimize the amount of transmission power

11.5.4.3 Broadcast and Multicast

Earlier protocols, gossiping and unicast, try to find efficient ways to send data between nodes, possibly over several hops. For this, many nodes must collect or distribute the information to every node in the network (broadcast). In fact, broadcast is a common operation in WSN applications.

Similarly, sometimes it is necessary to distribute data to a subset of previously known nodes. This process is called multicast.

11.5.4.4 Geographic Routing

This kind of routing appeared due to two main motivations:

1. Many applications need the node location as a reference address to allow destinations of the type: "every node in a given region" or "the closer node to a point". If these requirements are needed, an appropriated routing scheme must be provided. This first idea, sending data randomly to every node in a given region, is called geo-casting.
2. When the source and destination position is known and also the nodes among them, this information can be used to improve the routing process. For that, the destination node location must be specified geographically or relatively (with a location service). This second is called position-based routing.

11.5.4.5 Mobile Nodes

These aspects with motion ability should be considered for WSNs: mobile sensor nodes, mobile base station, mobile sensed phenomenon, or combination of these.

11.5.5 WSN Protocols Presented by Crossbow (Xbow)

There are many protocols proposed for WSNs, but most of them are at the developing stage or at the literature level, which means that all are not implemented. Therefore, attention is needed on routing techniques already implemented in the most popular operating system TinyOS installed on most of the wireless sensor nodes. Numerous proposals have been available along the different TinyOS contributions, but almost they are similar. Leading edge solutions provider in WSN Technology and the Largest Manufacturer of WSN, the Crossbow Technology, Inc., offers four kinds of routing [41,42]:

1. Basic Routing (normal and improved variants)
2. Reliable Routing
3. Low Power Routing
4. XMesh Routing

Main aspects of these routing techniques and lot of experimental results are presented and discussed by Castillo et al. [42].

11.5.6 WSN Protocols Based on Cooperative Behavior

Another class of WSN routing protocols, cooperative routing, is discussed by Castillo et al. [42] In this special class of routing, sensor nodes send data to a central node where data can be amassed and may be subject to further processing; this approach reduces route cost in terms of energy uses.

11.5.7 WSN Protocols Based on State

Based on state, WSNs routing protocols can be classified as follows [13]:

■ Stateful Ad Hoc Routing
■ Stateless Geometric Ad Hoc Routing

11.5.7.1 Stateful Ad Hoc Routing

The stateful ad hoc routing protocols require node to maintain some routing information that is collected using the routing protocol (e.g., through route request propagation or by reversing paths taken by the query). Stateful routing protocols need the routing information maintained at each intermediate node through the data-forwarding path. More specifically, state is kept at some nodes about nonlocal areas in the network (e.g., the path to reach some nonlocal node). Examples of stateful ad hoc routing algorithm/protocols are DSR, AODV, and DSDV.

11.5.7.2 Stateless Geometric Ad Hoc Routing

These kinds of protocols only track the position of their neighbors and select among them a neighbor that is likely to be closer to the destination. Stateful routing may not be efficient or even possible for very large networks with limited sensor node capabilities. Accordingly, stateless routing protocols that do not maintain per-route state have been proposed. They scale effectively in terms of routing overhead because the tracked routing information does not grow with the network size or the number of active sinks. Geographic (and more generally location based) routing protocols are the main type of stateless routing protocols. Examples of stateless geometric ad hoc routing algorithm/protocols are Greedy/Geographic Forwarding, Face Routing, GPSR, and COMPASS.

11.5.8 WSN Protocols Based on Epidemic Behavior

A classification of WSN routing protocol based on epidemic behavior is available in a research contribution [43]. Here routing protocol follows the model of nature to spread information and define simple rules for information to flow between nodes of a network. Epidemic algorithms can be differentiated from each other by their style of communication between neighboring nodes:

- Pull-based epidemic algorithm
- Push-based epidemic algorithm
- Pull–push-based epidemic algorithm

11.5.8.1 Pull-Based Epidemic Algorithm

A node asks a selected neighbor for new information. The node will receive new information only if the neighbor has new information.

11.5.8.2 Push-Based Epidemic Algorithm

A node with new information sends the information to a selected neighbor.

11.5.8.3 Pull–Push-Based Epidemic Algorithm

This algorithm is a combination of two models described above. A node employing such an algorithm sends information to a selected neighbor when it has some information available; it also asks and receives new information from the selected neighbor if the neighbor has new information.

11.5.9 WSN Protocols Based on Protocol's Initialization Point

Based on protocol's initialization point, WSNs routing protocols can be classified as [44]

- Source-initiated (Src-initiated)
- Destination-initiated (Dst-initiated)

11.5.9.1 Source-Initiated (Src-Initiated)

A source-initiated protocol sets up the routing paths upon the demand of the source node, and starting from the source node. Here source advertises the data when available and initiates the data delivery.

11.5.9.2 Destination-Initiated (Dst-Initiated)

A destination-initiated protocol, on the other hand, initiates path setup from a destination node.

11.5.10 WSN Protocols Based on "How the Source Finds the Destination"

Based on "how the source finds a route to the destination," WSN routing protocols can be classified into three categories [44]:

1. Proactive routing
2. Reactive routing
3. Hybrid routing

11.5.10.1 Proactive or Table-Driven Routing

In this type of routing protocols, all routes are computed before they are really needed. Protocol maintains fresh lists of destinations and respective routes by periodically distributing routing tables throughout the network. The main drawbacks of this approach are respective amount of data for maintenance and slow reaction on restructuring and failures.

11.5.10.2 Reactive Routing

In this type of routing protocols, routes are computed on demand. When sensor nodes are static, it is preferable to have table-driven routing protocols rather than reactive protocols. A significant amount of energy is used in route discovery and setup of reactive protocols. The main drawbacks of this approach are high latency time in route finding and excessive flooding that can lead to network clogging.

11.5.10.3 Hybrid Routing

This type of routing protocols uses a combination of the two above-discussed approaches (proactive and reactive) by considering advantages of these two. In this hybrid approach, routes are initially established with the help of some proactively prospected routes and then serve the demand from additionally activated nodes through reactive approach. The option for one or the other method requires predetermination for typical cases. The main drawbacks of this approach are advantages depend on amount of activated nodes and reaction to traffic demand depends on gradient of traffic volume.

11.5.11 WSN Protocols Based on "How to Reduce Useful Energy Consumption" [45]

- Protocols that control the transmission power level at each node by increasing network capacity while keeping the network connected
- Protocols that make routing decisions based on power optimization goals
- Protocols that control the network topology by determining which nodes should participate in the network operation (be awake) and which should not (remain asleep)

11.5.12 Classification Based on Application

On the basis of applications for which a particular routing protocol is very suitable, Al-Khdour and Baroudi [46] classify routing protocols as follows:

- Routing protocols suitable for Query-based WSN Applications
- Routing protocols suitable for Event-driven WSN Applications
- Routing protocols suitable for Periodic WSN Applications

11.5.13 Classification by Sha et al. [47]

Sha et al. classify routing protocols as follows:

- Greedy Geographic Routing
- Load Balanced Routing
- Energy Aware Routing
- Fault Tolerant Routing
- Information Exploiting Routing

11.5.14 Classification by Perillo and Heinzelman [48]

Apart from above classifications, Perillo and Heinzelman [48] classify network layer protocols as follows:

- Resource-Aware Routing
- Energy-Aware Routing
- Fidelity-Aware Routing
- Data-Centric Routing Protocols
- Geographic Routing
- Clustering

11.5.15 Protocols with Security Features for WSN

Current routing protocols optimize for the limited capabilities of nodes and the application-specific nature of networks, but do not consider security. Although these protocols have not been designed with security as a goal, it is important to analyze their security properties.

One aspect of sensor networks that complicates the design of a secure routing protocol is in-network aggregation. In WSNs, in-network processing makes end-to-end security mechanisms harder to deploy because intermediate nodes need direct access to the contents of the messages.

11.5.15.1 Ariadne: A Secure On-Demand Routing Protocol for Ad Hoc Networks [49]

It is a new secure reactive ad hoc network routing protocol that prevents attackers or compromised nodes from tampering with uncompromised routes consisting of uncompromised nodes, as well as prevents from a number of Denial-of-Service attacks. The design of Ariadne is based on basic operation of Dynamic Source Routing protocol (DSR). Ariadne is efficient, due to use of highly efficient symmetric cryptographic primitives. The major one is that it requires neither a trusted hardware nor powerful processors. It can authenticate routing messages using one of three schemes:

1. Shared secrets between each pair of nodes
2. Shared secrets between communicating nodes combined with broadcast authentication
3. Digital signatures

11.5.15.2 Secure Adaptive Routing Protocol (SARP) for WSNs [50]

SARP is a highly distributed, dynamic, efficient tree-based routing algorithm, with integrated trust framework that uses the multidimensional trust-metrics in order to find the most trusted paths. The main objective of this protocol is to detect an attack, avoid the attacked node(s) in routing, and warn other sensor nodes against the attack. SARP relies on updating trust of nodes based on performance of the nodes in transactions. Once a transaction completes, the data source begins a trust reporting process so that the nodes along the path can update the trust values of their neighbors.

11.5.15.3 Secure Implicit Geographic Forwarding (SIGF) [51]

A family of configurable and secure routing protocols for WSNs, which provide good enough security with high performance, SIGF comprises three protocols, which actually extend Implicit Geographic Forwarding (IGF) [52] and populate the gap between pure statelessness and traditional shared-state security; here each protocol form a base for the next:

■ SIGF-0 keeps no state, but provides probabilistic defenses.
■ SIGF-1 uses local history and reputation to protect against certain attacks.
■ SIGF-2 uses neighborhood-shared state to provide stronger security guarantees.

Each protocol encompasses the features of the earlier with additional mechanisms to defend against attacks that are more sophisticated. SIGF chooses the next hop dynamically and nondeterministically, rather than maintain routing tables. This approach contains the effect of compromise to a local neighborhood, increases robustness to node mobility and failure, and spreads energy drain more evenly across neighbors.

11.5.15.4 Secure Energy-Efficient Routing Protocol (SERP) for Densely Deployed WSNs [53]

SERP uses one-way hash chain and prestored shared secret keys for ensuring data transmission security. In this protocol, a sink-rooted tree structure is created as the backbone of the network. This energy-efficient network structure is used for authenticated and encrypted data delivery from the source sensors to the base station. Also with this energy-efficient network structure, maximum lifetime of the network could be achieved. To introduce data freshness, SERP includes an optional key refreshment mechanism that could be applied depending on the application at hand.

11.5.15.5 Secure and Energy-Efficient Protocol (SEEP) for Large-Scale WSNs [54]

This protocol operates on a clustered network and provides security at both intra- and intercluster levels. The strength of SEEP stems from its ability to meet conflicting goals of providing high-level security while maximizing energy efficiency.

11.5.15.6 Energy-Efficient Secure Routing Protocol for Sensor Networks (ESecRout) [55]

ESecRout protocol uses the symmetric cryptography to secure messages, and uses a small cache in sensor nodes to record the partial routing path (previous and next nodes) to the destination. It guarantees that the destination will be able to identify and discard the tampered messages and ensure that the messages received are not tampered with. Through experiments, authors show that ESecRout provides energy-efficient secure routing.

11.5.16 Energy-Efficient Protocols for WSNs

Recent advances in the area of WSNs led to many new protocols specifically designed for special class of network where energy awareness is a primary and essential consideration. Sensor nodes are frequently deployed in remote, antagonistic, and unattended environments, so it is vital to design algorithms and protocols that operate on minimal energy. Therefore, implementers must reduce communication between sensor nodes, abridge computations, and apply lightweight security solutions. In a research article [45], energy consumption by sensor node is attributed as either useful or wasteful sources (see Table 11.5).

Table 11.5 Energy Consumption in a Sensor Node

Useful energy consumption can be due to	1. Transmitting/receiving data 2. Processing query requests 3. Forwarding queries/data to neighboring nodes
Wasteful energy consumption can be due to	1. Idle listening to the media 2. Retransmitting due to packet collisions 3. Overhearing 4. Generating/handling control packets

A classification on energy efficient/aware routing protocols is available [10], which classifies this type of protocols into

- Energy Saver Protocols
- Energy Manager Protocols

Energy saver protocols decrease energy consumption totally, because most of them try to find the shortest path between source and destination to reduce energy consumption. The objective of energy manager protocols is to balance energy consumption in networks to avoid network partitioning. In the first approach, finding best route is totally based on energy balancing consideration; it may lead to a long path with high delay and decrease network lifetime, whereas in the later approach, finding best route only with the shortest distance consideration may lead to network partitioning. Many researches were conducted on the energy efficiency/awareness issues, and some are presented here with major theoretical aspects.

11.5.16.1 Threshold-Sensitive Energy-Efficient Sensor Network Protocol (TEEN) [56]

It was the first protocol for reactive networks with enhanced efficiency. Time critical data reach the user almost instantaneously. TEEN is eminently well suited for time critical data-sensing applications. Message transmission consumes much more energy than data sensing. Therefore, even though the nodes sense continuously, the energy consumption in this scheme can potentially be much less than in the proactive network, because data transmission is done less frequently. The soft threshold can be varied, depending on the criticality of the sensed attribute and the target application. A smaller value of the soft threshold gives a more accurate picture of the network, at the expense of increased energy consumption. Thus, the user can control the trade-off between energy efficiency and accuracy. At every cluster change time, the attributes are broadcast afresh, so the user can change them as required. The main drawback of this scheme is that, if the thresholds are not reached, the nodes will never communicate; the user will not get any data from the network at all and will not come to know even if all the nodes die. Thus, this scheme is not well suited for applications where the user needs to get data regularly. Another possible problem with this scheme is that a practical implementation would have to ensure that there are no collisions in the cluster.

11.5.16.2 Adaptive Periodic TEEN (APTEEN) [57]

It is a protocol for Hybrid network (inherit best characteristics of both proactive and reactive networks). It provides periodic data collection as well as near real-time warnings about critical events. By sending periodic data, it gives the user a complete picture of the network. It also responds immediately to drastic changes, thus making it responsive to time critical situations. Thus, it combines both proactive and reactive policies. It offers a flexibility of allowing the user to set the time interval (TC) and the threshold values for the attributes. Energy consumption can be controlled by the count time and the threshold values. The hybrid network can emulate a proactive network or a reactive network, by suitably setting the count time and the threshold values. The main drawback of this scheme is the additional complexity required to implement the threshold functions and the count time. However, this is a reasonable trade-off and provides additional flexibility and versatility.

11.5.16.3 Hybrid Energy-Efficient Distributed Clustering (HEED) [45]

It is an energy-efficient clustering protocol using residual energy as a primary parameter and network topology features such as node degree and distances to neighbors as secondary parameters. Here all nodes are assumed to be homogenous nodes (with same initial energy). It extends the basic scheme of LEACH. The clustering process is divided into a number of iterations, and in iteration nodes that are not covered by any cluster head, it doubles their probability of becoming a cluster head. It enables every node to decide independently and probabilistically on its role in the clustered network and thus cannot guarantee optimal elected set of cluster heads.

11.5.16.4 Heterogeneous HEED (H-HEED) [58]

It is a protocol for heterogeneous WSN. Cluster head selection is primarily based on the residual energy of each node. Since the energy consumed per bit for sensing, processing, and communication is typically known, and hence residual energy can be estimated. Intracluster communication cost is considered as the secondary parameter to break the ties; here tie means that a node might fall within the range of more than one cluster head. Different levels of heterogeneity are introduced: 2-level, 3-level, and multilevel in terms of the node energy. In 2-level H-HEED, two types of sensor nodes, that is, the advanced nodes and normal nodes, are used. In 3-level H-HEED, three types of sensor nodes, that is, the super nodes, advanced nodes, and normal nodes, are used. In this heterogeneous approach, all the sensor nodes are having different energy levels; as a result, nodes will die randomly. Multilevel H-HEED prolongs lifetime and shows better performance than other levels of H-HEED and HEED protocol.

11.5.16.5 Reactive Energy Decision Routing Protocol (REDRP) [59]

To solve the problem of limited energy, the loading of nodes has to be distributed as possible as it can. If energy consumption can be shared averagely by most nodes, then the lifetime of sensor networks will be enlarged. This protocol will create the routes in reactive routing method to transmit the data node gathered. It uses the residual energy of nodes as the routing decision for energy aware.

11.5.16.6 Power-Efficient Gathering in Sensor Information Systems (PEGASIS) [60]

It is a near-optimal chain-based protocol and an enhanced descendant of LEACH. It has two main objectives: to increase the lifetime of each node by using collaborative techniques and to permit only local coordination among nodes that are close together to reduce bandwidth consumed in communication. Nodes route data intended ultimately for the base station through intermediate nodes. While determining, the routes only consider the energy of the transmitter and neglect the energy dissipation of the receivers. It assumes that each sensor node can be able to communicate with the base station directly and that all nodes maintain a complete database about the location of all other nodes in the network. The method through which the node locations are obtained is not outlined. It also assumes that all sensor nodes have the same level of energy and that they are likely to die at the same time.

11.5.16.7 Hierarchical-PEGASIS [61]

The objective is to decrease the delay incurred for packets during transmission to the base station. In its concept, only spatially separated nodes are allowed to transmit at the same time. This chain-based protocol with CDMA-capable nodes constructs a chain of nodes that forms a tree-like hierarchy, and each selected node at a particular level transmits data to the node in the upper level of the hierarchy, ensuring data transmitting in parallel and reducing the delay significantly. Savvides et al. [61] show that the hierarchical extension of PEGASIS performs better than the regular PEGASIS scheme by a factor of about 60.

11.5.16.8 Scaling Hierarchical Power-Efficient Routing (SHPER) [62]

SHPER is an enhanced integration of a hierarchical reactive routing protocol. It supposes the coexistence of a base station and a set of homogeneous sensor nodes that are randomly distributed within a delimited area of interest. It consists of two phases: the initialization phase and the steady-state phase. Hard and soft thresholds are utilized in the SHPER protocol as with TEEN. It is best suited in real-life applications where imbalance in energy distribution is the common case. Network scalability is retained because it adopts both multihop routing and hierarchical architecture.

11.5.16.9 Low-Energy-Adaptive Clustering Hierarchy (LEACH) [63]

This is the most popular cluster-based protocol that includes distributed cluster formation. The idea is to form clusters of the sensor nodes based on the received signal strength and use local cluster heads as routers to the sink. It randomly selects a few sensor nodes as cluster-heads and rotates this role to distribute evenly the energy load among the sensors in the network. Its operation is separated into two phases: setup phase, where clusters are organized and CHs are selected, and steady-state phase, where the actual data transfer to the base station takes place. It uses a TDMA/CDMA MAC to reduce inter- and intracluster collisions. The optimal number of cluster heads is estimated to be 5% of the total number of nodes. This protocol is most appropriate for the applications where there is a need for constant monitoring.

11.5.16.10 Simple Energy-Efficient Routing (SEER) [64]

SEER protocol considers energy saving and balancing, but not has enough good idea about energy balancing. Once the network has been deployed in the area where it is to operate, the sink transmits a broadcast packet. Each node in the network is assumed to have a unique address within the network. When a node observes new data, it initiates the process of routing. Two types of data packets can be sent: normal data and critical data. When nodes receive a data message, they update the remaining energy value in the neighbor table for the neighbor that sent the message. Nodes that forward data messages follow the same process, except for minor differences. If a node's remaining energy falls below a certain threshold, it transmits an energy message to all of its neighbors to inform them its energy level. The sink node periodically sends a broadcast message through the network so that nodes can add new neighbors that joined the network to neighbor tables and remove neighbors that have failed from the neighbor tables. Nodes also update remaining energy values stored in the neighbor tables.

11.5.16.11 Balanced Energy-Aware Routing (BEAR) [10]

BEAR is an extended version of SEER protocol with some visible differences particularly in forwarding data mechanism that save and balance energy consumption in WSNs. It finds optimal route in energy level and hop count both. Routing decisions in BEAR are based on the distance to the base station as well as on the remaining battery energy level of nodes on the path toward the base station. BEAR is better than the SEER protocol in energy managing because BEAR sends data packet along a balanced path.

11.5.17 Routing Protocols under Real-Time Constraints

An-dong et al. [65] highlight the following points as to why real-time routing is very challenging in WSNs:

- A global addressing scheme is hard to build for the deployment of sheer number of sensor nodes.
- A careful resource management is required due to highly constrained nature of sensor nodes in terms of transmission power, on-board energy, processing capacity, and storage.
- New designs of real-time routings are necessary for offering real-time QoS in WSNs with guaranteed end-to-end delivery time, delay jitter, and other associated QoS metrics.

11.5.17.1 Timely, Reliable, Energy Efficient, and Dynamic (TREnD) [66]

It is a novel cross-layer WSN protocol for control applications. It is hierarchically subdivided into two parts: a static route at the interclusters level and a dynamical routing algorithm at the node level. This is supported at the MAC layer by hybrid TDMA/CSMA solution. The protocol parameters are adapted by an optimization problem, whose objective function is the network energy consumption, and the constraints are the reliability and latency of the packets. It uses a simple algorithm that allows the network to meet the reliability and latency while minimizing for energy consumption. It is a best fit for industrial environments.

11.5.17.2 Opportunistic Real-Time Routing (ORTR) in Multihop WSNs [67]

It is a novel real-time routing protocol designed to achieve guaranteed delivery of data under time constraints using efficient power consumption and balance overall power level. This heuristic method integrates routing-layer and MAC-layer functionalities to meet real-time requirements. To satisfy time requirements, a region of interest where real-time data must be delivered is defined with effective transmission power and a relay node within this region of interest is selected for balancing overall energy levels.

11.5.17.3 Stateless Protocol for Real-Time Communication in Sensor Networks (SPEED) [68]

SPEED protocol provides three types of real-time communication services: real-time uni-cast, real-time area-multicast, and real-time area-anycast. SPEED uses geographical location for routing. One key difference with other schemes of this genre is its spatiotemporal character. It takes into account timely delivery of the packets. End to-end soft real-time communication is achieved by maintaining a desired delivery speed across the sensor network through a combination of feedback control and nondeterministic geographic forwarding. It works satisfactorily under scarce resource conditions and can provide service differentiation. SPEED takes the first step in addressing the concerns of real-time routing in Wireless Multimedia Sensor Network (WMSN).

11.5.17.4 Multipath Multi-SPEED (MMSPEED) [69]

MMSPEED is an extension of SPEED protocol, a novel packet delivery mechanism that adopts a probabilistic approach to provide QoS guarantee of reliability and timeliness in WSNs. It is based on a cross-layer approach between the network and the MAC layer. The QoS provisioning in MMSPEED is performed in two quality domains, namely, timeliness and reliability. Between the two layers, a judicious choice is made over reliability and timeliness of packet arrival. It has the capabilities of differentiating between flows with different delays and reliability requirements. Multiple QoS levels are provided in the timeliness domain by guaranteeing multiple packet delivery speed options. In the reliability domain, various reliability requirements are supported by probabilistic multipath forwarding. Differentiation in reliability is argued as being an effective way of channeling resources from flows with relaxed requirements to flows with tighter requirements. This way, MMSPEED can guarantee end-to-end requirements in a localized way, which is desirable for scalability and adaptability to large-scale dynamic sensor networks.

11.5.17.5 Random Re-Routing (RRR) Algorithm [70]

It is a distributed and adaptive routing algorithm that can detect the occurrence of unusual events and provide better QoS for packets that carry information of these unusual events. In this algorithm, the sensor and forwarding nodes will change their routing policy adaptively according to the current traffic level. All decisions are taken in a distributed fashion at the sensor and forwarding nodes. It distinguishes packets of routine data and unusual events; packets from unusual events are routed along preferred paths, while the routine data are randomly shunted to slower and possibly longer secondary paths.

11.6 Theoretical Aspects of Data Link (MAC) Layer Protocols for WSNs

All communication issues between neighbor sensor nodes are handled by link layer. The link layer often bundles bytes to packets; these packets contain data and checksum bits that allow the receiver to detect transmission errors. Link layer assumes that bits can be transmitted to a neighbor over the lower physical layer, specifying the encodings and length limits on packets such that messages can be sent and received by the underlying physical layer and also responsible for ensuring reliable data transfer. Like other wireless networks, also in WSNs all nodes share a common communication medium, which needs control on it. This controlling mechanism is implemented through a sublayer of link layer called Medium Access Control (MAC). Thus, this section focuses on all issues related to MAC sublayer and protocols.

11.6.1 Fundamental Requirements, Design Constraints, and Performance Metrics of/for WSN MAC Protocols

Like other wireless networks, the first objective of the MAC protocol for WSN is the creation of the network infrastructure [46]. The second objective is to share the medium communication between the sensor nodes [71]. In designing a MAC protocol for a WSN, some of the unique features of WSN must be taken into consideration. For example, IEEE 802.11 [72] is a well-known MAC protocol for ad hoc network (IEEE working group 1999), but the energy constraints in the wireless sensor nodes make it inconvenient to apply the IEEE 802.11 protocol directly in WSN. IEEE 802.11 has a power save mode, but it is designed for a single-hop network, where all nodes can hear each other, which is not the case in WSN. A set of MAC protocols for the WSN were proposed. Most of the existing protocols aimed to save power consumption in the sensor nodes. The following are some of the design constraints:

■ Protocol must be able to consume low power.
■ Protocol has to minimize the coordination and synchronization between sensor nodes.
■ Protocol must be able to support a large number of sensor nodes.
■ Protocol must have high degree of scalability.
■ Protocol must take into account the limited bandwidth availability.

To design a good WSN-specific MAC protocol, Ye et al. [73] suggest the following attributes for consideration: Primarily, consider energy efficiency, scalability, and adaptivity on network changes in term of size, node density, and topology. Secondarily, consider fairness, latency, throughput, and bandwidth utilization.

In a research paper [74], authors compare performance of nine real-time WSN MAC protocols and presented in tabular format.

11.6.2 Energy-Saving Mechanisms for MAC Protocols

The major design parameter in MAC protocols for WSNs is energy efficiency. Ameen et al. focus on the various energy-saving mechanisms used in MAC protocols for energy efficiency: they include duty cycling, energy-efficient scheduling, scheduled rendezvous, on-demand wake-up scheme, energy efficiency through directional antennae, clustering, data rate adaptation, channel polling, and hybrid approaches [75]. They also focus on some external mechanisms that support energy efficiency in MAC; they are energy-efficient routing, energy efficiency through topology control, and data aggregation. According to Kaur and Mahajan [76], WSN MAC protocols must deal with the following energy-related issues: collisions, overhead, overhearing, and idle listening. These are the major sources of energy waste.

11.6.3 Classifications of WSN MAC Protocols

According to Kaur and Mahajan [76], MAC techniques proposed for WSNs are divided into two categories:

1. Schedule-based protocols can avoid collisions, overhearing, and idle listening by scheduling transmit and listen periods but have strict time synchronization requirements.
2. Contention-based protocols, on the other hand, relax time synchronization requirements and easily adjust to the topology changes as some new nodes may join and other may die few years after deployment. Protocols are based on the CSMA technique and have higher costs for message collisions, overhearing, and idle listening.

Akyildiz et al. [1] present a category of WSN-specific MAC protocols in tabular form; the categories are as follows:

■ Fixed allocation or Dedicated assignment based
■ Demand based
■ Random access or Contention based

In same research contribution [1], qualitative overview of some MAC protocols for sensor networks is also presented in tabular form.

Langendoen [77] provides a broad overview of the WSN-specific MAC protocols, surveys, and details of the historic development of the most common styles of medium access control for WSNs. According to this author, WSN-specific MAC protocols can be classified according to how nodes organize access to the shared radio channel:

- Random Access
- Slotted Access
- Frame-Based Access
- Hybrid Protocols

Some design considerations for energy-efficient MAC protocols are identified in a research contribution [75], which are as follows: network topology, deployment strategy, antenna mode, controlling mechanisms, delay, throughput, QoS requirements, and number of channels to be used in communication. Almost everyone is familiar with these terms. Also the in same research contribution [75], authors presented a classification tree of MAC Protocols, which classifies WSN-specific MAC protocols into three major categories and further classifies into several subcategories:

- Centralized MAC Protocols
  - Schedule-Based Protocols
    - Fixed Assignment
    - Demand Assignment
  - Contention-Based Protocols
- Distributed MAC Protocols
  - Schedule-Based Protocols
  - Contention Based Protocols
    - Contention Based with Scheduling
    - Contention Based with Reservation
      - Asynchronous Protocols
      - Synchronous Protocols
        - TDMA Based
    - Pure Contention Based
      - Receiver Initiated Protocols
      - Sender Initiated Protocols
        - Single Channel Protocols
        - Multi Channel Protocols
          - CDMA Based
          - FDMA Based
- Other MAC Protocols
  - Hybrid Protocols
  - Directional Antennas

In a research article [74], according to the difference of applications, real-time MAC protocols for WSNs are classified as hard real-time MAC protocol and soft real-time MAC protocol. To guarantee the constraint time for ensuring the validity of alarm messages in the time critical application, several proposed approaches are classified as branches from hard and soft real-time MAC protocols, which are as follows:

- Hard Real-time WSN-specific MAC Protocols
  - TDMA based
  - Dual Mode
  - Message Ordering
  - Traffic Management
- Soft Real-time WSN-specific MAC Protocols
  - S-MAC based
  - Spatial Channel Reuse
  - Energy Aware

11.6.4 WSN MAC Protocols

In this subsection, major theoretical aspects of some renowned MAC protocols are summarized.

11.6.4.1 nanoMAC [78]

It is a p-nonpersistent (p for probability) CSMA/CA-based WSN MAC protocol specifically designed for dense WSNs. It is intended to be highly scalable and supports IEEE-addressing schemes. The basic operation cycle of nanoMAC is defined by lightweight RTS-CTS-nDATA-ACK handshake. This mechanism not only coordinates the sleeping and synchronization mechanism but also prevents collisions, saves energy consumption, and adds latency associated with packet retransmissions. It is able to perform with high reliability, throughput, and efficient channel utilization. The idle channel listening and overhearing is reduced to a minimum by a sophisticated sleep algorithm employed in nanoMAC.

11.6.4.2 IEEE 802.15.4 MAC

IEEE 802.15.4 Wireless MAC specifications specifically target for low-rate, low-power WPANs. MAC characteristics of the IEEE 802.15.4 standard are available in a research contribution [79]. Important features of IEEE 802.15.4 include collision avoidance through CSMA/CA, real-time suitability with the help of reserving guaranteed time slots, and integrated secure communication support. Here devices also include power management functions like energy detection and link quality. In a research contribution [80], performance of the contention access period specified in IEEE 802.15.4 MAC standard is analyzed in terms of throughput and energy consumption that facilitated by a modeling of the contention access period as nonpersistent CSMA with backoff. The IEEE 802.15.4 MAC layer is fundamentally that of CSMA/CA system together with optional time slot structure and security functionality [81]. Network topologies supported by this specification are either star or peer-to-peer. This specification differentiates between a full function device (FFD) and a reduced function device (RFD), and, in each case, an FFD device acting as a coordinator manages the local network operations. The standard defines four frame types: beacon, data, acknowledgment, and MAC control.

11.6.4.3 Sensor-MAC (S-MAC) [73]

It is single-frequency contention based that uses few novel techniques to support self-configuration and reduce energy consumption. S-MAC enables low-duty-cycle operation in multihop networks. Nodes form virtual clusters are based on common sleep schedules to reduce control overhead and enable traffic-adaptive wake-up. It uses in-channel signaling to avoid overhearing unnecessary traffic. Finally, it applies message passing to reduce contention latency for applications that require in-network data processing. The basic idea behind this protocol is that time is divided into large frames. Every frame has two parts: an active part and a sleeping part. During the sleeping part, a node turns off its radio to preserve energy. During the active part, it can communicate with its neighbors and send any messages queued during the sleeping part. S-MAC needs some synchronization.

11.6.4.4 Traffic Aware, Energy-Efficient MAC (TEEM) [82]

It is a protocol inspired by S-MAC, based on the concept of listen/sleep mode cycle. However, unlike the S-MAC, where the duration of listen and sleep modes are fixed, our TEEM protocol makes these durations adaptive by utilizing traffic information of each node, achieving a significant decrease in power consumption. Thus, with the proposed scheme, the listen time of nodes can be reduced by putting them into sleep state earlier when they expect no data traffic to occur. Also, S-MAC has a long listen interval and divided into three parts, SYNC, RTS, and CTS packets, whereas TEEM has much smaller listen interval and divided into only two parts, SYNCdata and SYNCnodata. The duration of listen and sleep modes in TEEM is not fixed, but adaptive.

11.6.4.5 Timeout-MAC (T-MAC) [83]

It is an adaptive, energy-efficient, contention-based WSN MAC protocol with a novel idea to reduce idle listening by forwarding messages in bursts of variable length, and inducing sleeping between bursts. To handle load variations in time and location, it introduces an adaptive duty cycle in a novel way: by dynamically ending the active part of it. This reduces the amount of energy wasted on idle listening, in which nodes wait for potentially incoming messages, while still maintaining a reasonable throughput. In T-MAC, every node periodically wakes up to communicate with its neighbors, and then go to sleep again until the next frame. Meanwhile, new messages are queued. In this protocol frame synchronization is inspired by virtual clustering, same as S-MAC.

11.6.4.6 GANGS [84]

It is a hybrid, energy-efficient, self-organized, cluster-based MAC protocol proposed for WSNs. This is called GANGs because each cluster acts like a gang. Node with highest remaining energy will be elected as cluster/gang head, and it is cluster head's responsibility to controls its own gang/cluster, and negotiates with other clusters/gangs. The gangs construct a network in which a transmission can reach every cluster. It incorporates two medium access schemes: contention-free and contention-based. The contention-free (like TDMA) scheme is deployed by cluster head to construct a contention-free network backbone. The contention-based (like IEEE 802.11) scheme is deployed by ordinary nodes to communicate with cluster head nodes.

11.6.4.7 MAC with a Dynamic Duty Cycle for Sensor Network (DSMAC) [85]

The objective of this WSN MAC protocol is to achieve a good tradeoff between power consumption and latency. Unlike S-MAC, in this protocol duty cycle is changed based on average delay of the data packet and the power consumption. Duty cycle can be changed by changing the sleeping interval while fixing listening interval. Like S-MAC, the nodes in DSMAC form groups of peers and each set of neighbors follows a common schedule. Also, in this protocol one-hop packet latency is proposed, which is the time since a packet gets into the queue until it is successfully sent out.

11.6.4.8 BitMAC [86]

It is a deterministic, collision-free, and robust MAC protocol, designed for densely deployed data collection WSNs. It is based on a spanning tree of the sensor network with the sink at the root. BitMAC will use OOK (On-Off-Keying) communication model only for a limited number of control operations, but for other operations (like, payload data transmission), perhaps more efficient modulation schemes can be used if supported by the radio. Applications with real-time and robustness requirements are supported by BitMAC, because it is designed to support a large class of permanent or temporary node failures efficiently and without introducing contention or indeterminism. By the design of BitMAC, there are no collisions between nodes in communication range, but the most vulnerable part of BitMAC is the channel and ID assignment.

11.6.4.9 Traffic-Adaptive Medium Access (TRAMA) [87]

It is an energy-efficient, conflict-free channel access protocol for WSNs. Channel access is energy efficient while maintaining good throughput, acceptable latencies, and fairness. It consists of three components: Neighbor Protocol (NP), Schedule Exchange Protocol (SEP), and Adaptive Election Algorithm (AEA). Energy efficiency is attained by two ways: first, transmission schedules that avoid collisions of data packets at the receivers, and second, having nodes switch to low power radio mode when there are no data packets intended for those nodes. It derives collision-free transmission schedules based on the identifiers of nodes one and two hops away, the current time slot, and traffic information that specifies which node intends to transmit to which other node.

11.6.4.10 Wireless Sensor MAC (WiseMAC) [88]

It is an ultralow-power MAC Protocol for the downlink (traffic direction base station/sink € sensor node) of infrastructure WSNs based on CSMA with preamble sampling that minimizes idle listening. Exploit sensor nodes sampling schedules to minimize length of the wake-up preamble, and data frames are repeated in long preambles to mitigate overhearing. It is adaptive in nature to the traffic; per packet overhead decreases in high traffic conditions.

11.7 Theoretical Aspects of Physical Layer Protocols for WSNs

Data-link/media-access protocols get used to parameters of the underlying physical layer; this lowest layer of WSN protocol stack is responsible for signal detection, carrier frequency generation, frequency selection, modulation, and data encryption. WSN physical layer protocols are responsible for radio-related tasks and capable to perform computational electronics. The physical layer addresses the needs of a simple but robust modulation, transmission, and receiving techniques [1].

Bhat, author of a blog post [89], identified six important design issues, which must be considered while designing WSN-specific physical layer protocols:

1. Protocol must consume low power.
2. Protocol must support low transmission and reception range.
3. Protocol must be robust against interference from other systems, working in the same band.

4. Low complexity, in terms of used modulation scheme.
5. Normally support low duty cycle, because in most of the WSN applications sensing nodes are usually switched off in most of the time.
6. Normally support low data rates, because high data rates are required occasionally for short period.

On the same blog post [89], as per the author the most challenging aspects during design of WSN-specific physical layer protocols are to find out low-cost transceivers that consume less power, and choice of simple modulation schemes that are robust enough to provide required service.

Different schemes implemented at the physical layer, such as time division multiple access (TDMA) versus frequency division multiple access (FDMA) and binary modulation versus M-ary modulation are compared in a research contribution [90]:

From sensor node's point of view, TDMA scheme is preferred if receiver consumes less power, whereas FDMA is more suitable if the receiver uses more power; for average power consumption, hybrid TDM-FDM schemes are appropriate. In a TDMA scheme, the entire bandwidth of the channel is dedicated to a single sensor for communication purposes in a specified time, while in FDMA scheme, totally available entire bandwidth is divided among all sensor nodes equally; that is, all sensor nodes always equally share a single communication channel, whereas in hybrid TDM-FDM schemes both time and frequency are divided into available transmission slots.

Modulation scheme used by the radio is another important aspect requiring high rate data transfer at a low symbol rate. In M-ary modulation (M-ary PSK, M-ary FSK), radio sends multiple bits per symbol, but it will increase the circuit complexity and power consumption of the radio also reduces efficiency of the power amplifier and only efficient when startup time is small. This implies that more power will be needed to obtain reasonable levels of transmit output power. The data encoder parallelizes serially input bits and then passes the result to a digital-to-analog converter (DAC). The analog values produced serve as output levels for the in-phase (I) and quadrature (Q) components of the output signal. On the other hand, in a generic binary modulation, scheme modulation circuitry and frequency synthesizer are integrated. To transmit data using this architecture, the voltage-controlled oscillator can be either directly or indirectly modulated.

At the physical layer, three main components that contribute to energy loss in a wireless transmission, as discussed in a research contribution [91], are as follows:

1. Loss due to the channel and the fixed energy cost to run the transmission and reception circuitry.
2. Tradeoffs between the transmit power and the probability of error.
3. Modulation or coding of the system determines the probability of success of the transmission.

The main contribution of this research paper [91] is to present a method of finding the optimum physical layer parameters to minimize energy dissipation in a multihop WSNs.

Wireless sensor nodes have low bandwidth and low range of transceivers (such as 10 Kbps, 76 Kbps, 250 Kbps, 2.4 GHz, IEEE 802.15.4, Bluetooth, etc.). A discussion of existing radio standards (IEEE 802.11x, Bluetooth [IEEE 802.15.1 and.2], IEEE 802.15.4, ZigBee, and IEEE 1451.5) with focus on how they may or may not be suitable for WSNs is given in a book chapter [92].

In a technical report [93], technical specifications of some well-known wireless sensor nodes are presented in tabular format; a modified version focused on node's radio is presented in Table 11.6.

Most of the wireless sensor nodes are based upon RF circuit design that requires modulation, band pass filtering, demodulation, and multiplexing circuitry, but micro-Adaptive Multi-domain

Table 11.6 Radio Specifications of Some Wireless Sensor Nodes

Platform	MCU	RF Transceiver	Frequency	Radio Range (FEET)
MICA	Atmel ATMega128L	TR1000	433, 916 MHz	200
MICA2		CC1000	315, 433, 916 MHz	500
MICA2Dot		CC1000	315, 433, 916 MHz	500
MICAz		CC2420	2.4 GHz	410
Cricket		CC1000	433 MHz	500
MANTIS nymph		CC1000	315, 433, 868, 915 MHz	500
BTnode3		Zeevo-BT/ CC1000	2.4 GHz/868 MHz	328/500
Smart-Its mote	Atmel ATMega103L	Ericsson-BT/ TR1001	2.4 GHz/868 MHz	328/984
TelosA, TelosB	TIMSP430	CC2420	2.4 GHz	410
EYES, ECR, ESB		TR1001	868 MHz	984
Tmote Sky		CC2420	2.4 GHz	410
TinyNode 584		Xemics XE1205	868 MHz	200
ZebraNet H/W		9XStream	902–928 MHz	328
Intel mote2	PXA27x (Core)	CC2420	2.4 GHz	410
Intel mote	ARM7TDMI (Core)	Zeevo-BT	2.4 GHz	328
XYZ mote		CC2420		410

Source: Data from Manjunath, D., "A Review of Current Operating Systems for Wireless Sensor Networks," Technical Report, Department of ECE, Indian Institute of Science, Bangalore, India, 2007.

Power-aware Sensors (μAMPS) wireless sensor nodes use a Bluetooth compatible commercial single-chip 2.4-GHz transceiver with an integrated frequency synthesizer. Infrared may be another possible mode of communication in WSNs. Advantages of this mode include license-free communication, robustness from interferences induced by electrical devices, low cost of transceivers, and easier to build.

11.8 Theoretical Aspects of Cross-Layer Protocols for WSNs

Cross-layer is an emerging design trend that transcends traditional communication layers for performance gains in WSNs [94]. Most of the communication protocols for WSNs follow the traditional layered protocol architecture [2]. Although these protocols may achieve very high performance in terms of the metrics related to each of these individual layers, they are not jointly optimized to maximize the overall network performance while minimizing the energy consumption [95]. In view of the resource constraint nature of wireless sensor nodes in terms of power and computational capabilities, joint optimization and design of networking layers (i.e., cross-layer design) stand as the most promising alternative to inefficient traditional layered protocol architectures [2]. To improve the power efficiency and system throughput of WSNs, many researchers studied the necessity and possibility of taking advantages of cross-layer design.

The basic idea behind this approach is to optimize the control and exchange of information over more than two layers for achieving a significant performance improvement by exploiting the interactions among protocol layers. Safwati et al. [96] propos optimal cross-layer designs for energy-efficient wireless ad hoc and sensor networks. Sichitiu [97] worked on cross-layer scheduling for power efficiency in WSNs. A nice work is available on cross-layer design and optimization for WSNs in a research contribution [98]. The benefits of this approach include relaxation from the rigid traditional layering approach and associated design complexity and stability issues. An important question in the area of cross-layer design is what parameters need to be shared among different layers of the protocol stack and how each layer can be made robust to the changing network conditions [2].

Author of a textbook [94] presents a wider look on cross-layer design perspectives, from basics of cross-layer design, design guidelines, cross-layer architectures to applied cross-layer approaches (design coupling and information sharing) and also works on global performance goals, input aspects, configuration optimization, and implementations with case studies. Most of the research contributions related to cross-layer design focus on the interaction or modularity among physical, MAC, and routing layers [2]; only few researches, such as [98], cover all layers. Al-Khdour and Baroudi [46] present a summary of some cross-layer protocols designed for WSNs; a modified version is presented in Table 11.7.

Table 11.7 Summary of Some Cross-Layer Protocols for WSNs

Cross-Layer Objective	Approach	Layers	Protocols
Maximization of probability of sending packet to its D at N transmission	Mathematical model: Probabilistic dynamic programming	MAC, Network	Energy-Constrained Path Selection (ECPS) [96]
Minimize energy: Multiple simultaneous routes load distribution			Energy-Efficient Load Assignment (E2LA) [96]
Optimize performance of WSN	Framework (Optimization Agent)	All layers	Weilian Su [98]
Maximize sleep duration	Heuristic	MAC, Network	MAC CROSS [99]
Maximize throughput		Physical, MAC	Opportunistic ALOHA (O-ALOHA) [100]
Optimize number of cluster		Physical, MAC, Network	Power On With Elected Rotation (POWER) [101]

(Continued)

Table 11.7   Summary of Some Cross-Layer Protocols for WSNs (*Continued*)

Cross-Layer Objective	Approach	Layers	Protocols
Minimize total distortion	Game theory	Application, Physical	Game Theoretic Approach
Maximize network lifetime	Modeling as optimization problem	Routing, Link, MAC Layer	Shuguang Cui [102]
	Heuristic	MAC, Network	Cross-Layer Scheduling
		MAC, Physical, Network	Cross-Layer design for cluster formulation
Maximizing network lifetime, and monitoring coverage	Heuristic	MAC, Network	Sense-Sleep Trees (SS-Trees) [103]

Source: Data from Al-Khdour, T., and U. Baroudi, "Literature Review of MAC, Routing and Cross Layer Design Protocols for WSN, Chapter 2," in *Wireless Sensor Networks*, edited by S. Tarannum, Austria, 2011.

Apart from the above-presented cross-layer protocols, others include SERAN [104], which is a clustered two-layer protocol based on a semi-random approach that combines randomized and deterministic components to jointly define routing and MAC layer, and other one is Breath [105], which is based on a randomized routing, MAC, and duty cycling.

11.9 Radio Irregularities on WSNs

Impact and causes of radio irregularities on WSNs are discussed in a research contribution [106]. Studying these research materials stimulates to present some aspects about it here. Not many researches focused on radio irregularity in wireless communication. Only few researches, such as [106–110], focus on this common and nonnegligible phenomenon, which arises from multiple factors, such as variance in RF sending power and different path losses depending on the direction of propagation. It results in irregularity in radio range and variations in packet loss in different directions, and it is considered as a vital reason for asymmetric links as viewed by the upper layers in the WSN protocol stack.

Authors of research contribution [106] establish a radio model for simulation, called the Radio Irregularity Model (RIM), and presented the following facts about radio irregularity through analytical and experimental results:

■ General causes of radio irregularity include nonisotropic properties of the propagation media and the heterogeneous properties of devices.
■ Radio irregularity is a vital reason for asymmetric radio interference and asymmetric links in upper layers. Many aspects of upper layer performance can be directly or indirectly affected by radio irregularity.
■ Radio irregularity has a significant impact on routing protocols. Three most widely used techniques in routing protocols are path-reversal, multiround discovery, and neighbor

discovery. Routing protocols based on path-reversal and neighbor-discovery (such as geographic forwarding) are greatly influenced by radio irregularity; on the other hand, routing protocols based on multiround discovery technique (such as AODV and DSR) are able to deal with radio irregularity, but with relatively high overhead.

■ Radio irregularity has a relatively small impact on WSN-specific MAC protocols. Asymmetric radio interference between neighboring nodes affects the correctness of MAC layer functions. Contention-based MAC protocols are based on carrier sensing or handshaking techniques. Radio irregularity increases the chance for MAC protocols that use the carrier-sensing techniques to get involved in the hidden terminal problem. The handshaking technique is specially designed to resolve hidden and exposed terminal problems. However, they cannot resolve the hidden and exposed terminal problems due to asymmetry, which can be produced by radio irregularity.

In addition, the authors of the same research paper [106] propose six solutions to deal with radio irregularity.

11.10 Future Research Directions

WSN-specific application and physical layer protocols need more exploration and investigation. After evaluating a number of research contributions, it is also identified that performance comparison of WSN-specific application layer and physical layer protocols also has been required. Security at WSN-specific application and transport layer needs attention from the researchers; cross-layer security solution is also an unexplored field that needs attention. As per applications demand, if any WSN consists of all or few mobile sensor nodes, it opens new challenges for protocol design issues. A new type of WSN called WSN-HEAP [111], which consists wireless sensor nodes powered by ambient energy sources (such as water, light, wind, thermal, and RF), introduces a new set of design challenges in the area of protocol design for WSNs. A new research area, "Integration of WSNs with the existing TCP/IP-based networks," is gaining popularity. New classes of protocols are required for these new types of heterogeneous networks. "Role of WSNs in clean slate Future Internet Architectures" is also an ongoing challenge for researchers.

11.11 Educts of this Investigation

Work under the title "An Investigation on Protocols for Wireless Sensor Networks" is necessary in the current context of WSN technology, which has capability to change many of the information communication aspects in the upcoming era and in view of day-to-day suitability with the area of applications, increasing from domestic use to military use and from ground to space. The objective of this contribution has clearly brought forth important findings about each layer of the WSN protocol stack, cross-layer designing approaches, and, finally, radio irregularity perspectives that are very important in WSNs. This contribution contains very basic to high-level technical issues related to all the aspects fulfilling the objectives of title, obtained from highly cited research contribution in a concluding manner, but presenting whole aspects related to the respective field with some future research directions as well.

Acknowledgment

The authors gratefully acknowledge the immense contribution of the researchers whose work is thankfully cited in this paper. Their zest for scientific exploration and hard work in conducting various simulation studies has given a new insight to the community and paved the way for many product developments.

References

1. Akyildiz, I. F., W. Su, Y. Sankarasubramaniam, and E. Cayirci. 2002. "Wireless Sensor Networks: A Survey." *Computer Networks* 38: 393–422.
2. Marco, P. D. April 2008. "Protocol Design and Implementation for Wireless Sensor Networks." Masters' Degree Project Report-XR-EE-RT 2008:005, KTH Electrical Engineering Department, Stockholm, Sweden.
3. Saxena, M. 2007. "Security in Wireless Sensor Networks: A Layer-based Classification." CERIAS Tech Report 2007-04, Purdue University, West Lafayette.
4. Akkaya, K., and M. Younis. 2008. "A Survey on Routing Protocols for Wireless Sensor Networks." *Elsevier Journal of Ad Hoc Networks* 3 (3): 325–49. ISSN 1570-8705.
5. Chalak, A., V. Sivaraman, N. Aydin, and D. Turgut. May 9–12, 2006. "A Comparative Study of Routing Protocols in Wireless Sensor Networks." In *Proceedings of the 13th International Conference on Telecommunications*, Funchal, Portugal.
6. Becker, M., S. Schaust, and E. Wittmann. 2007. "Performance of Routing Protocols for Real Wireless Sensor Networks." In *Proceedings of the 10th International Symposium on Performance Evaluation of Computer and Telecommunication Systems, SPECTS'07*, San Diego, CA.
7. Boukerche, A., R. W. N. Pazzi, and R. B. Araujo. April 2006. "Fault-Tolerant Wireless Sensor Network Routing Protocols for the Supervision of Context-Aware Physical Environments." *Journal of Parallel and Distributed Computing*, Special Issue on Algorithms for Wireless and Ad-hoc Networks, 66 (4): 586–99. ISSN 0743-7315.
8. Li, Y., and T. Newe. December 4–6, 2006. "Wireless Sensor Networks-Selection of Routing Protocols for Applications." In *Proceedings of the Australian Telecommunication Networks and Applications Conference (ATNAC)*, 334–8. Melbourne, Australia.
9. Raman, B., and C. Kameswari. 2008. "Censor networks: A Critique of 'Sensor Networks' from a Systems Perspective." Newsletter, *ACM SIGCOMM Computer Communication Review* 38 (3): 75–8. New York: ACM.
10. Ahvar, E., and M. Fathy. October 2010. "BEAR: A Balanced Energy-Aware Routing Protocol for Wireless Sensor Networks." *Wireless Sensor Networks* 2: 793–800.
11. Saaranen, A., and C. A. Pomalaza-Ráez. 2004. "Comparison of Reactive Routing and Flooding in Wireless Sensor Networks." In *Proceedings of the Nordic Radio Symposium*, Oulu, Finland.
12. Dwivedi, A. K., and O. P. Vyas. 2010. "Network Layer Protocols for Wireless Sensor Networks: Existing Classifications and Design Challenges." *International Journal of Computer Applications* 8 (12, Article 6): 30–34.
13. Eriksson, J. April 2009. "Detailed Simulation of Heterogeneous Wireless Sensor Networks." Dissertation for the Degree of Licentiate of Philosophy in Computer Science at Uppsala University Sweden. ISSN 1404-5117.
14. Al-Karaki, J. N., and A. E. Kamal. December 2004. "Routing Techniques in Wireless Sensor Networks: A Survey." *IEEE Wireless Communications* 11 (6): 6–28.
15. Karl, H., and A. Willig. 2006. *Protocols and Architectures for Wireless Sensor Networks*. New York: John Wiley & Sons Ltd. ISBN: 978-0-470-09510-2.
16. Wachs, M., J. I. Choi, J. W. Lee, K. Srinivasan, Z. Chen, M. Jain, and P. Levis. 2007. "Visibility: A New Metric for Protocol Design." In *Proceedings of the 5th international conference on Embedded Networked Sensor Systems (ACM SenSys'07)*, 73–86. ISBN: 978-1-59593-763-6. doi: 10.1145/1322263.1322271.
17. Liu, C., and G. Cao. April 2011. "Spatial-Temporal Coverage Optimization in Wireless Sensor Networks." *IEEE Transaction on Mobile Computing* 10 (4): 465–78.
18. Al-Obaisat, Y., and R. Braun. March 2006. "On Wireless Sensor Networks: Architectures, Protocols, Applications, and Management." In *Proceedings of the AusWireless 2006 Conference*, Sydney, Australia.
19. Sharma, K., and M. K. Ghose. January 2011. "Cross Layer Security Framework for Wireless Sensor Networks." *International Journal of Security and Its Applications* 5 (1): 39–52.

20. Gawdan, I. S., C. O. Chow, T. A. Zia, and Q. I. Gawdan. September 2011. "Cross-Layer Based Security Solutions for Wireless Sensor Networks." *International Journal of the Physical Sciences* 6 (17): 4245–54.
21. Kalita, H. K., and A. Kar. December 2009. "Wireless Sensor Network Security Analysis." *International Journal of Next-Generation Networks* 1 (1): 1–10.
22. Wood, A. D., and J. A. Stankovic. October 2002. "Denial of Service in Sensor Networks." *IEEE Computer* 35 (10): 54–62. doi: 10.1109/MC.2002.1039518.
23. Karlof, C., and D. Wagner. 2003. "Secure Routing in Wireless Sensor Networks: Attacks and Countermeasures." *Ad Hoc Networks* 1: 293–315. doi: 10.1016/S1570-8705(03)00008-8.
24. Pathan, Al-S. K., H.-W. Lee, and C. S. Hong. February 2006. "Security in Wireless Sensor Networks: Issues and Challenges." In *Proceedings of the ICACT2006*, 1043–48.
25. Xiao, M., X. Wang, and G. Yang. June 21–23, 2006. "Cross-Layer Design for the Security of Wireless Sensor Networks." In *Proceedings of the 6th World Congress on Intelligent Control and Automation*, Dalian, China, 104–08.
26. Haenselmann, T. April 2006. *Sensor Networks, GFDL Wireless Sensor Network Textbook*. Available at http://pi4.informatik.uni-mannheim.de/~haensel/sn_book.pdf.
27. Sohrabi, K., J. Gao, V. Ailawadhi, and G. J. Pottie. September 1999. "Protocols for Self-Organization of a Wireless Sensor Network." In *Proceedings of the 37th Allerton Conference on Communication Computing and Control*.
28. Sangiovanni-Vincentelli, A., M. Sgroi, A. Wolisz, and J. M. Rabaey. 2004. "A Service-Based Universal Application Interface for Ad-Hoc Wireless Sensor Networks," Whitepaper, University of California, Berkeley.
29. Wang, C., K. Sohraby, Y. Hu, B. Li, and W. Tang. March 2005. "Issues of Transport Control Protocols for Wireless Sensor Networks." In *Proceedings of the International Conference on Communications Circuits and Systems*, ICCCAS, Hong-Kong, China.
30. Wang, C., K. Sohraby, B. Li, M. Daneshmand, and Y. Hu. May–June 2006. "A Survey of Transport Protocols for Wireless Sensor Networks." *IEEE Network* 20: 34–40.
31. Rahman, Md. A., A. E. Saddik, and W. Gueaieb. 2008. *Wireless Sensor Network Transport Layer: State of the Art, Sensors*, 221–44. Berlin Heidelberg: Springer-Verlag.
32. Iyer, Y. G., S. Gandham, and S. Venkatesan. 2005. "STCP: A Generic Transport Layer Protocol for Wireless Sensor Networks." In *Proceedings of the IEEE International Conference on Computer Communications and Networks*, ICCCN.
33. Tezcan, N., and W. Wang. "ART: An Asymmetric and Reliable Transport Mechanism for Wireless Sensor Networks." *International Journal of Sensor Networks*, Special Issue on Theoretical and Algorithmic Aspects in Sensor Networks 2 (3/4): 188–200. Geneva, Switzerland: Inderscience Publishers.
34. Pereira, P. R., A. Grilo, F. Rocha, M. S. Nunes, A. Casaca, Cl. Chaudet, P. Almström, and M. Johansson. December 2007. "End-to-End Reliability in WSNs: Survey and Research Challenges." In *Proceedings of the EuroFGI Workshop on IP QoS and Traffic Control*, Lisbon, Portugal.
35. Kim, S., R. Fonseca, and D. Culler. 2004. "Reliable Transfer on Wireless Sensor Networks." In *Proceedings of the First Annual IEEE Communications Society Conference on Sensor and Ad Hoc Communications and Networks*, 449–59. doi: 10.1109/SAHCN.2004.1381947.
36. Monteiro, D. M., B. Vaidya, and J. J. P. C. Rodrigues. 2010. "Performance Assessment of Congestion Control Transport Protocols for Wireless Sensor Networks." In *Proceedings of the International Conference on Wireless Information Networks & Business Information System*, Kathmandu, Nepal, 07–17. ISSN: 2091-0266.
37. Pedrosa, L. D., and R. M. Rocha. 2008. "WMTP—A Modular WSN Transport Protocol: The Fairness Module." In *Proceedings of the IEEE Conference on New Technologies Mobility and Security*, NTMS, Tangier, 1–6. doi: 10.1109/NTMS.2008.ECP.103.
38. Wan, C. Y., A. T. Campbell, and L. Krishnamurthy. September 2002. "PSFQ: A Reliable Transport Protocol for Wireless Sensor Networks." In *Proceedings of the WSNA'02*. Atlanta, GA.
39. Sankarasubramaniam, Y., Ö. B. Akan, and I. F. Akyildiz. June 2003. "ESRT: Event-to-Sink Reliable Transport in Wireless Sensor Networks." In *Proceedings of the MobiHoc'03*, Annapolis, MD.

40. Marchi, B., A. Grilo, and M. Nunes. July 2007. "DTSN—Distributed Transport for Sensor Networks." In *Proceedings of the 12th IEEE Symposium on Computers and Communications*, ISCC2007, Aveiro, Portugal.

41. Olivares, T., P. J. Tirado, F. Royo, J. C. Castillo, and L. Orozoco-Barbosa. 2007. "IntellBuilding: A Wireless Sensor Network for Intelligent Buildings." Poster of the 4th European Conference on Wireless Sensor networks (EWSN), Parallel and Distributed Systems Report Series, Report Number PDS-2007-00, Delft, Netherland. ISSN 1387-2109.

42. Castillo, J. C., T. Olivares, and L. Orozco-Barbosa. 2007. "Routing Protocols for Wireless Sensor Networks-Based Network." Technical Report, Albacete Research Institute of Informatics, University of Castilla, Spain.

43. Akdere, M., C. C. Bilgin, O. Gerdaneri, I. Korpeoglu, Ö. Ulusoy, and U. Cetintemel. 2006. "A Comparison of Epidemic Algorithms in Wireless Sensor Networks." *Elsevier Journal of Computer Communications* 29: 2450–57.

44. Biradar, R. V., V. C. Patil, S. R. Sawant, and R. R. Mudholkar. 2009. "Classification and Comparison of Routing Protocols in Wireless Sensor Networks." *UbiCC Journal*, Special Issue on Ubiquitous Computing Security Systems 4: 704–11.

45. Younis, O., and S. Fahmy. 2004. "HEED: A Hybrid, Energy-Efficient, Distributed Clustering Approach for Ad Hoc Sensor Networks." *IEEE Transactions on Mobile Computing* 3: 366–79.

46. [Al]-Khdour, T., and U. Baroudi. June 2011. "Literature Review of MAC, Routing and Cross Layer Design Protocols for WSN, Chapter 2." In *Wireless Sensor Networks*, edited by S. Tarannum. Austria: InTech. ISBN 978-953-307-325-5. doi: 10.5772/38724.

47. Sha, K., J. Du, and W. Shi. January 2005. "WEAR: A Balanced, Fault-Tolerant, Energy-Aware Routing Protocol for Wireless Sensor Networks." Technical Report MIST-TR-2005-001, Wayne State University.

48. Perillo, M., and W. Heinzelman. 2005. Chapter 36, in *Handbook of Algorithms for Wireless Networking and Mobile Computing*, edited by A. Boukerche, 790–813. Chapman and Hall/CRC.

49. Hu, Y. C., A. Perrig, and D. B. Johnson. 2005. "Ariadne: A Secure On-Demand Routing Protocol for Ad Hoc Networks." In *Journal of Wireless Networks* 11 (1–2): 21–38. Hingham, MA: Kluwer Academic Publishers.

50. DiPippo, L. C., Y. Sun, K. Rahn Jr., R. Anachi, and O. Savas. 2010. "Secure Adaptive Routing Protocol for Wireless Sensor Networks." Technical Report—TR10-329, Department of Computer Science, University of Rhode Island.

51. Wood, A. D., L. Fang, J. A. Stankovic, and T. He. October 2006. "SIGF: A Family of Configurable, Secure Routing Protocols for Wireless Sensor Networks." In *Proceedings of the SASN'06*, Alexandria, VA.

52. Blum, B., T. He, S. Son, and J. A. Stankovic. 2003. "IGF: A State-Free Robust Communication Protocol for Wireless Sensor Networks." Technical Report CS-2003-11, University of Virginia, Charlottesville, VA.

53. Pathan, A. S. K., and C. S. Hong. October 2008. "SERP: Secure Energy-Efficient Routing Protocol for Densely Deployed Wireless Sensor Networks." *Annals of Telecommunications*, 63 (9–10): 529–41. ISSN 0003-4347 (Print), 1958-9395 (Online). doi: 10.1007/s12243-008-0042-5.

54. Al-Karaki, J. N. December 2006. "Analysis of Routing Security-Energy Trade-offs in Wireless Sensor Networks." *International Journal of Security and Networks* 1 (3/4): 147–57. ISSN 1747-8405. doi: 10.1504/IJSN.2006.011774.

55. Yin, J., and S. K. Madria. 2008. "ESecRout: An Energy Efficient Secure Routing for Sensor Networks." *International Journal of Distributed Sensor Networks*, Special Issue: Advances on Heterogeneous Wireless Sensor Networks 4 (2) 67–82. doi: 10.1080/15501320802001101.

56. Manjeshwar, A., and D. P. Agarwal. April 2001. "TEEN: A Routing Protocol for Enhanced Efficiency in Wireless Sensor Networks." In *Proceedings of the 1st International Workshop on Parallel and Distributed Computing Issues in Wireless Networks and Mobile Computing*.

57. Manjeshwar, A., and D. P. Agarwal. 2002. "APTEEN: A Hybrid Protocol for Efficient Routing and Comprehensive Information Retrieval in Wireless Sensor Networks." In *Proceedings of the International Parallel and Distributed Processing Symposium, IPDPS*, 195–202.

58. Kour, H., and A. K. Sharma. 2010. "Hybrid Energy Efficient Distributed Protocol for Heterogeneous Wireless Sensor Network." *International Journal of Computer Applications* 4: 01–05.

59. Ying-Hong, W., L. Yi-Chien, F. Ping-Fang, and T. Chih-Hsiao. 2006. "REDRP: Reactive Energy Decisive Routing Protocol for Wireless Sensor Networks." In *Proceedings of the Ubiquitous Intelligence and Computing*, LNCS, Vol. 4159, 527–35. Berlin/Heidelberg: Springer.
60. Lindsey, S., and C. Raghavendra. 2002. "PEGASIS: Power-Efficient Gathering in Sensor Information Systems." In *Proceedings of the IEEE Aerospace Conference*, Vol. 3, 1125–30.
61. Savvides, A., C.-C. Han, and M. Srivastava. July 2001. "Dynamic Fine-Grained Localization in Ad-Hoc Networks of Sensors." In *Proceedings of the 7th ACM Annual International Conference on Mobile Computing and Networking, MobiCom*. 166–79.
62. Kandris, D., P. Tsioumas, A. Tzes, G. Nikolakopoulos, and D. D. Vergados. 2009. "Power Conservation through Energy Efficient Routing in Wireless Sensor Networks." *Sensors* 9: 7320–42. ISSN 1424-8220.
63. Heinzelman, W., A. Chandrakasan, and H. Balakrishnan. January 2000. "Energy-Efficient Communication Protocol for Wireless Microsensor Networks." In *Proceedings of the 33rd Hawaii International Conference on System Sciences, HICSS '00*.
64. Hancke, G. P., and C. J. Leuschner. 2007. "SEER: A Simple Energy Efficient Routing Protocol for Wireless Sensor Networks." *South African Computer Journal* 39: 17–24.
65. An-dong, Z., X. Tian-yin, C. Gui-hai, Y. Bao-liu, and L. Sang-lu. 2010. "A Survey on Real-time Routing Protocols for Wireless Sensor Networks." *Computer Communication* 2: 104–12.
66. Marco, P. D., P. Park, C. Fischione, and K. H. Johansson. June 2010. "TREnD: A Timely, Reliable, Energy-efficient and Dynamic WSN Protocol for Control Applications." In *Proceedings of the Information Communication Conference*.
67. Kim, J., and B. Ravindran. March 2009. "Opportunistic RealTime Routing in MultiHop Wireless Sensor Networks." In *Proceedings of the SAC'09*, Honolulu, HI.
68. He, T., J. A. Stankovic, C. Lu, and T. F. Abdelzaher. 2003. "Speed: A Stateless Protocol for Real-time Communication in Sensor Networks." In *Proceedings of the IEEE International Conference on Distributed Computing System, ICDCS'03*, 46–55.
69. Felemban, E., E. Ekici, and C. G. Lee. 2006. "MMSPEED: Multipath Multi-SPEED Protocol for QoS Guarantee of Reliability and Timeliness in Wireless Sensor Networks." *IEEE Transactions on Mobile Computing* 5 (6): 738–54. doi: 10.1109/TMC.2006.79.
70. Gelenbe, E., E. Ngai, and P. Yadav. April 2009. "Routing of High Priority Packets in Wireless Sensor Networks." In *Proceedings of the SPIE Defense, Security, and Sensing, SPIE DSS'09*, Orlando, FL.
71. Akylidiz, I. F., W. Su, Y. Sankarasubramaniam, and E. Cayirci. August 2002. "A Survey on Sensor Networks." *IEEE Personal Communications Magazine*, 102–14.
72. The Working Group for WLAN Standards. 1999. "IEEE 802.11 Standards, Part 11: Wireless Medium Access Control (MAC) and Physical Layer (PHY) Specifications." IEEE Technical Report.
73. Ye, W., J. Heidemann, and D. Estrin. June 2004. "Medium Access Control With Coordinated Adaptive Sleeping for Wireless Sensor Networks." *IEEE/ACM Transactions on Networking* 12 (3): 493–505. doi: 10.1109/TNET.2004.828953.
74. Teng, Z., and K. Kim. May 2010. "A Survey on Real-Time MAC Protocols in Wireless Sensor Networks." *Communications and Network* 2: 104–12. doi: 10.4236/cn.2010.22017.
75. Ameen, M. A., S. M. R. Islam, and K. Kwak. 2010. "Energy Saving Mechanisms for MAC Protocols in Wireless Sensor Networks." *International Journal of Distributed Sensor Networks*, Volume 2010 Article ID 163413. doi: 10.1155/2010/163413.
76. Kaur, S., and L. Mahajan. 2011. "Power Saving MAC Protocols for WSNs and Optimization of S-MAC Protocol." *International Journal of Radio Frequency Identification and Wireless Sensor Networks* 1 (1): 01–08.
77. Langendoen, K. 2008. "Medium Access Control in Wireless Sensor Networks." In *Medium Access Control in Wireless Networks, Vol. 2: Practice and Standards*, edited by H. Wu and Y. Pan, 535–60. Nova Science Publishers.
78. Ansari, J., J. Riihijarvi, P. Mahonen, and J. Haapola. 2006. "Implementation and Performance Evaluation of nanoMAC: A Low-Power MAC Solution for High Density Wireless Sensor Networks." *International Journal of Sensor Networks* 1 (4): 1–10. doi: 10.1504/IJSNET.2007.014361.
79. Callaway, E., P. Gorday, L. Hester, J. Gutierrez, M. Naeve, B. Heile, and V. Bahl. August 2002. "Home Networking with IEEE 802.15.4: A Developing Standard for Low-Rate Wireless Personal Area Networks." *IEEE Communication Magazine* 40: 70–77.

80. Das, I. R. A. K., and S. Roy. February 2006. "Analysis of the Contention Access Period of IEEE 802.15.4 MAC." UWEE Technical Report -UWEETR-2006-0003, University of Washington, Seattle, WA.
81. Petrova, M., J. Riihijöarvi, P. Möahöonen, and S. Labella. 2006. "Performance Study of IEEE 802.15.4 Using Measurements and Simulations." In *Proceedings of the IEEE WCNC.*
82. Suh, C., and Y. B. Ko. May 2005. "A Traffic Aware, Energy Efficient MAC Protocol for Wireless Sensor Networks." In *Proceedings of the IEEE International Symposium on Circuits and Systems,* ISCAS, Vol. 3, 2975–78.
83. Dam, T. V., and K. Langendoen. November 2003. "An Adaptive Energy-Efficient MAC Protocol for Wireless Sensor Networks." In *Proceedings of the ACM SenSys'03,* 171–80. Los Angeles, CA. New York: ACM.
84. Barowski, Y. D. May 2010. "GANGs: An Energy Efficient Medium Access Control Protocol." Dissertation for Degree of Doctor of Philosophy. Auburn University, AL.
85. Lin, P., C. Qiao, and X. Wang. March 2004. "Medium Access Control with a Dynamic Duty Cycle For Sensor Networks." In *Proceedings of the WCNC.*
86. Ringwald, M., and K. Röomer. January 2005. "BitMAC: A Deterministic, Collision-Free, and Robust MAC Protocol for Sensor Networks." In *Proceedings of the 2nd European Workshop on Wireless Sensor Networks,* EWSN, Istanbul, Turkey, 57–69.
87. Rajendran, V., K. Obraczka, and J. J. Garcia-Luna-Aceves. 2004. *Energy-Efficient, Collision-Free Medium Access Control for Wireless Sensor Networks.* Netherlands: Kluwer Academic Publishers.
88. El-Hoiydi, A., and J.-D. Decotignie. June–July 2004. "WiseMAC: An Ultra Low Power MAC Protocol for the Downlink of Infrastructure Wireless Sensor Networks." In *Proceedings of the 9th International Symposium Computers and Communications,* ISCC, Vol. 1 (28), 244–51.
89. Bhat, P. Blog posted on August 7, 2011. http://sensors-and-networks.blogspot.com/2011/08/physical-layer-for-wireless-sensor.html as on 27/12/2011.
90. Shih, E., S. Cho, N. Ickes, R. Min, A. Sinha, A. Wang, and A. Chandrakasan. July 2001. "Physical Layer Driven Protocol and Algorithm Design for Energy-Efficient Wireless Sensor Networks." In *Proceedings of the ACM MobiCom'01,* Rome, Italy, 272–86.
91. Holland, M. 2007. "Optimizing Physical Layer Parameters for Wireless Sensor Networks." Thesis submitted for Degree of Master of Science, University of Rochester, New York.
92. Townsend, C., and S. Arms. 2005. "Wireless Sensor Networks: Principles and Applications." In *Sensor Technology,* edited by J. S. Wilson, 575–89, Oxford: Elsevier-Newnes.
93. Manjunath, D. 2007. "A Review of Current Operating Systems for Wireless Sensor Networks." Technical Report, Department of ECE, Indian Institute of Science, Bangalore, India.
94. Jurdak, R. November 2006. *Wireless Ad Hoc and Sensor Networks: A Cross-Layer Design Perspective.* New York: Springer.
95. Melodia, T., M. C. Vuran, and D. Pompili. "The State of the Art in Cross-Layer Design for Wireless Sensor Networks, Broadband and Wireless Networking Laboratory." In *Proceedings of the Second international conference on Wireless Systems and Network Architectures in Next Generation Internet (EURO-NGI'05),* 78–92. Berlin, Heidelberg: Springer-Verlag. ISBN: 3-540-34025-4 978-3-540-34025-6. doi: 10.1007/11750673_7.
96. Safwati, A., H. Hassanein, and H. Mouftah. April 2003. "Optimal Cross-Layer Designs for Energy-Efficient Wireless Ad hoc and Sensor Networks." In *Proceedings of the IEEE International Conference of Performance, Computing, and Communications,* 123–28.
97. Sichitiu, M. L. 2004. "Cross-Layer Scheduling for Power Efficiency in Wireless Sensor Networks." In *Proceedings of the 23rd Annual Joint Conference of the IEEE Computer and Communications Societies,* Vol. 3, 1740–50.
98. Su, W., and T. L. Lim. June 2006. "Cross-Layer Design and Optimization for Wireless Sensor Networks." In *Proceedings of the 7th ACIS International Conference on Software Engineering, Artificial Intelligence, Networking, and Parallel/Distributed Computing,* 278–84.
99. Suh, C., Y. B. Ko, and D. M. Son. 2006. "An Energy Efficient Cross-Layer MAC Protocol for Wireless Sensor Networks." In *Proceedings of the APWeb,* Springer LNCS, Vol. 3842, 410–19.

100. Venkitasubramaniam, P., S. Adireddy, and L. Tong. October 2003. "Opportunistic ALOHA and Cross Layer Design for Sensor Networks." In *Proceedings of the IEEE Military Communications Conference*, MILCOM 2003, Vol. 1, 705–10.

101. Wang, L. C., and C. W. Wang. March 2004. "A Cross-layer Design of Clustering Architecture for Wireless Sensor Networks." In *Proceedings of the IEEE International Conference on Networking, Sensing & Control*, Taiwan, 547–52.

102. Shuguang, C., R. Madan, A. Goldsmith, and S. Lall. May 2005. "Joint Routing, MAC, and Link Layer Optimization in Sensor Networks with Energy Constraints." In *Proceedings of the IEEE International Conference on Communications*, ICC 2005, Vol. 2, 725–29.

103. Ha, R. W., P. H. Ho, and X. S. Shen. 2005. "Cross-Layer Application-Specific Wireless Sensor Network Design with Single-Channel CSMA MAC over Sense-Sleep Trees." *Computer Communications*, Special Issue on Energy Efficient Scheduling and MAC for Sensor Networks, WPANs, WLANs, and WMAN 29 (17): 3425–44. Elsevier. http://dx.doi.org/10.1016/j.comcom.2006.01.019.

104. Bonivento, A., C. Fischione, A. Sangiovanni-Vincentelli, F. Graziosi, and F. Santucci. November 2005. "SERAN: A Semi Random Protocol Solution for Clustered Wireless Sensor Networks." In *Proceedings of the IEEE International Conference on Mobile Adhoc and Sensor Systems Conference (MAHSS'05)*. ISBN: 0-7803-9465-8. doi: 10.1109/MAHSS.2005.1542819.

105. Park, P. G., C. Fischione, A. Bonivento, K. H. Johansson, and A. Sangiovanni-Vincentelli. June 2008. "Breath: A Self-Adapting Protocol for Wireless Sensor Networks." Accepted paper in IEEE SECON 2008, San Francisco, CA.

106. Zhou, G., T. He, S. Krishnamurthy, and J. A. Stankovic. June 2004. "Impact of Radio Irregularity on Wireless Sensor Networks." In *Proceedings of the 2nd International Conference on Mobile Systems, Applications, and Services (ACM MobiSys'04)*, 125–38. New York: ACM. ISBN: 1-58113-793-1. doi: 10.1145/990064.990081.

107. Cerpa, A., N. Busek, and D. Estrin. September 2003. "SCALE: A Tool for Simple Connectivity Assessment in Lossy Environments." CENS Technical Report 0021 Center of Embedded Networked Systems (CENS). Los Angeles, CA: University of California.

108. Ganesan, D., B. Krishnamachari, A. Woo, D. Culler, D. Estrin, and S. Wicker. 2002. "Complex Behavior at Scale: An Experimental Study of Low-Power Wireless Sensor Networks." Technical Report UCLA/CSD-TR 02-0013 UCLA Computer Science Division.

109. Woo, A., T. Tong, and D. Culler. November 2003. "Taming the Underlying Challenges of Reliable Multihop Routing in Sensor Networks." In *Proceedings of the ACM SenSys*, Los Angeles, CA.

110. Zhao, Y. J., and R. Govindan. November 2003. "Understanding Packet Delivery Performance in Dense Wireless Sensor Network." In *Proceedings of the ACM SenSys-03*, Los Angeles, CA.

111. Seah, W. K. G., Z. A. Eu, and H. P. Tan. 2009. "Wireless Sensor Networks Powered by Ambient Energy Harvesting (WSN-HEAP)—Survey and Challenges." In *Proceedings of the 1st International Conference on Wireless VITAE*, 1–5.

Chapter 12

Data Gathering and Data Management Techniques in Wireless Sensor Networks

---

M. Bala Krishna and M. N. Doja

Contents

12.1 Introduction

Wireless sensor networks (WSNs) are spatially distributed autonomous nodes with constraints in computing and storage features. Sensor nodes (SNs) are deployed on a large scale to monitor events in weather forecasts, agriculture and animal habitats, manufacturing industry, military fields, and so on. With recent emerging technology and applications in Mobile Communications, 3G/4G Communications, Wi-Fi, WiMax, Ad hoc and Sensor Networks, Internet, and so on, the main challenge is the *data*, which need to be organized, stored, and processed efficiently. In sensor networks, the data transmission between the nodes is dependent on external factors like environmental conditions, density of mobile nodes, network congestion, and so on. In contrast to the conventional data management applications, the sensor data applications need to be monitored continuously, based on event and time triggers. In *event-based* queries, the input streams are bounded and user queries are processed by using event-based attributes. In *time-based* queries, the input streams are unbounded and user queries are processed by using the most recent inputs. Wireless Sensor Networks [1] is applied in terrestrial, underwater, underground, and real-time multimedia communications.

Data gathering techniques are primarily aimed to collect the data and are used extensively in data mining, web mining, data monitoring applications, and so on. Data management techniques are used to organize and store the collected data for efficient query processing. Data gathering techniques in sensor networks is based on data diffusion and data dissemination. Data management techniques in sensor networks [2] is based on data aggregation, data fusion, data storage, and query processing. Data aggregation with robust data dissemination [3] is applied for fault-tolerant applications. Data-centric networking is used to manage the data across various clusters in the network. The salient features of data-centric networking are given as follows:

1. Data packets from the SNs are sent to the cluster head (CH) to minimize the packet overhead and save the node energy.
2. Query processing is based on the single-valued or multivalued attributes of the data-centric CH node or base station (BS) rather than the SNs.
3. Support multi-hop and peer-to-peer communication.
4. Data-centric protocols are used in routing and data query applications.

This chapter is organized as follows. Section 12.2 explains the features of data gathering, data diffusion, and data dissemination. This section explains the flow of sensor data using data gathering techniques that are combined and stored at the CH or BS using data management techniques. Section 12.3 expounds the data gathering process based on data diffusion and data dissemination techniques with their respective protocols. Pull and push diffusion techniques are explained with parametric analysis. Section 12.4 describes the data combining process based on data fusion and data aggregation techniques with their respective protocols. Section 12.5 explain the properties and metrics of sensor network database. This section explains various types of queries used in sensor networks. Section 12.6 elucidates data storage techniques based on External Storage (ES), Internal Storage (IS), and CH-Centric Storage (CH-CS) and gives their respective analysis. Section 12.7 elucidates the metrics used in data gathering and data management for sensor networks. Section 12.8 concludes the chapter with focus on data gathering techniques in sensor networks, which save the node energy, increase the throughput rate, and increase the network life cycle. Finally, Section 12.9 gives an insight into the future research and challenges in data gathering and data management techniques for sensor networks.

12.2 Features of Data Gathering and Data Management

12.2.1 Data Gathering

Data gathering in sensor networks differ from the traditional database systems with respect to the size of data attributes. The data-centric storage and real-time query processing are the main features of data gathering in sensor networks. Data gathering protocols operate on (1) communication and networking attributes, (2) event-time attributes, (3) SN attributes, and (4) environmental attributes. Since the sensor networks are dense by nature, the data gathering protocols are attributed with features like (1) scalability, (2) adaptability, (3) robustness, (4) energy-efficient design, and (5) dynamic topology. Figure 12.1 illustrates the flow of data from the SNs to CH and BS. The data-gathering phase shown in the upper layer of Figure 12.1 collects the data from SNs using, data diffusion and data dissemination techniques. The data management phase shown in the lower layer of Figure 12.1 uses the data fusion and data aggregation techniques to combine the sensor data and store it in the database. CHs and BS use efficient query processing techniques to monitor and retrieve the sensor data.

12.2.2 Data Diffusion

Data diffusion is defined as the process of sending or receiving the data based on the data-centric node, which is located at the center of adjacent nodes. Source node or sink node function as the

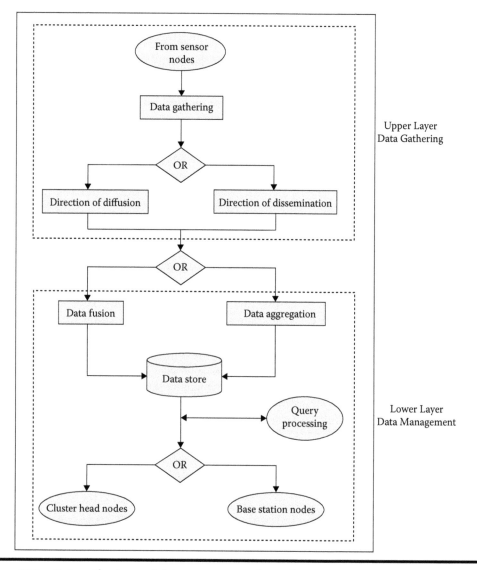

Figure 12.1 Data gathering and data management layers in WSNs.

active participant and initiate the data diffusion session. Data diffusion protocols are classified based on the direction of source or sink node. In sensor networks, data gathering protocols collect the data based on the direction of diffusion. Data diffusion protocols [4] in sensor networks are classified as one-phase pull diffusion, two-phase pull diffusion, directed diffusion, and push diffusion.

12.2.3 Data Dissemination

Data dissemination is defined as the broadcasting process in which a monitor head node sends or receives the data from the neighboring CHs or BS. In a large-scale sensor network, when congestion occurs, the data packets are split into small-sized packets and disseminated to the neighboring SNs or CHs. Data gathering protocols collect the data based on the direction of dissemination.

Based on the event-time and data packet attributes, the data dissemination [5] protocols are classified into proactive, reactive, and hybrid methodologies.

12.2.3.1 Proactive Dissemination

In this method, SNs publish the information during the dissemination time interval, even though there is no variation in the node data attributes. The node data attributes are updated after the expiry of dissemination time interval. This method is suitable for dense networks and avoids the network congestion. Proactive disseminate rate, $\text{PDissemination}_{\text{rate}}$, in the time interval [t1, t2] (where t1 and t2 define the dissemination time interval) to publish the node information across z clusters with n nodes per cluster and k adjacent nodes per SN is given as follows:

$$\text{PDissemination}_{\text{rate}} = \int_{t1}^{t2} \sum_{j=1}^{z} \sum_{i=1}^{n} \text{PublishInformation}_j (\text{SN}_i, \text{AdjSN}_i^k) \qquad (12.1)$$

12.2.3.2 Reactive Dissemination

In this method, SNs publish the information when an event or variation in node attribute occurs. Due to less frequent updates, this method minimizes the communication overhead. When the network traffic is high, congestion is resolved by using event scheduling and event priority techniques. Reactive dissemination rate based on packet variation, $R_{\text{Pkt}}\text{Dissemination}_{\text{rate}}$, in the time interval [t1, t2] (where t1 and t2 define the dissemination time interval) to publish the node information across z clusters with n nodes per cluster and k adjacent nodes per SN is given as follows:

$$R_{\text{Pkt}}\text{Dissemination}_{\text{rate}} = \int_{t1}^{t2} \sum_{j=1}^{z} \sum_{i=1}^{n} \text{PublishInformation}_j \left(\text{SN}_i, \text{AdjSN}_i^k, \text{PktVariation}_i\right) \qquad (12.2)$$

Reactive dissemination rate based on event, $R_{\text{Event}}\text{Dissemination}_{\text{rate}}$ for node i with p events Event_p in the time interval [t1, t2] (where t1 and t2 define the dissemination time interval) to publish the node information across z clusters with n nodes per cluster and k adjacent nodes per SN is given as follows:

$$R_{\text{Event}}\text{Dissemination}_{\text{rate}} = \int_{t1}^{t2} \sum_{j=1}^{z} \sum_{i=1,p=1}^{n,m} \text{PublishInformation}_j \left(\text{SN}_i, \text{AdjSN}_i^k, \text{Event}_p\right) \qquad (12.3)$$

12.2.3.3 Hybrid Dissemination

In this method, SNs publish the information based on the combination of proactive and reactive dissemination attributes. Probability and random-based methods are used to publish the node information.

12.2.4 Data Fusion

In sensor networks, data redundancy exists in both the temporal and the spatial domains. Data collected from the SNs and neighboring nodes at different time intervals is correlated and may contain redundant information. Data fusion is the process of compressing correlated data and eliminate data redundancy. Data fusion is performed at the source node, intermediate nodes, or the CHs. As the size of sensor network increases, data fusion uses the distributed data combining techniques [6] as compared to the centralized data combining techniques.

12.2.5 Data Aggregation

Data aggregation is the process of combining the related data from the SNs using aggregation functions [7] like sum, average, maximum, minimum, and so on. Data aggregation process is executed at the source node, intermediate node, or the CHs. Data aggregation nodes (DANs) are used to collect the data from neighboring nodes, balance the network load, and increase the throughput rate.

Data fusion and data aggregation are seen as separate entities in the data management phase irrespective of their similarities in data combining process. Data fusion is defined as the process of compressing the correlated data from SNs and forwarding it to the neighboring CHs and BS. In data fusion, the data is compressed at the source SN and/or intermediate nodes. If the data is not fused at regular intervals, then it leads to network congestion. At the abstract level, data fusion uses decision-based logic and aggregation functions. Data fusion is primarily used in dense network applications. Data aggregation is the process of combining the correlated data from SNs (using aggregation functions) and forwarding it to the neighboring CHs and BS. Data aggregation is used in relatively less denser network applications as compared to the data fusion technique. In secure data management, the data fusion and data aggregation are attributed with secured keys to combine the data packets.

12.3 Data Gathering Process Based on Packet Forwarding

Data gathering process [8] based on the packet forwarding mechanism is classified as follows:

- Diffusion
- Dissemination

12.3.1 Data Gathering Mechanism Based on Diffusion

Data gathering is initiated with a set of control signals between the source node and the sink node. The sequence of control messages is (1) query, (2) wait, and (3) reply. Data gathering node or CH sends the query message to the SN, waits for the reply message, and confirms the data gathering session based on time or event attributes. If a nonparticipating node wishes to join the data gathering session, it waits in the queue and receives the confirmation from CH. Flooding of sensor query messages can be high in data gathering based diffusion due to (1) a large number of sink or source nodes and (2) multievent applications.

12.3.1.1 Data Diffusion Protocols in Wireless Sensor Networks

Data diffusion protocols [8] in sensor networks are classified into pull and push techniques based on the number of source and sink nodes and the operation mode (broadcast or flooding). Table 12.1 gives the characteristic features of pull and push diffusion techniques for packet forwarding in sensor networks.

Table 12.1 Characteristic Features of Pull and Push Diffusion for Packet Forwarding in Wireless Sensor Networks

Diffusion Type	Active Initiator	Node Size	Operation	Control Overhead
One-phase pull diffusion [4,8]	Sink	Many source nodes and few sink nodes	Only query is broadcasted	Control overhead is less due to the elimination of exploratory data
Two-phase pull diffusion [8,9]	Sink	Many source nodes and few sink nodes	Both query and reply messages are broadcasted	Control overhead is more due to the inclusion of exploratory data
Push diffusion [8,10]	Source	Few source nodes and many sink nodes	Flooding is used to send the data	Control overhead is more due to the inclusion of exploratory data

12.3.1.1.1 One-Phase Pull Diffusion

In one-phase pull diffusion [4], the routes are established based on the gradient factor. Sink node sends the query across multiple route paths and the reply is based on minimum latency from the neighboring nodes. Since the source node does not send the empirical data to the sink node, the communication control overhead is less as compared to other diffusion techniques. One-phase pull diffusion is used in applications with many source nodes and few sink nodes. One-phase pull diffusion, Pull Diffusion$_{\text{One-Phase}}$ across z clusters between the sink node, SinkSN, and the source node, SrcSN, based on query message and the shortest paths with minimum latency is given as follows:

$$\text{Pull Diffusion}_{\text{One-Phase}} = \sum_{c=1}^{z} \sum_{i=1, j=1}^{m,n} \left(\text{Query}_c \left(\text{SinkSN}_i, \text{SrcSN}_j \right) + \text{FormShortPath}_c \left(\text{SrcSN}_j, \text{SinkSN}_i \right) \right) \quad (12.4)$$

12.3.1.1.2 Two-Phase Pull Diffusion

In two-phase pull diffusion [9], the multiple routes are established between the source node and the sink node. Sink node sends the query across multiple route paths and receives the empirical data from the source node through multiple paths. Sink node uses reinforced gradient (minimum cost path) with neighbor nodes to confirm the route path, and the process is repeated until the source node is reached. The sink node reinforces the best route with no faulty node, thereby ensuring a guaranteed route with minimum latency and increase in the throughput rate. Two-phase pull diffusion is used in applications with many source nodes and few sink nodes. Two-phase pull diffusion, Pull Diffusion$_{\text{Two-Phase}}$ across z clusters between the sink node, SinkSN, and the source

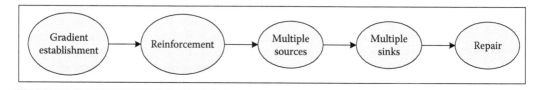

Figure 12.2 Sequence of steps in directed diffusion.

node, SrcSN based on the query message and reply message, where multiple paths are pruned to find the shortest path, is given as follows:

$$\text{Pull Diffusion}_{\text{Two-Phase}} = \sum_{c=1}^{z} \sum_{i=1, j=1}^{m, n} \left(\begin{array}{l} \left(\text{Query}_c \left(\text{SinkSN}_i, \text{SrcSN}_j \right) + \text{Reply}_c \left(\text{SrcSN}_j, \text{SinkSN}_i \right) \right) + \\ \left(\text{PruneMultiplePaths}_c \left(\text{SinkSN}_i, \text{SrcSN}_j \right) + \text{FormShortPath}_c \left(\text{SrcSN}_j, \text{SinkSN}_i \right) \right) \end{array} \right) \quad (12.5)$$

12.3.1.1.3 Directed Diffusion

In directed diffusion [9], the sink node sends control message that is distributed throughout the network. Source nodes transmit the data to the intermediate nodes, which are then forwarded to the sink node. Directed diffusion applies two-phase pull routing in which the sink node searches for the data source and decides the best optimal route path. The sequence of steps in directed diffusion is shown in Figure 12.2.

12.3.1.1.4 Push Diffusion

In push diffusion [10], the source node floods the query message and the sink nodes collect the data based on query–reply messages. Each query is attributed with a set of SNs per session. Consider query $q1$, where the sink node collects data from a set of source nodes $s(s1, s2, \ldots, sk)$, and query $q2$, where the intermediate node collects data from different set of source nodes $r(r1, r2, \ldots, rm)$. Here, the queries $q1$ and $q2$ are attributed with the sensor node set s and r respectively. Push diffusion is used for applications with multiple sink nodes and few source nodes. Sensor protocol for information through negotiation (SPIN) [10] uses push diffusion based on a set of negotiation messages (ADV, REQ, and DATA) between the source node, intermediate node, and the sink node.

12.3.2 Data Gathering Mechanism Based on Dissemination

The performance of data gathering techniques can be improved by using effective dissemination topology structures. The data dissemination protocols in sensor networks use chain, tree, cluster, grid, and hybrid topology structures. For static SNs, the network topology is fixed, and for node mobility, the network topology is varied. Data dissemination protocols [8] in sensor networks are based on the topological structures given in Table 12.2.

12.3.2.1 Data Dissemination Protocols in Wireless Sensor Networks

12.3.2.1.1 Two-Tier Data Dissemination

In two-tier data dissemination (TTDD) [15], the source node constructs the grid structure and transmits the data to multiple mobile nodes. When a sink node moves from one cell position to

Table 12.2   Characteristic Features of Network Topology used in Data Dissemination Protocols

Chain	To collect the data from a set of nodes in the route path. Power-efficient data gathering protocol for sensor information system (PEGASIS) [11] uses the chain structure based on greedy technique
Tree	Sink node constructs the tree with root as itself and the source node as the leaf node. Minimum spanning tree [12] in data dissemination protocols establish and maintain the energy-efficient route
Cluster	CH collects the data from SNs and forwards it to the BS. Hierarchical clustering [13] based on data dissemination uses mobile sink nodes
Grid	For multiple source nodes and multiple sink nodes, grid [14] structure is used to collect the sensor data from multiple alternate paths and establish a fault-tolerant network
Hybrid	Hybrid data dissemination combines the above topology structures based on application-specific and network-specific attributes

another position, a query message is flooded to the source node and the route paths are updated. The tier I or higher tier operates on the dissemination nodes of the grid and the tier II or lower tier operates on the current position of the sink node. TTDD between the souce node, SrcSN, and the sink node, SinkSN, based on p geographic positions *GPos* and adjacent sink node, SinkAdjSN, is given as follows:

$$\text{TTDD} = \underbrace{\sum_{GPos=1}^{p} \text{ReqQuery}(\text{SrcSN}_{GPos}, \text{SinkSN}_{GPos})}_{\text{Tier I}} \left(\underbrace{\sum_{i=1,j=1}^{n,m} \text{ReqQuery}(\text{SinkSN}^i_{GPos}, \text{SinkAdjSN}^j_{GPos})}_{\text{Tier II}} \right) + \tag{12.6}$$
$$\sum_{GPos=1}^{n} \text{DataDisseminate}(\text{SrcSN}_{GPos}, \text{SinkSN}_{GPos})$$

12.3.2.1.2 Scalable and Robust Data Dissemination

In scalable and robust data dissemination [16], the publish/subscribe model is based on the disjoint properties of space, time, and node synchronization. In robust data dissemination, the network is partitioned into several groups and region-specific routing is used in each group.

12.3.2.1.3 Dynamic Multiresolution Data Dissemination

In dynamic multiresolution data dissemination [17], minimum incremental dissemination tree for data dissemination request is based on lightweight trees. Real-time data dissemination is dependent on multiple temporal attributes and node attributes.

Data dissemination methodologies [18] in sensor networks use mobile sink nodes for sparse networks and mobile SNs.

12.4 Data Combining Techniques

Emerging research and development in the data management techniques indicate variations in the data fusion and the data aggregation protocols for sensor networks. This section explains the characteristic features of data fusion and data aggregation. Data fusion is used in applications with large magnitude data to perform complex queries and is based on compression functions like compress(Attribute1), compress(Attrribute[1-k]) and so on. Data aggregation is used in applications with less magnitude data to perform simple queries and is based on aggregation functions [7] like sum, count, average, minimum, maximum, and so on. Figure 12.3 gives an illustration of data fusion [19] used in sensor networks.

12.4.1 Data Fusion

The sensor data transmitted to the CH is forwarded to the neighboring CH and finally to the BS for data storage and query processing. Data fusion reduces the number of packet transmissions,

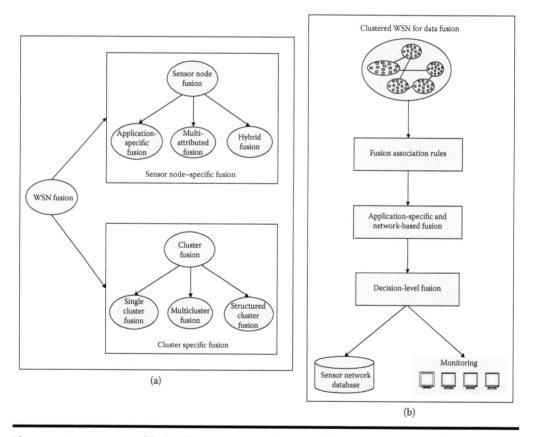

Figure 12.3 **(a) Types of fusion in sensor networks and (b) levels of sensor data fusion [19].**

minimize data redundancy, and thus saves the node energy and bandwidth [6]. The data fusion functions used in sensor networks are start fusion, increase fusion rate, decrease fusion rate, prioritize fusion rate, and stop fusion. Various levels of data fusion [20] used in sensor networks are given as follows:

- *Low-level data fusion*: Data collected from the SNs in primitive format is processed at CH or BS.
- *Middle-level data fusion*: Data collected from the SNs in template format (with attributes like distance, signal strength, node energy, temperature, pressure, etc.) is processed at CH or BS. Based on sensor application and node density, combination of the primitive data and the template data is used in middle-level data fusion.
- *High-level data fusion*: Data collected from the SNs is fused based on *decision functions* and resolved using the node density and network conditions. Soft computing techniques like artificial intelligence and fuzzy logic are used in high-level data fusion applications.
- *Multilevel fusion*: In wireless and ad hoc sensor networks, the data packets are transmitted from the source node to the sink node through CHs. The packet traversal depth is maintained at each CH. Single-level data fusion is constructed and maintained within each cluster and multilevel data fusion is constructed and maintained across various clusters.

12.4.1.1 Real-Time Data Fusion

Hard data fusion: Data fusion performed exactly at time *t* for real-time critical applications is known as *hard data fusion*. If the data fusion is done before time *t* or after time *t*, then the sensor application will not be able to give the desired response result. For real-time and event-based critical applications, hard data fusion is used.

Soft data fusion: Data fusion performed within the time range [*t*1–*t*2] for real-time applications is known as *soft data fusion*. If the data fusion is done before time *t*1 or after time *t*2, then the sensor application will not be able to give the desired response result. For time interval and event-based applications, soft data fusion is used.

12.4.1.1.1 Data Fusion Metrics in Sensor Networks

Compression: Combine and eliminate redundant sensor data.

Error free data: Parity bits are maintained in the data packet so that the received data can be verified and corrected if necessary. The number of parity bits depend on the protocol complexity. Distributed optimization [6] techniques are used to minimize the error rate in data fusion.

Node and network attributes: Data fusion is a function of node attributes (like adjacent node distance, signal strength, and battery power) and network attributes (like bandwidth, node density, topology, cluster size, transmitter, and receiver signal strength [21]).

Fusion performance parameters: Fusion accuracy and fusion latency primarily define the performance metrics in data fusion. Fusion accuracy is defined as the percentage of data accuracy in the fusion process, and fusion latency is defined as the time taken to fuse the sensor data from a set of nodes.

12.4.1.2 Data Fusion Attributes

12.4.1.2.1 Node Density-Based Data Fusion

Data fusion is a function of SN set, packet size, and fusion function. Data fusion to combine the sensor data from n SNs based on packet size, pktsize, is given as follows:

$$\mathrm{DFusion}_n\big((SN_1, SN_2, SN_3, \ldots, SN_n), \text{pktsize}, n\big) \qquad (12.7)$$

Data fusion to compress the sensor data from n SNs based on packet size, pktsize, and fusion function, $f_{compress}$, is given as follows:

$$\mathrm{DFusion}_{compress}\big((SN_1, SN_2, SN_3, \ldots, SN_n), \text{pktsize}, f_{compress}\big) \qquad (12.8)$$

12.4.1.2.2 Time-Based Data Fusion

Data fusion is a function of time (specific time or time range) and SN set. Data fusion to compress the sensor data from n SNs at time t based on packet size, pktsize, is given as follows:

$$\mathrm{DFusion}_{time(t)}\big((SN_1, SN_2, SN_3, \ldots, SN_n), \text{pktsize}, t\big) \qquad (12.9)$$

Data fusion to compress the sensor data from n SNs in the time interval range $[t1-t2]$ based on packet size, pktsize, is given as follows:

$$\mathrm{DFusion}_{time[t1-t2]}\big((SN_1, SN_2, SN_3, \ldots, SN_n), \text{pktsize}, t1, t2\big) \qquad (12.10)$$

12.4.1.2.3 Distance-Based Data Fusion

Data fusion is a function of coverage area and SN set. Data fusion to compress the sensor data from n SNs in the coverage area CArea is given as follows:

$$\mathrm{DFusion}_{coveragearea(a)}\big((SN_1, SN_2, SN_3, \ldots, SN_n), \text{pktsize}, \text{CArea}\big) \qquad (12.11)$$

12.4.1.2.4 Packet Arrival Rate-Based Data Fusion

Data fusion is a function of packet arrival rate and SN set. The packet arrival rate is defined by the hop count, hopcnt, and time traversal, t. Data fusion to compress the sensor data from n SNs at packet arrival rate of k bits per second and hop count, hopcnt, is given as follows:

$$\mathrm{DFusion}_{PktArrivalRate(k)}\big((SN_1, SN_2, SN_3, \ldots, SN_n), \text{PktArrivalRate}(k, \text{hopcnt}, t)\big) \qquad (12.12)$$

12.4.1.3 Data Fusion Protocols in Wireless Sensor Networks

Data fusion in WSNs is used to save the node energy, and achieve the desired reliability in network performance.

12.4.1.3.1 Adaptive Energy-Efficient Data Fusion

Data fusion results in significant energy consumption. The routing protocols use effective data fusion methods to decrease the network traffic. Adaptive data fusion for energy-efficient routing [22] uses Minimum Fusion Steiner Tree (MFST) that is based on the principle of intersection node and combine (fuse) the data packets that arrive at the intersection node. Adaptive Fusion Steiner Tree (AFST) performs the fusion task at the precise node instead of the intersection node to avoid network congestion.

12.4.1.3.2 Data Fusion with Desired Reliability

Data fusion based on in-network processing [23] uses minimum energy estimated from the reliable information calculated from redundant transmissions in the fusion path. This method uses optimal chain and tree topologies to maintain the desired reliability.

12.4.2 Data Aggregation

Data aggregation is the process of joining the data packets at SNs, intermediate SNs, or CHs in small-scale or medium-scale sensor networks. Data aggregation is used if the packet arrival rate is less than the threshold value. If the packet arrival rate is more than the threshold value, it leads to *data exploding* (packet loss and redundant transmissions due to network congestion). Therefore, solution would be to increase the buffer size or use additional buffers at the Data Aggregation Node. Data aggregation is dependent on the geographical attributes of the node. The preferred modes of data aggregation are as follows:

- Temporal—Per node
- Spatial—Per cluster

12.4.2.1 Data Aggregation Functions

Data aggregation techniques [11] in sensor networks are based on hierarchical and flat networks. WSNs use data aggregation functions [7] like COUNT, SUM, AVERAGE, MIN, MAX, and so on. Data aggregation functions use the parameter SNAttributes, which is a function of adjacent node size, aggregation time t, aggregation distance d, and packet arrival rate per second. Data aggregation functions used in small-scale and medium-scale sensor networks are given in Table 12.3.

Table 12.3 Data Aggregation Functions in WSNs

COUNT	fcnt($s1, s2, s3, ..., sn$, SNAttributes)	Total number of n SNs
SUM	fsum($s1, s2, s3, ..., sn$, SNAttributes)	Sum of n SN attributes
AVERAGE	favg($s1, s2, s3, ..., sn$, SNAttributes)	Average of n SN attributes
MAX	fmax($s1, s2, s3, ..., sn$, SNAttributes)	Maximum value from n SNs
MIN	fmin($s1, s2, s3, ..., sn$, SNAttributes)	Minimum value from n SNs

12.4.2.2 Data Aggregation Protocols in WSNs

12.4.2.2.1 Power-Efficient Data Gathering and Aggregation

The sink nodes or CHs perform data gathering and data aggregation based on the energy levels and hop count of the neighboring SNs. Data aggregated spanning tree is constructed with edge parameters like residual energy of SN, packet size, hop count, number of neighboring nodes, and the residual energy level of neighboring nodes [24].

12.4.2.2.2 Local Minimum Spanning Tree-Based Data Aggregation

Data aggregation process is based on the tree structure. Data aggregation tree is constructed using source-initiated or sink-initiated methods and maintained by the Data Aggregation Node or CH. Local Minimum Spanning Tree (LMST) [25] is constructed such that the edges with longer paths are minimized to reduce the network load.

12.4.2.2.3 Delay-Constrained Optimal Data Aggregation

Delay-Constrained Optimal Data Aggregation (DeCODA) uses hierarchical structure [26], where clustering and gateway nodes are based on signal strength and node distance. Architectural components of DeCODA [26] are DeCODA scheduler, channel management, buffer management, and data aggregation. DeCODA is executed periodically, when (1) the data packets reach the threshold or (2) the network delay is more in congestion-free traffic. DeCODA achieves energy efficiency based on the channel management and the aggregate packet threshold level.

Delay-efficient data aggregation [27] uses packet scheduling in distributed approach to generate a collision-free network. The packet scheduling is attributed with adjacent node distance and radius of the aggregator node. The data gathering techniques based on capacity and latency [28] are used in resource constrained applications to enhance the performance of WSNs.

12.4.2.2.4 Latency-Constrained Aggregation in Sensor Networks

Latency-Constrained Aggregation in Sensor Networks (LCASN) [29] consider the latency constraint to estimate the aggregation time, $t_{aggregation}$ and data arrival time, $t_{arrivaldata}$ at the sink node. Energy-efficient data aggregation based on hierarchy [30] uses a number of aggregators to minimize the node energy consumption and use a distributed network topology.

12.4.2.3 Comparison between Data Fusion and Data Aggregation

Even though few similarities exist between data fusion and data aggregation methods, the protocol design primarily differs in node density and query frequency. Data fusion is applied in large-scale sensor networks to fuse the data packets using compression functions like compress (Attribute1), compress (Attribute[1-k]), and so on. Data aggregation is applied in small-scale sensor networks to combine the data packets using aggregation functions like count, sum, average, compress, and so on. If a set of SNs generate data continuously, then data fusion is invoked periodically, else it would lead to *data exploding* (packet loss and redundant transmissions due to network congestion).

12.5 Sensor Network Database

Sensor generated data (for applications like weather forecast, air traffic controller, water level in irrigation dams, seismic conditions, chemical radiation levels, military field, medical and health care, vehicular traffic, retail market, etc.) are stored in the database for monitoring and query processing. The WSN design protocol components for sensor database applications [31] are in-network aggregation, data storage methodologies, and range queries. Table 12.4 specifies various types of sensor queries and their respective features.

12.5.1 Properties of Sensor Database

The properties of data-centric storage [31,32] for distributed sensor networks are essentially based on scalability and robustness. The primary attributes used in scalable sensor database are as follows:

Persistence: Data stored in the sensor database must be persistent irrespective of the changes in network topology or node failure.

Consistency and uniformity: Query must be sent to the CH in prescribed template format. If the CH is replaced by an alternate CH, the routing tables need to be updated to maintain a consistent database.

Query load balancing: The sensor data need to be stored uniformly across various CHs in the network. If a CH is loaded with multiple queries, then subset of queries are to be assigned to adjacent CHs to balance the network load. For sensor queries beyond a threshold limit $Query_{Th}$, additional CHs are added in the network.

Scalability: The number of SNs are proportional to the magnitude of query data base. Sensor database applications support node scalability to (1) increase the query processing rate and (2) minimize the query latency.

Table 12.4 Types of Queries [8] in the Sensor Network

Query Type	Feature
Previous state queries	Queries based on the data collected from previous history (time, $t_{previous}$; date, $date_{previous}$; attribute, $attribute_{previous}$)
Current state queries	Queries based on the data collected from current state (time, $t_{current}$; date, $date_{current}$; attribute, $attribute_{current}$)
Extensive queries	Queries based on the sensor data collected over a time period (tp–tq)
Regular interval queries	Queries based on the data collected at regular intervals (every 10 minutes in an hour, every 15 minutes in an hour, etc.)
Random queries	Queries based on the data using random and probabilistic based methods
Event-based queries	Queries based on the data using event triggers
Multivalue ranged queries	Queries based on the data collected from multiple node attributes

Secure data access: Encryption and decryption techniques are used to support secure the data access and secure query processing in sensor applications.

Concurrency: The sensor data is distributed across various CHs. Concurrency is used to support multiple queries across SNs, CHs, and BS.

12.5.2 Metrics Used in Sensor Database and Query Processing

12.5.2.1 Network Metrics

1. *Network bandwidth*: Maximum number of queries supported by the network.
2. *Node density*: Maximum number of nodes per query.
3. *Hot spot usage*: Number of alternate paths found in the sensor query when a node failure occurs. Hot spot usage affects the performance of network lifetime.
4. *Query throughput*: Average number of queries processed per unit time.

12.5.2.2 Sensor Node Metrics

1. *CPU processing time*: Processing time of the SNs to store and retrieve the data.
2. *Memory storage*: Memory needed to processes the sensor queries.
3. *Signal strength*: Transmitter and receiver signal strength to process the sensor queries.
4. *Mobility*: Speed of the SNs in data gathering and query processing protocols.
5. *Location*: Distance of the SN from CH.

12.5.2.3 Cluster Head Node and Storage Metrics

1. *CPU processing time*: Query processing time of the CH node.
2. *Memory storage*: Memory needed to process the sensor queries by CH.
3. *Index creation time*: Time taken to create the index table which is used in hashing.
4. *Storage space requirement*: Physical storage needed to save the sensor data.
5. *Update and maintenance cost*: Update the sensor database. Delete the redundant and unused long-standing data.

12.6 Data Storage Techniques Based on Storage Location

The sensor data aggregated or fused is stored in the database for query processing. The parameters considered in query processing are the number data attributes, size of each attribute, number of nodes, and node distance. Query rate defines the performance metric in query processing. As the query rate increases, it leads to energy consumption. Therefore, energy efficient sensor query protocols are used to balance the query load across various clusters in the network. The data is forwarded to the CH based on (1) the senor node identity and (2) the node General Packet Radio Service (GPRS) metrics. The data store is centralized or distributed based on the storage location. Data storage locations [8,32] for efficient query processing and data management are given as follows:

- External Storage
- Internal Storage
- Cluster Head Node-centric Storage

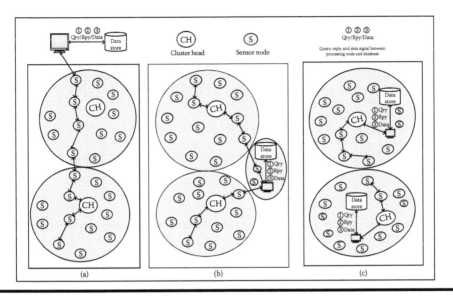

Figure 12.4 Data storage techniques in sensor networks. (a) External storage, (b) internal storage, and (c) cluster head–centric storage [8].

The data-centric storage protocols [32] in sensor networks are based on Geographic Hash Table (GHT). Figure 12.4 illustrates various types of data storage locations used in sensor networks.

12.6.1 Data Storage Location Used in Sensor Networks

12.6.1.1 External Storage

Data collected from the source nodes in the network are sent to the CH or sink node and further forwarded to the external data store [32]. The data is organized using single-dimensional and multi-dimensional attributes. Due to the limitations in sensor node energy, CHs and BS support features like store, retrieve, and query the sensor data. For a sensor network with n nodes and external storage, the query cost is based on events and query size [32]. Figure 12.4a gives an illustration of external data storage for query processing in sensor networks. Sensor query processing based on external data storage $DS_{External}$ for SN_i with j adjacent nodes and minimum hop, MinHop, is given as follows:

$$QueryDS_{External} = \left(\sum_{c=1}^{z} \left(\sum_{i=2}^{n} \sum_{j=1}^{n-1} Query_c(SN_i, SN_j, MinHop_{i,j}) \right) + Query_{CH_c} \right) +$$
$$Query_{External}(SN_j, CH_c, SNDataStore) \tag{12.13}$$

where $Query_c(SN_i, SN_j, MinHop_{i,j})$ is the sensor data query for c clusters between SN_i and SN_j, $Query_{CH_c}$ is the query by CH in cluster c, and $Query_{External}(SN_j, CH_c, SNDataStore)$ is the external query between SN, CH, and external data store. The total query time within a cluster is the weighted average of node distance from the ES. Equation 12.13 indicates the external data storage for SNs for *nonfaulty* query route paths. If a faulty node occurs in the query path, then an alternate route is found from the adjacent SNs with minimum hop count. Sensor query processing based on external data storage due to a faulty node k is given as follows:

$$
\begin{aligned}
&\text{QueryDS}_{\text{External}}\text{FaultyNode}_k = \\
&\text{QueryDS}_{\text{External}} + \text{FindAlternateRoute}\left(\text{SN}_k, \text{AdjSN}_k, \text{MinHop}_{k,\text{Adj}k}\right)
\end{aligned}
\tag{12.14}
$$

12.6.1.2 Internal Storage

The internal data store is within the coverage area of the sensor network and accessed through CHs or intermediate nodes. Internal storage [32] reduces the complexity in routing and query processing as compared to the external or CH-centric storage. Figure 12.4b gives an illustration of internal data storage for query processing in sensor networks. Sensor query processing based on internal data storage $\text{DS}_{\text{Internal}}$ for SN_i with j adjacent nodes, minimum hop, MinHop, and coverage area, CArea is given as follows:

$$
\begin{aligned}
\text{QueryDS}_{\text{Internal}} = {}&\left(\sum_{c=1}^{z}\left(\sum_{i=2}^{n}\sum_{j=1}^{n-1}\text{Query}_c\left(\text{SN}_i, \text{SN}_j, \text{MinHop}_{i,j}\right)\right) + \text{Query}_{\text{CH}_c}\right) + \\
&\text{Query}_{\text{Internal}}\left(\text{SN}_j, \text{CH}_c, \text{CArea}, \text{SNDataStore}\right)
\end{aligned}
\tag{12.15}
$$

The main limitation of internal storage is query flooding, when a set of nodes try to access the internal data store simultaneously. For a sensor network with n nodes and internal storage, the query cost is based on events and query size [32].

12.6.1.3 Cluster Head Node-Centric Storage

The data store is located within each cluster and managed by the CH [32]. For complex queries, CHs work in coordination and use query-controlled multicluster attributes. For a sensor network with n nodes and CH node-centric storage, the query cost is based on events and query size [32]. Figure 12.4c gives an illustration of CH data-centric storage for query processing in sensor networks. Sensor query processing based on CH data storage DS_{CH} for SN_i with j adjacent nodes and minimum hop, MinHop is given as follows:

$$
\text{QueryDS}_{\text{CH}} = \left(\sum_{c=1}^{z}\left(\sum_{i=2}^{n}\sum_{j=1}^{n-1}\text{Query}_c(\text{SN}_i, \text{SN}_j, \text{MinHop}_{i,j})\right) + \text{Query}_{\text{CH}_c}\right)
\tag{12.16}
$$

Query processing by the BS from the data collected from CH and data storage type (DSType: external or internal or CH node-centric) across c clusters is given as follows:

$$
\text{Query}_{\text{BS}} = \sum_{c=1}^{z}\text{Query}_{\text{CH}_c} + \text{Query}_{\text{DSType}}
\tag{12.17}
$$

12.6.2 Query Processing in Sensor Networks

Query processing [33] is the procedure of data retrieval from data store in the sensor network based on query attributes. Query processing is classified based on (1) centralized and (2) distributed data storage methods. In centralized data storage method, the sensor data is processed independent of the node and network attributes. This method is used in dense networks to query the historical data. The main limitation is, this method is suitable for static network conditions and not for dynamic network conditions. In distributed data storage method, the query data is based on SNs and CHs that are distributed across various clusters in the network. The CHs retrieve the data from SNs based on node attributes and time-event attributes. This method is used in sparse and dense networks. The main limitation of this method is query latency, which increases with node density and distributed data storage methods. The query processing (with single-valued and multivalued attributes) is efficient and faster in centralized storage methods as compared to distributed storage methods. SNs distributed across geographical regions use spatiotemporal attributes in query processing. The classification of distributed query processing [34] in sensor networks is based on in-network processing, data acquisition processing, data-centric processing and cross-layer data query processing.

12.6.2.1 Query Processing Protocols in Wireless Sensor Networks

12.6.2.1.1 Dynamic Multiroot and Multiquery Processing

Dynamic multiroot and multiquery processing [35] is based on the naive, static, and heuristic multiqueue models. Naive, a primitive query model, is used for small-scale sensor networks. The static queue model is used when the query regions are known in advance. Heuristic queue model is used for large-scale queries and overlap queries.

12.6.2.1.2 Energy-Efficient Query Processing

Energy-efficient query processing protocols save the node energy and extend the network life cycle. The data query with application specific attributes and accuracy constraints [36] minimize the node energy consumption as compared to the data query with generic attributes and range-based constraints.

12.7 Metrics in Data Gathering and Data Management

Data gathering process in sensor networks using approximation techniques and probabilistic methods [37] is given as follows.

12.7.1 Size of Source Nodes and Sink Nodes

Sink and source nodes in the sensor network are classified as follows:

- One sink – one source
- One sink – many sources
- Many sinks – one source
- Many sinks – many sources

One sink–one source: Body sensor application for a patient is an example of one sink – one source. Sensor devices are attached to the patient's body and the data is transmitted to the remote hospital for monitoring the patient's condition.

One sink–many sources: Weather forecast attributed with parameters like heat, humidity, air pressure, wind direction, and so on is an example of one sink – many sources. Sensors in the weather application send multiple monitored data to the remote meteorological work station.

Many sinks–one source: Vehicular traffic is an example of many sinks – one source. One source traffic light controls and directs many sink vehicles in each direction. The dimension of multiple sink traffic can be two-way traffic, three-way traffic, four-way traffic, and so on.

Many sinks–many sources: Satellite and mobile communications, GPRS applications, and so on are the examples of many sinks – many sources.

12.7.2 Energy and Bandwidth

To save the node energy, data is transmitted at a slower rate and additional CHs are placed in the neighborhood to save the routing time. An approximate aggregation technique [38] for sensor databases gives an illustration to save the node energy. CHs and SNs use channel access techniques like time division multiple access (TDMA) and frequency division multiple access (FDMA) to transmit the data in network congestion paths and optimize the network bandwidth.

12.7.3 Accuracy and Latency

Network density and node distance are the main performance attributes in fusion/aggregation accuracy. For dense and distributed networks, the data is partitioned into multiple sets and fusion accuracy is calculated using parity bits in each data, which is similar to error detection process. For a network with relatively less denser nodes, the aggregation accuracy is calculated using parity bits. The trade-off between static and dynamic query attributes in WSN is maintained by using fusion/aggregation accuracy methods.

The amount of time taken to join the sensor data attributes using data fusion/data aggregation protocols by CH or DAN is known as fusion/aggregation latency. Local-based minimum spanning tree and shortest path algorithms are used to minimize the latency and increase the throughput rate.

12.8 Conclusions

Data gathering and data management techniques are extensively used in sensor networks to aggregate the data and provide query processing services. This chapter elucidates various phases of data gathering and data management techniques in WSNs. Data gathering based on data diffusion and data dissemination protocols are explained in detail. Data management based on data fusion and data aggregation protocols are given for sensor networks. Sensor network protocols use (1) data gathering–based data diffusion to collect the data, and data fusion methods to combine the data and (2) data gathering–based data dissemination to collect the data, and data aggregation methods to combine the data. Data gathering and data management protocols are implemented in the upper layer of the network protocol stack in sensor networks. The data flow sequence explained in Section 12.2 illustrates the flow of sensor data based on diffusion and dissemination methods. The data is further aggregated or fused and stored in the database for query processing. The data storage locations for efficient query processing in sensor networks is based on external storage, internal storage and cluster node centric storage methods. The sequence of process for data management in WSNs is (1) data gathering, to collect the data from SNs; (2) data fusion/data aggregation, to organize and combine the data; (3) data store, to save the data attributes in database; and (4) query processing, to retrieve and update the sensor data attributes.

12.9 Future Research and Challenges

Data gathering and data management protocols based on energy-efficient and security attributes for event-based and time-based critical applications in WSNs need to be explored and resolved. Scheduling and priority methods need to be incorporated in data gathering protocols for real-time applications. Secure mobile-based data gathering protocols need to be enhanced to support the features like network density and varying topology structure. Robust and scalable data gathering protocols based on load balancing (network and node) and fault-tolerant attributes are yet to be developed. Data gathering applications based on IPv6 over low-power wireless personal area networks (6LoWPAN) and node identity need to be designed and integrated with wireless local area network (WLAN) and wireless metropolitan area network (WMAN). Sensor query processing with mobile-based data store need to be explored for sensor applications.

References

1. Akyildiz, I. F., and M. C. Vuran. 2010. *Wireless Sensor Networks, Series in Communication and Networking.* Chichester, West Sussex, UK: John Wiley & Sons Ltd.
2. Jinbao, L., C. Zhipeng, and L. Jianzhong. 2008. "Chapter 12: Data Management in Sensor Networks." In *Wireless Sensor Networks and Applications, Springer Series on Signals and Communication Technology,* edited by L. Yingshu, M. T. Thai, and W. Weili. New York: Springer Science+Business Media.
3. Waltenegus, D., and P. Christian. 2010. *Sensor Networks: Where Theory Meets Practice, Springer Series on Signal and Communication Technology.* Berlin, Heidelberg: Springer-Verlag.
4. Heidemann, J., F. Silva, and D. Estrin. 5–7 November 2003. "Matching Data Dissemination Algorithms to Application Requirements." In *Proceedings of ACM First International Conference on Embedded Networked Sensor Systems (Sensys),* 218–229. Los Angeles, CA.
5. Avinash, S., T. Joshua, W. Jie, and C. Mihaela. 2009. "Reputation-and-Trust-Based Systems for Ad Hoc Networks." In *Algorithms and Protocols for Wireless and Mobile Ad Hoc Networks, Wiley Series on Parallel and Distributed Computing,* edited by A. Boukerche. Hoboken, NJ: John Wiley & Sons Inc.
6. Rabbat, M., and R. Nowak. 26–27 April 2004. "Distributed Optimization in Sensor Networks." In *Proceedings of IEEE Third International Symposium on Information Processing in Sensor Networks (IPSN),* 20–27. Berkeley, CA.
7. Jen-Yen, C., P. Gopal, and X. Dongyan. September 2006. "Robust Computation of Aggregates in Wireless Sensor Networks: Distributed Randomized Algorithms and Analysis." *IEEE Transactions on Parallel and Distributed Systems* 17 (9): 987–1000.
8. Wei-Peng, C., and C. H. Jennifer. 2005. "Chapter 15: Data Gathering and Fusion in Sensor Networks." In *Handbook of Sensor Networks: Algorithms and Architectures,* edited by I. Stojmenovic, Hoboken, NJ: John Wiley & Sons Inc.
9. Intanagonwiwat, C., R. Govindan, D. Estrin, J. Heidemann, and F. Silva. February 2003. "Directed Diffusion for Wireless Sensor Networks." *IEEE/ACM Transactions on Networking* 11 (1): 2–16.
10. Kulik, J., W. Heinzelman, and H. Balakrishnan. March–May 2002. "Negotiation-Based Protocols for Disseminating Information in Wireless Sensor Networks." *Springer Wireless Networks* 8 (2/3): 169–185.

11. Rajagopalan, R., and P. K. Varshney. Fourth Quarter 2006. "Data Aggregation Techniques in Sensor Networks: A Survey." *IEEE Communications Surveys & Tutorials* 8 (4): 48–63.

12. Tan, H. O., I. Korpeoglu, and I. Stojmenovic. 21–23 May 2007. "A Distributed and Dynamic Data Gathering Protocol for Sensor Networks." In *Proceedings of IEEE Twenty First International Conference on Advanced Information Networking and Applications (AINA)*, 220–227. Niagara Falls, Canada.

13. Ching-Ju, L., C. Po-Lin, and C. Cheng-Fu. 3–6 July 2006. "HCDD: Hierarchical Cluster Based Data Dissemination in Wireless Sensor Networks with Mobile Sink." In *Proceedings of ACM International Conference on Wireless Communications and Mobile Computing (IWCMC)*, 1189–1194. Vancouver, British Columbia, Canada.

14. Sharma, T. P., R. C. Joshi, and M. Misra. 14–17 December 2008. "GBDD: Grid Based Data Dissemination in Wireless Sensor Networks." In *Proceedings of IEEE Sixteenth International Conference Advanced Computing and Communications (ADCOM)*, 234–240. MIT Campus, Anna University, Chennai, India.

15. Haiyun, L., Y. Fan, C. Jerry, L. Songwu, and Z. Lixia. January 2005. "TTDD: Two-Tier Data Dissemination in Large-Scale Wireless Sensor Networks." *Springer Wireless Networks* 11 (1–2): 167–175.

16. Soochang, P., L. Euisin, Y. Fucai, and K. Sang-Ha. August 2010. "Scalable and Robust Data Dissemination for Large-Scale Wireless Sensor Networks." *IEEE Transactions on Consumer Electronics* 56 (3): 1616–1624.

17. Guoliang, X., L. Minming, L. Hongbo, and J. Xiaohua. September 2009. "Dynamic Multi-Resolution Data Dissemination in Wireless Sensor Networks." *IEEE Transactions on Mobile Computing* 8 (9): 1205–1220.

18. Hamida, E. B., and G. Chelius. December 2008. "Strategies for Data Dissemination to Mobile Sinks in Wireless Sensor Networks." *IEEE Wireless Communications* 15 (6): 31–37.

19. Guang-Zhong, Y., and H. Xiaopeng. 2006. "Chapter 8: Multi-Sensor Fusion." In *Body Sensor Networks*, edited by Y. Guang-Zhong, New York: Springer-Verlag London Limited; Springer Science+Business Media.

20. Nakamura, E. F., A. F. Loureiro, and A. C. Frery. September 2007. "Information Fusion for Wireless Sensor Networks: Methods, Models, and Classifications." *ACM Computing Surveys (CSUR)* 39 (3): 1–55.

21. Hongli, X., H. Liusheng, Z. Yindong, H. He, J. Shenglong, and L. Gang. September 2010. "Energy-Efficient Cooperative Data Aggregation for Wireless Sensor Networks." *Elsevier Journal of Parallel and Distributed Computing* 70 (9): 953–961.

22. Hong, L., L. Jun, L. Yonghe, and K. D. Sajal. October 2006. "Adaptive Data Fusion for Energy Efficient Routing in Wireless Sensor Networks." *IEEE Transactions on Computers* 55 (10): 1286–1299.

23. Hong, L., T. Huixiang, M. Huadong, and K. D. Sajal. March 2011. "Data Fusion with Desired Reliability in Wireless Sensor Networks." *IEEE Transactions on Parallel and Distributed Systems* 22 (3): 501–513.

24. Tan, H. O., I. Korpeoglu, and I. Stojmenovic. March 2011. "Computing Localized Power-Efficient Data Aggregation Trees for Sensor Networks" *IEEE Transactions on Parallel and Distributed Systems* 22 (3): 489–500.

25. Ovalle-Martinez, F. J., I. Stojmenovic, F. Garcia-Nocetti, and J. Solano-Gonzalez. February 2005. "Finding Minimum Transmission Radii and Constructing Minimal Spanning Trees in Ad Hoc and Sensor Networks." *Elsevier Journal of Parallel and Distributed Computing* 65 (2): 132–141.

26. Kui, W., L. Chong, Y. Xiao, and L. Jiangchuan. October 2009. "Delay-Constrained Optimal Data Aggregation in Hierarchical Wireless Sensor Networks." *Springer Mobile Network Applications* 14 (5): 571–589.

27. Xiang-Yang, L., M. Xufei, T. Shaojie, and W. Shiguang. January 2011. "A Delay-Efficient Algorithm for Data Aggregation in Multihop Wireless Sensor Networks." *IEEE Transactions on Parallel and Distributed Systems* 22 (1): 163–175.

28. Santi, P. September 2010. "On the Data Gathering Capacity and Latency in Wireless Sensor Networks." *IEEE Journal on Selected Areas in Communications* 28 (7): 1211–1221.

29. Becchetti, L., P. Korteweg, A. Marchetti-Spaccamela, M. Skutella, L. Stougie, and A. Vitaletti. 11–13 September 2006. "Latency Constrained Aggregation in Sensor Networks." In *Proceedings of Springer Fourteenth Annual European Symposium on Algorithms (Algorithms ESA) (LNCS 4168)*, 88–99. ETH Zürich, Zürich, Switzerland.

30. Chen, Y. P., A. L. Liestman, and J. Liu. May 2006. "A Hierarchical Energy-Efficient Framework for Data Aggregation in Wireless Sensor Networks." *IEEE Transactions on Vehicular Technology* 55 (3): 789–796.

31. Feng, Z., and G. Leonidas. 2004. "Chapter 6: Sensor Network Database." In *Wireless Sensor Networks: An Information Processing Approach*, edited by F. Zhao and L. Guibas, New Delhi, India: Elsevier Inc.; Morgan Kaufmann Publishers; Reed Elsevier India Private Limited.

32. Ratnasamy, S., B. Karp, S. Shenker, D. Estrin, R. Govindan, L. Yin, and F. Yu. August 2003. "Data-Centric Storage in Sensornets with GHT, A Geographic Hash Table." *Springer Mobile Networks and Applications* 8 (4): 427–442.

33. Banerjee, T., and D. P. Agrawal. 2009. "Chapter 7: Query Processing and Data Aggregation." In *Wireless Sensor Networks: A Networking Perspective*, edited by J. Zheng and A. Jamalipour, 215–240. Hoboken, NJ: IEEE and John Wiley & Sons Inc.

34. Chatterjea, S., and P. Havinga. July 2007. "A Taxonomy of Distributed Query Management Techniques for Wireless Sensor Networks." *Wiley International Journal of Communication Systems* 20 (7): 889–908.

35. Zhiguo, Z., D. K. Ajay, and S. M. Shatz. June 2010. "Dynamic Multiroot Multiquery Processing Based on Data Sharing in Sensor Networks." *ACM Transactions on Sensor Networks* 6 (3): Article 25, 1–38.

36. Jun-Zhao, S. 23–25 June 2008. "An Energy-Efficient Query Processing Algorithm for Wireless Sensor Networks." In *Proceedings of Springer Fifth International Conference on Ubiquitous Intelligence and Computing (UIC) (LNCS 5061)*, 373–385. Oslo Norway: Oslo University College.

37. Chu, D., A. Deshpande, J. Hellerstein, and W. Hong. 3–7 April 2006. "Approximate Data Collection in Sensor Networks Using Probabilistic Models." In *Proceedings of IEEE International Conference on Data Engineering (ICDE)*, 1–12. Atlanta, GA.

38. Jeffrey, C., H. Marios, L. Feifei, B. John, and K. George. April 2009. "Robust Approximate Aggregation in Sensor Data Management Systems." *ACM Transactions on Database Systems (TODS)* 34 (1): Article 6, 1–35.

Chapter 13

Data Gathering Algorithms for Wireless Sensor Networks

Natarajan Meghanathan

Contents

13.1 Introduction

A wireless sensor network encompasses several smart sensor nodes that can gather data about the surrounding environment and process them before propagating to a control center called the sink, typically located far away from the network field. End users of the application access the network nodes and the gathered data through the sink. The sensor nodes are constrained with limited battery charge, memory, and processing capability as well as operate under a limited transmission range. Two sensor nodes that are outside the transmission range of each other cannot directly communicate. A wireless sensor network also has a limited bandwidth, and the communication medium is shared by the sensor nodes that are in the transmission range of each other. Due to all the above resource and operating constraints, it will not be a viable solution to require every sensor node to directly transmit their data to the sink over a longer distance. Also, if several signals are transmitted at the same time over a longer distance, it could lead to lot interference and collisions. Thus, there is a need for employing energy-efficient data gathering algorithms that can effectively combine the data collected at these sensor nodes and send only the aggregated data (that is a representative of the entire network) to the sink.

In this chapter, we consider the problem of periodically gathering the data from the sensor nodes in the network. Accordingly, the data gathering algorithms operate in several rounds, and during each round, data from the sensor nodes are collected, aggregated, and forwarded to the sink through a communication topology determined on an underlying network graph based on the transmission range of the sensor nodes. The data gathering algorithms differ in the communication topology used for data aggregation and forwarding: clusters [1], chain [2], grid [3], connected dominating set (CDS), [4] and spanning trees [5]. The data gathering algorithms target to optimize one or more of these performance metrics: (1) node lifetime, (2) delay per round, (3) energy per round, and (4) energy*delay per round. In this chapter, the node lifetime will be referred to as the number of rounds the network can run before sustaining the first node failure due to exhaustion of battery charge; this definition for node lifetime is also sometimes interpreted as the network lifetime in the literature. Throughout the chapter, the terms "node" and "vertex," "link" and "edge," and "gathering" and "aggregation" are used interchangeably. They mean the same. The rest of the chapter is organized as follows:

- Section 13.2 explains the system model and the performance metrics used in the description and analysis of the different data gathering algorithms presented in this chapter.
- Section 13.3 explains the cluster-based LEACH (Low-Energy Adaptive Clustering Hierarchy) data gathering algorithm [1], which still serves as one of the two standards (the other one being the PEGASIS algorithm discussed in Section 13.4) to which the performance of any data gathering algorithm newly proposed in the literature is compared to.
- Section 13.4 describes the chain-based PEGASIS (Power-Efficient Gathering in Sensor Information Systems) algorithm [2] and analyzes the different procedures available in the literature to form the chain [6] as well as select the leader node [7] and their impact on the performance of PEGASIS.
- Section 13.5 presents two grid block energy based hierarchical data gathering (GBE-DG) algorithms [3] that differ from each other depending on whether the nodes within a grid are arranged as a cluster or as a chain.
- Section 13.6 presents an algorithm to construct an energy-aware connected dominating set (ECDS)-based data gathering tree [4] that prefers to include nodes with higher energy level as the intermediate aggregating nodes of the tree.

- Section 13.7 presents a spanning-tree-based energy-aware maximal leaf nodes data gathering (EMLN-DG) algorithm [5] that prefers to include only nodes that have a relatively larger number of uncovered neighbors as well as a higher energy level as the intermediate nodes of the tree and in turn maximizes the number of leaf nodes of the data gathering tree. This section also presents an algorithm to calculate the delay per round of data gathering, which is also applicable to the ECDS-based data gathering algorithm described in Section 13.6.
- Section 13.8 presents the individual and comparative performance study of the above data gathering algorithms through extensive simulation studies with respect to metrics such as the node lifetime, delay per round, energy lost per round, and the energy*delay per round.
- Section 13.9 presents the future work, and Section 13.10 presents the conclusions.

13.2 System Model and Performance Metrics

The system model that is common to the different data gathering algorithms presented in this chapter can be summarized as follows:

- The underlying network graph considered in the construction of the communication topology used for data gathering is a unit disk graph wherein there exists a link between any two nodes if and only if the physical distance between the two end nodes of the link is less than or equal to R, the default transmission range that is the same for every sensor node in the network.
- Sensor nodes do transmission power control. In other words, a sensor node can adjust its transmission power according to the distance to the receiver node; this helps a sensor node to reach out to nodes that may not lie within its default transmission range, thus enhancing the network connectivity. Some data gathering algorithms exclusively work based on this assumption, while other algorithms require a node to transmit data to nodes outside of its transmission range only if needed and by default, operate the nodes with a limited transmission range.
- Sensor nodes are either only TDMA (time division multiple access) enabled or CDMA (code division multiple access) enabled or sometimes both TDMA and CDMA enabled, depending on the assumptions of the individual algorithms. With TDMA-only systems, data can move from one node to another, one node at a time, whereas in CDMA-based systems [8], each node is assigned a unique CDMA code that is also known to every other node so that there can be simultaneous communication between any two nodes in the network. For systems that assume sensor nodes to be both TDMA and CDMA enabled, data gathering in the CDS- and tree-based algorithms proceeds as follows: Every upstream node broadcasts a time schedule (for data transmission) to its immediate downstream nodes; a downstream node transmits its data to the upstream node according to this schedule. Such a TDMA-based communication between every upstream node and its immediate downstream child nodes can occur in parallel, with each upstream node using a unique CDMA code.
- We assume the size of the aggregated data packet is the same as the size of the individual data packets sent by the sensor nodes. In other words, aggregation at any node does not result in increase in the size of the data packets transmitted from the sensor nodes toward the sink. Data aggregation helps to reduce the uncorrelated noise in several signals and produce a more accurate signal that is considered to be a representative of the network condition and is transmitted to the sink.

The performance metrics evaluated for and sometimes optimized by the different data gathering algorithms are as follows:

■ Energy lost per round: This is the sum of the energy lost at all the nodes for the transmission, reception, and fusion of the data, averaged across several rounds of data gathering.
■ Delay per round: This is measured in terms of the number of time units per round of data aggregation and transmission to the sink, averaged across several rounds of data gathering.
■ Energy*delay per round: This is simply the product of the average energy lost per round and the average delay incurred per round of data aggregation and transmission to the sink.
■ Node lifetime: This is the round of first node failure due to the exhaustion of energy at a node.

13.3 Cluster-Based Data Gathering

In this section, we discuss a well-known clustering-based data gathering algorithm, abbreviated as LEACH. The LEACH algorithm [1] considers the network to be divided into multiple clusters with a clusterhead for each. A sensor node directly transmits its data to the closest clusterhead. A clusterhead gathers data from the sensor nodes in its cluster, aggregates the collected data, and directly transmits to the sink. To ensure fairness with regard to energy usage, clusterheads are rotated for every round of data gathering. The optimal number of clusterheads that will maximize the node lifetime and reduce the overall energy consumption depends on several parameters, including the network topology and the relative costs of computation and communication. Note that having 0 clusterheads or 100% clusterheads means the same and both scenarios correspond to having every sensor node directly communicating with the sink (Figure 13.1).

The decision of a node to become a clusterhead for a round depends on (1) the percentage P of clusterheads for the network (an input parameter to the LEACH algorithm) and (2) whether the node has been a clusterhead in the last $1/P$ rounds. If r is the round number and G is the set of nodes that have not been clusterheads in the last $1/P$ rounds, the probability of a node $n \in G$ to become a clusterhead is given by

$$T(n) = \frac{P}{1 - P * \left(r \bmod \frac{1}{P} \right)} \tag{13.1}$$

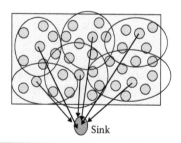

Figure 13.1   LEACH Clusters.

To decide whether or not to become a clusterhead for a round, each node n generates a random number between 0 and 1 and if the number is less than T(n), then the node becomes the clusterhaed for the round. Under this model, a node will become a clusterhead for some round within the last $1/P$ rounds. After the clusterheads are selected, every non-clusterhead node associates itself to the cluster of a clusterhead that is closest to it. A clusterhead then broadcasts a time schedule to the member nodes within its cluster; a member node uses the time slot assigned to it to transmit its data to the clusterhead. Each clusterhead uses a unique CDMA code for communication within its cluster; thus, there can be simultaneous data gathering within each cluster. A clusterhead aggregates the data collected from all its member nodes with its own data, if any, and then directly transmits the aggregated data to the sink using the unique CDMA code assigned to it so that there will not be much interference with transmissions from other clusterheads.

The delay incurred for data gathering using LEACH is computed as follows: Since data gathering within a cluster is TDMA based, the number of time units it takes to complete data gathering within a cluster is simply the number of member nodes for the cluster. The delay incurred for data gathering for the whole network is then one plus the maximum of the number of time units encountered in the clusters. The "plus one" is to account for the transmission from the clusterhead to the sink.

It is difficult to find an optimal number of clusterheads to simultaneously minimize the energy consumption and delay in data gathering. If we choose more clusterheads, there could be fewer transmissions within a cluster and with each cluster operating using a unique CDMA code, the overall delay might be lower. But several clusterheads would be transferring the locally aggregated data over long distances to the sink. On the other hand, if we operate with fewer clusters, the regular sensor nodes end up spending more energy in transmitting their data to a distantly located clusterhead and owing to a larger cluster size, there will be more delay incurred within each cluster to gather data from all the member nodes. The above design problem and the lack of a concrete model to select clusterheads based on the available energy level at the nodes have been the motivation for the subsequent development of more energy-efficient data gathering algorithms. Nevertheless, LEACH is one of the popular and still commonly used data gathering algorithms for performance comparison studies.

13.4 Chain-Based Data Gathering

In this section, we describe the well-known PEGASIS algorithm [2] that uses a network-wide chain of sensor nodes for data gathering. The chain of sensor nodes is constructed in a greedy fashion, starting from the node farthest to the sink; the node that is closest to the last added node to the chain is the next node included in the chain. This procedure is repeated until the chain comprises all the sensor nodes. Note that a node cannot be located at more than one position in the chain. PEGASIS uses the same chain for every round of data gathering.

For every round of data gathering, a leader node is randomly chosen among the nodes in the chain. The sink could select the leader node and notify it to all the nodes in the network. The leader node is responsible for aggregating data from both the ends of the chain and forwarding the data to the sink. Data aggregation starts from the two ends of the chain, with each node (other than the two end nodes and the leader node) receiving and transmitting the aggregated data to the next node in the chain, until the data reaches the leader node. The leader node combines the data received from both sides of the chain with its own data and transmits the

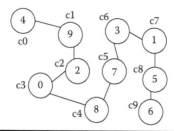

Figure 13.2 PEGASIS Chain.

aggregated data to the sink node. The working of the chain-based data gathering in PEGASIS is illustrated for a 10-node network in Figure 13.2. Note that the position of a node in the chain (called the chain index) is different from the node ID. In Figure 13.2, node 3 with chain index 6 (i.e., c6) is selected as the leader node; as a result, the flow of data would be in the following order: $c0 \rightarrow c1 \rightarrow c2 \rightarrow c3 \rightarrow c4 \rightarrow c5 \rightarrow c6 \leftarrow c7 \leftarrow c8 \leftarrow c9$. Note that a node has to wait for data to arrive from one of the two ends of the chain to aggregate and further transmit its own data. The leader node will have to wait for data to arrive from both sides of the chain and with one transmission per time unit (TDMA), the delay incurred with the original version of PEGASIS (hereafter, referred to as PEGASIS-TDMA) would be significantly high and actually equal to the number of nodes in the network.

13.4.1 PEGASIS for CDMA Systems

With PEGASIS-TDMA, data have to propagate through the complete chain of sensor nodes, in a linear fashion, before getting transmitted to the sink. For CDMA systems, the delay incurred with PEGASIS could be optimized by letting the data to propagate through the chain in a binary hierarchical fashion [9] wherein a round of data gathering comprises log N levels (N corresponds to the number of nodes in the network). Each node transmits the data to a nearby node in a given level of the hierarchy; only the nodes that receive the data rise to the next level. The leader node will be the only node at the topmost level and it transmits the aggregated data to the sink. At a particular level of the hierarchy, there can be simultaneous aggregation of data between as many pairs of nodes as possible, thus minimizing the delay as well as the energy*delay per round. Figure 13.3 illustrates an example for data aggregation at different levels of a 10-node chain with the node at chain index 3 serving as the leader node.

13.4.2 Use of Tree Traversal Algorithms for PEGASIS Chain Formation

The original version of PEGASIS uses the network-wide chain formed using the greedy distance-based heuristic. Greedy strategies rarely yield optimal solution and in our simulation studies involving PEGASIS (e.g., [6]), we observe that the distance between consecutive nodes in the chain gets larger as we progress away from the starting node of the chain, especially in the second half of the chain. This motivated us to consider alternate strategies to form the chain. In [6], we showed that the network-wide chain could be constructed using the traditional graph theory tree

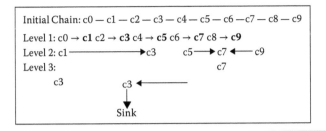

Figure 13.3   PEGASIS Chain data aggregation for CDMA systems.

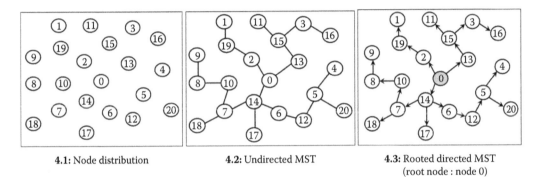

| 4.1: Node distribution | 4.2: Undirected MST | 4.3: Rooted directed MST (root node : node 0) |

Pre-order chain: [0, 2, 19, 1, 13, 15, 3, 16, 11, 14, 6, 12, 5, 4, 20, 7, 10, 8, 9, 18, 17]

In-order chain: [2, 19, 1, 0, 13, 3, 16, 15, 11, 6, 12, 4, 5, 20, 14, 10, 8, 9, 7, 18, 17]

Post-order chain: [1, 19, 2, 16, 3, 11, 15, 13, 4, 20, 5, 12, 6, 9, 8, 10, 18, 7, 17, 14, 0]

Figure 13.4   Example to illustrate the generation of node chain using the tree traversal algorithms.

traversal algorithms (such as the pre-order, in-order, and post-order traversals) [10]. The procedure to generate the chain (refer to Figure 13.4 for an example) is as follows: We first form an undirected minimum-weight spanning tree (*ud-MST*) of the entire sensor network with the objective of minimizing the sum of the Euclidean distances of the tree edges. We then transform the *ud-MST* to a rooted directed minimum-weight spanning tree (*rd-MST*) by conducting a Breadth First Search on the *ud-MST*, starting from the node located closest to the center of the network. The tree traversal algorithms can now be executed on the *rd-MST*, and the node sequence resulting from each of the traversals is used as the chain of nodes for the PEGASIS algorithm. Simulation results [6] illustrate that the node lifetime increases as large as by 30% (for PEGASIS-TDMA) and 19% (for PEGASIS-CDMA) as well as the energy consumed per round reduces by 35% (for PEGASIS-TDMA) and 19% (for PEGASIS-CDMA) compared to the greedy distance-based heuristic, with the post-order and the pre-order tree traversal strategies appearing, respectively, to be the most and least effective.

13.4.3 Impact of Leader Node Selection and Network Topology on the Performance of PEGASIS

In [7], we evaluated the impact of the strategy used to select a leader node (of the chain for data aggregation per round) and the shape of the network topology on the performance of the original PEGASIS algorithm. We investigated the following leader node selection strategies: random (leader of a round is selected randomly), shuffle (a node is selected as leader only once in N rounds in a network of N nodes), high energy (the node with the highest energy is selected as the leader of the round), and 2-block and 4-block (the network is divided into two blocks or four blocks, and for each round a random block is selected; the node with the highest energy in the selected block is the leader of the round).

To maximize the node lifetime, we observed the following: For any topology, when the sink is located outside the network field, the energy-aware leader selection strategies should be used. For the rectangular topologies and any sink location, the energy-aware selection should be done for the leader node. For square and circular topologies, when the sink is located at the center of the network field, the random and shuffle strategies are preferable, whereas we can opt for any leader node selection strategy if the sink is located at the origin of the network field. To minimize the energy*delay per round, we observe the following: For rectangular topologies, the energy-aware leader selection strategies (especially the 2-block and 4-block strategies) are preferable. For square and circular network topologies, we can opt for any leader node selection strategy.

13.5 Grid Block Energy-Based Data Gathering

In this section, we present two GBE-DG algorithms [3] that differ from each other depending on whether the nodes within a grid are arranged as a cluster or as a chain. The entire sensor network is divided into grid blocks of equal size. The following computations/selections are executed for every round of data gathering: The energy level of a grid block is the sum of the available energy levels of all the sensor nodes within the grid block. The grid block that has the maximum energy level is called the leader grid block (LGB), and the sensor node having the largest available energy in the LGB is called the global cluster leader (GCL) serving as the root of the data gathering tree for the round. For every grid block, the sensor node having the largest available energy is called the local cluster leader (LCL). For the LGB, the GCL and LCL roles are assigned to the same sensor node. The LCL nodes (except the GCL node) form the intermediate nodes, and the non-LCL nodes within each grid block form the leaf nodes of the data gathering tree. The leaf nodes within a grid block either directly transmit their data to the LCL node according to a time schedule (GBE-Cluster-DG-tree) like in a LEACH cluster or form a grid block-wide chain of sensor nodes involving the LCL node as the leader of the chain (GBE-Chain-DG-tree) as in PEGASIS.

For the GBE-Chain-DG-tree, the grid block-wide chain is constructed using the same greedy distance-based heuristic of the original PEGASIS algorithm; the only differences are that the leader node (i.e., the LCL node) is preselected before the chain construction and only the nodes within a grid block (not network wide) are part of the chain. Data aggregation starts simultaneously from either ends of the chain and proceeds toward the leader LCL node. We assume that the data aggregation occurring to the left of the leader node will not interfere with that occurring at the right of the leader node. Such an assumption is realistic as a non-LCL node will only be receiving data from a node to its left (or right) in the chain, aggregate with its own data, and then transmit the aggregated data to a node to its right (or left) in the chain. To avoid any collisions,

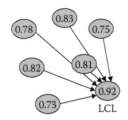

Figure 13.5 GBE-Cluster-DG for sample grid block.

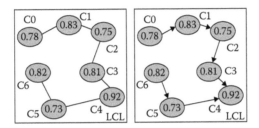

Figure 13.6 GBE-Chain-DG: Chain formation in a sample grid block and data aggregation toward the LCL node.

if the nodes to the immediate left and right of the LCL node are ready with their aggregated data and could transmit them simultaneously, the LCL node sends them a time schedule to serialize their transmissions in a sequential order.

In the case of the GBE-Cluster-DG-tree, for each grid block, the LCL node broadcasts a time schedule to all the non-LCL nodes so that there can be only one transmission at any time instant within a grid block. However, the LCL nodes use a unique CDMA code for communication within a grid block; hence, data aggregations can simultaneously occur within each grid block.

For both GBE-Cluster-DG and GBE-Chain-DG, the LCL nodes serve as intermediate nodes of the data gathering tree rooted at the GCL node. The upstream node for an LCL node m is the closest LCL node n such that the distance between n and the GCL node is less than the distance between m and the GCL. The upstream node for an LCL node could be the GCL node itself if no other LCL nodes are nearby. Figures 13.5 and 13.6 illustrate an example for the execution of the cluster and chain formation and data aggregation within a grid block; the label inside the circle indicates the available energy at the node. In Figure 13.6, the label outside the circle indicates the position of the node in the chain (chain index), with C4 and C0 being, respectively, the chain indices of the LCL node and the node farthest away from the LCL node.

13.6 CDS-Based Data Gathering

A CDS of a network graph comprises a subset of the vertices in the graph such that for every vertex in the graph, the vertex is also in the CDS or has a neighbor node that is in the CDS. The CDS has been widely used in the MANET (Mobile Ad hoc Network) literature (e.g., [11]) for efficient broadcasting because a CDS is said to cover all nodes in the network, that is, a broadcast message (initiated by a node) forwarded by the nodes in the CDS reaches all the nodes in the network. Recently, we had investigated the use of CDS for data gathering in wireless sensor networks and in this pursuit, we had proposed an ECDS-based data gathering (referred to as ECDS-DG) algorithm [4] that considers the residual energy available at the nodes as the criteria to include a node into the CDS (i.e., the node with a larger residual energy gets included into the ECDS).

The objective of the ECDS-DG algorithm is to simultaneously minimize the energy lost per round and maximize the node lifetime (round of first node failure due to exhaustion of battery charge). In this pursuit, for every round of data gathering, the ECDS-DG algorithm forms a CDS of the sensor nodes in the network with nodes having higher residual energy (i.e., available energy) preferred for inclusion into the ECDS. The leader node for a round is the sensor node with the largest available energy and is obviously the first node to be part of the ECDS.

13.6.1 Stage 1: Construction of the ECDS

The ECDS-DG algorithm is executed for every round of data aggregation. The following four data structures are maintained by the algorithm:

1. ECDS-List—includes all the nodes that are part of the ECDS
2. Uncovered-Nodes-List—includes all the nodes that are not covered by a node in the ECDS-List
3. Covered-Nodes-List—includes nodes that are either in ECDS-List or covered by a node in ECDS-List
4. Priority-Queue—includes nodes that are in the Covered-Nodes-List and are probable candidates for addition to the ECDS-List. This list is sorted based on the decreasing order of energy level of the nodes. A dequeue operation on this queue returns the node with the highest energy level in the list

For a given unit disk graph $G = (V, E)$ of the underlying network (where V is the set of vertices and E is the set of edges), the Uncovered-Nodes-List is initialized to the vertex set V, and the ECDS-List and the Covered-Nodes-List are set to be empty. In the first iteration, the leader node (the node with the largest available energy) is removed from the Uncovered-Nodes-List and added to the ECDS-List and the Covered-Nodes-List. All the neighboring nodes (within the transmission range) of the leader node are now said to be covered, and these nodes are also removed from the Uncovered-Nodes-List and added to the Priority-Queue and the Covered-Nodes-List. The nodes are stored in the Priority-Queue in the decreasing order of their available energy.

For subsequent iterations, we first check if the Uncovered-Nodes-List is still empty. If so, we execute a dequeue operation on the Priority-Queue and extract a node s (having the highest available energy among the nodes in the Covered-Nodes-List and not yet in the ECDS-List). If the extracted node s has at least one neighbor node that is in the Uncovered-Nodes-List, then s is added to the ECDS-List and the uncovered neighbors of s are removed from the Uncovered-Nodes-List and added to the Covered-Nodes-List and the Priority-Queue. The Priority-Queue is resorted so that it stores the list of nodes in the decreasing order of their available energy. If the node extracted from the Priority-Queue has no uncovered neighbor node, then the node is not added to the ECDS-List and is simply removed from the Priority-Queue; the next node in the Priority-Queue is considered for prospective inclusion into the ECDS-List. We continue the above procedure until the Uncovered-Nodes-List gets empty and the Covered-Nodes-List contains all the vertices in the network graph. The algorithm returns the ECDS-List accumulated until then as the set of ECDS nodes for the network graph. At any stage of the algorithm if the Priority-Queue gets empty, while the Uncovered-Nodes-List remains nonempty, it implies that the underlying network graph is not connected and the algorithm stops without returning the ECDS-List.

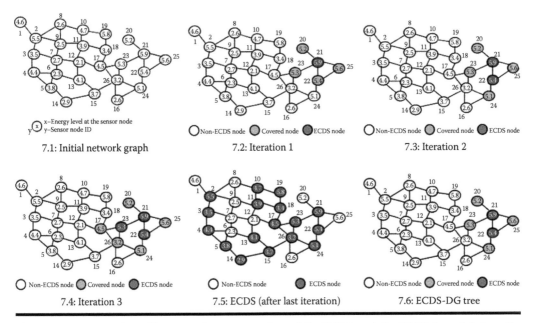

Figure 13.7   Example to illustrate the construction of the ECDS and the ECDS-based data gathering tree.

Stage 1 of the algorithm can be run in $O(|V| + |E| + |V|*\log|V|) = O(|E| + |V|*\log|V|)$ time if the Priority-Queue is implemented as a binary heap [10]. If implemented using arrays, this stage of the algorithm takes $O(|V| + |E| + |V|^2) = O(|E| + |V|^2)$ time. Figure 13.7 illustrates an example for the construction of the ECDS and the ECDS-DG tree. The integer outside a circle represents the node ID, and the real number inside the circle represents the available energy at the node. As can be seen in the iteration 2 of this example, a covered node that has the largest remaining energy (node 25 with 5.6 J) may not be still included into the ECDS if the node has no uncovered neighbors; instead, we consider the covered node that has the next largest remaining energy (node 22 with 5.4 J) and include it into the ECDS as it can cover at least one uncovered node.

13.6.2 Stage 2: Construction of the ECDS-DG Tree

After the ECDS is constructed, we form a data gathering tree (the ECDS-DG tree) with the leader node as the root of the tree. The rest of the ECDS nodes are the intermediate nodes of the tree and the non-ECDS nodes form the leaf nodes. The upstream node of a non-ECDS node is the closest ECDS node. The upstream node for an ECDS node m is the closest ECDS node n such that node n is closer to the leader than node m; the leader node itself is the upstream node if no such closer ECDS node exists for ECDS node m. Data gathering in the ECDS-DG tree starts from the leaf nodes and propagates all the way to the leader node through the intermediate ECDS nodes responsible for receiving, fusing, and transmitting the aggregated data to the upstream node in the tree. The leader node forwards the aggregated data to the sink. Since the links of the ECDS-DG tree are formed with a leaf node preferring a nearby ECDS node and an ECDS node preferring a closer ECDS node as the upstream node, the energy lost due to transmission is significantly reduced leading to an extended node lifetime. Stage 2 of the algorithm can be run in $O(|V|^2)$ time as it would take $O(|V|)$ time to find a suitable upstream node for every non-root node (a non-ECDS or an ECDS node). As $\log|V| < |V|$, when we combine the run-time complexity of the two stages, the overall run-time complexity of the ECDS-DG algorithm can be concluded as $O(|E| + |V|^2)$.

13.7 Spanning Tree-Based EMLN Data Gathering

In this section, we describe another energy-aware data gathering strategy, one based on maximizing the number of leaf nodes in a data gathering tree, for wireless sensor networks. The motivation for this strategy stems from the observation that a leaf node in a data gathering tree needs to turn on itself only to periodically sense and transmit data to its parent, and the leaf node can remain asleep for the rest of the time. On the other hand, an intermediate node would need to be turned on and stay in the active and listening modes much longer than a leaf node because the intermediate node would need to receive data from all its child nodes, aggregate with its own sensed data, and transmit to its parent. On these lines, we had recently proposed an algorithm to determine the EMLN-DG tree [5] that spans the entire sensor network and the tree is formed based on the residual energy level at the nodes and the number of uncovered neighbors of a node.

The problem of determining a maximal leaf spanning tree has been also proven (in [12]) to be polynomially equivalent to the problem of approximating a minimum connected dominating set (MCDS). A MCDS [10] is a CDS with the minimum number of constituent nodes covering the rest of the nodes in the network graph. A common approach (e.g., [13]) for approximating an MCDS is to consider inclusion of nodes that have a larger number of uncovered neighbors. With EMLN-DG, we provide energy awareness to the MCDS strategy by computing the weight of a node as the product of the number of uncovered neighbors and the available residual energy at the node. The EMLN-DG tree is formed for every round of data gathering, starting from a root-Node, the node that has the largest weight value. Only nodes that have a relatively larger number of neighbors and a higher energy level are included as intermediate nodes in the EMLN-DG tree. By maximizing the number of leaf nodes in a data gathering tree and considering the energy level available at the nodes while forming the tree, we reduce energy consumption per round and balance the energy level across all the nodes in the network. This contributes significantly to maximizing the node lifetime as only fewer nodes (i.e., the intermediate nodes of the tree) are involved in the more energy consuming trio of reception, transmission, and fusion.

13.7.1 Algorithm to Construct the EMLN-DG Tree

The EMLN-DG algorithm, run for every round of data gathering, uses the following four principal data structures to facilitate the computation of the DG tree:

1. Intermediate-Nodes-List—the list of all the intermediate nodes, including the root node of the DG tree
2. Leaf-Nodes-List—the list of all the leaf nodes of the DG tree
3. Covered-Nodes-List—the list of all nodes that are spanned (i.e., covered) by the DG tree
4. Priority-Queue—a queue whose entries are tuples of the form ($Weight(u)$, u), and the entries are stored in the decreasing order of the node weights. A dequeue operation on this queue extracts the node with the maximum weight from the queue. The weight of a node is the product of the number of uncovered neighbors of a node and the residual energy available at the node. If two or more nodes have the same maximum weight, a node is randomly chosen and extracted from the queue

To start with, the Intermediate-Nodes-List, Leaf-Nodes-List, and Covered-Nodes-List are all set to empty. The Priority-Queue is populated in decreasing order of the node weights. We keep track of the list of nodes at each Level of the data gathering tree through a data structure referred to as Nodes-All-Levels. The rootnode is at Level 0; the maximum value for Level corresponds to the Height of the tree. The rootnode (with the largest value for the node weight) is the first node to be extracted from the Priority-Queue using a dequeue operation and added to the Intermediate-Nodes-List and Covered-Nodes-List. Upon adding a node to the Intermediate-Nodes-List, all its neighboring nodes (if not yet covered) are added to the Covered-Nodes-List; the entries of the nodes in the Priority-Queue are resorted taking into consideration the changes in the values of the weight of a node due to the inclusion of new nodes into the Covered-Nodes-List. A node that is newly added to the Covered-Nodes-List is also added to the Leaf-Nodes-List. If a covered node gets selected as the intermediate node, then the node is removed from the Leaf-Nodes-List and added to the Intermediate-Nodes-List. The Level of a newly covered node is one more than the Level of its immediate predecessor intermediate node through which the former was covered. Until the Covered-Nodes-List contains all the vertices in the network graph, we repeat the above procedure by extracting the node with the largest weight from the Priority-Queue, adding it to the Intermediate-Nodes-List, and removing from the Leaf-Nodes-List. Any of the uncovered neighbors of this intermediate node are added to the Covered-Nodes-List and the Leaf-Nodes-List; their Level value is set to one more than the Level value of the intermediate (predecessor) node. When we assign the Level value of a node, we also include that node in the Nodes-All-Levels data structure into the list corresponding to the particular Level value.

If the algorithm runs until all the nodes in the network are covered, it returns the following five data structures/variables that are used to compute the delay per round associated with the EMLN-DG tree: Intermediate-Nodes-List, Leaf-Nodes-List, Nodes-All-Levels, Height-DG-Tree, and rootNode. The time complexity of the algorithm is $O(|V|*(|V|+|E|))$, as it takes $O(|V|+|E|)$ time per iteration and at most there could be $O(|V|)$ intermediate nodes in the tree. For each of the iterations, we have to recompute the number of uncovered neighbors for every node in the network. There are $|V|$ nodes in the network, and we have to process each of the $|E|$ edges twice, once for each vertex on which the edge is incident.

13.7.2 Algorithm to Compute the Delay per Round of Data Gathering

The delay at a node indicates the number of time slots it takes for the node to gather data from all its immediate child nodes. The delay of the data gathering tree is one plus the delay at the root node. We assign one time slot per child node to transfer data to its immediate predecessor node in the tree. We start calculating the delay at the intermediate nodes from the bottom of the data gathering tree. The delay associated with each of the leaf nodes is 0. For each intermediate node u at a particular level, we prepare a sorted list of the delay associated with each of its immediate child nodes. The delay associated with the intermediate node is computed through a temporary running variable, Temp-Delay (initialized to zero), as we explore the delay associated with each of the child nodes in the sorted list. For every child node v in the sorted list of the delay, Temp-Delay is set to the maximum of Temp-Delay + 1 and Delay(v) + 1 as we assume it takes one time slot for a child node to transfer its aggregated data to the immediate predecessor intermediate node. The delay associated with the intermediate node u, Delay(u), is the final value of Temp-Delay after we run through the sorted list of the delays associated with the Child-Nodes(u). The above procedure is repeated for all the intermediates nodes, from levels one less than the Height of the tree all the way to zero (i.e., the root node).

The delay per round on a $|V|$-node data gathering tree can be computed in $O(|V|*\log|V|)$ time as there are only $|V| - 1$ edges in the tree, and it takes $O(\log|V|)$ time to update additions to a sorted list that is implemented as a heap. The algorithm described here to compute the delay per round of data gathering can also be applied to the ECDS-DG tree.

13.7.3 Example to Illustrate the Construction of the EMLN-DG Tree and Calculation of Delay

Figure 13.8 illustrates an example to demonstrate the working of the EMLN-DG algorithm and the computation of the delay for the data gathering tree. In Figure 13.8 (8.1 through 8.9), each circle represents a node. The integer outside the circle represents the node ID, and the integer inside the circle represents the number of uncovered neighbors of the corresponding node. The real

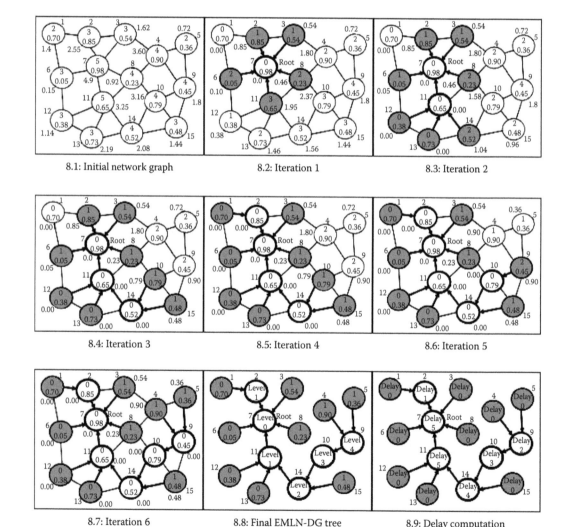

Figure 13.8 Example illustrating the construction of the EMLN-DG tree and the computation of its delay.

number inside the circle represents the residual energy (in Joules) currently available at the node, and the real number outside the circle represents the weight (product of the residual energy and the number of uncovered neighbors) for the particular node. The intermediate nodes that are part of the EMLN-DG tree have their circles bolded. We shade the circles of leaf nodes that are covered by the intermediate nodes of the EMLN-DG tree. The circles of nodes that are not yet part of the EMLN-DG tree (i.e., nodes that are neither intermediate nodes nor leaf nodes) are neither shaded nor made bold.

On the 15-node example illustrated in Figure 13.8, we observe that it takes six iterations to compute the final EMLN-DG tree. The EMLN-DG tree has only six intermediate nodes, including the root node. The remaining nine nodes (i.e., 60% of the nodes) are leaf nodes. The height of the tree is 5 (nodes 4 and 5 are at level 5), and the delay at the root node (i.e., number of time slots it takes for the root node 7 to receive the aggregated data from all its immediate downstream child nodes) is six time slots.

13.8 Simulation Studies

In this section, we present the results from simulation studies evaluating the performance of the data gathering algorithms discussed in the earlier sections, listed as follows: (1) LEACH, (2) PEGASIS-TDMA (the original version of PEGASIS), (3) PEGASIS-CDMA (the improved binary-chain-based version supporting aggregation at different levels), (4) GBE-Cluster-DG, (5) GBE-Chain-DG, (6) ECDS-DG, and (7) EMLN-DG. All our simulations are conducted in a discrete event simulator developed by us in Java. This simulator has been successfully used to develop and study data gathering algorithms in some of our recent studies [3–5]. The link layer model is assumed to be an ideal MAC (medium access control) model without any interference and collisions as this will facilitate us to obtain the best possible performance from each data gathering algorithm.

The network dimension is 100×100 m². The number of nodes in the network is 100, and the nodes are uniformly and randomly distributed throughout the network. The sink node is located at (50, 300), outside of the sensor network field. In the case of ECDS-DG and EMLN-DG, the transmission range per sensor node adopted for forming the network graph in the algorithm is varied from 15 to 50 m. There exists an edge between any two vertices in the graph if the distance between the corresponding nodes in the network is less than or equal to the transmission range. Sensor nodes running the GBE-Cluster-DG, GBE-Chain-DG, LEACH, PEGASIS-TDMA, and PEGASIS-CDMA algorithms are assumed to be able to conduct transmission power control (i.e., vary their transmission range) depending on the distance to the receiver node. For the GBE-Cluster-DG and GBE-Chain-DG algorithms, the grid block length values used are 20, 25, 33.33, 50, and 100 m. The performance metrics evaluated are those defined in Section 13.2, which are again listed as follows: (1) delay per round, (2) energy lost per round, (3) energy*delay per round, and (4) node lifetime, measured as the round of first node failure. In addition to the above four performance metrics, we also measure the network connectivity, which is a measure of the probability that the communication topology envisioned by these algorithms spans all the sensor nodes when operating under a limited transmission range. The initial energy provided per node during a simulation session is 1 J.

The rest of this section is organized as follows: We first present the energy consumption model used in the simulations. Next, we present the performance of the ECDS-DG and EMLN-DG data gathering trees with respect to different operating transmission ranges of the sensor nodes as well as illustrate the performance of the GBE-Cluster-DG and GBE-Chain-DG algorithms with respect to the different grid block lengths. For the chosen network dimensions and node density, we observe that both the ECDS-DG and EMLN-DG algorithms give optimal performance for a transmission range of 25 m; likewise, we observe both the GBE-Cluster-DG and GBE-Chain-DG algorithms to give optimal performance for a grid block length of 25 m. Hence, we use the performance metric values obtained for the transmission range, grid block length value of 25 m for these four energy-aware algorithms to compare with the classical energy-unaware data gathering algorithms in Section 13.8.4. For the LEACH algorithm, we use a value of 0.05 for the probability of a node becoming a clusterhead during a round. In our earlier simulations for several works (e.g., [3–5]) on data gathering algorithms, we observed LEACH to give the best performance when operated with this value of 0.05 for the probability of a node becoming a clusterhead.

13.8.1 Energy Consumption Model

We use the following first-order radio model [14] that has been also used to model energy consumption in several previous works (e.g., [1,2,9]) in the literature. According to this model, the energy expended by a radio to run the transmitter or receiver circuitry is $E_{elec} = 50$ nJ/bit and $\in_{amp} = 100$ pJ/bit/m^2 for the transmitter amplifier. The radios are turned off when a node wants to avoid receiving unintended transmissions. An r^2 energy loss model is used to compute the transmission costs. The energy lost in transmitting a k-bit message over a distance d is given by $E_{TX}(k, d) = E_{elec} * k + \in_{amp} * k * d^2$. The energy lost in receiving a k-bit message is $E_{RX}(k) = E_{elec} * k$.

13.8.2 Impact of Transmission Range on the Performance of the ECDS-DG and EMLN-DG Trees

Figures 13.9 (9.1) and 13.10 (10.1) illustrate that the probability of network connectivity reaches 0.995 (99.5%) when the transmission range of the sensor nodes is fixed to be 25 m. Lower values of transmission range result in poor network connectivity, and as the transmission range increases, the connectivity increases. After the transmission range value reaches 25 m, the probability of

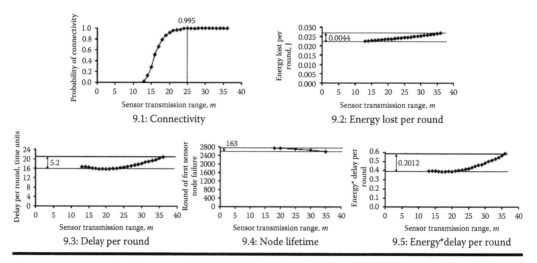

Figure 13.9   Impact of transmission range per sensor node on the performance of the ECDS-DG tree.

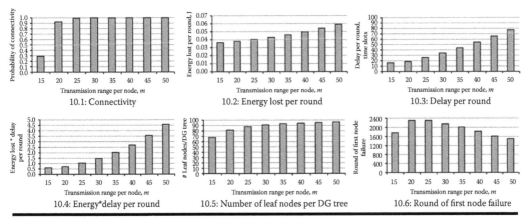

Figure 13.10 Impact of transmission range per sensor node on the performance of the EMLN-DG tree.

network connectivity does not change much, increases only from 0.995 to 0.999 as we increase the transmission further.

The height of the EMLN-DG tree decreases with increase in the transmission range per node, and this can be attributed to the corresponding increase in the number of leaf nodes (and a decrease in the number of intermediate nodes). As the transmission range increases, the number of nodes that can be covered by the inclusion of a node as an intermediate node in the EMLN-DG tree increases. Hence, fewer intermediate nodes needed to be added in the EMLN-DG tree. On a 100-node network, we observe that only less than 35% of the nodes serve as intermediate nodes of the EMLN-DG tree at a very low transmission range per node value of 15 m and as the transmission range per node increases, the percentage of nodes serving as intermediate nodes significantly decreases. At transmission ranges of 25 m and 50 m per sensor node, less than 15% and 4% of the nodes, respectively, serve as intermediate nodes of the tree.

For low and moderate values of transmission range, more nodes are added to the ECDS and likewise, more intermediate nodes are added to the EMLN-DG tree. This leads to fewer downstream nodes per upstream node in the data aggregation trees. On the other hand, for higher transmission range values, fewer nodes are included into the ECDS and likewise, fewer intermediate nodes are added to the EMLN-DG tree; each upstream node in the data aggregation tree has several downstream nodes. There can be simultaneous transmissions across the different upstream node–downstream node sets, with each upstream node assigning a unique CDMA code to its set of downstream nodes. In order to avoid too much of interference with simultaneous transmissions, we assume that all the downstream nodes of an upstream node send their data to the upstream node using a single CDMA code according to a time schedule and only one transmission is allowed per time unit (this is also the approach adopted for communication within a cluster in LEACH). An upstream node has thus to wait to receive data from all its downstream nodes before propagating the data further. The more the number of downstream nodes, more the delay incurred at an upstream node to collect, aggregate, and further transfer the data.

The energy lost per round increases with increase in the transmission range of the sensor nodes as more energy is expended with increase in the distance. But the increase in the energy lost per round is not proportional to the increase in the transmission range. As the transmission range is doubled, the energy lost per round only increases by at most 20%. This can be attributed to the reduction in the number of nodes in the ECDS and the number of intermediate nodes in the

EMLN-DG tree, with increase in the transmission range per sensor node and not all sensor nodes need to do data aggregation, all the time.

For the ECDS-DG tree, as we increase the sensor transmission range (say, it is doubled), we do notice that the node lifetime (in terms of the number of rounds) decreases, but only slightly, within 10%. This can be attributed to only a slight increase in the energy lost per round. For the EMLN-DG tree, the node lifetime is maximum when the transmission range per node is at 20 and 25 m. As we increase the transmission range per node from 25 m to 35 and 50 m, the node lifetime decreases by about 12% and 35%, respectively. This can be attributed to a slight to moderate increase in the energy lost per round. For both ECDS-DG and EMLN-DG, when we compute the energy*delay values, it remains low for smaller and moderate transmission range values and increases rapidly as the transmission range gets high.

13.8.3 Performance of the GBE-DG Algorithms and the Impact of Grid Block Length

For both the GBE-Cluster-DG and GBE-Chain-DG algorithms, the probability of network coverage (Figure 13.11 [11.1]) is the same as this metric depends only on the distribution of the nodes and the grid block length and not specifically on the data aggregation principle. We define a network to be completely covered if, after the random deployment of the sensor nodes and the construction of the grid blocks, at least one sensor node can be located in each grid block. Lower values of the grid block length result in poor network coverage and as the grid block length increases, the network coverage increases. For grid block lengths of 33.33, 50, and 100 m, the probability of network coverage remains at 1. For grid block length values of 25 m or more, the probability of network coverage does not change much, increases only from 0.978 to 1.

The node lifetime (round of first node failure) for both the GBE-DG algorithms is higher for smaller grid block lengths of 20 and 25 m, and as the grid block lengths increase, the node lifetime decreases. This can be attributed to the longer transmission distance encountered by the sensor nodes to transmit their data to the LCLs and also to the longer transmission distance encountered by the LCLs when they forward the aggregated data to the peer LCLs. The energy lost per round increases significantly as the grid block length increases.

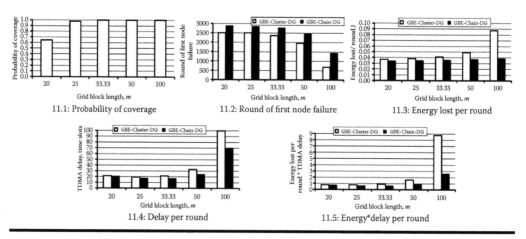

11.1: Probability of coverage   11.2: Round of first node failure   11.3: Energy lost per round

11.4: Delay per round   11.5: Energy*delay per round

Figure 13.11   GBE-DG algorithms: impact of the grid block length on the performance metrics.

For all grid block lengths, GBE-Chain-DG trees yield longer node lifetimes compared to GBE-Cluster-DG trees. The magnitude of the difference in the node lifetimes between the two GBE-DG trees increases with increase in the grid block length. For a grid block length of 100 m, GBE-Cluster-DG tree basically turns out to be a direct polling-based data gathering and GBE-Chain-DG transforms to a PEGASIS-based approach where the leader for a round is the node with the highest available residual energy and the chain construction starts with the node that is farthest away from the leader of the round, rather than a static chain starting farthest away from the sink, as in traditional PEGASIS. The relatively poor performance of GBE-Cluster-DG tree with increase in the grid block length can be attributed to the transmission of the data packets over a longer distance, compared to the distance transmitted in the case of the GBE-Chain-DG tree. With GBE-Chain-DG, the transmissions and receptions occur over the chain that is formed according to a greedy distance-based heuristic wherein the next node added to the chain is the node closest to the most recently added node to the chain. Such an approach is bound to yield a lower transmission energy loss, compared to the direct transmissions of data packets from each leaf sensor node to its LCL node, as in the case of GBE-Cluster-DG. Overall, as the grid block length increases, the difference in the energy consumed per round for the two GBE-DG algorithms increases and GBE-Chain-DG yields significantly lower energy loss per round—contributing to a relatively longer node lifetime.

The data aggregation delay is lower at smaller grid block length values as there will be less number of sensor nodes per grid block and fewer time slots are sufficient to transfer data to the LCL nodes in each grid block. For larger grid block lengths, there will be more sensor nodes per grid block and relatively more time slots are needed to transfer data from all the sensor nodes to a LCL node within each grid block. There would be more interference if simultaneous data transfers occur within a grid block and hence we require data transfer to occur sequentially within a grid block. But there would be relatively less interference if simultaneous data transfers occur across the different grid blocks. Hence, we assume data transfer can occur at the same time in each of the grid blocks. When we compute the energy*delay per round for TDMA systems, we notice that it remains almost the same at lower grid block length values (of 20, 25, and 33.33 m) and then increases significantly.

We notice that GBE-Chain-DG trees incur a lower delay per round compared to GBE-Cluster-DG trees, and the difference in the delay per round increases with increase in the grid block length. The relatively lower delay per round incurred with GBE-Chain-DG can be attributed to the provision for simultaneous propagation of the aggregated data on either sides of the chain within a grid block (i.e., to the left and right of the LCL node in the chain). In the case of GBE-Cluster-DG, as all the leaf sensor nodes have to directly transmit their data to the LCL node, no more than one transmission is possible at any time slot. GBE-Chain-DG trees incur 10%–25% relatively lower delay, compared to GBE-Cluster-DG, as the grid block length is varied from 25 to 50 m. The relatively lower energy consumed and delay incurred per round for GBE-Chain-DG enables the algorithm to return a significantly lower energy*delay per round compared to GBE-Cluster-DG. The difference in the magnitudes of energy*delay per round increases significantly as the grid block length is increased.

13.8.4 Performance Comparison of the Data Gathering Algorithms

With ECDS-DG and EMLN-DG, the non-CDS nodes and leaf nodes, respectively, chose an intermediate node that lies within the transmission range and hence spend relatively less energy in transmitting to an upstream node in the data gathering tree. With these two algorithms, it is only the sensor nodes which are part of the ECDS or the intermediate nodes of the tree that end up spending more energy to reach out to upstream nodes located beyond their transmission range. Similarly, with the grid block based data gathering algorithms (GBE-Cluster-DG and GBE-Chain-DG), the non-LCL sensor nodes within a grid block are expected to transmit to distances only within the block; it is only the LCL sensor nodes and the GCL node that are expected to transmit to any distance. With all these four data gathering algorithms (ECDS-DG, EMLN-DG, GBE-Cluster-DG, and GBE-Chain-DG), the choice of the sensor nodes that are expected to transmit to any distance is carefully made for every round of data aggregation—only the sensor nodes that have a relatively larger residual energy are assigned the responsibility of gathering data from one or more sensor nodes, aggregating them (i.e., fusion), and then forwarding them to another sensor node that may be at any distance, depending on the underlying principle used to construct the topology for data gathering. Thus, energy awareness is built into the design of the above four data gathering algorithms and this contributes to the significantly longer node lifetime achieved with these algorithms. On the other hand, with the classical energy-unaware LEACH and PEGASIS (both CDMA and TDMA versions) algorithms, every sensor node is expected to reach out to any other sensor node, if needed. Such an energy-unaware design flaw significantly contributes to the relatively lower node lifetime obtained with these classical data gathering algorithms.

The strategy of both the GBE-DG algorithms to choose the LCL nodes and the GCL node for each round of data communication helps to equally rotate the data aggregation and forwarding load across all the sensor nodes. For moderate values of the grid block length (25 and 33.33 m), the sensor nodes are almost equally distributed across the grid blocks and they are fairly used. This is also vindicated by the significantly lower percentage of the initial energy that is left unutilized at the time of first node failure. On the other hand, we notice that LEACH and PEGASIS are not that much fair to all the sensor nodes in the network. For PEGASIS and LEACH, the percentage of initial energy left unutilized at the nodes at the time of first node failure is in the 55%–65% and 28%–30% range, respectively.

The unfairness of node usage in PEGASIS is due to the random selection of the leader for the chain (without any consideration of the energy levels at the nodes or the position of the nodes vis-à-vis the sink location) and also due to the unequal placement of the sensor nodes in the chain. The distance between the successive nodes in the chain increases as we traverse from the beginning to the end of the chain. As a result, nodes in the beginning of the chain spend less energy to transfer data to the next node in the chain, whereas nodes in the second half of the chain spend more energy to transfer data to the next node in the chain. PEGASIS prefers to start the chain with the node that is farthest from the sink location to facilitate this node to spend most of its energy as a chain leader rather than spending much energy to transmit to the next node in the chain. With random selection of the chain leader for every round, all nodes are almost equally selected as leaders. Since the chain is not reconstructed for every round, nodes in the second half of the chain lose more energy as they continue to transfer the aggregated data over a relatively longer distance to the next node in the chain for every round and lose more energy serving as the chain leaders.

In the case of LEACH, if P is the probability of a node serving as a clusterhead (also translates to the percentage of clusterheads), each sensor node is elected as the clusterhead exactly once within $1/P$ rounds. So, over a given time period, every sensor node serves as the clusterhead for almost an equal number of times. But a clusterhead located far away from the sink has to spend more energy to transfer the aggregated data compared to the nodes that are closer to the sink. Thus, LEACH incurs significantly larger energy consumption per round; as on average, the sensor nodes have to transmit over a relatively longer distance to reach the clusterhead and also each of the clusterheads have to transfer the aggregated data directly to the sink.

EMLN-DG incurs a relatively longer delay than the other three energy-aware data gathering algorithms as well as that of PEGASIS-CDMA, and this can be attributed to the use of time-division multiplexing during the aggregation of the data packets at an upstream node from its immediate downstream nodes. All the downstream nodes of an upstream node forward data to the upstream node according to a time schedule, with only one transmission allowed per time slot. Data aggregation at different upstream nodes can however occur in parallel, using different CDMA codes. Also, an intermediate node in the EMLN-DG tree cannot forward the aggregated data to its upstream node until it receives data from each of its immediate downstream nodes. Since EMLN-DG aims for maximizing the number of leaf nodes, an upstream node has to wait for a longer time to receive data from all its downstream nodes (especially if they are leaf nodes); the delay incurred in gathering data from all the leaf nodes manifests itself as the data propagates to the root of the data gathering tree.

As noted in Figures 13.12 (13.1 and 13.4), respectively, ECDS-DG incurs the lowest energy per round and the GBE-Chain-DG algorithm incurs the largest node lifetime in terms of the number of rounds before the first sensor node dies due to the exhaustion of battery charge. ECDS-DG follows suit and yields the second largest value for the node lifetime. We also observe ECDS-DG to yield a relatively lower delay per round (in terms of the number of time units incurred to complete data gathering), and its delay is only more than the delay incurred with PEGASIS-CDMA, which is the chain-based binary version of the PEGASIS algorithm. Note that PEGASIS-CDMA assumes that at a given level in the binary hierarchy, any two nodes in the chain can communicate simultaneously with each other using a unique CDMA code. Even though CDMA codes help to

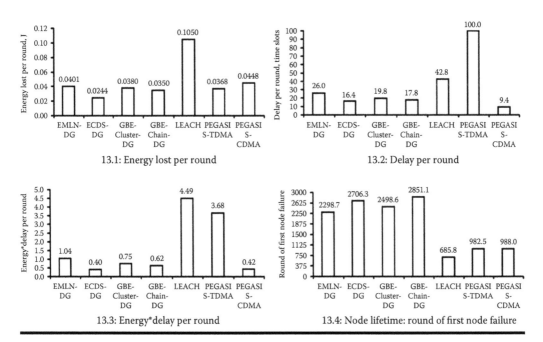

13.1: Energy lost per round

13.2: Delay per round

13.3: Energy*delay per round

13.4: Node lifetime: round of first node failure

Figure 13.12   Performance comparison of the data gathering algorithms.

Table 13.1   Ranking of the Data Gathering Algorithms with Respect to Different Performance Metrics

Performance Metric	1 (Best)	2	3	4	5	6	7 (Worst)
Node lifetime	GBE-Chain-DG	ECDS-DG	GBE-Cluster-DG	EMLN-DG	PEGASIS-CDMA	PEGASIS-TDMA	LEACH
Delay per round	PEGASIS-CDMA	ECDS-DG	GBE-Chain-DG	GBE-Cluster-DG	EMLN-DG	LEACH	PEGASIS-TDMA
Energy per round	ECDS-DG	GBE-Chain-DG	PEGASIS-TDMA	GBE-Cluster-DG	EMLN-DG	PEGASIS-CDMA	LEACH
Energy*delay per round	ECDS-DG	PEGASIS-CDMA	GBE-Chain-DG	GBE-Cluster-DG	EMLN-DG	PEGASIS-TDMA	LEACH

avoid mixing up of the data signals, there can be still some interference among the signals if all the simultaneous communications are targeted to a single node.

Considering the practical difficulty associated with operating PEGASIS-CDMA and the fact that ECDS-DG incurs the lowest value for the energy*delay per round, we opine that ECDS-DG would be a better choice over PEGASIS-CDMA to simultaneously minimize both the delay per round and the energy incurred per round of data gathering. In addition, the node lifetime for ECDS-DG (2706 rounds) is only 5% less than the node lifetime incurred for GBE-Chain-DG (2851 rounds), whereas the energy*delay per round for GBE-Chain-DG (0.62 Joules-time units) is 45% more than the energy*delay per round for ECDS-DG (0.40 Joules-time units). Hence, we can very well conclude that ECDS-DG is the best data gathering algorithm with respect to minimizing both the energy*delay per rounds and maximizing the node lifetime—measured in terms of the number of rounds before the first node failure due to exhaustion of battery charge.

Table 13.1 presents a comprehensive ranking (from best to worst) of the seven data gathering algorithms with respect to the different performance metrics. Note that ECDS-DG occupies the first two rankings for all the four performance metrics.

13.9 Future Research Directions

We identify the following as possible extensions of the performance comparison study in future:

1. Compare the performance of the data gathering algorithms with respect to different definitions of network lifetime: (1) number of rounds for first node failure; (2) number of rounds for the network to get disconnected, and (3) number of rounds for a certain percent of the sensor nodes to have failed.
2. Compare the performance of the data gathering algorithms for different locations of the sink node as well as in the presence of multiple sinks. In the current simulation, we assumed there is one sink node and it is located outside the network field. One could plan to compare the performance of the data gathering algorithms when the sink is located in the center of the network or at the network boundary. Similarly, one could also compare the performance of the data gathering algorithms when the aggregated data has to be forwarded to the closest sink among multiple sinks evenly distributed around the network field.

3. Compare the performance of the data gathering algorithms in the presence of a mobile sink that moves around the network field and makes random stops at different locations for every round of data gathering. The leader node selection for a round could be then either based on the sensor node that lies closest to the sink or the sensor node that has the largest available energy. One could compare the pros and cons of both these leader node selection strategies in the presence of a mobile sink.

13.10 Conclusions

The high-level contribution of this chapter is a comprehensive description of the various data gathering algorithms for wireless sensor networks thereby illustrating the use of different communication topologies (cluster, chain, grid, CDS, and spanning trees) for data aggregation and forwarding as well as of the potential benefits associated with energy-aware data gathering over energy-unaware data gathering. We have described in detail the working of seven data gathering algorithms (LEACH, PEGASIS-TDMA, PEGASIS-CDMA, GBE-Cluster-DG, GBE-Chain-DG, ECDS-DG, and EMLN-DG) and analyzed their individual as well as comparative performance in a simulation environment. We observe the modern energy-aware data gathering algorithms (GBE-Cluster-DG, GBE-Chain-DG, ECDS-DG, and EMLN-DG) to give optimal performance over the classical energy-unaware algorithms (LEACH, PEGASIS-TDMA, and PEGASIS-CDMA) with respect to node lifetime (measured as the round of first node failure), delay per round (measured as the number of time units required to collect data from all the sensor nodes and transfer them to the sink), energy per round, and the energy*delay per round (a measure of the tradeoff between energy and delay incurred per round of data gathering). Among the modern energy-aware algorithms, we observe the ECDS-based data gathering algorithm to be the ideal choice for simultaneously maximizing the node lifetime and minimizing the energy*delay metrics without any significant tradeoff. We also observe that the choice of the communication topology used for data gathering as well as the strategy used to construct these topologies (for example, the flat-chain-based approach for the original PEGASIS algorithm vis-à-vis a hierarchical grid-based chain approach for the GBE-Chain-DG algorithm) does have a significant impact on the performance of the data gathering algorithms.

References

1. Heinzelman, W., A. Chandrakasan, and H. Balakrishnan. 2000. "Energy-Efficient Communication Protocols for Wireless Microsensor Networks." Paper presented at the Proceedings of the Hawaiian International Conference on Systems Science, Maui, Hawaii January 2000.
2. Lindsey, S., C. Raghavendra, and K. M. Sivalingam. September 2002. "Data Gathering Algorithms in Sensor Networks using Energy Metrics." *IEEE Transactions on Parallel and Distributed Systems* 13 (9): 924–35.
3. Meghanathan, N. December 2010. "Grid Block Energy based Data Gathering Algorithms for Wireless Sensor Networks." *International Journal of Communication Networks and Information Security* 2 (3): 151–61.
4. Meghanathan, N. July–September 2010. "A Data Gathering Algorithm based on Energy-Aware Connected Dominating Sets to Minimize Energy Consumption and Maximize Node Lifetime in Wireless Sensor Networks." *International Journal of Interdisciplinary Telecommunications and Networking* 2 (3): 1–17.

5. Meghanathan, N. May 2010. "An Algorithm to Determine Energy-Aware Maximal Leaf Nodes Data Gathering Tree for Wireless Sensor Networks." *Journal of Theoretical and Applied Information Technology* 15 (2): 96–107.

6. Meghanathan, N. December 2009. "Use of Tree Traversal Algorithms for Chain Formation in the PEGASIS Data Gathering Protocol for Wireless Sensor Networks." *KSII Transactions on Internet and Information Systems* 3 (6): 612–27.

7. Shukla, I., and N. Meghanathan. December 2009. "Impact of Leader Selection Strategies on the PEGASIS Data Gathering Protocol for Wireless Sensor Networks." *Ubiquitous Computing and Communication Journal* 4 (5): 20–9.

8. Viterbi, A. J. April 1995. *CDMA: Principles of Spread Spectrum Communication*, 1st ed. Upper Saddle River, NJ: Prentice Hall.

9. Lindsey, S., C. Raghavendra, and K. M. Sivalingam. April 2001. "Data Gathering in Sensor Networks using the Energy*Delay Metric." Paper presented at the Proceedings of the 15th International Parallel and Distributed Processing Symposium, pp. 2001–08. San Francisco, CA.

10. Cormen, T. H., C. E. Leiserson, R. L. Rivest, and C. Stein. July 2009. *Introduction to Algorithms*, 3rd ed. Cambridge, MA: MIT Press.

11. Meghanathan, N., and A. Farago. July 2008. "On the Stability of Paths, Steiner Trees and Connected Dominating Sets in Mobile Ad Hoc Networks." *Ad hoc Networks* 6 (5): 744–69.

12. Caro, Y., D. B. West, and R. Yuster. April 2000. "Connected Domination and Spanning Trees with Many Leaves." *SIAM Journal of Discrete Mathematics* 13 (2): 202–11.

13. Meghanathan, N. February 2006. "An Algorithm to Determine the Sequence of Stable Connected Dominating Sets in Mobile Ad Hoc Networks." Paper presented at the Proceedings of 2nd Advanced International Conference on Telecommunications, Guadeloupe, French Caribbean.

14. Rappaport, T. S. January 2002. *Wireless Communications: Principles and Practice*, 2nd ed. Upper Saddle River, NJ: Prentice Hall.

SECURITY ISSUES IN WIRELESS SENSOR NETWORKS

Chapter 14

Privacy in Wireless Sensor Networks: Issues, Challenges, and Solutions

Arijit Ukil

Contents

14.1 Introduction

The groundbreaking experiment conducted by Milgram [1] to find the average path length in social networking found that everyone in this world is just six steps away from one another (six degrees of separation). This amazing finding makes us believe that we are living in a small world, and this small world is not strictly privacy protected. Security is required by the individual and the organization to survive, and privacy is needed for sustaining in this small world. With the advent of smart cities, smart homes, and smart cars, and eventually the world becoming a smart planet, sensors possess a good amount of computational power, are networked, and work in collaboration to implement some applications and offer some services. This milieu of six degrees of separation and smart planet make our world very comfortable to live in and connect with each other with remarkable ease. However, every good thing comes with some demerits. The demerit of this small world with wireless sensor network (WSN) web is weak protection of privacy. Research on preserving privacy has started mostly from data mining perspective after witnessing huge business loss in e-commerce [2,3]. Over the years, the trend of privacy research shifted from privacy preserving data mining (PPDM) to other important problems like anonymous authentication, private data aggregation, and contextual privacy, which find appropriate use cases in WSN. Another factor needs to be considered in WSN is the low computational power of legacy or tiny legacy sensor nodes, which should not be overburdened to carry out the computation for privacy preservation algorithms. In this context, we would like to differentiate between privacy and security. In nontechnical or semitechnical literature, privacy and security aspects and issues are mostly found in the same brackets and people intend to think same solution exist to counter the problems of security and privacy. However, security and privacy are not at all same with only some amount intersection between the challenges and techniques used in these two domains. Security has five pillars or services: confidentiality, authentication, integrity, nonrepudiation, and availability. Privacy consists of the following primitives:

- *Anonymity:* It means hiding the information source.
- *Untraceability:* It is defined as the degree of difficultness for an adversary to identify that a given set of actions were performed by the same subject.
- *Unlinkability:* It is hiding the relationship between the items of the data set.
- *Unobservability:* It is hiding of the items, which means that the adversary cannot recognize the items at all.
- *Pseudonymity:* It is camouflaging the identity of the source and the destination.

Similar to security, two types of privacy algorithms exist.

1. *Information-theoretic perfect privacy:* In this case, a priori and a posteriori probability distribution functions are identical so that information leakage to any adversary is zero.
2. *Computationally perfect privacy:* Here, a probabilistic polynomial-time algorithm cannot distinguish between prior and posterior probability distribution functions so that no information is leaked to realistic adversary.

In the context of WSNs, we consider computationally perfect privacy as information-theoretic perfect privacy algorithms are mostly Non-Polymonial (NP)-hard in nature. Our main concerns will be location or contextual privacy, PPDM (k-anonymity, l-diversity, utility-privacy tradeoff, data sanitizer), secure multiparty computation (SMC), and anonymous authentication, which take care of the challenges of almost all WSN privacy issues. Location or context privacy ensures the node's contextual information is not leaked to the unintended parties. In PPDM, the data available at the different nodes are kept in statistically perturbed way so that raw data that contains sensitive information about different nodes or the network per se can be transformed to protect privacy but ensures the statistical properties are not modified beyond the prescribed threshold limit. SMC is useful when multiple nodes require computing collaboratively without revealing the actual content of their data. Suppose a WSN node launches some queries to private statistical databases, which aims to maximize the accuracy of queries while minimizing the chances of identifying its records. This is handled by the concept of differential privacy. In anonymous authentication, nodes can send or report private data without revealing its own identity. Accordingly, we organize the chapter as follows. In Section 14.2, we describe the considered system model and common privacy model. In Section 14.3, we discuss the different algorithms and state-of-the-art proposals in WSN privacy, which include contextual privacy, PPDM, SMC, differential privacy, and anonymous authentication. In Section 14.4, we study two related and elegant schemes for solving privacy preserving data aggregation (PPDA) in WSN. Finally, we conclude and illustrate future research scope along with the existing open research problems.

14.2 System Model and Common Privacy Threat Model

We consider a WSN with large number of sensor nodes reporting directly (single-hop) or indirectly (multi-hop) to one or more servers, which process the data and send through Internet or directly to the concerned client systems. Let N number of sensor nodes be arbitrarily distributed and they communicate with each other and also to M number of server nodes through ad hoc wireless links. This is depicted in Figure 14.1. We consider a very generic WSN architecture with sensor nodes that are connected wirelessly with other sensor nodes and some of them to the server. Servers can be connected to client machines and/or to Internet to post the processed sensor data. We consider all the communication links as wireless, except servers to the Internet, which may be wired. Another important thing we consider is in-network processing, where few of the sensor nodes form cluster

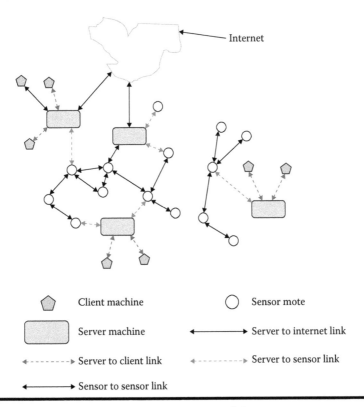

Figure 14.1   Wireless sensor network (WSN) system model.

and cluster head processes the data and routes those data to the server or another cluster head. This invariably cuts down communication cost by saving bandwidth and energy. Secure in-network processing for finding exact aggregation (SUM) results while satisfying the required security concerns using homomorphic encryption and secret sharing is proposed Papadopoulos et al. [4]. To facilitate our study on privacy in WSN, we investigate the following five features:

1. Secure multiparty computation
2. Anonymous authentication
3. Privacy preserving data mining
4. PPDA and homomorphic encryption
5. Contextual privacy

Privacy threat model for WSN can be broadly classified into four categories (data privacy at transit, data privacy at storage, data privacy at publishing, and contextual privacy):

1. *Data privacy at transit:* When the private data are communicated over wireless link, adversary (malicious WSN nodes inside or outside the network) may intercept the data over wireless link. For data privacy in transit, SMC [5] and anonymous authentication [6] are few of the solutions.
2. *Data privacy at storage:* Private data are stored at WSN nodes or at the servers. External agents may intend to access the stored private data. Even finding partial list sometimes becomes useful as in most cases strong correlation exists between the different attributes. PPDA is a good candidate when aggregation data are more important than individual sensor data.
3. *Data privacy at publishing:* When private data are published or given access to external agents, sensitive part of the data which may be hidden can get revealed using background knowledge, multiple database analysis, and other data correlating techniques. For both data privacy at storage and data privacy at publishing, PPDM [7] concepts are useful.
4. *Contextual privacy:* Apart from data privacy, sensitive data can be disclosed by monitoring the contextual information associated with the database. Contextual privacy preservation addresses the issue of revelation of private data from traffic analysis [8]. Knowledge of simple contextual information like spatial and temporal data sometimes becomes enough to infer about the sensitive data [41]. For an example, the following shows how inference is a strong tool to know with high probability the private information.

Let $E1$ be the event that a patient PA has some kind of disease and $E2$ be the event that PA sees a specific physician PH. By statistics, the following conditions are available. If PA has the disease, then the probability that he will see the specific PH is $\rho1$, that is, $P_r [E2|E1] = \rho1 = 98\%$. If PA does not have the disease, then the probability that he will see the specific PH is $\rho2$, that is, $P_r [E2|\neg E1] = \rho2 = 1\%$. Suppose that only 5% of the population in a region has the disease, that is, $P_r [E1] = 5\%$.

Inference: When the personal health information of the patient PA reaches the specific physician, based on the Bayesian inference, we have

$$P_r[E1 \mid E2] = \frac{P_r[E2 \mid E1].P_r[E1]}{P_r[E2 \mid E1].P_r[E1] + P_r[E2 \mid \neg E1].P_r[\neg E1]} = .84$$

The collaborative and distributive nature of WSN makes trust management a challenging task. Trust can be defined as the shelter when an entity faces uncertainty and risk [9]. WSN nodes are deployed in unknown areas and remote terrains so that the hostility of the environment and uncertainty and unpredictability of the threats are the reasons of a robust trust management in WSN installation and operation. Reputation-based trust modeling is a suitable modeling for trust management in such kind of environment [10]. The subjective nature of trust and its goal of finding a reputation framework helps WSN privacy to utilize the robust trust management to address some of its issues [11,12]. The distributed trust model [13] utilizes the concept of recommendation protocol to exchange, revoke, and refresh recommendations about other network entities or particularly the target node. By executing recommendation-based trust management framework, WSN node can compute the trust level of the target node while requesting for a certain service. When the trust score of target node is above some preset threshold value, the initiator node transacts with the target node. Another interesting trust management scheme is threshold management, where partial keys are available with the nodes of a group. When a threshold number of nodes of that group participate in executing some events, they can only do it together with complete cooperation [14]. It employs Shamir's t-out-of-n scheme based on Lagrange's interpolation. Apart from communication or link-level trust management, trust establishment at the device level using trusted computing is very important [15]. For providing security and trust establishment at the physical or execution level or computing level, solution built around secure execution environment (SEE) is becoming the de facto condition [16]. An SEE is a part of the overall computing system to execute applications in a protected manner so that the attacks originating from outside the sensor node cannot tamper with the code and data belonging to the SEE. In fact, trust, security, and privacy are related and they have few intersected features [17], as depicted in Figure 14.2, where we broadly classify and distinguish the features of trust, security, and privacy. We observe that few of the features like integrity, nonrepudiation, recommendation, threshold cryptography, SMC, PPDM, PPDA, homomorphic encryption, and contextual privacy are remotely dependent for security, trust, and privacy, whereas confidentiality, authentication, reputation, anonymity, access control, and confidence score are interdependent factors for trust, security, and privacy.

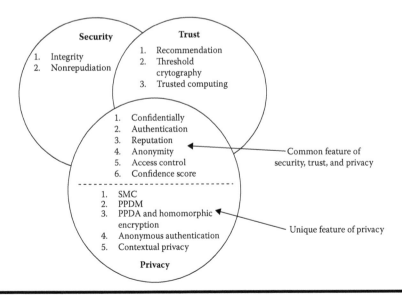

Figure 14.2 Trust–security–privacy.

14.3 Privacy in WSN: Algorithm and Schemes

In this section, we discuss about the privacy algorithms and schemes appropriate in WSN. The schemes are considered based on the different threat models that WSN devices, networks, and applications encounter.

14.3.1 Secure Multiparty Computation

SMC is defined as the cooperative secure computation of a computable function. In more generic terms, "multiparty secure computation allows N parties to share a computation, each learning only what can be inferred from their own inputs and the output of the computation. For example, the parties can compute summary statistics on their shared transaction logs, including cross-checking of the logs against counterparties to a transaction, without revealing those logs. Consider N number of interconnected nodes in a WSN and they like to send the aggregated result of the data to the server at each period T. Let $P(t) = \{ p_1, p_2, ..., p_N \}$ be the data set of the N sensors at time instant t. The objective is that the server will get know about $\sum_{n=1}^{N} p_n$ without getting information about the individual values, that is, $p(t)$ is unknown to the server. Another condition is that any node cannot know the values of other nodes. Another example is that two sensor nodes like to compare the values held by them and the larger value is reported to the server. This problem was conceived by Yao in his famous work [18], which he termed as the millionaire problem. The problem is summarized as follows: A and B are two millionaires who want to find out who is richer without revealing the precise amount of their wealth. This problem is analogous to a more general problem where there are two numbers a and b, and the goal is to solve the inequality $a \geq b$ without revealing the actual values of a and b. We consider the first problem of privacy preserving summation.

To simplify the discussion, we consider that the sensor nodes are interconnected. One node is at least connected with other and no loop is formed in routing data to the server. Consider the simple architecture in Figure 14.3, where three nodes, N1, N2, and N3 are serially connected and are directly (may be indirectly, where the connection is robust and trusted) connected with the server. We follow the scheme depicted by Ukil [19].

Suppose that the sum is in the range [0, M]. Our objective is to find out the sum Z privately without revealing the private data p_n, $i = 1, 2, ..., N$ to each other as well as to the server.

$$Z = \sum_{n=1}^{N} p_n$$

The process is initiated by the server. The server randomly chooses one of the source nodes and signals it to initiate the process. The source node first chosen by the server is denoted by N_1. This

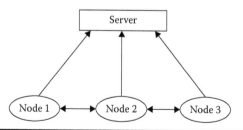

Figure 14.3 Simplified architecture for secure multiparty computation.

node possesses its private data p_1 and it generates one random number ε between the range $[0, M]$ and then computes Γ_1, where X is very large number $> \Gamma_N$:

$$\Gamma_1 = (\varepsilon + p_1) \, modX$$

After computing Γ_1, the source node N_1 performs neighborhood discovery to find out the other source nodes it is connected to. This information N_1 passes to the server. Server keeps the knowledge of the nodes already participated. If the source nodes connected to N_1 is not already participated, the server randomly chooses one of those nonparticipated source nodes and sends that message to N_1. Let this next source node be N_2. Now, accordingly N_1 passes Γ_1 to N_2. The source node N_2 computes Γ_2:

$$\Gamma_2 = (\Gamma_1 + p_2) \, modX$$

The source node follows the same procedure as N1 and sends Γ_2 to N_3. This way $N_N = 3$ is reached, which computes $\Gamma_{N=3}$:

$$\Gamma_N = (\Gamma_{N-1} + p_N) \, modX$$

The server when finds out that all the nodes are participated, it asks the last node to send Γ_N to it. Server now directs the first source node N_1 to compute the summation as

$$Z = (\Gamma_N - \varepsilon) \, modX = \sum_{n=1}^{N} p_n$$

The described scheme can be extended in multiplicative domain also, where

$$Z_m = \prod_{n=1}^{N} p_n$$

$$\Gamma_{1m} = (\varepsilon. \ p_1) \, modX$$

$$\Gamma_{Nm} = (\Gamma_{N-1}. \ p_N) \, modX$$

$$Z_m = (\Gamma_{Nm} / \varepsilon) \, modX = \prod_{n=1}^{N} p_n$$

The scheme shown above works only if all the parties are honest but curious and intends to complete the aggregation function, which may be beneficial to them. When one or more parties cheat, the whole scheme fails. For example, one node could compute everything according to the scheme until a certain point. After that, it decides to change its inputs or begin sending random values. Another problem is when any of the nodes becomes defective and forwards random values. Interested readers can refer to Kamvar et al. and Wan et al. [20,21] to understand how some of the security issues of SMC are resolved using trusted third party and gradient descent or other methods [22]. However, recent research directions are focusing on PPDA techniques using homomorphic encryption, which is much more secure to address the above-stated problems.

14.3.2 Anonymous Authentication

In SMC, user data are hidden, but user can be identified. There is another class of problem in WSN, where user data are public, but user should not be identifiable. In certain WSN, the nodes are capturing sensitive information. Consider participating sensor kind of applications (like crowdsourcing), where sensor nodes capture information of a city through street cameras and the process is helped by few registered volunteers, thus forming a larger WSN to provide different functions for making the city smart. In this process, the volunteers may capture some sensitive photos or information, which need to be sent to the server or to the administrator for (say) safety purposes. In such kind of scenarios, volunteers may not like to disclose its identity. However, they need to authenticate to the server to restrict information sharing by the authorized volunteers only. So the problem is that the verifier cannot know who the actual signer is, but only that the signer belongs to a certain set of possible signers. This is called seemingly paradoxical "anonymous authentication" [23]. One of the earliest work on anonymous authentication is done by Chaum [24], which is the famous "dining cryptographers problem." Let two WSN nodes ϖ_1 and ϖ_2 intend to broadcast message $M1$ and $M2$ without revealing their identity. It is assumed that ϖ_1 and ϖ_2 possess two secret keys $k1$, $k2$, and one binary bit b. The nodes broadcast $M1$ and $M2$ using the following policy:

1. If $b = 0$, ϖ_1 broadcasts the pair $(M1 \oplus k1, k2) = \vartheta_1^0$ and ϖ_2 broadcasts the pair $(k1, M2 \oplus k2) = \vartheta_2^0$.
2. If $b = 1$, ϖ_1 broadcasts the pair $(k1, M1 \oplus k2) = \vartheta_1^1$ and ϖ_2 broadcasts the pair $(M2 \oplus k1, k2) = \vartheta_2^1$.
3. Decipher $M1$ and $M2$ by $\vartheta_1^0 \oplus \vartheta_1^1$ and $\vartheta_2^0 \oplus \vartheta_2^1 \oplus \vartheta_2^1$.
4. If $b = 0$, the broadcast message is $(M1, M2)$, and for $b = 1$, it is $(M2, M1)$.
5. As long as $k1$, $k2$, and b are kept secret, adversary cannot learn which party broadcast which message.
6. This can be extended into N-party protocol.

Research on anonymous authentication paves a long way after the introduction of dining cryptographer problem, elicited earlier. In group signature–based schemes [25], a trusted party (group manager) predetermines a group of users and distributes keys to the members, allowing each member to sign messages anonymously on behalf of the group. However, this scheme suffers from scalability and initial setup procedure, which may not be suitable for WSN kind of applications. In 2001, Rivest, Shamir, and Tauman [6] proposed simplified group signature scheme called "ring signature." The disadvantage of group signature is that members need to fully cooperate, and expanding group means revoking the initial key establishment process. In this scheme, possible signers constitute the ring. Each ring member A_i has a RSA public key $P_i = (N_i, e_i)$. This defines trap-door one way permutation f_i on $\mathbf{Z}_{N_i}$ such that: $f_i(x) = x^{e_i} \bmod N_i$. For each f_i, generate extended trap-door permutation g_i over $\{0,1\}^b$ such that

$$g_i(m) = \begin{cases} q_i N_i + f_i(r_i), & \text{if } (q_i + 1) \, N_i \leq 2^b \\ m, & \text{otherwise} \end{cases}$$

where $m = q_i N_i + r_i, 0 \leq r_i < N_i, 0 \leq r_i < N_i$
b is a 160 bits larger than any N_i for 1024 bit RSA.

Now consider a symmetric key encryption algorithm E such that it takes any key k as input and outputs a permutation function E_k over b-bit strings. From extended trap-door permutation g_i, key k, initialized value v, and symmetric encryption, we can obtain the combining function $C_{k,v}(g_1(x_1), \ldots, g_r(x_r))$ whose range is $\{0,1\}^b$. The combining function $C_{k,v}(y_1, \ldots, y_r)$ where $y_i = g_i(x_i)$ has the following properties:

- For each s, $1 \leq s \leq r$ and for any fixed values of all the other inputs y_i, $i \neq s$, $C_{k,v}$, y is a one-to-one mapping from y_s to z.
- Given z and all inputs y_i, except y_s, it is possible to find efficiently y_s such that $C_{k,v}$ $(y_1, \ldots, y_r) = z$ for $x_{1, \ldots,} x_r$ if the adversary cannot invert $g_1, \ldots, g_r$.
- For example, the combining function can be

$$y_1 = g_1(x_1), \, t_1 = E_k(v \oplus y_1)$$

$$y_2 = g_2(x_2), \, t_2 = E_k(t_1 \oplus y_2)$$

$$\ldots$$

$$y_r = g_r(x_r), \, z = t_r = E_k(t_{r-1} \oplus y_r)$$

Now we can generate the ring signature as follows:

- Given the message m to be signed, signer's secret key S_s, and the sequence of public keys $P_1, \ldots, P_r$ of all the ring members, signer computes ring signature as follows:
  - Compute symmetric key $k = h(m)$ as hah of m.
  - Pick a value v uniformly at random from $\{0,1\}^b$.
  - Pick random x_i for all other ring members $1 \leq i \leq r$, $i \neq s$ uniformly and independently from $\{0,1\}^b$ and computes $y_i = g_i(x_i)$.

Solve y_s from $C_{k,v}(y_1, \ldots, y_r) = v$
Calculate $x_s = g_s^{-1}(y_s)$
Ring signature: $(P_1, \ldots, P_r, v, x_1, \ldots, x_r)$.

- At the receiver, ring signature is verified as
  Compute $y_i = g_i(x_i)$, for $1 \leq i \leq r$
  - Verifier hashes the message m to find the encryption key k.
  - Verifier checks whether y_i's satisfy $C_{k,v}(y_1, \ldots, y_r) = v$. If y_i's satisfy, verifier accepts the signature as valid, else rejects.

This ring signature scheme can be viewed in Figure 14.4.

14.3.3 Privacy Preserving Data Mining

It is a challenging task to protect personally identifiable information while sharing information in a distributed system like WSN. With the immergence of sophisticated data mining tools based on statistical and semantic analysis, data privacy against data mining is becoming an interesting research area. In recent years, data mining has been viewed as a threat to privacy because of the

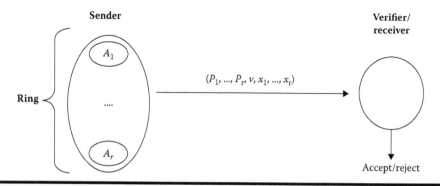

Figure 14.4   Rivest ring signature.

widespread proliferation of electronic data maintained by corporations. This has led to increased concerns about the privacy of the underlying data. In literature, a number of techniques have been proposed for modifying or transforming the data in such a way so as to preserve privacy [26,41]. To design a privacy preserving data publishing system, we must first quantify the very notion of privacy or information loss. From data privacy perspective, privacy can be defined as the measure of probability of leakage of information to unintended sources. Another possible parameter is assessing anonymity of the sensitive information, which is measured by the amount of obfuscation at the point of releasing data. Another lever is data perturbation, which refers to a data transformation process typically performed at the data source before publishing. Perturbation techniques are often evaluated with two basic metrics: level of privacy guarantee and level of model-specific data utility preserved, which are often measured by the loss of accuracy for data classification and data clustering [27]. From WSN PPDM perspective, we consider data perturbation technique, where data (mostly the sensitive attributes) are perturbed before publishing so that private information release becomes as minimum as possible under the utility constraints. Data perturbation techniques are one of the preferred models for PPDM. It is especially useful for applications where data owners want to participate in cooperative mining while simultaneously preventing the leakage of privacy sensitive information in their published database or data sets. Data perturbation refers to a transformation or mapping process mostly performed at the data source before transmitting, recording, or storing the data. By performing the transformation act, the sensitive attributes are masked at the public or published domain. However, the mapping process (original data set mapped to perturbed data set) can be such that use case or requirement-specific properties can be retained. Data perturbation consists of two methods: (1) data randomization and (2) data generalization.

14.3.3.1 Data Randomization

In a large sample of sensor database, almost any combination of continuous scaled variable has unique property or is more specifically unique in nature. When an attacker uses statistical matching techniques, sensitive data can be retrieved from masked data set. To prevent such kind of attacks, data randomization method is used to distort the sensitive data set and quasi-identifiers, which provides randomness bias to the overall data base.

Let us consider a set of data records denoted by $X = \{x_1, x_2, \ldots, x_N\}$ with the property $X \sim (\mu, \Sigma)$ Let us add additive noise ε, which is drawn from the probability distribution $f_p(\varepsilon)$ with the property $\varepsilon \sim \aleph(0, \Sigma_\varepsilon)$ and $\Sigma_\varepsilon = \gamma.\mathrm{diag}\left(\sigma_1^2, \sigma_2^2, \ldots, \sigma_N^2\right)$, where $\gamma > 0$ and $Cov\left(\varepsilon_i \neq \varepsilon_j\right)$,

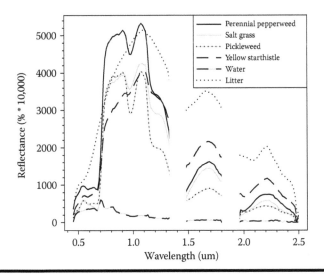

Figure 14.5   Data distribution after additive noise addition.

$\forall$ i $\neq$ j. The noise components are drawn independently and are denoted as ε_1, ε_2, ..., ε_n [28]. Therefore, the new set of distorted records obtained is denoted by

$$Z = \left\{ z_1, z_2, ..., z_N \right\}$$

and

$$z_1 = x_1 + \varepsilon_1$$
$$z_2 = x_2 + \varepsilon_2$$
$$....$$
$$z_N = x_N + \varepsilon_N$$

We assumed that the variance of the additive noise is large enough so that the original record values cannot be easily guessed from the distorted data. So we can retrieve the distribution pattern of the added noise, but the original records cannot be recovered. This is depicted in Figure 14.5, which clearly shows that adding random additive noise distorts the original value but retains the data pattern. For simplicity, we have used statistically sorted input data.

One key advantage of the randomization method is that it is relatively simple and does not require knowledge of the distribution of other records in the data. However, when data sets of related attributes is to be perturbed, additive randomization does not yield good result and containing the statistical property (approximately) in published database is not feasible. So, in such scenario, data generalization methods like *k*-anonymity and *l*-diversity are to be used.

14.3.3.2 Data Generalization

In data generalization method, sensitive data or quasi-identifiers are mapped to another domain, which enables statistical disclosure control. There exists fair distinction between sensitive data and quasi-identifiers. Data sets or the attributes that should not be disclosed in public is called sensitive data or sensitive attribute. Quasi-identifiers are those attributes or combination of attributes which on their own are nonsensitive in nature, but on combination with external data, they are capable of identifying private records. Another important term in reference to PPDM is "equivalence class," which is defined as the set of tuples that cannot be distinguished from each other with respect to quasi-identifiers. The motivating factor behind the generalization method is that many attributes in the data set are pseudo-identifiers. When they are analyzed with respect to the publicly available records, sensitive data (which is masked or randomized) can be retrieved back. For example, if in a certain medical database, patient's name is obfuscated, attacker can identify the patient by analyzing the attributes like zip code, birth date, doctors attended, and so on. In fact, k-anonymity was developed to address the problem of indirect inference to identify private data from public record [29].

Let $T = \{t_1, t_2, ..., t_n\}$ be a table with attributes $\{A_1, A_2, ..., A_m\}$. We assume that T is a subset of some larger population Ω, where each tuple represents an individual from the population. Let Λ denote the set of all attributes $\{A_1, A_2, ..., A_m\}$ and $t[A_i]$ denote the value of attribute A_i for tuple t. Now if $C = \{C_1, C_2, ..., C_p\} \subseteq \Lambda$, then we use the notation $t[C]$ to denote the tuple $(t[C_1], ..., t[C_p])$, which is the projection of t onto the attributes in C while S denote the set of all sensitive attributes and N denote the set of all nonsensitive attributes.

k-anonymity [29] is a property that captures the protection of released data against possible reidentification of the respondents to whom the released data refer. It reduces the granularity of the data representation such that any given record maps onto at least k other records in the data. A table T satisfies k-anonymity if for every tuple $t \in T$ there exist $k - 1$ other tuples $t_{i_1}, t_{i_2}, ..., t_{i_{i-1}} \in T$ such that $t[C] = t_{i_1}[C] = t_{i_2}[C] = ..., t_{i_{i-1}}[C]$ for all $C \in QI$. We have shown how to construct k-anonymized table from normal data set in Tables 14.1 and 14.2. Here, we have shown two anonymized representations.

k-anonymity suffers from homogeneity and background knowledge attack. Therefore, the technique of l-diversity was proposed, which not only maintains the minimum group size of k but also focuses on maintaining the diversity of the sensitive attributes. l-diversity puts constraints on minimum number of distinct values seen within a equivalence class for any sensitive attribute. This property of l-diversity nullifies background attack, predominant in k-anonymity. When there are l or more well-represented values for the sensitive attribute in an equivalence class, then the table is said to contain l-diversity. Let a q^*-block be a set of tuples such that its nonsensitive values generalize to q^*. A q^*-block is l-diverse if it contains l "well-represented" values for the sensitive attribute S [30]. It is known that identity disclosure is successfully handled by l-diversity. However, it does not prevent attribute disclosure when the overall distribution is skewed. In l-diversity, all values of a given attribute are treated in a similar way even if they are semantically related. Similarity attacks take place when the sensitive attributes are semantically similar. Similarity attacks are the main motivation for the researchers to look for more robust technique like t-closeness [31].

Table 14.1 Normal Data Representation

Sensor Serial No.	Sensor Location (Zip Code)	Sensor Data (X)	Sensor Data (Y)
1	100011	3	11
2	100012	4	14
3	100013	3	12
4	100014	7	10
5	100015	4	19
6	100016	6	15

Table 14.2   Privacy Preserved Data Representation with Generalization Using *k*-anonymity (*k* = 2)

Sensor Serial No.	Sensor Location (Zip Code)	Sensor Data (X)	Sensor DATA (Y)
1	100011–10012	3–4	11–14
2	100011–10012	3–4	11–14
3	100013–10014	3–7	10–12
4	100013–10014	3–7	10–12
5	100015–10016	4–6	15–19
6	100015–10016	4–6	15–19

Another approach traditionally used is cluster-based data generalization. It is assumed that quasi-identifiers can be represented in an arbitrary metric space. The idea is to cluster the points within a predefined threshold distance with minimum number of points and publish only the cluster centers. Instead of publishing individual or generalized values, cluster center of a cluster of *r* points can be beneficial in many applications.

14.3.4 Privacy Preserving Data Aggregation

In variety of WSN applications, aggregated data are of importance, whereas individual source data are confidential in nature. So the objective is to aggregate the sensor data without getting to know the content of the data. Consider the application of smart meters, where smart meters are distributed within a certain region and the energy provider bills the consumer periodically. But there are two problems:

1. Smart meter data are very much sensitive in nature, analyzing which pattern of users' presence, absence, and other activities in the residence can be derived. So the user may not like to share his/her consumption data without reaching certain threshold granularity. Likewise, only monthly consumption for billing purpose can be shared.
2. Smart grid is responsible for prediction, computation, and distribution of energy to the household, and it requires granular information from the smart meter, contradicting the user concern. However, smart grid is interested in getting aggregated energy consumption value of its concerned area.

So the solution is PPDA, where aggregation is done without knowing the smart meter reading. The problem can be modeled as depicted in Figure 14.6. Consider N number of sources (smart meter) and one aggregator (smart grid). The set of data at the sources at time instant T is

$$D\,|_T = \{d_1, d_2, \ldots, d_N\}$$

Sources to preserve privacy send another set of data to the aggregator:

$$X\,|_T = \{x_1, x_2, \ldots, x_N\}$$

where $x_i \neq d_i$, $\forall i$

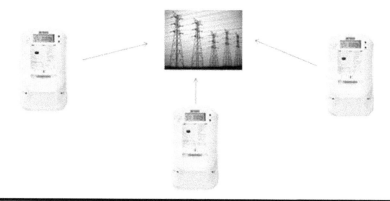

Figure 14.6 Privacy preserving data aggregation.

Aggregator, using privacy preserving algorithm, can compute the aggregation (for simplicity, we assume SUM as the aggregation function):

$$Y\mid_T = f_p\left\{x_1, x_2, ..., x_N\right\} = \sum_{n=1}^{N} d_n$$

$$\sum_{n=1}^{N} d_n$$

There are different solutions to this problem. Like SMC as described earlier, SMART (slice-mixed aggregation) scheme [32], which slices original data into pieces and then recombines them randomly, is one of the solutions to this problem. However, these solutions are vulnerable to different types of threats and some are very complex (both computational and communication overhead). One of the most elegant solution is homomorphic encryption [33] in which the researchers are showing immense interest. Homomorphic encryption scheme allows to efficiently compute arbitrary functions over encrypted data without the decryption key (i.e., given encryptions $E(d_1)$, ..., $E(d_N)$ of d_1, ..., d_N for any computable function f). This problem was posed by Rivest et al. in 1978 [34]. For normal encryption process, data needs to be decrypted to plain text to perform some computations on the plain text data, as shown in Figure 14.7.

However, in this method, the plain text that contains private data is revealed after decryption. Homomorphic encryption eliminates the step of decryption for computing purpose (Figure 14.8). It is required that the homomorphic public key cryptosystem needs to support secure (k, k)-threshold decryption so that the corresponding private key is shared by a group of n players. Successful decryption is possible only when all the entities are acting together. This results in a decryption protocol, which is secure against malicious players. This is normally performed when all the entities to prove in zero-knowledge and to follow the threshold decryption protocol.

Homomorphic encryption as defined above is called fully homomorphic, which is a challenging research topic [34]. Here, we will consider additive and multiplicative homomorphic encryption schemes in the range of the integer, where f = SUM/MULTIPLY. We formally define homomorphic encryption scheme as follows:

Let R and S be sets and we regard an encryption function E: R → S

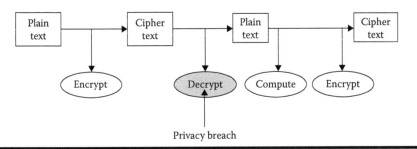

Figure 14.7 Normal encryption process.

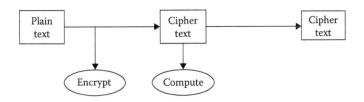

Figure 14.8 Homomorphic encryption process.

1. Additively homomorphic when $E(x \text{ ADD } y) = \text{Add } (E(x), E(y))$
2. Multiplicatively homomorphic when $E (x \text{ MULTI } y) = \text{Multi } (E(x), E(y))$

Basic RSA is a multiplicatively homomorphic encryption scheme with public key $p_k = \{N, e\}$, then the cipher text ξ of plain text ϕ is $\xi = \varepsilon(\phi) = \phi^e \, mod \, N$.

So

$$\varepsilon(\phi_1) . \, \varepsilon(\phi_2) = (\phi_1 . \, \phi_1)^e \, mod \, N = \varepsilon(\phi_1 . \, \phi_2)$$

More generally,

$$\prod_{k=1}^{K} \varepsilon(\phi_k) = \varepsilon \left(\prod_{k=1}^{K} \phi_k \right)$$

For additive homomorphic encryption,

$$\sum_{k=1}^{K} \varepsilon(\phi_k) = \varepsilon \left(\sum_{k=1}^{K} \phi_k \right)$$

Paillier cryptosystem is a case for additive homomorphic encryption [35]. This scheme applies composite residuosity class problem in public key cryptography and consist of three encryption algorithms: one trap-door permutation and two homomorphic probabilistic encryption algorithms. The decisional composite residuosity assumption is the intractability hypothesis upon which this cryptosystem is based. We are interested in the homomorphic property of this scheme. The scheme is an additive homomorphic cryptosystem; this means that, given only the public key and the encryption of x_1 and x_2, one can compute the encryption of $x_1 + x_2$. Let us consider public key is the modulus N and the base e, then the encryption of a message ϕ is

$$\varepsilon(\phi) = \left(e^{\phi} r^N\right) \ mod \ N$$

$$\varepsilon(\phi_1) \cdot \varepsilon(\phi_2) = \left(e^{(\phi_1 + \phi_2)} \left(r_1 r_2\right)^N\right)$$

$$= \varepsilon(\phi_1 + \phi_2 \ mod \ N)$$

14.3.5 Contextual Privacy Preservation

With the advent of ubiquitous computing and Internet of Things (IoT), information does not reside entirely on the data; contextual information also contributes enormously for complete understanding of the system, service, and helps to do the analytics. An adversary might deduce sensitive information by observing and analyzing the contextual data. The derived knowledge then compared and correlated with its background knowledge and published records. Attacker may obtain the sensitive information (approximately) which are supposedly hidden. For simplicity, we assume context information in the form of time and location. Context and location awareness turn devices like mobile phones into ubiquitous tools for quantifying personal patterns and habits. For context privacy, we mainly consider the protection of location and temporal information. We consider an adversary that attempts to obtain information about mobile nodes based on observed messages. In practice, the adversary can be a rogue individual that deploys its own infrastructure (e.g., by placing eavesdropping devices in the network) or a set of malicious legitimate nodes. In the worst case, the adversary is global: it has a complete coverage and eavesdrops communications throughout the entire network. The goal of the adversary is to track users' locations from which it can implicitly obtain the true identity of the owner of the device. Similarly, temporal privacy also needs to be protected. Linking location and temporal information of a device, attacker can acquire number of private information of the user. It is observed that context information is spread throughout the network layer from physical layer, link layer, network layer to application layer. The basic approach is to introduce privacy wrapper at middleware, which uses contextual information to control the sharing of private data or to obfuscate the contextual information altogether. Privacy wrapper dynamically determines when to share personal information. Intuitively, the privacy wrapper monitors the network, application type, nature of the private data and determines whether the context provides enough privacy for users to share their personal information or obfuscates the private data according to the context (Figure 14.9).

To preserve and control contextual information privacy, confusion distance and decorrelation time–based approach can be considered. Confusion distance is the minimum distance in which two similar device's location is indistinguishable. For example, in smart home use case, it may be the distance between two consecutive smart homes. In decorrelation time–based scheme, the adversary cannot figure out the relation of the sender and the sent data. One approach of decorrelation is to divide the total estimated transmission time T into n slots and data are also sliced in n parts and then to send the data in random fashion within the time slot T. This method may not be applicable for hard real-time systems.

Consider that the contextual information regarding any of the sensor nodes $(c_i, i \in N)$ needs to be protected. These information are $c_i = \{c_{it}, c_{il}\}$, which are for temporal and location identity of the nodes, respectively. The requirement of contextual privacy is that c_i, which is implicitly

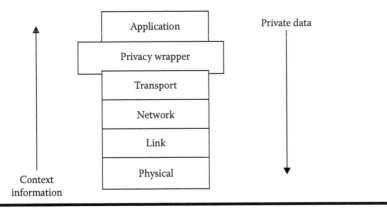

Figure 14.9 Privacy wrapper for context privacy.

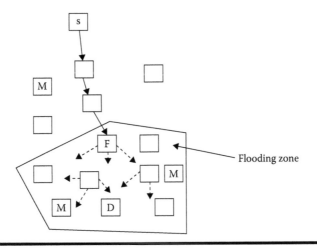

Figure 14.10 Location privacy preservation.

or explicitly associated with the data d_i, is to be protected; that is, the attacker should find it extremely difficult to analyze c_i and the process is computationally expensive. c_{it} contains the time when sensitive data are created at data source, collected by a sensor node, and then transmitted to the base station [36].

Location privacy issue is addressed by Ukil [37], where the objective is to send data d_i to a phantom source that is far from the original source and to find a flooding zone such that it is as minimum (with respect to number nodes message needs to be flooded to) as possible, which will make traceback low probabilistic. Finding this optimality condition can be explained by the following example.

Let the probability of original source detection be P_r, which is a very small number, $P_r \rightarrow 0$. Typically, $P_r < 10^{-2}$. To find the destination (assuming that our first phase has been broken) D, sensor source node S and flooding node F are to be identified. M be the attacker nodes present inside and outside the flooding zone. To derive the location of origin (S), the attacker has to try out each of the flooded message from F. The system model is shown in Figure 14.10. Let N number of nodes are available in the flooding zone and average H number of hops possible in that zone, then message is broadcast to K number of nodes, where

$$K = \frac{N!}{H!(N-H)!}$$

So finding S from K possible nodes is of probability: $1/K$. So the optimal condition is

$$P_r < \frac{1}{K} \rightarrow K > \frac{1}{P_r}$$

Now let us consider it numerically for practical case. Consider

$$P_r = 10^{-2}$$

$$H = 3$$

N turns out to be 10 to satisfy the condition $K > \frac{1}{P_r}$.

Privacy preservation technique is heavily dependent on properly evaluating the privacy metric. Privacy metric evaluation scheme outcome can be used as a scale to measure the amount of privacy of an application. For example, certain applications like financial or insurance transaction, certain medical data require more privacy than data like users retail purchasing profile. This kind of privacy sensitive communication enables users (or their devices) to make communication decisions based on their privacy level, which allows users to regulate the sharing of their information by making dynamic privacy based decisions, or the application can dynamically regulate the release of private information. Privacy preserving scheme sets the privacy of the data or application as per the measurement of the privacy scale. It is to be remembered that privacy preservation has strong correlation with the data or context utility. Privacy preservation results in utility loss. It is obvious more the data is obfuscated, more confusion arises and more information is lost. Privacy preservation is a hard requirement that must be satisfied, while utility is to be maximized. This is an optimization problem. The objective is to

1. Find solution for utility measure.
2. Optimize point between utility and privacy.
3. Attain efficiency and scalability: even simple restriction of optimized k-anonymity is NP-hard [7].

14.4 Case Study

As discussed earlier, PPDA is an important research issue, particularly in WSN where sensors nodes are deployed in hostile environment, remotely accessed and maintained infrequently and due to in-network processing, where cluster head aggregates (processes) the data sent by source nodes to improve energy and bandwidth efficiency [32,38,42]. He et al. [32] proposed CPDA (Cluster-based Private Data Aggregation) scheme, which performs privacy preserving data aggregation in low communication overhead in a self-organized multi-hop WSN. Let us consider that two source nodes $S1$ and $S2$ report to the cluster head A, which does in-network aggregation and

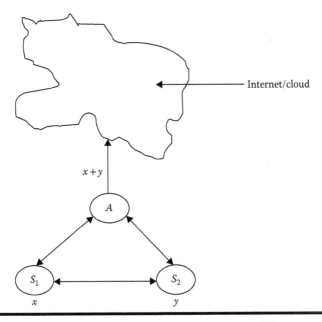

Figure 14.11   Cluster-based Private Data Aggregation system model.

forward the aggregated result, as described in Figure 14.11. We assume that $S1$, $S2$, and A do not trust each other and they possess a communication link among them. Data of $S1$ and $S2$ are x and y, respectively, and private in nature. However, aggregated value of the data of $S1$ and $S2$ is important for all three nodes.

CPDA algorithm consists of two parts:

1. Value distortion: Let the data values in the sink node $S1$ and $S2$ be x and y, and z be the dummy variable at the aggregator node "A." In the first step, the server/aggregator sends three seeds a, b, and c to the friend pairs. Based on that A computes

$$\alpha_{S1}^A = z + R_1^A b + R_2^A b^2$$

$$\alpha_{S2}^A = z + R_1^A c + R_2^A c^2$$

$$\alpha_A^A = z + R_1^A a + R_2^A a^2$$

where R_1^A and R_2^A are two random numbers generated by A.

Similarly, $S1$ computes

$$\alpha_{S1}^{S1} = x + R_1^{S1} b + R_2^{S1} b^2$$

$$\alpha_A^{S1} = x + R_1^{S1} a + R_2^{S1} a^2$$

$$\alpha_{S2}^{S1} = x + R_1^{S1} c + R_2^{S1} c^2$$

Similarly $S2$ computes

$$\alpha_A^{S2} = y + R_1^{S2}a + R_2^{S2}a^2$$

$$\alpha_{S1}^{S2} = y + R_1^{S2}b + R_2^{S2}b^2$$

$$\alpha_{S2}^{S2} = y + R_1^{S2}c + R_2^{S2}c^2$$

where R_1^{S1} and R_2^{S1} are two random numbers generated by sink node $S1$; R_1^{S2} and R_2^{S2} are other two random numbers generated by sink node $S2$. After that, the calculated α_{S1}^A and α_{S2}^A are sent to sink node $S1$ and sink node $S2$ by A, securely as described earlier. Similarly, α_A^{S1} and α_{S2}^{S1} are sent to sink node $S2$ and A by sink node $S1$ and α_A^{S2} and α_A^{S2} and α_{S1}^{S2} are sent to A and sink node $S1$ by sink node $S2$.

2. Value aggregation: After the private data values (x and y) are distorted, all the nodes aggregates the values available to them and generates aggregated result. Sink node $S1$ calculates Ψ_{S1}, sink node $S2$ calculates Ψ_{S2}, and A calculates Ψ_A:

$$\Psi_A = \alpha_A^A + \alpha_A^{S1} + \alpha_A^{S2} = (x + y + z) + R_1 a + R_2 a^2$$

$$\Psi_{S1} = \alpha_{S1}^A + \alpha_{S1}^{S1} + \alpha_{S1}^{S2} = (x + y + z) + R_1 b + R_2 b^2$$

$$\Psi_{S2} = \alpha_{S2}^A + \alpha_{S2}^{S1} + \alpha_{S2}^{S2} = (x + y + z) + R_1 c + R_2 c^2$$

where $R_1 = R_1^A + R_1^{S1} + R_1^{S2}$ and $R_2 = R_2^A + R_2^{S1} + R_2^{S2}$. These aggregated results from sink node $S1$ and sink node $S2$ are securely sent to the aggregator A. Now, the aggregator has the simple task to solve the above equation for $(x + y + z)$ with the knowledge of the values of a, b, c and Ψ_A, Ψ_{S1}, and Ψ_{S2}. After solving for $D = x + y + z$, node A internally knows its own data z, so it can find out the result $(x + y)$.

However, CPDA suffers from two critical limitations:

1. Computation of the privacy preservation algorithm increases with the number of sink nodes.
2. In most of the practical scenarios, the sink nodes cannot communicate directly with each other in a peer-to-peer mode. In such cases, usefulness of CPDA is doubtful.

SPPDA (Simplified Privacy Preserving Data Aggregation) scheme proposed by Ukil [38] eliminates these two limitations. SPPDA introduces intelligent cluster formation and robust key management with the privacy preservation scope of CPDA. Instead of probabilistic cluster formation in CPDA, SPPDA uses deterministic cluster formation with in-built intelligence. SPPDA adapts to newer configuration in the event that nodes join or participated nodes leave the network. The key management part consists of source to aggregator key establishment and source to source key establishment. The advantage of less computational time results in smooth running of real-time applications. However, SPPDA is developed for single-hop sensor networks and it is assumed that n-hop WSN is virtual n-layer single-hop WSN. For approximate aggregation, use the generic privacy preserving algorithm [39].

14.5 Future Research Direction

With the advent of cloud computing, users' data is hosted by a third party cloud service provider like Google or Amazon. This is a paradigm shift in the computing world heralding the era of thin client. Cloud computing comes with various advantages. However, the most important thing that will determine the success of cloud computing is the amount of confidence cloud service provider can assure to its customers in the domain of privacy preservation. With the inevitable redundancy of infrastructure and backup storage in the traditional cloud architecture, the pertinent questions are

- How can cloud service provider assure the clients or tenants that the data privacy is protected as per the requirement and demand of the customers?
- How can data owner control the privacy of its data? The customer should decide the access control policy by themselves.

The giant resource management and enormous computing obligation of cloud infrastructure, meeting the above-mentioned conditions are challenging in nature. In fact, private multiclient applications are important part of overall cloud architecture as they characterize cloud standards and protocols. There are different proposals, and research directions are given by researchers and scientists. The common objective is to find an optimal but efficient trustworthy computation for private multiclient applications. One of the pioneering work is done by Gentry [40] by proposing fully homomorphic encryption. However, this elegant scheme is far complex to realize and implement. Finding a computationally efficient fully homomorphic encryption is a exigent and "hot" research topic in today's privacy computing research particularly when dealing with cloud and other distributed computing like IoT paradigm.

Another challenging research problem is privacy encoding, where different encoding schemes can be made privacy preserved. For example, in case of efficient in-network processing, networking coding technique is used in recent WSNs. Network coding–based privacy preserving scheme against traffic analysis in multi-hop wireless networks is to be considered as a pioneering direction for privacy research in WSN. This kind of research should focus on packet flow untraceability, bidirectional message confidentiality, and source–destination unlinkability with the objective of thwarting traffic analysis and flow tracing attacks.

In PPDM, t-closeness is found be the most advanced scheme to nullify the strongest data mining techniques. However, in t-closeness, correlation between different attributes is lost as each attribute is generalized separately terminating their dependence and association. T-closeness also affects the utility when t is very small. And consequently, small t will result in increase in computational time. So finding a new PPDM technique that provides strong resistance against different data mining attacks and also at least offers near optimal privacy-utility tradeoff is an interesting research problem.

With the advent of IoT, WSN is becoming ubiquitous and penetrating the fabric of human life by allowing tiniest, most common, and trivial devices (sensors) to be connected to Internet. To protect privacy in this kind of ecosystem and provide a holistic privacy feature to such systems, a separate privacy layer needs to be incorporated between application layer and application program interface as shown in Figure 14.12. Privacy layer mainly consists of three components:

1. Privacy middleware
2. Privacy policy language
3. Privacy in database

Designing and developing privacy layer for WSN-specific applications is very challenging. The basic architecture is similar as depicted in Figure 14.13. The main challenge lies in defining

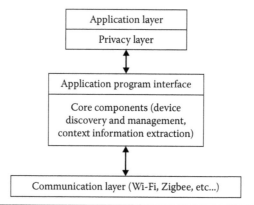

Figure 14.12 Privacy layer in WSN (Internet of Things) software stack.

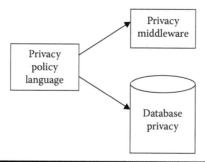

Figure 14.13 Privacy layer architecture.

an appropriate privacy policy and designing the middleware optimized for that privacy policy. Another important feature needs to be considered is privacy violation detection, which needs to be an integral part of privacy preserving policy. The existing IoT middlewares do not provide a uniform and comprehensive suite of security and privacy requirements for various data communication across heterogeneous devices using different protocol for communications in the domain of IoT.

When publishing or sharing sensitive data with a third party, the probability of privacy breach from publicly available information becomes very high due to the availability of powerful analytic tools and large storage systems. So, it is prudent to evaluate the risk of privacy breaching probability while sharing sensitive data to third parties by estimating the trustworthiness of the third party [43]. Another interesting issue is to design PPDM algorithms as games. Most of the existing PPDM algorithms assume that each participant acts cooperatively and behaves harmlessly. However, in practical systems, this may be the ideal scenario, not the real one. Every entity in real life likes to maximize its own benefit by launching different types of attacks like collusion attack. It is an open problem to design PPDM algorithms, which can resist such malpractices or can achieve optimal solution in presence of such rogue nodes. Game theoretic approach is best known tool to design such algorithms.

14.6 Conclusion

In this chapter, we have presented the motivation, requirements, objectives, and schemes for privacy preservation in WSN. We discussed that due to widespread use of social network, surveillance, and mining tools, data privacy becomes a challenging task. In WSN, the situation is grimmer due its ad hoc, infrastructure-less, and decentralized nature. We have shown the different aspects of privacy preservation, which on the other hand finds good amount of intersection with the security and trust management primitives. We described methods for privacy preservation of data and context addressing the problems particularly unique in WSN. Finally, we have investigated two related schemes for PPDA as a case study to show how privacy preservation in WSN research is evolving. Another important observation is that privacy research in WSN is relatively new, compared to trust and security research. However, privacy research is enriched with good amount of ongoing research activities diversifying privacy research from application-specific issues to functional models to architectural descriptions.

Acknowledgment

I would like to thank Prof. Subhamoy Maitra of the Indian Statistical Institute in Kolkata, India for many useful discussions, particularly for the section on anonymous authentication.

References

1. Milgram, S. "Small World Experiment." http://en.wikipedia.org/wiki/Small_world_experiment. Accessed on July 18, 2012.
2. European Union Directive. "Consumer Privacy." http://www.cdt.org/issue/consumer-privacy. Accessed July 18, 2012.
3. Agrawal, R., and R. Srikant. 2000. "Privacy Preserving Data Mining." In *ACM SIGMOD Conference Management of Data*, pp. 439–50. New York: ACM.
4. Papadopoulos, S., A. Kiayias, and D. Papadas. 2011. "Secure and Efficient In-Network Processing of Exact SUM Queries." *IEEE ICDE*, 528, pp. 517–28.
5. Goldreich, O. 2001."Secure Multi-Party Computation." Working Draft, Version 1.3. http://www.wisdom.weizmann.ac.il/~oded/pp.html. Accessed on 18 July, 2012.
6. Rivest, R. L., A. Shamir, and Y. Tauman. 2001. "How to Leak a Secret." *ASIACRYPT*, pp. 552–65.
7. Bayardo, R. J., and R. Agrawal. 2005. "Data Privacy through Optimal k-anonymization." In *Proceedings of the 21st International Conference on Data Engineering*, pp. 217–8. IEEE.
8. Deng, J., R. Han, and S. Mishra. 2006. "Decorrelating Wireless Sensor Network Traffic to Inhibit Traffic Analysis Attacks." *Pervasive and Mobile Computing Elsevier*, 2(2): 159–86.
9. Artz, D., and Y. Gil. 2007. "A Survey of Trust in Computer Science and the Semantic Web." *Journal of Web Semantics*, 5 (2): 58–71.
10. Marsh, S., and M. R. Dibben. 2005. "Trust, Untrust, Distrust and Mistrust—An Exploration of the Dark(er) side." In *iTrust 2005*, LNCS vol. 3477.
11. Srinivasany, A., J. Teitelbaumy, H. Liangz, J. Wuyand, and M. Cardei. "Reputation and Trust-Based Systems for Ad Hoc and Sensor Networks." In *On Trust Establishment in Mobile Ad-Hoc Networks*, edited by A. Boukerche. New York: Wiley & Sons.
12. Boukerche, A., L. Xu, and K. El-Khatib. 2007. "Trust-Based Security for Wireless Ad Hoc and Sensor Networks." *Computer Communications*, 30(11–12): 2413–27.
13. Abduhl-Rahman, A., and S. Hailes. 1997. "A Distributed Trust Model." In *Workshop on New Security Paradigms*, pp. 48–60. Langdale, Cumbria, United Kingdom: IEEE.

14. Ertaul, L., and N. Chavan. 2005. "Security of Ad Hoc Networks and Threshold Cryptography." In *International Conference on Wireless Networks, Communications, and Mobile Computing*, pp. 69–74.
15. Trusted Computing Group. https://www.trustedcomputinggroup.org. Accessed July 18, 2012.
16. Ukil, A., J. Sen, and S. Koilakonda. 2011. "Embedded Security for Internet of Things." *IEEE NCETAC*, pp. 1–6.
17. Ukil, A. 2010. "Trust and Reputation Based Collaborating Computing in Wireless Sensor Networks." *CIMSIM*, pp. 464–9.
18. Yao, A. 1982. "Protocols for Secure Computations." In *Proceedings of the 23rd Annual Symposium on Foundations of Computer Science*, pp. 160–4.
19. Ukil, A. 2010. "Privacy Preserving Data Aggregation in Wireless Sensor Networks." *ICWMC*, pp. 435–40.
20. Kamvar, S. D., M. T. Schlosser, and H. G. Molina. 2003. "The Eigentrust Algorithm for Reputation Management in p2p Networks." In *International World Wide Web Conference*, pp. 640–41. New York: ACM.
21. Wan, L., W. K. Ng, S. Han, and V. C. S. Lee. 2007. "Privacy Preservation for Gradient Descent Methods." In *ACM SIGKDD International Conference on Knowledge Discovery and Data Mining*, pp. 775–83. New York: ACM.
22. Lindell, Y., and B. Pinkas. 2007. "An Efficient Protocol for Secure Two-Party Computation in the Presence of Malicious Adversaries." *EUROCRYPT*.
23. Lindell, Y. 2010. "Anonymous Authentication." *Journal of Privacy and Confidentiality*, 2(2–4): 35–63.
24. Chaum, D. L. 1988. "The Dining Cryptographers Problem: Unconditional Sender and Recipient Untraceability." *Journal of Cryptology*, 1(1): 65–75.
25. Chaum, D., and E. van Heyst. 1991. "Group Signatures." In *Advances in Cryptology—EUROCRYPT*, edited by D. W. Davies, pp. 257–65. Springer-Verlag.
26. Verykios, V. S., E. Bertino, I. N. Fovino, L. P. Provenza, Y. Saygin, and Y. Theodoridis. 2004. "State-of-the-Art in Privacy Preserving Data Mining." In *ACM SIGMOD* 33 (1), 50–57.
27. Aggarwal, G., T. Feder, K. Kenthapadi, S. Khuller, R. Panigrahy, D. Thomas, and A. Zhu. 2006. "Achieving Anonymity via Clustering." *ACM PODS*, 153–62.
28. Brand, R. 2004. "Microdata Protection through Noise Addition." *Privacy in Statistical Databases*, Springer LNCS vol. 3050, pp. 347–59. Springer.
29. Samarati, P., and L. Sweeney. 1998. "Protecting Privacy When Disclosing Information: k-Anonymity and its Enforcement through Generalization and Suppression." *IEEE Symposium on Security and Privacy*. IEEE.
30. Liu, J., and K. Wang. 2010. "On Optimal Anonymization for l-Diversity." *IEEE ICDE Conference*, pp. 213–24. IEEE.
31. Li, N., T. Li, and S. Venkatasubramanian. 2007. "t-Closeness: Privacy Beyond k-Anonymity and l-Diversity." *ICDE Conference*.
32. He, W., X. Liu, H. Nguyen, K. Nahrstedt, and T. T. Abdelzaher. 2007. "PDA: Privacy-Preserving Data Aggregation in Wireless Sensor Networks." In *IEEE INFOCOM*, pp. 2045–53.
33. Gentry, C., and S. Halevi. 2011. "Implementing Gentry's Fully-Homomorphic Encryption Scheme." *EUROCRYPT*, pp. 129–86.
34. Rivest, R., L. Adleman, and M. Dertouzos. 1978. "On Data Banks and Privacy Homomorphisms." In *Foundations of Secure Computation*, edited by R. DeMillo, D. Dobkin, A. Jones, and R. Lipton, pp. 169–80. New York: Academic Press.
35. Paillier, P. 1999. "Public-Key Cryptosystems Based on Composite Degree Residuosity Classes." *EUROCRYPT*, pp. 223–38.
36. Kamat, P., W. Y. Xu, W. Trappe, and Y. Y. Zhang. 2007. "Temporal Privacy in Wireless Sensor Networks." In *27th International Conference on Distributed Computing Systems*, p. 23. IEEE.
37. Ukil, A. 2010. "Context Protecting Privacy Preservation in Ubiquitous Computing." *IEEE CISIM*, pp. 273–8.
38. Ukil, A. 2011. "A Secure Privacy Preserved Data Aggregation Scheme in Non Hierarchical Networks." *ICCSA*, LNCS vol. 6785, pp. 436–51. Springer.

39. Zhang, W. S., C. Wang, and T. M. Feng. 2008. "GP^2S: Generic Privacy-Preservation Solutions for Approximate Aggregation of Sensor Data, Concise Contribution." *6th Annual IEEE International Conference on Pervasive Computing and Communications*, pp. 179–84.

40. Gentry, C. 2009. "Fully Homomorphic Encryption Using Ideal Lattices." In *STOC*, pp. 169–78. New York: ACM.

41. Li, N., N. Zhang, S. K. Das, and B. Thuraisingham. 2009. "Privacy Preservation in Wireless Sensor Networks: A State-of-the-Art Survey." *Elsevier Ad Hoc Networks* 9, 1501–14.

42. Ukil, A. 2010. "Security and Privacy in Wireless Sensor Networks." *Smart Wireless Sensor Networks*, Intechweb.

43. Ukil, A. 2012. "Connect With Your Friends and Share Private Information Safely." In *IEEE 9th International Conference on Information Technology: New Generations (ITNG)*, pp. 367–72.

Chapter 15

Security in Wireless Sensor Networks

Jaydip Sen

Contents

15.1 Introduction

Wireless sensor networks (WSNs) consist of hundreds or even thousands of small devices each with sensing, processing, and communication capabilities to monitor the real-world environment. They are envisioned to play an important role in a wide variety of applications ranging from critical military surveillance applications to forest fire monitoring and building security monitoring in the near future [1]. In these networks, a large number of sensor nodes are deployed to monitor a vast field in which the operational conditions are most often harsh or even hostile. However, the nodes in WSNs have severe resource constraints due to their lack of processing power, limited memory, and limited energy. Since these networks are usually deployed in remote places and left unattended, they should be equipped with security mechanisms to defend against attacks such as node capture, physical tampering, eavesdropping, denial-of-service (DoS) attacks, etc. Unfortunately, traditional security mechanisms with high overheads are not feasible for resource-constrained sensor nodes. Researchers in WSN security have proposed various security schemes that are optimized for these networks with resource constraints. A number of secure and efficient routing protocols [2–5], secure data aggregation protocols [6–11], etc., have been proposed by several researchers in WSN security.

In addition to resolving traditional security issues such as secure routing and secure data aggregation, security mechanisms deployed in WSNs should involve collaborations among nodes due to the decentralized nature of the networks and absence of any infrastructure. In real-world WSNs, the nodes cannot be assumed to be trustworthy a priori. Researchers have therefore focused on building a sensor trust model to solve the problems that are beyond the capabilities of traditional cryptographic mechanisms [12–19]. Since in most cases the sensor nodes are unattended and physically insecure, vulnerability to physical attack is an important issue in WSNs. A number of propositions exist in the literature for defense against physical attack on sensor nodes [20–29].

In this chapter, we present a comprehensive overview of various security issues in WSNs. First, we outline the constraints of WSNs, security requirements in these networks, and various possible attacks and their corresponding countermeasures. Then, a holistic view of the security issues is presented. These issues are classified into six categories: (1) Cryptography, (2) key management, (3) secure routing, (4) secure data aggregation, (5) intrusion detection, and (6) trust management. The advantages and disadvantages of various security protocols are discussed, compared, and evaluated. Some open research issues in each of these areas are also discussed.

The remainder of the chapter is organized as follows: In Section 15.2, various constraints in WSNs are discussed. Section 15.3 presents the security requirements in WSNs. Section 15.4 discusses various attacks that can be launched on WSNs. Section 15.5 presents the numerous countermeasures for all possible attacks on WSNs. Finally, Section 15.6 concludes the chapter, highlighting some future directions of research in WSN security.

15.2 Constraints in Wireless Sensor Networks

A WSN consists of a large number of sensor nodes that are inherently resource-constrained devices. These nodes have limited processing capability, very low storage capacity, and constrained communication bandwidth. These constraints are due to limited energy and small physical size of the sensor nodes. Due to these constraints, it is difficult to directly employ the conventional security mechanisms in WSNs. In order to optimize the conventional security algorithms for WSNs, it is necessary to be aware of the constraints of sensor nodes [30]. Some of the major constraints of a WSN are listed as follows:

Energy constraints: Energy is the biggest constraint for a WSN. In general, energy consumption in sensor nodes can be categorized into three categories: (1) Energy for the sensor transducer, (2) energy for communication among sensor nodes, and (3) energy for microprocessor computation. The study by Hill et al. [31] found that each bit transmitted in WSNs consumes about as much power as executing 800–1000 instructions. Thus, communication is more costly than computation in WSNs. Any message expansion caused by security mechanisms comes at a significant cost. Further, higher security levels in WSNs usually correspond to more energy consumption for cryptographic functions. Thus, WSNs could be divided into different security levels depending on energy cost [32,33].

Memory limitations: A sensor is a tiny device with only a small amount of memory and storage space. Memory in a sensor node usually includes flash memory and random access memory (RAM). Flash memory is used for storing downloaded application code, and RAM is used for storing application programs, sensor data, and intermediate results of computations. There is usually not enough space to run complicated algorithms after loading the operating system (OS) and application code. In the SmartDust project, for example, TinyOS consumes about 4K bytes of instructions, leaving only 4500 bytes for running security algorithms and applications [31]. A common sensor type—TelosB—has a 16-bit, 8-MHz reduced instruction set computing (RISC) central processing unit (CPU) with only 10K RAM, 48K program memory, and 1024K flash storage [34]. The current security algorithms are, therefore, infeasible in these sensors [35].

Unreliable communication: Unreliable communication is another serious threat to sensor security. Normally, packet-based routing of sensor networks is based on connectionless protocols and thus inherently unreliable. Packets may get damaged due to channel errors or may get dropped at highly congested nodes. Furthermore, the unreliable wireless communication channel may also lead to damaged or corrupted packets. High error rates mandate robust error-handling schemes to be implemented, leading to higher overhead on sensor nodes. In some situations, even if the channel is reliable, communication may not be so. This is due to the broadcast nature of wireless communication, as packets may collide in transit and need retransmission [1].

High latency in communication: In a WSN, multihop routing, network congestion, and processing in intermediate nodes may lead to a high latency in packet transmission. This makes synchronization very difficult to achieve. Synchronization issues are sometimes critical in security as some security mechanisms may rely on critical event reports and cryptographic key distribution [36].

Unattended operation of networks: In most cases, the nodes in a WSN are deployed in remote regions and are left unattended. The likelihood that a sensor encounters a physical attack in such an environment is therefore very high. Remote management of a WSN makes it virtually impossible to detect physical tampering. This makes security in WSNs a particularly difficult task.

15.3 Security Requirements in Wireless Sensor Networks

A WSN is a special type of network. It shares some commonalities with a typical computer network but also exhibits many characteristics that are unique to it. The security services in a WSN should protect information communicated over the network and network resources from attacks and misbehavior of nodes. The most important security requirements in a WSN are listed as follows:

Data confidentiality: The security mechanism in a WSN should ensure that no message in the network is understood by anyone except the intended recipient. In a WSN, the issue of confidentiality should address the following three requirements [30,35]: (1) A sensor node should not allow its readings to be accessed by its neighbors unless they are authorized to do so, (2) key distribution mechanism should be extremely robust, and (3) public information such as sensor identities and public keys of nodes should also be encrypted in certain cases to protect against traffic analysis attacks.

Data integrity: The mechanism should ensure that no message is altered by an entity as it traverses from the sender to the recipient.

Availability: This requirement ensures that the services of a WSN is always available even in the occurrence of an internal attack or external attacks such as a DoS attack. Different approaches have been proposed by researchers to achieve this goal. Whereas some mechanisms make use of additional communication among nodes, others propose the use of a central access control system to ensure successful delivery of every message to its recipient.

Data freshness: It implies that the data sent is recent and ensures that no adversary replays old messages. This requirement is especially important when the WSN nodes use shared keys for message communication and a potential adversary can launch a replay attack using the old key as the new key is being refreshed and propagated to all the nodes in the WSN. A "nonce" or time-specific counter may be added to each packet to check the freshness of the packet.

Self-organization: Every node in a WSN should be self-organizing and self-healing. This feature of a WSN also poses a great challenge to security. The dynamic nature of a WSN sometimes makes it impossible to deploy any preinstalled shared-key mechanism among the nodes and the base station [37]. A number of key predistribution schemes have been proposed in the context of symmetric encryption [37–40]. However, for applying public key cryptographic techniques, an efficient mechanism for key distribution is essential. It is desirable that the nodes in a WSN self-organize among themselves not only for multihop routing but also for carrying out key management and developing trust relations.

Secure localization: In many situations, it becomes necessary to accurately and automatically locate each sensor node in a WSN. For example, a WSN designed to locate faults requires accurate locations of sensor nodes to identify the faults. A potential adversary can easily manipulate and provide false location information by reporting false signal strength, replaying messages, etc., if the location information is not secured properly. Capkun et al. [41] have described a technique called "verifiable multilateration" (VM). In multilateration, the position of a device is accurately computed from a series of known reference points. The authors have used authenticated ranging and distance bounding to ensure accurate location of a node. Because of the use of distance bounding, an attacking node can only increase its claimed distance from a reference point. However, to ensure location consistency, the attacker has to prove that its distance from another reference point is shorter. As it is not possible for the attacker to prove this, the attacker can be detected. Lazos et al. [42] have described a scheme called "secure range-independent localization" (SeRLoC). The scheme is a decentralized range-independent localization scheme. It is assumed that the locators are trustworthy and cannot be compromised by any attacker. A sensor computes its location by listening to the beacon information sent by each locator, which includes the locator's location information. The beacon messages are encrypted using a shared, global symmetric key that is predistributed in the sensor nodes. Using the information from all the beacons received by a sensor node, the sensor node computes its approximate location based on the coordinates of the locators. The sensor node then computes an overlapping antenna region using a majority voting scheme. The final location of the sensor node is determined by computing the center of gravity of the overlapping antenna region.

Time synchronization: Most of the applications in sensor networks require time synchronization. Any security mechanism for WSNs should also be time synchronized. A collaborative WSN may require synchronization among a group of sensors. In one study [43], a set of secure synchronization protocols have been proposed.

Authentication: It ensures that the communicating node is the one that it claims to be. An adversary can not only modify data packets but also change a packet stream by injecting fabricated packets. Therefore, it is essential for a receiver to have a mechanism in place to verify that the received packets have indeed come from the actual sender node. In case of communication between two nodes, data authentication can be achieved through a "message authentication code" (MAC) computed from the shared secret key. A number of authentication schemes for WSNs have been proposed by researchers, most of which are for secure routing.

15.4 Security Vulnerabilities in Wireless Sensor Networks

The WSNs are vulnerable to various types of attacks. These attacks can be broadly categorized as follows [44]:

- *Attacks on secrecy and authentication*: Standard cryptographic techniques can protect the secrecy and authenticity of communication channels from outsider attacks such as eavesdropping, packet replay attacks, and modification or spoofing of packets.
- *Attacks on network availability*: Attacks on availability are often referred to as DoS attacks. The DoS attacks may target any layer of a sensor network.
- *Stealthy attack against service integrity*: In a stealthy attack, the goal of the attacker is to make the network accept a false data value. For example, an attacker compromises a sensor node and injects a false data value through that sensor node.

In these attacks, keeping the sensor network available for its intended use is essential. The DoS attacks against WSNs may permit real-world damage to the health and safety of people [29]. A DoS attack usually refers to an adversary's attempt to disrupt, subvert, or destroy a network. However, a DoS attack can be any event that diminishes or eliminates a network's capacity to perform its expected functions [29].

15.4.1 Denial-of-Service Attacks

Wood et al. have defined a DoS attack as an event that diminishes or attempts to reduce a network's capacity to perform its expected function [29]. Several standard techniques are described in the literature to cope with some of the more common DoS attacks, although, in a broader sense, the development of a generic defense mechanism against DoS attacks is still an open problem. Moreover, most of the defense mechanisms require high computational overhead and, hence, they are not suitable for resource-constrained WSNs. Since DoS attacks in WSNs can sometimes prove very costly, researchers have spent a great deal of effort in identifying various types of such attacks and devising strategies to defend against them. Some important types of DoS attacks in WSNs are discussed in Sections 15.4.1.1 through 15.4.1.4.

15.4.1.1 Physical Layer Attacks

The physical layer is responsible for frequency selection, carrier frequency generation, signal detection, modulation, and data encryption [1]. As with any radio-based medium, there exists the possibility of jamming in WSNs. There are two broad categories of attacks on WSNs in the physical layer: (1) Jamming and (2) tampering. They are described as follows:

Jamming: It is a type of attack that interferes with the radio frequencies that the nodes use in a WSN for communication [29,44]. A jamming source may be powerful enough to disrupt the entire network. Even with less powerful jamming sources, an adversary can potentially disrupt communication in the entire network by strategically distributing the jamming sources. Intermittent jamming attacks may also prove detrimental to the network [29].

Tampering: Sensor networks typically operate in outdoor environments. Due to their unattended and distributed nature, the nodes in a WSN are highly susceptible to physical attacks [45]. Physical attacks may cause irreversible damage to the nodes. An adversary can extract cryptographic keys from a captured node, tamper with its circuitry, modify the program codes, and even replace the node with a malicious sensor [28]. It has been shown that sensor nodes such as MICA2 motes can be compromised in less than 1 minute [22].

15.4.1.2 Link Layer Attacks

The link layer is responsible for multiplexing of data streams, data frame detection, medium access control, and error control [1]. Attacks at this layer include purposefully created collisions, resource exhaustion, and unfairness in allocation. A collision occurs when two nodes attempt to transmit simultaneously on the same frequency [29]. When packets collide, they are discarded and need to be retransmitted. An adversary may strategically cause collisions in specific packets such as ACK control messages. A possible result of such collisions is the costly exponential backoff. The adversary may simply violate the communication protocol and continuously transmit messages in an attempt to generate collisions. Repeated collisions can also be used by an attacker to cause resource exhaustion [29]. For example, a naive link layer implementation may continuously attempt to retransmit the corrupted packets. Unless these retransmissions are detected early, the energy levels of the nodes would be exhausted quickly. Unfairness is a weak form of DoS attack [29]. An attacker may cause unfairness by intermittently using the aforementioned link layer attacks. In this case, the adversary causes degradation of real-time applications running on other nodes by intermittently disrupting their frame transmissions.

15.4.1.3 Network Layer Attacks

The network layer of WSNs is vulnerable to different types of attacks such as spoofed routing information, selective packet forwarding, sinkhole, Sybil, wormhole, blackhole and grayhole, HELLO flood, Byzantine, information disclosure, and acknowledgment spoofing. These attacks are briefly described as follows:

Spoofed routing information: The most direct attack against a routing protocol is to target the routing information in the network. An attacker may spoof, alter, or replay routing information to disrupt traffic in the network [46]. These disruptions include creation of routing loops, attracting or repelling network traffic from selected nodes, extending or shortening source routes, generating fake error messages, causing network partitioning, and increasing end-to-end latency.

Selective forwarding: In a multihop network such as a WSN, all the nodes need to forward messages accurately for message communication. An attacker may compromise a node in such a way that it selectively forwards some messages and drops others [46].

Sinkhole: In a sinkhole attack, an attacker makes a compromised node more attractive to its neighbors by forging routing information [29,46,47]. The result is that the neighbor nodes choose the compromised node as the next-hop node to route their data through. This type of attack makes selective forwarding very simple as all traffic from a large area in the network flows through the compromised node.

Sybil attack: It is an attack in which one node presents more than one identity in a network. It was originally described as an attack intended to defeat the objective of redundancy mechanisms in distributed data storage systems in peer-to-peer networks [48]. Newsome et al. describe this attack from the perspective of a WSN [47]. In addition to defeating distributed data storage systems, the Sybil attack is effective against routing algorithms, data aggregation, voting, fair resource allocation, and misbehavior detection. Regardless of the target (voting, routing, aggregation), the Sybil algorithm functions similarly. All the techniques involve utilizing multiple identities. For instance, in a sensor network voting scheme, a Sybil attack might utilize multiple identities to generate additional "votes." Similarly, to attack the routing protocol, the Sybil attack can rely on a malicious node taking on the identity of multiple nodes and thus route multiple paths through a single malicious node.

Wormhole: A wormhole is low-latency link between two portions of a network over which an attacker replays network messages [46]. This link may be established either by a single node forwarding messages between two adjacent, but otherwise nonneighboring nodes or by a pair of nodes in different parts of a network communicating with each other. The latter case is closely related to the sinkhole attack as an attacking node near the base station can provide a one-hop link to that base station via the other attacking node in a distant part of the network.

Blackhole and Grayhole attacks: In the blackhole attack, a malicious node falsely advertises good paths (e.g., the shortest path or the most stable path) to the destination node during the path-finding process (in reactive routing protocols) or in the route update messages (in proactive routing protocols). The intention of the malicious node could be to hinder the path-finding process or to intercept all data packets sent to the concerned destination node. A more delicate form of this attack is known as the grayhole attack, in which the malicious node intermittently drops data packets thereby making its detection more difficult.

HELLO flood: Most of the protocols that use HELLO packets have the naive assumption that receiving such a packet implies that the sender is within the radio range of the receiver. An attacker may use a high-powered transmitter to fool a large number of nodes and make them believe that they are within its neighborhood [46]. Subsequently, the attacker node falsely broadcasts a shorter route to the base station and all the nodes that received the HELLO packets attempt to transmit to the attacker node. However, these nodes are out of the radio range of the attacker.

Byzantine attack: In this attack, a compromised node or a set of compromised nodes works in collusion and carries out attacks such as creating routing loops, forwarding packets in non-optimal routes, and selectively dropping packets [49]. Byzantine attacks are very difficult to detect since under such attacks networks usually do not exhibit any abnormal behavior.

Information disclosure: A compromised node may leak confidential or important information to unauthorized nodes in a network. Such information may include information regarding the network topology, geographic location of nodes, or optimal routes to authorized nodes in the network.

Resource depletion attack: In this type of attack, a malicious node tries to deplete resources of other nodes in a network. The typical resources that are targeted are battery power, bandwidth, and computational power. Attacks could be in the form of unnecessary requests for routes, very frequent generation of beacon packets, or forwarding of stale packets to other nodes.

Acknowledgment spoofing: Some routing algorithms for WSNs require transmission of acknowledgment packets. An attacking node may overhear packet transmissions from its neighboring nodes and spoof the acknowledgments, thereby providing false information to the nodes [46]. In this way the attacker is able to disseminate wrong information in the network about the status of the nodes, since some acknowledgment may arrive from nodes that are not alive in reality.

In addition to the aforementioned categories of attacks, there are various types of possible attacks on the routing protocols in WSNs. Most of the routing protocols in WSNs are vulnerable to attacks such as routing table overflows, routing table poisoning, packet replication, route cache poisoning, and rushing attacks. A comprehensive discussion on these attacks is given by Sen [50].

15.4.1.4 Transport Layer Attacks

The attacks that can be launched on the transport layer in a WSN are flooding attacks and desynchronization attacks:

Flooding: Whenever a protocol is required to maintain state at either end of a connection, it becomes vulnerable to memory exhaustion through flooding [29]. An attacker may repeatedly make new connection requests until the resources required by each connection are exhausted or reach a maximum limit. In either case, further legitimate requests are ignored.

Desynchronization: Desynchronization refers to the disruption of an existing connection [29]. An attacker may, for example, repeatedly spoof messages to an end host causing the host to

Table 15.1 Attacks on Various Layers of a WSN and Their Countermeasures

Layer	Attacks	Defense
Physical	Jamming	Spread-spectrum communication, priority messages, lower duty cycle, region mapping, mode change
Link	Collision	Error-correcting code
	Exhaustion	Rate limitation
	Unfairness	Small frames
Network	Spoofed routing information and selective forwarding	Egress filtering, authentication, monitoring
		Redundancy probing
	Sinkhole	Authentication, monitoring, redundancy
	Sybil	Authentication, probing
	Wormhole	Authentication, packet leashes by using geographic and temporal information
	HELLO flood	Authentication, verifying the bidirectional link
	Acknowledgment flooding	Authentication
Transport	Flooding	Client puzzles
	Desynchronization	Authentication

Source: Data from Wang, Y. et al., *IEEE Comm. Surveys and Tutorials*, 8, 2, 2–23, 2006.

request retransmission of missed frames. If timed correctly, an attacker may degrade or even prevent the ability of end hosts to successfully exchange data, causing them to waste energy instead attempting to recover from errors that never really exist. The possible DoS attacks and their corresponding countermeasures are listed in Table 15.1.

15.4.2 Attacks on Secrecy and Authentication

There are different types of attacks under this category. They are described in Sections 15.4.2.1 through 15.4.2.2.

15.4.2.1 Node Replication Attack

In node replication attack, an attacker attempts to add a node to an existing WSN by replicating (i.e., copying) the node identifier of an already existing node in the network [51]. A node replicated and joined to the network in this manner can potentially cause severe disruption in message communication in the WSN by corrupting the packets and forwarding them to wrong routes. This may also lead to network partitioning and communication of false sensor readings. In addition, if the attacker gains physical access to the entire network, it is possible for him or her to copy the cryptographic keys and use these keys for message communication from the replicated node. The attacker can also place the replicated node in strategic locations in the network so that he or she could easily manipulate a specific segment of the network, possibly causing a network partitioning.

15.4.2.2 Attacks on Privacy

Since WSNs are capable of automatic data collection through efficient and strategic deployment of sensors, these networks are also vulnerable to potential abuse of these vast data collections. Privacy preservation of sensitive data in a WSN is a particularly difficult challenge [52]. Moreover, an adversary may gather seemingly innocuous data to derive sensitive information if he or she knows how to aggregate data collected from multiple sensor nodes. This is in analogy to the "panda hunter problem," in which the hunter can accurately estimate the location of the panda by systematically monitoring the traffic [53].

Privacy preservation in WSNs is even more challenging since these networks make large volumes of information easily available through remote access mechanisms. Since the adversary need not be physically present to carry out the surveillance, the information gathering process can be done anonymously with very low risk. In addition, remote access allows a single adversary to monitor multiple sites simultaneously [54].

Some of the common attacks on sensor data privacy [52,54] are as follows:

- *Eavesdropping and passive monitoring*: This is the most common and easiest form of attack on data privacy. If messages are not protected by cryptographic mechanisms, an adversary can easily understand their contents. Packets containing control information in a WSN convey more information than what is accessible through the location server. Eavesdropping on these messages prove very effective for an adversary.
- *Traffic analysis*: In order to make an effective attack on privacy, eavesdropping is combined with traffic analysis. Through an effective analysis of traffic, an adversary can identify some sensor nodes with special roles and activities in a WSN. For example, a sudden increase in message communication between certain nodes signifies that these nodes have some specific activities and events to monitor. Deng et al. have demonstrated two types of attacks that can identify the base station in a WSN without even underrating the contents of the packets being analyzed in traffic analysis [55].
- *Camouflage*: An adversary may compromise a sensor node in a WSN and later on use that node to masquerade as a normal node in the network. This camouflaged node may advertise false routing information and attract packets from other nodes for further forwarding. After the packets start arriving at the compromised node, it starts forwarding them to strategic nodes where privacy analysis on the packets may be carried out systematically.

It may be noted from this discussion that WSNs are vulnerable to a number of attacks at all layers of the Transmission Control Protocol/Internet Protocol (TCP/IP) stack. However, as pointed out by Perrig et al. [56], there may be other types of attacks possible that are not yet identified. Securing a WSN against all these attacks is quite a challenging task.

15.5 Security Mechanisms for Wireless Sensor Networks

In this section, defense mechanisms for combating various types of attacks on WSNs are discussed. First, different cryptographic mechanisms for WSNs are presented. Both public key cryptography and symmetric key cryptographic techniques are discussed for WSN security. A number of key management protocols for WSNs are discussed next. Various methods of defending against DoS attacks, secure broadcasting mechanisms, and various secure routing mechanisms are also discussed. In addition, various mechanisms for defending Sybil attacks, node replication attacks, traffic analysis attacks, and attacks on sensor privacy are also presented. Finally, intrusion detection mechanisms for WSNs, secure data aggregation mechanisms, and various trust management schemes for WSN security are discussed.

15.5.1 Cryptography in Wireless Sensor Networks

Selecting the most appropriate cryptographic method is vital in WSNs as all security services are ensured by cryptography. Cryptographic methods used in WSNs should meet the constraints of sensor nodes and be evaluated by code size, data size, processing time, and power consumption. In this section, we focus on the selection of cryptographic methods in WSNs. We discuss public key cryptography first, which is followed by a discussion on symmetric key cryptography.

15.5.1.1 Public Key Cryptography in Wireless Sensor Networks

Many researchers believe that the code size, data size, processing time, and power consumption of public key algorithm techniques, such as "Diffie–Hellman key agreement protocol" [57] or RSA signatures [58], make them undesirable to be used in WSNs.

Public key algorithms such as RSA are computationally intensive and usually execute thousands or even millions of multiplication instructions to perform a single cryptographic operation. Further, a microprocessor's public key algorithm efficiency is primarily determined by the number of clock cycles required to perform a multiplication instruction [30]. Brown et al. found that public key algorithms such as RSA usually require on the order of tens of seconds up to minutes to perform encryption and decryption operations in resource-constrained wireless devices, which exposes a vulnerability to DoS attacks [59]. On the other hand, Carman et al. found that it usually takes a microprocessor thousands of nanojoules to do a simple multiplication function with a 128-bit result [30]. In contrast, symmetric key cryptographic algorithms and hash functions consume much less computational energy than public key algorithms. For example, the encryption of a 1024-bit block consumes approximately 42 mJ on MC68328 DragonBall processor using RSA, and the estimated energy consumption for a 128-bit AES block is much lower at 0.104 mJ [30].

Studies have shown that it is feasible to apply public key cryptography to sensor networks by using the right selection of algorithms and associated parameters, optimization, and low-power techniques [60–62]. The investigated public key algorithms include Rabin's scheme [63], Ntru-Encrypt [64], RSA [58], and "elliptic curve cryptography" (ECC) [65,66]. Most studies in the literature focus on RSA and ECC algorithms. The attraction of ECC is that it offers equal security for a far smaller key size compared to RSA, thereby reducing processing and communication overheads. For example, RSA with 1024-bit keys (RSA-1024) provides a currently accepted level of security for many applications and is equivalent in strength to ECC with 160-bit keys (ECC-160) [67]. To protect data beyond the year 2010, RSA security recommends RSA-2048 as the new minimum key size, which is equivalent to ECC with 224-bit keys (ECC-224) [68]. Table 15.2 summarizes the execution of ECC and RSA on an Atmel ATmega128 processor (used by MICA2 mote) [23]. The execution time is measured on average for a point multiplication in ECC and a modular exponential operation in RSA. The ECC secp160r1 and secp224r1 are two standardized elliptic curves defined in "Recommended Elliptic Curve Domain Parameters" [69]. As shown in Table 15.2, by using the small integer $e = 2^{16} + 1$ as the public key, RSA public key operation is slightly faster than ECC point multiplication. However, ECC point multiplication outperforms

Table 15.2   Public Key Cryptography: Average ECC and RSA Execution Times

Algorithm	Operation Time (s)
ECC secp160r1	0.81
ECC secp224r1	2.19
RSA-1024 public key, $e = 2^{16} + 1$	0.43
RSA-1024 private key (with Chinese remainder theorem)	10.99
RSA-2048 public key, $e = 2^{16} + 1$	1.94
RSA-2048 private key (with Chinese remainder theorem)	83.26

Source: Data from Wang, Y. et al., *IEEE Comm. Surveys and Tutorials*, 8, 2, 2–23, 2006.

Table 15.3   Public Key Cryptography: Average Energy Costs of Digital Signature and Key Exchange in Millijoules (mJ)

Algorithm	Signature		Key Exchange	
	Sign	Verify	Client	Server
RSA-1024	304	11.9	15.4	304
ECDSA-160	22.82	45.09	22.3	22.3
RSA-2048	2302.7	53.7	57.2	2302.7
ECDSA-224	61.54	121.98	60.4	60.4

Source: Data from Wang, Y. et al., *IEEE Comm. Surveys and Tutorials*, 8, 2, 2–23, 2006.

RSA private key operation by one order of magnitude. Since the RSA private key operations are too slow, they have limited use in sensor network. The ECC has no such issues because both public key operation and private key operation use the same point multiplication operations.

Wander et al. investigated the energy cost of authentication and key exchange based on RSA and ECC on an Atmel ATmega128 processor [62]. The result is shown in Table 15.3. The ECC-based signature is generated and verified with the "elliptic curve digital signature algorithm" (ECDSA) [70]. The key exchange protocol is a simplified version of SSL handshake, which involves two parties: (1) A client initiating the communication and (2) a server responding to the initiation [71]. The WSN is assumed to be administered by a central point with each sensor having a certificate signed by the central point's private key using an RSA or ECC signature. In the handshake process, the two parties verify each other's certificate and negotiate the session key to be used in the communication. As Table 15.3 shows, compared with RSA cryptography at the same security level, ECDSA signatures are significantly cheaper than RSA signatures. Further, the ECC-based key exchange protocol outperforms the RSA-based key exchange protocol at the server side, and there is almost no difference in energy cost for these two key exchange protocols at the client side. In addition, the relative performance advantage of ECC over RSA increases as the key size increases in terms of execution time and energy cost. Tables 15.2 and 15.3 indicate that ECC is more appropriate than RSA for use in sensor networks.

The implementation of RSA and ECC on MICA2 [31] nodes further proved that a public key–based protocol is viable for WSNs. Watro et al. [72] have described a system named TinyPK in which RSA system is implemented on MICA2 motes using TinyOS development environment. The authors demonstrated that authentication and key agreement protocol can be efficiently realized by this scheme in resource-constrained sensor nodes. Another scheme, TinyECC [73], based on ECC have been designed and implemented on MICA2. A similar study was conducted by Malan et al. on ECC using a MICA2 mote [67]. In their work, ECC was used to distribute a single symmetric key for the link layer encryption provided by the TinySec module [74].

Although public key cryptography is possible in sensor nodes, private key operations are still expensive. The assumptions mentioned in the literature [57–61] may not be satisfied in some applications. For example, some studies [57–61] concentrated on the public key operations only, assuming the private key operations will be performed by a base station or a third party. By selecting appropriate parameters, for example, using the small integer $e = 2^{16} + 1$ as the public key, the public key operation time can be reduced significantly, although the private key operation time does not change. The limitation that private key operation occurs only at a base station makes many security services using public key algorithms unavailable under these schemes. Such services include peer-to-peer authentication and secure data aggregation.

In contrast, Tables 15.4 and 15.5 show the execution time and energy cost of two symmetric cryptography protocols on an Atmet ATmega128 processor. In Table 15.4, the execution time was measured on a 64-bit block using an 80-bit key. From Table 15.4, we can see that symmetric key cryptography is faster and consumes less energy compared to public key cryptography.

Table 15.4   Symmetric Key Cryptography: Average RC5 and Skipjack Execution Times

Algorithm	Operation Time (s)
Skipjack (C) [75]	0.38
RC5 (C, assembly) [76]	0.26

Source: Data from Wang, Y. et al., *IEEE Comm. Surveys and Tutorials*, 8, 2, 2–23, 2006.

Table 15.5   Symmetric Key Cryptography: Average Energy for AES and SHA-1

Algorithm	Operation Time (s)
SHA-1 (C) [77]	5.9 µJ/byte
AES-128 Encryption/Decryption (assembly) [78]	1.62/2.49 µJ/byte

Source: Data from Wang, Y. et al., *IEEE Comm. Surveys and Tutorials*, 8, 2, 2–23, 2006.

15.5.1.2 Symmetric Key Cryptography in Wireless Sensor Networks

Since most of the public key cryptographic mechanisms are computationally intensive, most of the research studies for WSNs focus on the use of symmetric key cryptographic techniques. Symmetric key cryptographic mechanisms use a single shared key between two communicating hosts, which is used for both encryption and decryption. However, one major challenge for deploying symmetric key cryptography is how to securely distribute the shared key between the two communicating hosts. This is a nontrivial problem since predistributing the key may not always be feasible.

Five popular encryption schemes, RC4 [79], RC5 [76], IDEA [79], SHA-1 [77], and MD5 [79,80], were evaluated on six different microprocessors ranging in word size from 8-bit (Atmel AVR) to 16-bit (Mitsubishi M16C) to 32-bit widths (StrongARM, XScale) in one study [81]. Execution time and code memory size were measured for each algorithm and platform. The experiments indicated uniform cryptographic cost for each encryption class and each architecture class. The impact of caches was negligible, whereas instruction set architecture (ISA) support is limited to specific effects on certain algorithms. Moreover, hashing algorithms (e.g., MD5 and SHA-1) incur almost an order of magnitude higher overhead than encryption algorithms (e.g., RC4, RC5, and IDEA).

Law et al. [82] evaluated two symmetric key algorithms: (1) RC5 and (2) TEA [83]. They further evaluated six block ciphers, RC5, RC6 [84], Rijndael [78], MISTY1 [85], KASUMI [86], and Camellia [87], on IAR Systems' MSP430F149 [82]. The benchmark parameters were code, data memory, and CPU cycles. The evaluation results are presented in Table 15.6, in which the algorithms are ranked based on the key setup and encryption mode used. In both cases, the algorithms are optimized for speed of execution and memory space requirement and then ranked on the basis of their speed of execution, code size, and data size in memory. The evaluation results showed that Rijndael is suitable for high security and energy efficiency requirements and MISTY1 is suitable for good storage and energy efficiency.

The performance of symmetric key cryptography is mainly decided by the following factors:

Embedded data bus width: Many encryption algorithms prefer 32-bit word arithmetic, but most embedded processors usually use an 8-bit- or 16-bit-wide data bus.

Instruction set: The ISA has specific effects on certain algorithms. For example, most embedded processors do not support the variable-bit rotation instruction such as *rotate bit left* (ROL) of the Intel architecture, which greatly improves the performance of RC5.

Selecting the appropriate cryptography method for sensor nodes is fundamental to providing security services in WSNs. However, the decision depends on the computation and communication capability of the sensor nodes in a WSN. Open research issues range from cryptographic algorithms to hardware design as described here:

■ Recent studies on public key cryptography have demonstrated that public key operations may be practical in sensor networks. However, private key operations are still too expensive in terms of computation and energy cost to accomplish in a sensor node. The application of private key operations to sensor nodes needs to be studied further.

■ Symmetric key cryptography is superior to public key cryptography in terms of speed and low energy cost. However, the key distribution schemes based on symmetric key cryptography are not perfect. Efficient and flexible key distribution schemes need to be designed.

■ It is also likely that more powerful motes need to be designed to support the increasing requirements on computation and communication in sensor nodes.

Table 15.6 Summary of Cipher Performance on Sensor Nodes

	By Key Steps					
	Size Optimized			Speed Optimized		
Rank	Code Memory	Data Memory	Speed	Code Memory	Data Memory	Speed
1	RC5-32	MISTY1	MISTY1	RC6-32	MISTY1	MISTY1
2	KASUMI	Rijndael	Rijndael	KASUMI	Rijndael	Rijndael
3	RC6-32	KASUMI	KASUMI	RC5-32	KASUMI	KASUMI
4	MISTY1	RC6-32	Camellia	MISTY1	RC6-32	Camellia
5	Rijndael	RC5-32	RC5-32	Rijndael	Camellia	RC5-32
6	Camellia	Camellia	RC6-32	Camellia	RC5-32	RC6-32
	By Encryption (CBC/CFB/OFB/CTR)					
	Size Optimized			Speed Optimized		
Rank	Code Memory	Data Memory	Speed	Code Memory	Data Memory	Speed
1	RC5-32	RC5-32	Rijndael	RC6-32	RC5-32	Rijndael
2	RC6-32	MISTY1	MISTY1	RC5-32	MISTY1	Camellia
3	MISTY1	KASUMI	KASUMI	MISTY1	KASUMI	MISTY1
4	KASUMI	RC6-32	Camellia	KASUMI	RC6-32	RC5-32
5	Rijndael	Rijndael	RC6-32	Rijndael	Rijndael	KASUMI
6	Camellia	Camellia	RC5-32	Camellia	Camellia	RC6-32

Source: Data from Ganesan, P. et al., in *Proceedings of the 2nd ACM International Conference on Wireless Sensor Networks and Applications,* ACM Press, New York, 151–59, 2003.

15.5.2 Key Management Protocols

The area that has received maximum research attention in WSN security is key management. Key management is a core mechanism to ensure security in network services and applications in WSNs. The goal of key management is to establish keys among nodes in a secure and reliable manner. In addition, the key management scheme must support node addition and revocation in a network. Since the nodes in WSNs have computational and power constraints, the key management protocols for these networks must be extremely lightweight. Most of the existing key management protocols for WSNs are based on symmetric key cryptography because public key cryptographic techniques are in general computationally intensive. Figure 15.1 presents a taxonomy of key management protocols in WSNs as described in [171]. In Sections 15.5.2.1 and 15.5.2.2, a brief overview of some of the most important key management protocols is given.

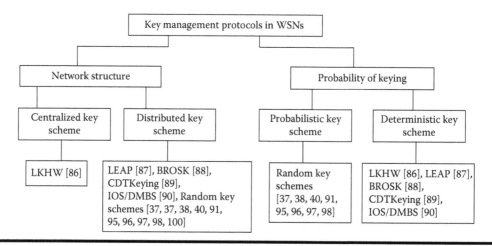

Figure 15.1 **Key management protocols in wireless sensor networks: A taxonomy. (From Wang, Y. et al.,** *IEEE Comm. Surveys and Tutorials,* **8, 2, 2–13, 2006.)**

15.5.2.1 Key Management Protocols Based on Network Structure

Depending on the underlying network structure, the key management protocols in WSNs may be centralized or distributed. In a centralized key management scheme, there is only one entity that controls the generation, regeneration, and distribution of keys. This entity is called "key distribution center" (KDC). The only protocol existing in the literature that is based on centralized key distribution is the LKHW scheme [88]. The LKHW scheme is based on a "logical key hierarchy" (LKH). In this scheme, the base station is treated as a KDC and all keys are logically distributed in a tree rooted at the base station. The main drawback of this scheme is its single point of failure. If the central controller fails, the entire network and its security will be affected. The lack of scalability is another issue. Moreover, it does not provide data authentication. In distributed key management protocols, different controllers are used to manage key-related activities. These protocols do not have the vulnerability of single point of failure, and they allow better scalability. Most of the key management protocols existing in the literature are distributed in nature. These schemes fall into either deterministic or probabilistic categories and are discussed in Section 15.5.2.2.1 and Section 15.5.2.2.2 respectively.

15.5.2.2 Key Management Protocols Based on Probability of Key Sharing

The key management protocols for WSNs may be classified on the probability of key sharing between a pair of sensor nodes. Depending on this probability, the key management schemes may be either deterministic or probabilistic.

15.5.2.2.1 Deterministic Key Distribution Schemes

The "localized encryption and authentication protocol" (LEAP) proposed by Zhu et al. [89] is a key management protocol for WSNs based on symmetric key algorithms. It uses different keying mechanisms for different packets depending on their security requirements. Four types of keys are established for each node in a WSN: (1) An individual key shared with the base station (predistributed), (2) a group of keys shared by all the nodes in the network (predistributed), (3) a pairwise key shared with immediate neighbor nodes, and (4) a cluster key shared with multiple neighbor nodes. The pairwise keys shared with immediate neighbor nodes are used to protect peer-to-peer communication and the cluster key is used for local broadcasting.

It is assumed that the time required to attack a node is greater than the network establishment time, during which a node can detect all its intermediate neighbors. A common initial key is loaded into each node before deployment. Each node derives a master key, which depends on the common key and its unique identifier. Nodes then exchange HELLO messages, which are authenticated by the receivers (since the common key and its identifier are known, the master key of the neighbor can be computed). The nodes then compute a shared key based on their master keys. The common key is erased in all nodes after the completion of the key distribution process and, by assumption, no node has been compromised up to this point. Since no adversary can get the common key, it is impossible to inject false data or decrypt the messages exchanged earlier. Also, no node can later forge the master key of any other node. In this way, pairwise shared keys are established between all immediate neighbors. The cluster key is established by a node after the pairwise key is established. A node generates a cluster key and sends it encrypted to each neighbor with its pairwise shared key. The group key can be preloaded, but it should be updated once any compromised node is detected. This could be done, in a naive way, by the base station, which sends the new group key to each node using its individual key, or in a hop-by-hop basis using cluster keys. Other sophisticated algorithms have been proposed for the same. Further, the authors [89] have proposed methods for establishing shared keys between multihop neighbors.

Lai et al. have proposed a "broadcast session key" (BROSK) negotiation protocol [90]. The BROSK assumes a master key shared by all the nodes in a network. To establish a session key with its neighbor node B, a sensor node A broadcasts a key negotiation message and both A and B arrive at a shared session key. The BROSK is a scalable and energy-efficient protocol.

Cametepe et al. have proposed a deterministic key distribution scheme for WSNs using combinatorial design theory [91]. The combinatorial design theory–based pairwise key predistribution (CDTKeying) scheme is based on block design techniques in combinatorics. It uses symmetric and generalized quadrangle design techniques. The scheme uses a finite projective plane of order n (for prime power of n) to generate a symmetric design with parameters $n^2 + n + 1$, $n + 1$, and 1. The design supports $n^2 + n + 1$ nodes and uses a key pool of size $n^2 + n + 1$. It generates $n^2 + n + 1$ key chains of size $n + 1$ in which every pair of key chains has exactly one key in common, and every key appears in exactly $n + 1$ key chains. After deployment, every pair of nodes finds exactly one common key. Thus, the probability of key sharing among a pair of sensor nodes is unity. The disadvantage of this proposition is that the parameter n has to be a prime power. Therefore, all network sizes can be supported for a fixed key chain size.

Lee et al. have proposed two combinatorial design theory–based deterministic schemes: (1) The "ID-based one-way function scheme" (IOS) and (2) "deterministic multiple space Bloms' scheme" (DMBS) [92]. They further discussed the use of combinatorial set systems in the design of deterministic key predistribution schemes for WSNs [93].

Chan et al. have proposed a deterministic key management protocol to facilitate key establishment between every pair of neighboring nodes in a WSN [94]. In this mechanism, known as "peer intermediaries for key establishment" (PIKE), all N sensor nodes are organized into a two-dimensional space, as shown in Figure 15.2, where the coordinate of each node is (x, y) for $x, y \in \{0, 1, \ldots \sqrt{N} - 1\}$. Each node shares unique pairwise keys with $2(\sqrt{N} - 1)$ nodes that have the same x or y coordinate in the two-dimensional space. For two nodes with no common coordinate, an intermediate node, which has a common x or y coordinate with both nodes, is used as a router to forward a key from them. However, the communication overhead of this scheme is rather high

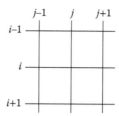

Figure 15.2   The peer intermediaries for key establishment scheme: Sensor nodes are organized in a two-dimensional space.

because "secure connectivity" is only $2/\sqrt{N}$, which means that each node must establish a key for almost each of its neighbors through multilink paths.

Huang et al. [95] have proposed a hybrid key establishment scheme that exploits the difference in computational power and energy between a sensor node and the base station in a WSN. The authors argue that an individual sensor node possesses far less computational power and energy than a base station. In light of this argument, they propose placing the major cryptographic computations on the base station. On the sensor side, lightweight symmetric key operations are deployed. Every sensor node and the base station mutually authenticate each other based on ECC. The proposed mechanism also uses certificates to establish the legitimacy of a public key. The certificates are based on an elliptic curve scheme. Such certificates are useful in verifying the authenticity of sensor nodes.

Zhou and Fang [96] have developed a scalable key agreement protocol that uses a t-degree $(k + 1)$-variate symmetric polynomial to establish keys in a deterministic way.

15.5.2.2.2 Probabilistic Key Distribution Schemes

Most of the key management protocols for WSNs are probabilistic and distributed schemes. Eschenauer et al. have proposed a "random key predistribution" scheme for WSNs that relies on probabilistic key sharing among the nodes of a "random graph" [37]. The mechanism has three phases: (1) Key predistribution, (2) shared key discovery, and (3) path key establishment. In the key predistribution phase, each sensor is equipped with a key ring stored in its memory. The key ring consists of k keys, which are randomly drawn from a large pool of P keys. The association information of the key identifiers in the key ring and the sensor identifier is also stored at the base station. Each sensor node shares a pairwise key with the base station. In the shared key discovery phase, each sensor discovers its neighbors with which it shares keys. The authors have suggested two methods for this purpose. The simplest method is for each node to broadcast a list of identifiers of the keys in their key rings in plaintext, allowing neighboring nodes to check whether they share a key. However, an adversary may observe the key-sharing patterns among sensors in this way. The second method uses the challenge–response technique to hide key-sharing patterns among nodes from an adversary. Finally, in the path key establishment phase, a path key is assigned for the sensor nodes that are within the communication range and not sharing a key, but are connected by two or more links at the end of the second phase. If a node is compromised, the base station can send a message to all other sensors to revoke the compromised node's key ring. Rekeying follows the same procedure as revocation. The messages from the base station are signed by the pairwise key shared by the base station and sensor nodes, thus ensuring that no adversary can forge a station. If a node is compromised, the attacker has a probability of approximately k/P to attack any link successfully. Since $k << P$, it only affects a small number of sensor nodes.

Eschenauer et al.'s work can be considered as the basic random key management scheme. A number of additional key predistribution schemes have been proposed [38,40,97–100].

In the basic random key management scheme, any two neighbor nodes need to find a single common key from their key rings to establish a secure link in the key setup phase. However, Chan et al. observed that increasing the amount of key overlap in a key ring can increase the resilience of the network to node capture [38]. The authors have proposed the "*q*-composite random key predistribution" scheme. It is required to share at least *q* common keys in the key setup phase to build a secure link between any two neighbor nodes. Further, they introduced a key update phase to enhance the basic random key management scheme. Suppose *A* has a secure link to *B* after the key setup phase and the secure key is *k* from the key pool *P*. Since *k* may be residing in the key ring memory of some other nodes in the network, the security of the link between *A* and *B* is jeopardized if any of these nodes are captured. Thus, it is better to update the communication key between *A* and *B* instead of using a key in the key pool. To address this problem, the authors have presented a multipath key reinforcement mechanism for the key update. An adversary in this case has to eavesdrop on all the disjoint paths between node *A* and node *B* if he or she wants to reconstruct the communication key. The security of the scheme is further augmented by a random pairwise key management scheme for node-to-node authentication.

To discover whether the key sets of two nodes have an intersection, usually both nodes need to broadcast their key indices or find common keys through a challenge–response procedure. Such methods have very high communication overheads. Pietro et al. [98] improved the basic random key management scheme by associating the key indices of a node with its identity. For example, each node is assigned a pseudorandom number generator $g(x, y)$ and the key indices for the node are computed as $g(\text{ID}, i)$ for $i = 1, 2, \ldots N$, where ID is the node identity. In this way, other nodes can find out which key is in its key set by checking its node identity.

Blundo et al. presented a "polynomial-based key predistribution" protocol for group key predistribution that can be adapted to WSNs [101]. The key setup server randomly generates a bivariate *t*-degree polynomial defined as follows:

$$f(x, y) = \sum_{i=0}^{t} \sum_{j=0}^{t} a_{ij} x^i y^j \tag{15.1}$$

The *t*-degree polynomial is defined over a finite field $\mathbf{F}_q$, where *q* is a prime value that is large enough to accommodate a cryptographic key. By choosing $a_{ij} = a_{ji}$, a symmetric polynomial is obtained, that is, $f(x, y) = f(y, x)$. Each sensor node is assumed to have a unique, integer-valued, nonzero identity. For each sensor node *u*, a polynomial share $f(u, y)$ is assigned, which means the coefficients of univariate polynomials $f(u, y)$ are loaded into node *u*'s memory. When nodes *u* and *v* need to establish a shared key, they broadcast their IDs. Subsequently, node *u* computes $f(u, v)$ by evaluating $f(u, y)$ at $y = v$, and node *v* can also compute $f(v, u)$ by evaluating $f(v, y)$ at $y = u$. Due to polynomial symmetry, the shared key between nodes *u* and *v* is established as $K_{uv} = f(u, v) = f(v, u)$. A *t*-degree bivariate polynomial is also $(t + 1)$ secure. Therefore, an adversary must compromise not less than $(t + 1)$ nodes holding the shares of the same polynomial to reconstruct it.

Liu et al. have proposed a "polynomial pool–based key predistribution" (PPKP) scheme in [40]. The scheme involves three phases: (1) Setup, (2) direct key establishment, and (3) path key establishment. In the setup phase, the setup server randomly generates a set *F* of bivariate *t*-degree polynomials over the finite field $\mathbf{F}_q$. For each sensor node, the setup server picks a subset of polynomials $F_i \subseteq F$ and assigns the polynomial shares of these polynomials to node *i*. In the [172] direct key establishment stage, a sensor node finds a polynomial that is shared with other sensor nodes and then establishes a pairwise key using the polynomial-based key predistribution scheme discussed by Blundo et al. [101]. The path key establishment phase in this scheme is similar to that in the basic random key management scheme. The general framework based on polynomial pool-based pairwise key predistribution can be applied in various ways. The authors [40] have provided two examples. In the random subset assignment strategy, during the setup phase, a setup server selects a random subset of polynomials and assigns their polynomial shares to each sensor. In the second strategy—the grid-based key predistribution strategy—the setup server assigns a polynomial share to each node that is determined based on a grid-structure. The grid-based key predistribution scheme is more resilient to a possible node compromise attack.

Du et al. have presented a "multiple-space key predistribution" (MSKP) scheme [97], which uses Blom's method [172]. The key difference between the schemes proposed by Liu et al. [40] and Du et al. [97] is that the scheme described by the former is based on a set of bivariate t-degree polynomials, whereas the scheme described by the latter is based on Blom's method. The proposed scheme allows any pair of nodes in a network to be able to find a pairwise secret key. As long as no more than λ nodes are compromised, the network is perfectly secure. To use Blom's method, during the predeployment phase a base station first constructs a $(\lambda + 1) \times N$ matrix G over a finite field $GF(q)$, where N is the size of the network and G is considered to be public information. Then, the base station creates a random $(\lambda + 1) \times (\lambda + 1)$ symmetric matrix D over $GF(q)$ and computes an $N \times (\lambda + 1)$ matrix $A = (D \cdot G)^T$, where $(D \cdot G)^T$ is the transpose of $D \cdot G$. Matrix D needs to be kept secret and should not be disclosed to adversaries. It is easy to verify that $A \cdot G$ is a symmetric matrix as follows:

$$A \cdot G = (D \cdot G)^T \cdot G = G^T \cdot D^T \cdot G = G^T \cdot D \cdot G = (A \cdot G)^T \qquad (15.2)$$

Therefore, $K_{ij} = K_{ji}$. The idea is to use K_{ij} (or K_{ji}) as the pairwise key between node i and node j. To perform the aforementioned computation, in the predistribution phase for any sensor node k the following two steps are carried out: (1) The kth row of matrix A is stored at node k, and (2) the kth column of matrix G is stored at node k. Then nodes i and j need to find the pairwise key between them; i and j first exchange their columns of G and then compute K_{ij} and K_{ji}, respectively, using their private rows of A.

In the proposed scheme, each sensor node is loaded with G and τ distinct D matrices drawn from a large pool of ω symmetric matrices, $D_1, \dots D_\omega$ of size $(\lambda + 1) \times (\lambda + 1)$. For each D_i, calculate the matrix $A_i = (D_i \cdot G)^T$ and store the jth row of A_i at this node. After deployment, each node needs to discover whether it shares any space with its neighbors. If they find that they have a common space, the nodes can follow Blom's method to build a pairwise key. The scheme is scalable and flexible. Moreover, it is substantially more resilient against node capture as compared to the scheme proposed by Liu et al. [40].

In the aforementioned scheme, each sensor node needs to keep many key materials such that a pair of nodes shares a key with a probability that can guarantee that the entire network is almost connected. This causes a large storage overhead on memory-constrained sensor nodes. Hwang et al. [39] proposed to enhance the basic random key management protocol [37] by reducing the amount of key-related materials required to be stored at each node while guaranteeing a certain probability of sharing a key between two nodes. Their idea is to guarantee secure connectivity in the largest subcomponent of a network rather than the entire network. The probability that two nodes have a key in common is reduced, but it is still large enough for the largest network component to be connected.

Hwang et al. extended the basic random key management scheme and proposed a cluster key grouping scheme [100]. They further analyzed the trade-offs involved among energy, memory, and security robustness.

In all the key management schemes discussed so far, the key materials are uniformly distributed in the entire terrain of a network. This uniform distribution makes the probability that two neighbor nodes share a direct key, called secure connectivity, rather small. Therefore, a lot of communication overhead is inevitable for the establishment of indirect keys. If some location information is known, two nearby sensor nodes can be preloaded with the same set of key materials. In this way, secure connectivity may be improved to a large extent.

In the "location-based key predistribution" (LBKP) scheme [102], the entire WSN is divided into many square cells. Each cell is associated with a unique t-degree bivariate polynomial. Each sensor node is preloaded with shares of the polynomials of its home cell and four other cells horizontally and vertically adjoining its home cell. After deployment, any two neighbor nodes can establish a pairwise key if they have shares of the same polynomial. For example, in Figure 15.3 the polynomial of cell C_{33} is also assigned to cells C_{32}, C_{34}, C_{23}, and C_{43}. The polynomials of other cells are assigned in the same way. As a result, a node in C_{33} has some polynomial information in common with other nodes in the shaded areas.

Du et al [99]. have also proposed a key predistribution scheme that uses network deployment knowledge. In the proposed scheme, the entire network is divided into many square cells. Each cell is assigned a subset key pool S_{ij}, $i = 1, \ldots u$ and $j = 1, \ldots v$, out of a global key pool S. These subset key pools are set up such that the key pools of two neighbor cells share a portion of keys. In each cell, the basic random key management scheme [37] is applied. Using the deployment knowledge—the information about the manner in which the nodes are deployed in the network—the scheme ensures that the value of the probability that a pair of neighboring nodes share a secret key is very high. The high value of the probability signifies that any pair of nodes in the network can establish secure communication sessions between them. The intelligent use of the deployment knowledge also ensures that the size of the key ring (i.e., the set of keys) held by a given node in the proposed scheme [99] is much smaller than that in the basic key management scheme proposed by Eschenauer and Gligor [37]. Hence, the scheme is very memory-efficient.

Some of the aforementioned key management schemes for WSNs are classified and compared in Table 15.7. Although a number of key management protocols have been proposed for WSNs, the design of key management protocols for WSNs is still largely open to research. Some of the open research issues are discussed as follows:

Memory: High security and low overhead are two objectives that a key management protocol needs to achieve. Although there have been several proposals for key establishment in sensor networks, they can hardly address these two requirements. Strong security protocols

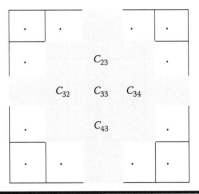

Figure 15.3 The location-based key distribution scheme.

Table 15.7 Classification and Comparison of Key Management Protocols in WSNs

Protocol Type	Protocol Name	Reference	Master Key	Pairwise Key	Path Key	Cluster Key	Scalability	Robustness	Processing Load	Communication Load	Storage Load
Deterministic	All pairwise		NA	Yes	No	No	Low	Low	Low	Low	High
	LEAP	[89]	Yes	Yes	Yes	Yes	Good	Low	Low	Low	Low
	BROSK	[90]	Yes	Yes	No	No	Good	Low	Low	Low	Low
	LKHW	[88]	Yes	Yes	No	Yes	Fair	Low	Low	Low	Low
	CDTKeying	[91]	NA	Yes	No	No	Good	Good	Medium	Medium	High
	IOS and DMBS	[92]	NA	Yes	No	No	Good	Good	Medium	Medium	High
Probabilistic	Basic	[37]	NA	Yes	Yes	No	Good	Good	Medium	Medium	High
	q-Composite	[38]	NA	Yes	No	No	Good	Good	Medium	Medium	High
	Polynomial based	[40]	NA	Yes	No	No	Good	Good	Medium	Medium	High
	Blom based	[97]	NA	Yes	No	No	Good	Good	Medium	Medium	High
	Deployment knowledge based	[99]	NA	Yes	No	No	Good	Good	Medium	Medium	High
	Cluster key grouping	[100]	NA	Yes	No	No	Good	Good	Medium	Medium	High
	Location based	[102]	NA	Yes	No	No	Good	Good	Medium	Medium	Medium

Source: Data from Zhang, Y. et al., *Security in Wireless Mesh Networks*, CRC Press, Boca Raton, FL, 2008.

usually require large amounts of memory, as well as high-speed processors and large power consumption. However, they cannot be easily supported due to the constraints on hardware resources of the sensor platform. It is well known that in a wireless environment, transmitting 1 bit can consume more energy than computing 1 bit. In key management protocols, direct key establishment does not require communication or it requires only a few rounds of one-hop communication, whereas indirect key establishment is performed over multi-hop communication. To reduce the multihop communication overhead, the probability of a direct key establishment between a pair of nodes should be as high as possible so that a secure connectivity among the nodes can be guaranteed. However, highly secure connectivity requires more key materials in each node, which is usually impractical, especially when the network size is large. Considering the aforementioned two issues, memory cost can be a major bottleneck in designing key management protocols in a WSN. How to reduce memory cost while still maintaining a certain level of security is a very important issue.

End-to-end security: The major merit of symmetric key cryptography is its computational efficiency. However, most current symmetric key schemes for WSNs aim at link layer security—not transport layer security—because it is impractical for each node to store a transport layer key for each of the other nodes in a network due to the huge number of nodes in the network. However, end-to-end communication at the transport layer is very common in many WSN applications. For example, to reduce unnecessary traffic a fusion node can aggregate reports from many source nodes and forward a final report to the sink node. During this procedure, the reports between source nodes and the fusion node and the report between the fusion node and the sink node should be secured. In hostile environments, however, any node can be compromised. If one of the intermediate nodes along a route is compromised, the message delivered along the route can be exposed or modified by the compromised node. Employing end-to-end security can effectively prevent message tampering by a malicious intermediate node. Compared with symmetric key technology, public key cryptography is expensive, but it has flexible manageability and supports end-to-end security. A more promising approach to key establishment in WSNs is to combine the merits of both symmetric key and public key techniques so that each node is equipped with a public key system and relies on it to establish end-to-end symmetric keys with other nodes. To achieve this goal, a critical issue is to develop more efficient public key algorithms and their implementations so that they can be widely used on sensor platforms. How to prove the authenticity of public keys is another important problem. A malicious node can otherwise impersonate any normal node by claiming its public key. Identity-based cryptography is a shortcut to avoid the problem. Currently, most identity-based cryptographic algorithms operate on elliptic curve fields, and pairing over elliptic curves is widely used in the establishment of identity-based symmetric keys. However, the pairing operation is very costly, with costs comparable to or even more than those of RSA. Therefore, fast algorithms and implementations are the major areas of research.

Efficient symmetric key algorithms: There is still a demand for the development of more efficient symmetric key algorithms because encryption and authentication based on symmetric keys are very frequent in the security operations of sensor nodes. For example, in the link layer security protocol TinySec [74], each packet must be authenticated and encryption can also be triggered if critical packets are transmitted. Therefore, fast and cost-efficient symmetric key algorithms should be developed.

Key update and revocation: Once a key is established between two nodes, the key can act as a master key and be used to derive different subkeys for many purposes (e.g., encryption and authentication). If a key is used for a long time, it may be exposed due to cryptanalysis over the ciphertexts intercepted by adversaries. To protect the master key and subkeys from cryptanalysis, it is wise to update keys periodically. The period of update, however, is difficult to choose. Because the cryptanalysis capability of adversaries is unknown, it is very difficult to estimate how long it takes for adversaries to expose a key by cryptanalysis. If the key update period is too long, the corresponding key may also be exposed. If it is too short, frequent updates can incur large overheads. A related problem is key revocation. If one node is detected to be malicious, its key must be revoked. However, key revocation has not been thoroughly investigated yet. Although Chan et al. [104] propose a distributed revocation protocol, it is only based on the random pairwise key scheme [38] and cannot be easily generalized into other key establishment protocols.

Node compromise: Node compromise is the most detrimental attack on sensor networks. Because compromised nodes have all the authentic key materials, node compromise can result in very severe damage to WSN applications and cannot be detected easily. How to counteract node compromise remains an open problem. Most of the current security protocols attempt to minimize the adverse impact on the network due to a possible node compromise through careful protocol design such that the impact of node compromise can be restricted to a small area of the network. However, a hardware approach is more promising. With advances in hardware design and manufacturing techniques, stronger, tamper-resistant, and cheaper devices can be installed on the sensor platform to counteract node compromise.

15.5.3 Defense against Denial-of-Service Attacks

Various types of DoS attacks in WSNs are discussed in Sections 15.4.1.1 through 15.4.1.4. In Sections 15.5.3.1 through 15.5.3.3, defense mechanisms for each of these attacks are presented in detail.

15.5.3.1 Defense Mechanisms in the Physical Layer

Jamming attack may be defended by employing variations of spread-spectrum communication such as frequency hopping and code spreading [29]. "Frequency-hopping spread spectrum" (FHSS) is a method of transmitting signals by rapidly switching a carrier among many frequency channels using a pseudorandom sequence that is known to both the transmitter and the receiver. As a potential attacker would not be able to predict the frequency selection sequence, it will be impossible for him or her to jam the frequency being used at a given point of time. Code spreading is another technique for defending a network against jamming. However, it requires greater design complexity and more energy consumption than FHSS and is thus not suitable for WSNs. In general, sensor devices are limited to single-frequency use and are highly susceptible to jamming attacks. One approach to tolerate jamming attacks in a WSN is to identify the jammed part of the network and effectively avoid it by routing around it. Wood et al. [29] have proposed an approach in which nodes along the perimeter of a jammed region report their status to their neighbors and the affected region is identified collectively and packets are routed around it.

15.5.3.2 Defense Mechanisms in the Link Layer

A typical defense against collision attack uses error-correcting codes [29]. Most codes work best with low levels of collisions such as those caused by environmental or probabilistic errors. However, these codes also add additional processing and communication overheads. It is reasonable to assume that an attacker will always be able to corrupt more than what can be corrected. Although it is possible to detect these malicious collisions, no complete defense mechanism against them is known today.

A possible solution for energy exhaustion attack is to apply rate-limiting MAC admission control. This allows a network to ignore requests that intend to exhaust the energy reserves of a node. A second technique is to use time-division multiplexing so that each node is allotted a time slot in which it can transmit [29]. This eliminates the need of arbitration for each frame and can solve the indefinite postponement problem in a backoff algorithm. However, it is still susceptible to collisions.

The effect of unfairness caused by an attacker who intermittently launches link layer attacks can be lessened by the use of small frames since this reduces the amount of time an attacker gets to capture the communication channel [29]. However, this technique often reduces efficiency and is susceptible to further unfairness such as an attacker trying to retransmit quickly instead of randomly delaying.

15.5.3.3 Defense Mechanisms in the Network Layer

A countermeasure against spoofing and alteration is to append a MAC to a message. By adding a MAC to a message, receivers can verify whether messages have been spoofed or altered. To defend the network against replayed information, counters or time stamps may be introduced in messages [35]. A possible defense against selective forwarding attack is to use multiple paths to send data [46]. A second defense is to detect the malicious node or assume it has failed and seek an alternative route.

Sen et al. have presented a cooperative detection scheme for identifying malicious packet-dropping nodes in an ad hoc network [105]. The scheme exploits the redundancy in routing information in an ad hoc network to build a robust detection framework so that it works even in the presence of transient network partitioning and Byzantine failure of nodes.

Hu et al. have proposed a novel and generic mechanism called "packet leashes" for detecting and defending against wormhole attacks [106]. In a wormhole attack, a malicious node eavesdrops on a series of packets, tunnels them through a path in the network, and replays them. This is done in order to make a false representation of the distance between two colluding nodes. It is also used, more generally, to disrupt the routing protocol by misleading the neighbor discovery process [46]. Hu et al. have presented a mechanism that employs a directional antenna to combat wormhole attacks [23]. Wang et al. have used a visualization approach to detect wormholes in a WSN [107]. In the mechanism proposed by the authors, a distance estimation is made between all the sensor nodes in a neighborhood. Using multidimensional scaling, a virtual layout of the network is then computed, and a surface-smoothing strategy is used to adjust the roundoff errors. Finally, the shape of the resulting virtual network is analyzed. If any wormhole exists, the shape of the network will bend and curve toward the wormhole; otherwise, the network will appear flat. Sen et al. have presented a security mechanism that can detect cooperative grayhole attacks in a wireless ad hoc sensor network [108]. In this scheme, every node monitors the packet-forwarding behavior of each of its neighbors and a global detection algorithm is used to detect any routing misbehavior.

To defend against flooding DoS attacks at the transport layer, Aura et al. have proposed using "client puzzles" [109] in which each client should demonstrate its commitment to the connection by solving a puzzle. As an attacker does not have infinite resources, it will be impossible for him or her to create new connections fast enough to cause resource starvation on the serving node. A possible defense against desynchronization attacks is to enforce a mandatory requirement of authentication of all packets communicated between nodes [29]. If the authentication mechanism is secure, an attacker will be unable to send any spoofed messages.

Some mechanisms for ensuring secure multicasting and broadcasting in WSNs are discussed in Section 15.5.3.3.1.

15.5.3.3.1 Secure Broadcasting and Multicasting Protocols

Multicasting and broadcasting techniques are used primarily to reduce the communication and management overheads of sending a single message to multiple receivers. In order to ensure that only legitimate group members receive the multicast and broadcast communication, appropriate authentication and encryption mechanisms must be in place. To handle this problem, several key management schemes have been devised such as centralized group key management protocols, decentralized key management protocols, and distributed key management protocols [110]. First, we discuss some generic security mechanisms for multicast and broadcast communication in wireless networks. Then we present some of the well-known propositions specific to WSNs.

In the case of centralized group key management protocols, a central authority is used to maintain a group. Decentralized management protocols, however, divide the task of group management among multiple nodes. In distributed key management protocols, the key management activity is distributed among a set of nodes rather than being assigned to a single node. In some cases, the entire group of nodes is responsible for key management [110].

An efficient way to distribute keys in a network is to use a logical key tree. Such techniques essentially fall under the category of centralized key management protocols. Some schemes have been developed for WSNs based on the logical key tree technique [88,111,112]. Although centralized solutions are not always the most efficient ones, these mechanisms may sometimes be very effective for WSNs, as relatively heavier computations can usually be carried out in powerful base stations.

Di Pietro et al. have proposed a directed diffusion–based multicast mechanism for WSNs that utilizes an LKHW [88]. In the logical hierarchy, a central key distributor is at the root of a tree and the nodes in the network are at the leaf level. The internal nodes of the tree contain keys that are used in the rekeying process. Directed diffusion is an energy-efficient data dissemination technique for WSNs [113]. In directed diffusion, a query is transformed into an interest and then diffused throughout the network. The source node then starts collecting data from the network based on the propagated interest. The dissemination technique also sets up certain gradients designed to draw events toward the interest. The collected data is then sent back to the source along the reverse path of interest propagation. The directed diffusion–based LKHW scheme as proposed by Di Pietro et al. allows nodes to join and leave groups. The key hierarchy is used to effectively reestablish keys for the nodes below the node that has left the group. When a node declares its intension to join a group, a key set is generated for the new node based on the keys within the existing key hierarchy.

Kaya et al. discuss the problem of multicast group management [114], in which the nodes in a network are grouped based on their locality and a security tree is constructed on the groups.

Lazos et al. have presented a tree-based key distribution scheme that is similar to the directed diffusion–based LKHW proposed by Di Pietro et al. [112]. In their proposed scheme, a routing-aware tree is constructed in which the leaf nodes are assigned keys based on all the relay nodes above them. As the scheme takes advantage of routing information for constructing the key hierarchy, it is more energy efficient than the routing schemes that arbitrarily arrange nodes into a routing tree.

In one study [111], the authors propose a mechanism that uses geographic location information for constructing an LKHW for secure multicast communication. Based on geographical location information, the nodes in a network are grouped into different clusters. The nodes within a cluster are able to reach each other with single-hop communication. Using the cluster information, a key hierarchy is constructed in a manner similar to the scheme proposed by Lazos et al. [112].

15.5.4 Defense against Attacks on Routing Protocols

Many routing protocols have been proposed for WSNs. These protocols can be divided into three broad categories according to network structure: (1) Flat structure-based routing, (2) hierarchical structure-based routing, and (3) location-based routing [115]. In flat-based routing, all nodes are typically assigned equal roles or functionalities. In hierarchical structure-based routing, nodes play different roles in a network. In location-based routing, sensor node positions are used to route data in the network. One common location-based routing protocol is "greedy perimeter stateless routing" (GPSR) [3]. It allows nodes to send packets to a region rather than a particular node. All these routing protocols are vulnerable to various types of attacks such as selective forwarding, sinkhole attacks, etc., as mentioned in Section 15.4. Elaborate discussions on various types of attacks on routing protocols in WSNs s are given by Karlof et al. [46] and Sen [50]. A comparative analysis of some of the well-known existing secure routing protocols for WSNs has been presented by Sen [50].

The goal of a secure routing protocol for a WSN is to ensure integrity, authentication, and availability of messages. Most of the existing secure routing algorithms for WSNs are based on symmetric key cryptography except the one described by Du et al. [116], which is based on public key cryptography. In this section, a number of security mechanisms for routing in WSNs are discussed in detail:

The μTESLA [9], which is the "micro" version of the timed, efficient, streaming, loss-tolerant authentication protocol, and its extensions [117,118] have been proposed to provide broadcast authentication for sensor networks. The μTESLA is a broadcast authentication protocol proposed by Perrig et al. for the "security protocols for sensor networks" (SPINS) protocol [35]. The μTESLA introduces asymmetry through the delayed disclosure of symmetric keys, resulting in an efficient broadcast authentication scheme. For its operation, it requires the base station and the sensor nodes to be loosely synchronized. In addition, each node must know an upper bound on the maximum synchronization error.

To send an authenticated packet, the base station simply computes a MAC on the packet with a key that is secret at that point of time. When a node gets a packet, it can verify that the corresponding MAC key was not yet disclosed by the base station. Because a receiving node is assured that the MAC key is known only to the base station, it can be sure that no adversary could have altered the packet in transit. The node stores the packet in a buffer. At the time of key disclosure, the base station broadcasts the verification key to all its receivers. When a node receives the disclosed key, it can easily verify the correctness of the key. If the key is correct, the node can now use it to authenticate the packet stored in its buffer. Each MAC is a key from the key chain and is generated by a public one-way function F. To generate the one-way key chain, a sender chooses the last key K_n from the chain and repeatedly applies F to compute all other keys: $K_i = F(K_{i+1})$.

Figure 15.4 shows an example of μTESLA. The receiver node is loosely time synchronized and knows K_0 in an authenticated way. Packets P_1 and P_2 sent in interval 1 contain a MAC with a key K_1. Packet P_3 has a MAC using key K_2. If P_4, P_5, and P_6 are all lost, as well as the packet that disclosed the key K_1, the receiver cannot authenticate P_1, P_2, and P_3. In interval 4, the base station broadcasts the key K_2, which is authenticated by the nodes by verifying $K_0 = F(F(K_2))$; hence, they

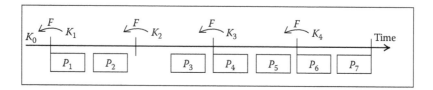

Figure 15.4 Illustration of time-released key chain for source authentication. (From Wang, Y. et al., *IEEE Comm. Surveys and Tutorials*, 8, 2, 2–23, 2006.)

also know that $K_1 = F(K_2)$ and so they can authenticate packets P_1 and P_2 with K_1, and P_3 with K_2. The SPINS limits the broadcasting capability to only the base station. If a node wants to broadcast authenticated data, it has to broadcast the data through the base station. The data is first sent to the base station in an authenticated way. It is then broadcasted by the base station.

To bootstrap a new receiver, µTESLA depends on a point-to-point authentication mechanism in which a receiver sends a request message to the base station and the base station replies with a message containing all the necessary parameters. It may be noted that µTESLA requires the base station to unicast initial parameters to individual sensor nodes and thus incurs a long delay to boot up a large-scale sensor network. Liu et al. propose a multilevel key chain scheme for broadcast authentication to overcome this deficiency [117,118].

The basic idea in the studies by Liu et al. [117,118] is to predetermine and broadcast the initial parameters required by µTESLA instead of using unicast-based message transmission. The simplest way is to predistribute the µTESLA parameters with a master key during the initialization of sensor nodes. As a result, all sensor nodes have the key chain commitments and other necessary parameters once they are initialized, and they are ready to use µTESLA as long as the staring time has passed. Furthermore, the authors have introduced a multilevel key chain scheme in which the higher-level key chains are used to authenticate the commitments of the lower-level ones. However, the multilevel key chain suffers from possible DoS attacks during commitment distribution stage. Further, none of the µTESLA or multilevel key chain schemes is scalable in terms of the number of senders. In one study [119], a practical broadcast authentication protocol has been proposed to support a potentially large number of broadcast senders using µTESLA as a building block.

The µTESLA provides broadcast authentication for base stations, but it is not suitable for local broadcast authentication. This is because µTESLA does not provide immediate authentication. For every received packet, a node has to wait for one µTESLA interval to receive the MAC key used in computing the MAC for the packet. As a result, if µTESLA is used for local broadcast authentication, a message traversing one hop will take at least one µTESLA interval to arrive at the destination. In addition, a sensor node has to buffer all unverified packets. Both the latency and the storage requirements limit the scheme for authenticating infrequent messages broadcast by the base station. Zhu et al. have proposed a one-way key chain scheme for one-hop broadcast authentication [89]. The mechanism is known as LEAP. In this scheme, every node generates a one-way key chain of certain length and then transmits the commitment (i.e., first key) of the key chain to each neighbor, which is encrypted with their pairwise shared key. Whenever a node has a message to send, it attaches to the message the next authenticated key in the key chain. The authenticated keys are disclosed in reverse order to their generation. A receiving neighbor can verify the message based on the commitment or an authenticated key that it received from the sending node more recently.

Deng et al. have proposed an "intrusion-tolerant routing protocol in wireless sensor networks" (INSENS) that adopts a routing-based approach to security in WSNs [2]. It constructs routing tables in each node, bypassing malicious nodes in the network. The protocol cannot totally rule out attack on nodes, but it minimizes the damage caused to the network. The computation, communication, storage, and bandwidth requirements at the nodes are reduced, although at the cost of greater computation and communication at the base station. To prevent DoS attacks, individual nodes are not allowed to broadcast to the entire network. Only the base station is allowed to broadcast, and the base station is authenticated using a one-way hash function so as to prevent any possible masquerading by a malicious node. Control information pertaining to routing is authenticated by the base station in order to prevent injection of false routing data. The base station computes and disseminates routing tables, since it does not have computational and energy constraints. Even if an intruder takes over a node and this node does not forward packets, INSENS uses redundant multipath routing so that the destination can still be reached without passing through the malicious node.

The INSENS has two phases: (1) Route discovery and (2) data forwarding. During the route discovery phase, the base station sends a request message to all the nodes in a network by multihop forwarding. Any node receiving a request message records the identity of the sender and sends the message to all its immediate neighbors if it has not already done so. Subsequent request messages are used to identify the senders as neighbors, but repeated flooding is not performed. The nodes respond with their local topology by sending feedback messages. The integrity of the messages is protected using encryption by a shared key mechanism. A malicious node can inflict damage only by not forwarding packets, but the messages are sent through different neighbors; so, it is likely that a message reaches a node by at least one path. Hence, the effect of malicious nodes is not totally eliminated but is restricted to only a few downstream nodes in the worst case. Malicious nodes may also send spurious messages and cause battery drain for a few downstream nodes. Finally, the base station calculates forwarding tables for all nodes, with two independent paths for each node, and sends them to the nodes. The second phase of data forwarding takes place based on the forwarding tables computed by the base station.

The SPINS is a suite of security protocols optimized for sensor networks [35]. The SPINS includes two building blocks: (1) Secure network encryption protocol (SNEP) and (2) the micro version of the timed, efficient, streaming, loss-tolerant authentication protocol (μTESLA). The SNEP provides data confidentiality, two-party data authentication, and data freshness for peer-to-peer communication (node to base station), whereas μTESLA provides authenticated broadcast as discussed already.

The SPINS assumes that each node is predistributed with a master key K, which is shared with the base station at its time of creation. All the other keys, including a key K_{encr} for encryption, a key K_{mac} for MAC generation, and a key K_{rand} for random number generation, are derived from the master key using a string one-way function. The SPINS uses RC5 protocol for confidentiality. If A wants to send a message to base station B, the complete message sent by A to B is as follows:

$$A \lozenge B: D_{<K_{encr}C>}, \text{MAC}\left(K_{mac}, C \mid D\right)_{<K_{encr}C>} \tag{15.3}$$

In the aforementioned expression, D is the transmitted data and C is a shared counter between the sender and the receiver for the block cipher in counter mode. The counter C is incremented after each message is sent and received by the sender and the receiver respectively. The SNEP also provides a counter exchange protocol to synchronize the counter value at both sides.

The SNEP provides the following properties:

Semantic security: The counter value is incremented after each message and, thus, the same message is encrypted differently each time.

Data authentication: A receiver can be sure that a received message originated from the claimed sender if the MAC verification produces positive results.

Replay protection: The counter value in the MAC prevents the replay of old messages by an adversary.

Weak freshness: The SPINS identifies two types of freshness: (1) Weak freshness provides partial message ordering and carries no delay information. (2) Strong freshness provides a total order on a request–response pair and allows delay estimation. In SNEP, the counter maintains message ordering in the receiver side and yields weak freshness. The SNEP guarantees only weak freshness, since there is no guarantee to node A that a message was created by node B in response to an event in node A.

Low communication overhead: The counter state is kept at each end point and need not be sent in each message.

Inspired by the work on public key cryptography [60–62,72], Du et al. investigated the public key authentication problem [116]. The use of public key cryptography eases many problems in secure routing, for example, issues associated with authentication and integrity. However, before a node A uses the public key from another node B, A must verify that the public key is actually B's, that is, A must authenticate B's public key; otherwise a "man-in-the-middle" attack is possible. In general networks, public key authentication involves signature verification on a certificate signed by a trusted third-party certificate authority (CA) [120]. However, the signature verification operations are very expensive for sensor nodes. Du et al. have proposed an efficient alternative that uses only a one-way hash function for public key authentication. The proposed scheme can be divided into two stages: In the predistribution stage, A Merkle tree R is constructed with each leaf L_i corresponding to a sensor node. Let pk_i represent node i's public key, V be an internal tree node, and V_{left} and V_{right} be V's two children. The value of an internal tree node is denoted by Φ. The Merkle tree can then be constructed as follows:

$$
\begin{aligned}
\Phi(L_i) &= h(id_i, pk_i) \text{ for } i = 1, \dots N \\
\Phi(V) &= h(\Phi(V_{\text{left}}) \,\|\, \Phi(V_{\text{right}}))
\end{aligned}
\tag{15.4}
$$

In the aforementioned expression, "$\|$" represents the concatenation of two strings and h is a one-way hash function such as MD5 or SHA-1. Let R be the root of the tree. Each sensor node v needs to store the root value $\Phi(R)$ and the sibling node values $\lambda_1, \dots \lambda_H$ along the path from v to R. If node A wants to authenticate B's public key, B sends its public key pk along with the value of $\lambda_1, \dots \lambda_H$ to node A. Then, A can use the same procedure to reconstruct the Merkle tree $R`$ and calculate the root value $\Phi(R`)$. A trusts B to be authentic if $\Phi(R`) = \Phi(R)$. A sensor node only needs $H + 1$ storage units for the extra hash values. Based on this scheme, Du et al. further extended the idea to reduce the height of the Merkle tree to improve the communication overhead of the scheme. The proposed scheme is more efficient than signature verification on certificates. However, the scheme requires that some hash values are distributed in a predistribution stage. This results in some scalability issues when new sensors are added to an existing WSN.

Tanachaiwiwat et al. have presented a novel secure routing protocol named "trusted routing for location aware sensor networks" (TRANS) [5]. It is primarily meant for use in data-centric networks.

It makes use of an asymmetric cryptographic scheme that relies on a loose-time synchronization mechanism to ensure message confidentiality. The authors have used μTESLA to ensure message authentication and confidentiality. Using μTESLA, TRANS is able to ensure that a message is sent along a path of trusted nodes utilizing location-aware routing. The base station broadcasts an encrypted message to all its neighbors. Only the trusted neighbors will possess the shared key necessary for decrypting the message. The trusted neighbors then add their locations (for the return trip), encrypt the new message with their shared key, and forward the message to their neighbors closest to the destination. Once the message reaches the destination, the recipient is able to authenticate the source (base station) using the MAC corresponding to the base station. To acknowledge or reply to the message, the destination node can simply forward a return message along the same trusted path along which the original message was received [5].

Sen et al. propose a routing protocol that is resilient against packet-dropping attack by malicious nodes in a WSN [121]. It essentially utilizes a single-path routing concept and hence saves energy compared to multipath routing protocols. If a malicious node is detected in the next hop on a routing path, the node is efficiently bypassed and packets are routed around the node to the base station still in a single path. The protocol is based on a robust "neighborhood-monitoring system" (NMS) that works on promiscuous monitoring of the neighborhood of a node and detection of possible malicious packet-dropping attacks by a cooperative algorithm using neighbor list checking.

One particular challenge to secure routing in WSNs is that it is very easy for a single node to disrupt the routing process by disrupting the route discovery process. Papadimitratos et al. have proposed a secure route discovery protocol that guarantees correct topology discovery in an ad hoc sensor network [4]. The protocol relies on MAC and an accumulation of the node identities along the route traversed by a message. In this way, a source node discovers the sensor network topology as each node along the route from the source to a destination appends its identity to the message. In order to ensure that the message has not been tampered with, a MAC is constructed and it can be verified at both the destination and the source nodes (for the return message from the destination).

15.5.5 *Defense against Sybil Attacks*

Any defense mechanism against the Sybil attack must ensure that a framework is in place in the network to validate that a particular identity is the only identity being held by a given physical node [47]. Newsome et al. have described three orthogonal dimensions of the Sybil attack taxonomy [47]. The three dimensions are as follows: (1) Direct versus indirect communication, (2) fabricated versus stolen identities, and (3) simultaneity. In direct communication, the Sybil nodes communicate directly with legitimate nodes. In this attack, when a legitimate node sends a radio message to a Sybil node, one of the malicious devices listens to the message. In indirect communication, no legitimate nodes are able to communicate directly with the Sybil nodes. Messages sent to a Sybil node are routed through one or more malicious nodes, which pretend to pass the message on to the Sybil node. In case of fabricated identities, the attacker creates new Sybil identities arbitrarily. However, if a mechanism is in place to detect false identities, an attacker cannot fabricate new identities. In this case, the attacker needs to assign other legitimate identities to Sybil nodes. This identity theft may go undetected if the attacker destroys or temporarily disables the impersonated nodes. In case of simultaneous attacks, the attacker tries to have all the Sybil identities participate in the network simultaneously. Alternatively, the attacker may present a large number of identities over a period of time while deploying a small number of identities at a given point of time. Newsome et al. primarily describe direct validation techniques, including a radio resource test. In the radio test, a node assigns to each of its neighbors a different channel and listens to each neighbor. If the node detects a transmission on the channel, it is assumed that the node transmitting on the channel is a physical node. Similarly, if the node does not detect a transmission on the specified channel, the node assumes that the identity assigned to the channel is not a physical identity.

To defend against the Sybil attack, random key predistribution techniques [37–38,97] can be used. In random key predistribution, a random set of keys or key-related information is assigned to each sensor node so that in the key setup phase, each node can discover or compute the common keys shared by it with its neighbors. The common keys are used as shared secret session keys to ensure node-to-node secrecy. Newsome et al. propose that the identity of each node be associated with the keys assigned to the node [47]. With a limited set of captured keys, there is some probability that an arbitrarily generated identity will work.

15.5.6 Detection of Node Replication Attack

Parno et al. have proposed a mechanism for distributed detection of node replication attacks in WSNs [51]. To address the fundamental limitations of currently existing mechanisms, for example, single point of failure in centralized schemes or neighborhood voting protocols that fail to detect distributed replications, the authors propose two algorithms that work through the collective actions of multiple nodes in a WSN. The algorithms are as follows: (1) Randomized multicast and (2) line-selected multicast algorithms. The randomized multicast algorithm distributes location information of a node to randomly selected witnesses and exploits the birthday paradox to detect replicated nodes. The line-selected multicast algorithm uses the network topology to detect replication as discussed here.

Randomized broadcasting evolved from traditional node-to-node broadcasting. In traditional node-to-node broadcasting, each node in a network uses an authenticated broadcast message to flood the network with its location information. Each node stores the location information of its neighbors, and if it receives a conflicting claim it revokes the offending node. This protocol can achieve 100% detection of all duplicate location claims if the broadcasts reach all the nodes. However, the total communication cost for the protocol is $O(n^2)$, which is too high for a large WSN. To reduce the communication cost of node-to-node broadcasting, deterministic multicast mechanism may be applied where a node's location claim is shared with a limited subset of deterministically chosen "witness nodes." The witnesses are chosen as a function of the node's ID. If an adversary replicates a node, the witnesses will receive two different location claims for the same node ID. The conflicting location claims trigger the revocation of the replicated node. The randomized multicast approach suggested by Parno et al. improves the robustness of the deterministic multicast mechanism. It randomizes the witnesses for a given node's location claim so that the adversary cannot anticipate their identities. When a node announces its location, each of its neighbors sends a copy of the location claim to a set of randomly selected witness nodes. If the adversary replicates a node, then two sets of witnesses will be selected. In a network of n nodes, if each location produces $\sqrt{n}$ witnesses, then the birthday paradox predicts at least one collision with high probability, that is, at least one witness will receive a pair of conflicting location claims. Two conflicting location claims form sufficient evidence to revoke a node; so the witness can flood the pair of location claims through the network, and each node can independently confirm the revocation decision. Unfortunately, however, the communication and storage overheads for randomized multicast are too high, which are $O(n^2)$ and $O(\sqrt{n})$, respectively. The authors have suggested some enhancements to the scheme for improving communication and storage overheads:

To reduce the communication cost of the randomized multicast approach, Parno et al. have proposed an alternative algorithm, that is, line-selected multicast. It is based on the "rumor routing protocol" [122]. The idea is that a location claim traveling from a source s to a destination d will also travel through several intermediate nodes. If each of these nodes records the location claims, then the path of the location claim through the network can be thought of as a line segment. The destination of the location claim is one of the randomly chosen witnesses. As the location claim routes through the network toward a witness node, the intermediate sensors check the claim. If a conflicting location claim crosses the line segment, then the node at the intersection detects the conflict and initiates a revocation broadcast. The line-selected multicast algorithm has a communication overhead of $O(n\sqrt{n})$ as long as each line segment is of $O(\sqrt{n})$ nodes. The storage overhead of the algorithm is $O(\sqrt{n})$.

15.5.7 Defense against Traffic Analysis Attacks

Deng et al. have proposed a mechanism for defending against traffic analysis attacks in a WSN [55]. The authors have argued that since the base station is a central point of failure, once the location of the base station is discovered an adversary can disable or destroy it, thereby rendering the data-gathering functionalities of the entire WSN ineffective. Two classes of traffic analysis attacks in WSNs are identified: (1) Rate monitoring attack and (2) time correlation attack. In a time correlation attack, an adversary monitors the packet-sending rate of nodes near it and moves closer to the nodes that have a higher packet-sending rate. In a time correlation attack, an adversary observes the correlation in sending times between a node and its neighbor node that is assumed to be forwarding the same packet and deduces the path by following the sound of each forwarding operation as the packet propagates toward the base station. The mechanism proposed by the authors prevents rate monitoring and time correlation attacks. The mechanism involves four techniques: First, a multiple-parent routing scheme is introduced that allows a sensor node to forward a packet to one of its multiple parents. This makes the patterns less pronounced in terms of routing packets toward the base station. Second, a controlled random walk is introduced into the multihop path traversed by a packet through the WSN toward the base station. This distributes packet traffic, thereby rendering rate monitoring attacks less effective. Third, random fake paths are introduced to confuse an adversary from tracking a packet as it moves toward a base station. This mitigates the effectiveness of time correlation attacks. Finally, multiple, random areas of high communication activities are created to deceive an adversary as to the true location of the base station, which further increases the difficulty of initiating rate monitoring attacks. The combination of these four strategies makes the proposed mechanism extremely robust to any traffic analysis attack.

15.5.8 Defense against Attacks on Sensor Privacy

The attacks on information privacy in WSNs are discussed in Section 15.4. In Sections 15.5.8.1 through 15.5.8.3, some schemes for protecting information privacy in WSNs are discussed.

15.5.8.1 Anonymity Mechanisms

Precise location information enables accurate identification of a user. This is a serious threat to privacy. One way to handle this problem is to make data sources anonymous. An anonymity mechanism depersonalizes the data before it is released from a source. Gruteser et al. have presented an analysis on the feasibility of anonymizing location information in location-based services in an automotive telematics environment [123]. Beresford et al. have proposed anonymity techniques for an indoor location system based on the Active Bat [124]. In one study [125], an efficient and reliable routing protocol for wireless ad hoc and mesh networks has been presented that ensures anonymity of the user. User anonymity, authentication, and data privacy is achieved by applying a novel protocol that is based on Rivest's ring signature scheme [126].

Since ensuring total anonymity is almost an impossible proposition, in nearly all practical scenarios a trade-off is made between anonymity and disclosure of public information in most of the privacy protection mechanisms. Four approaches have been proposed by researchers in this direction [52,127–129] for WSNs. These approaches are as follows: (1) Decentralization of storage of sensitive data, (2) establishment of secure channels for communication, (3) changing the pattern of data traffic, and (4) exploiting the mobility of nodes. The sensitive location data is stored in a spanning tree of nodes so that no single node holds a complete view of the location information. Communication using secure protocols such as SPINS [35] makes eavesdropping and active attack on a WSN extremely difficult. The data traffic pattern may be changed by selectively inserting some bogus data in network traffic so that traffic analysis by an external entity is not successful. Mobile sensor nodes make attacks on location privacy very difficult. The Cricket system [128] is a location-support system for mobile and location-dependent applications inside large buildings. It allows applications running on mobile and static nodes to learn their physical locations from a set of listeners. The listeners hear and analyze information from beacons in a building. Location sensors are placed on the mobile devices instead of some static locations in the building, and the location information is not disclosed during the process of position determination.

15.5.8.2 Policy-Based Approaches

In policy-based defense mechanisms, decisions on access control and authentication techniques are made on the basis of a specified set of privacy policies. Molnar et al. have presented the concept of private authentication and demonstrated its application in the radio frequency (RF) identification (RFID) domain [130]. Duri et al. propose a policy-based framework for protecting sensor information, in which a computer inside a car acts as a trusted agent for location privacy [131]. Snekkenes introduces various parameters for access control that enable specifying policies in the context of a mobile network [132]. Some of the parameters are time of request, location, speed, and identity of the located object. Myles et al. describe the architecture of a centralized location server that controls access requests from client applications through a set of validator modules based on a set of XML-coded privacy policies [133]. Hengartner et al. discuss various challenges that arise in the specification and implementation of policies that control access to location information [134]. The authors also present a design framework of an access control mechanism in their study.

15.5.8.3 Information Flooding

Ozturk et al. proposed various modifications to WSN routing protocols for protecting the location information of a source node [53]. In particular, the authors discuss a set of flooding protocols. Randomized data routing and a phantom traffic generation mechanism are used so that it is difficult for an adversary to track any data source. For ensuring privacy of source location, the authors discuss four types of flooding-based routing protocols: (1) Baseline flooding, (2) probabilistic flooding, (3) flooding with fake messages, and (4) phantom flooding. They are described as follows:

Baseline flooding: In baseline flooding, every node in a network forwards a message once and no node retransmits a message that was previously transmitted by it. When a message reaches an intermediate node, the node first checks whether it has received and forwarded the message before. If this is its first time, the node broadcasts the message to all its neighbors. Otherwise, the node just discards the message.

Probabilistic flooding: In probabilistic flooding, only a subset of nodes in the entire network participates in data forwarding and the other nodes simply discard the messages they receive. One possible weakness of this approach is that some messages may get lost in the network and, as a result, overall network connectivity is affected. However, Ozturk et al. [53] have proved analytically that this is not a significant problem.

Flooding with fake messages: Flooding cannot provide privacy protection because an adversary can easily identify the shortest path between a source and a sink and use it to backtrack to the source location. The main reason for lack of location privacy is that there is only one source node. One approach that can alleviate the risk of source-location privacy breaching is to augment the flooding protocols so that more sources can be introduced that inject fake messages into the network. If the fake messages have the same length as real messages and the fake messages are also encrypted, it will be impossible for an adversary to distinguish between fake and real messages.

Phantom flooding: Phantom flooding follows the same principle as probabilistic flooding. It too attempts to direct messages to different locations of a network so that an adversary cannot receive a steady stream of messages to track the message source. However, probabilistic flooding is not very effective since shorter paths are more likely to deliver more messages. Phantom flooding entices an attacker away from the real source toward a fake source, which is called the "phantom source." In phantom flooding, every message experiences two phases: (1) A walking phase, which may be a random walk or a directed walk, and (2) subsequent flooding, which is meant to deliver the message to the sink. When the source sends out a message, the message is unicast in a random fashion within the first h_{walk} hops. This is called the random walk phase. After h_{walk} hops, the message is flooded using the baseline flooding technique. This is the flooding phase. Phantom flooding significantly improves privacy and network safety period because every message may take a different (shortest) path to reach any node in the network.

Deng et al. address the problem of defending a base station against physical attacks by concealing the geographic location of the base station [55]. The authors have investigated several countermeasures against traffic analysis techniques aimed at disguising the location of a base station. In the proposed mechanism, a degree of randomness is introduced when selecting the multihop route to the base station. Then, random fake packets are introduced as a packet moves toward a base station. Metrics such as total entropy of the network, total energy consumed, and the ability to guard against heuristic-based techniques to locate the base station are evaluated analytically as well as using extensive simulations.

Xi et al. [135] have described a successful attack on the flooding-based phantom routing approach proposed by Ozturk et al. [53]. The authors have also proposed the "greedy random walk" (GROW) protocol, which involves a two-way random walk, that is, from both source and sink, to reduce the chance that an eavesdropper can collect the communicated location information. In the proposed mechanism, the sink first initiates an N-hop random walk; the source then initiates an *M*-hop random walk. Once the source packet reaches an intersection of these two paths, it is forwarded through the path created by the sink. Local broadcasting is used to detect when the two paths intersect each other. In order to minimize the chance of backtracking along a random walk, the nodes are stored in a bloom filter as the walk progresses. At each stage, the intermediate nodes are checked against the bloom filter to ensure that backtracking is minimized.

15.5.9 Intrusion Detection

The security mechanisms implemented in secure routing protocols and secure data aggregation protocols are configured beforehand to prevent an attacker from challenging the security of a network. However, these security mechanisms alone cannot ensure that a WSN is secure. Since it is possible for an attacker to compromise a sensor node, it is easy for him or her to inject false data into a WSN. Authentication and data encryption are not enough for ensuring data security. Another approach to protect WSNs involves mechanisms for detecting and reacting to intrusions.

An "intrusion detection system" (IDS) monitors a host or a network for suspicious activity patterns that are outside normal and expected behavior [29]. It is based on the assumption that there exists a noticeable difference in behavior between an intruder and a legitimate user of a network such that an IDS can match it with preprogrammed or possibly learned rules. Based on the analysis model used for analyzing audit data to detect intrusions, IDSs are usually classified into two types: (1) Rule-based IDSs and (2) anomaly-based IDSs [136]. Rule-based IDSs are used to detect known patterns of intrusions [137,138]. Anomaly-based systems are used to detect new or unknown intrusions [139,140]. A rule-based IDS has a low false-alarm rate compared to an anomaly-based system, and an anomaly-based IDS has a high intrusion detection rate compared to a rule-based system. Distributed IDSs have high detection efficiency. Sen et al. have presented the model of a distributed IDS that consists of a large number of autonomous and cooperating agents [141]. The architecture exploits interagent communication and distributed computation to achieve high detection efficiency with very low rates of false positives and false negatives.

However, WSNs are generally application specific and lack basic information on topology, normal usage, expected communication patterns, etc. It is impractical to preinstall some fixed patterns in sensors before they are deployed. Moreover, due to constraints in sensors, it is both time consuming and energy consuming to learn and detect these parameters after deployment. Thus, existing intrusion detection schemes in ad hoc networks may not be adapted to WSNs.

Research on intrusion detection in WSNs is still in its preliminary stage. Current research focuses on how to detect and eliminate injected false information. Thus, cooperation among sensors, especially neighboring nodes, is necessary to decide the validity of a report. Section 15.5.9.1 discusses some existing mechanisms of intrusion detection for WSNs.

15.5.9.1 Intrusion Detection in Wireless Sensor Networks

Brutch et al. discuss various types of possible attacks against WSNs in their study and present three different architectures for intrusion detection [142]: The first is a stand-alone architecture. In this case, each node functions as an independent IDS and is responsible for detecting attacks directed toward it. The nodes do not exchange intrusion data and no cooperative detection mechanisms are deployed. The second architecture is a distributed and cooperative architecture. In this architecture, an intrusion detection agent is deployed on each node. The local agents are responsible for detecting local attacks on nodes; they also cooperate among themselves by exchanging intrusion-related data to detect global intrusion attempts. The third architecture proposed by the authors is a hierarchical architecture. This is suitable for a multilayered WSN, in which the network is divided into clusters with the cluster-head node being responsible for routing within a cluster. Multilayered networks are primarily used for event correlation.

Zhu et al. propose an "interleaved hop-by-hop" authentication (IHOP) scheme [143]. The IHOP scheme guarantees that the base station will detect any injected false data packets when no more than a certain number, t, of nodes are compromised. The sensor network is organized in a cluster-based hierarchy. Each cluster-head node builds a route to the base station, and each intermediate node has an upper associate node and a lower associate node that are $t + 1$ hops away from it. The IHOP uses a number of shared keys: Every node shares a master key with the base station, each node knows its one-hop neighbors and establishes a pairwise key with each of them, and a node can establish a pairwise key with another node that is multiple hops away if needed.

Further, IHOP also assumes that the base station has a mechanism to authenticate broadcast messages, for example, μTESLA. A cluster-head node collects information from the members of its cluster and sends a report to the base station only when at least $t + 1$ sensors observe the same result. Meanwhile, a cluster-head node also collects the MACs from detecting nodes. Each detecting node sends two MACs to the cluster-head node: (1) A MAC using the key shared with the base station, referred to as individual MAC, and (2) a MAC using the key shared with its upper associate nodes, referred to as pairwise MAC. The cluster-head node then compresses the $t + 1$ individual MACs by XORing them to reduce the size of the report. However, the pairwise MACs are not compressed for transmission. If they were, a node replaying a message would not be able to extract the pairwise MACs and a compressed MAC for the base station. When an intermediate node receives a report, it verifies the MAC of its lower associate node. If it fails, the report is eliminated. Otherwise, the node removes the MAC, generates a new MAC using its upper associate node pairwise key, and appends the new MAC to the report. However, the pairwise MACs are not compromised for transmission. If they were, a node relaying the message would not be able to extract the pairwise MACs of interest to it. Thus, a legitimate report includes $t + 1$ pairwise MACs and a compressed MAC for the base station. When an intermediate node receives a report, it verifies the MAC of its lower associate node. If it fails, the report is eliminated. Otherwise, the node removes the MAC, generates a new MAC using its upper associate node pairwise key, and appends the new MAC to the report.

The IHOP scheme ensures that the base station can detect false data packets when no more than t nodes are compromised. However, the authors [143] have not shown how to select the parameter t for a sensor network.

Wang et al. proposed a scheme to detect whether a node is faulty or malicious with the collaboration of neighbor nodes [144]. In the proposed scheme, when a node suspects that one of its neighbors is faulty, it sends messages requesting opinions on the behavior of the suspected node to other neighbors of the suspect. After collecting the results, the node analyzes the results to diagnose whether the suspect has a fault. The authors formalized the problem as how to construct a dominating tree to cover all the neighbors of a suspect node and further proposed two tree-based propagation collection protocols to construct a dominating tree and collect information via the tree structure.

Albers et al. have proposed an intrusion detection architecture based on a local IDS (LIDS) on each node in a wireless ad hoc network [145]. In order to detect a network-wide intrusion, the LIDSs on the nodes collaborate with each other and exchange two types of data: (1) Security data and (2) intrusion alerts. Security data is used to exchange information with other network hosts, whereas intrusion alerts are used to inform LIDSs in neighboring nodes to exchange intrusion-related information. Although the framework is for an ad hoc network, its approach of local anomaly detection and cooperatively detecting any network-wide intrusion can be used to develop an intrusion detection mechanism for a WSN [6].

Intrusion detection in WSNs is still largely open to research. Key research issues include the following:

- Due to the constraints of WSNs, intrusion detection in WSNs has many aspects that are not of concern in other network types. The problem of intrusion detection needs to be well defined in WSNs.
- The IDS protocols proposed in the literature focus on filtering injected false information only [2,11,143]. These protocols need to be improved to address scalability issues.

15.5.10 Secure Data Aggregation

Data communication constitutes an important share of total energy consumption in a sensor network. Simulation [35] shows that data transmission accounts for 71% of the energy cost of computation and communication for SNEP protocols. An efficient data aggregation mechanism can greatly help in optimizing energy consumption.

In a WSN, there are certain nodes called "aggregators," which are responsible for carrying out data aggregation operations. If an aggregator node is compromised, it is easy for an adversary to inject false data into the network. Another possible attack is to compromise a sensor node and inject forged data through it. Without authentication, attackers may fool aggregators into reporting false data to the base station. Secure data aggregation requires authentication, confidentiality, and integrity of data to be maintained. Moreover, secure data aggregation also requires cooperation among the sensor nodes in identifying compromised sensors.

Before discussing some secure aggregation (SA) mechanisms detailed in the literature, an overview of some well-known aggregation techniques are presented.

Estrin et al. [6] propose a clustering-based algorithm that uses directed diffusion to gather a global perspective utilizing only the local nodes in each cluster. The nodes are assigned different levels with level 0 being assigned to the nodes lying at the lowest level. While the nodes at the highest level can communicate across the clusters, the nodes at the lower levels communicate among each other in the same cluster via the cluster head node. This effectively enables localized cluster computation, whereas the higher-level nodes communicate the local information of clusters to get a global picture.

Madden et al. [8] propose a mechanism called "tiny aggregation" (TAG) service. It is a generic data aggregation mechanism that involves a language similar to SQL to generate queries in a WSN. The base station generates a query using this language. The sensor nodes send the reply using routes constructed based on a routing tree. At each point in the tree, data is aggregated using some aggregation function that was defined in the initial query.

Shrivastava et al. have proposed a summary structure for supporting fairly complex aggregate functions, such as median and range queries [10]. In addition, computation of relatively easier functions such as min/max, sum, and average are also supported in the proposed framework. However, more complex aggregate functions, such as the most frequently reported data value, are not supported. The computed aggregate functions are approximate, but the estimated errors are statistically bounded.

A number of SA protocols for WSNs are detailed in the literature. However, fundamentally there is a conflict of interests in data confidentiality and data aggregation. Confidentiality requires data to be transmitted in ciphertext mode, whereas data aggregation is usually done on plaintext

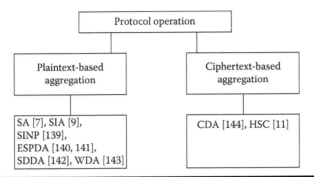

Figure 15.5 Secure data aggregation in wireless sensor networks: A taxonomy. (From Wang, Y. et al., *IEEE Comm. Surveys and Tutorials*, 8, 2, 2–23, 2006.)

content. A straightforward method is to invoke end-to-end encryption before executing data aggregation. This strategy, however, has a shortcoming. The encryption and decryption operations involve substantial computation overhead. An alternative method is to provide data aggregation on concealed data, which requires a particular class of encryption transformation. However, this method usually reduces the security level [103].

Figure 15.5 shows the taxonomy of secure data aggregation protocols for WSNs. There are two categories of SA protocols: (1) Plaintext-based protocols and (2) ciphertext-based protocols.

15.5.10.1 Secure Aggregation on Plaintext Data

Hu et al. have proposed an SA protocol that uses the µTESLA protocol [7]. The protocol is resilient to both intruder devices and single device key compromises. In the proposition, sensor nodes are organized into a tree in which the internal nodes act as aggregators. However, the protocol is vulnerable if a parent and one of its child nodes are compromised, since due to the delayed disclosure of symmetric keys the parent node will not be able to immediately verify the authenticity of the data sent by its children nodes.

Przydatek et al. have presented a "secure information aggregation" (SIA) framework for sensor networks [9]. The framework consists of three categories of nodes: (1) A home server, (2) the base station, and (3) sensor nodes. A base station is a resource-enhanced node that is used as an intermediary between the home server and the sensor nodes, and it is also the candidate to perform the aggregation task. The SIA assumes that each sensor has a unique identifier and shares a separate secret cryptographic key with both the home server and the aggregator. The keys enable message authentication and encryption if data confidentiality is required. Moreover, SIA assumes that the home server and the base station can use a mechanism, such as µTESLA, to broadcast authenticated messages. The proposed solution follows "aggregate-commit-prove" approach.

In the first phase, that is, aggregate phase, the aggregator collects data from sensors and locally computes the aggregation result using some specific aggregate function. Each sensor shares a key with the aggregator. This allows the aggregator to verify whether the sensor reading is authentic. However, there is a possibility that a sensor may have been compromised and an adversary has captured the key. In the proposed scheme, there is no mechanism to detect such an event.

In the second phase, that is, commit phase, the aggregator commits to the collected data. This phase ensures that the aggregator actually uses the data collected from the sensors and the statement to be verified by the home server about the correctness of computed results is meaningful.

One efficient mechanism for committing is a Merkle hash-tree construction [146]. In this method, data collected from the sensors is placed at the leaves of a tree. The aggregator then computes a binary hash tree starting with the leaf nodes. Each internal node in the hash tree is computed as the hash value of the concatenation of its two children nodes. The root of the tree is called the commitment of the collected data. As the hash function in use is collision-free, once the aggregator commits to the collected values it cannot change any of the collected values.

In the third and final phase, the aggregator and the home server engage in a protocol in which the aggregator communicates the aggregation result. In addition, the aggregator uses an interactive proof protocol to prove the correctness of the reported results. This is done in two logical steps: In the first step, the home server ensures that the committed data is a good representation of the sensor data readings collected. In the second step, the home server checks the reliability of aggregator output. This is done by checking whether the aggregation result is close to the committed results. The interactive proof protocol varies depending on the aggregation function used. Moreover, the authors also presented efficient protocols for secure computation of the median and average of the measurements, for the estimation of network size, and for finding the minimum and maximum sensor readings.

Deng et al. proposed a collection of mechanisms for "securing in-network processing" (SINP) for WSNs [147]. A security mechanism was proposed to address the downstream requirement that sensor nodes authenticate commands disseminated from their parent aggregators and the upstream requirement that aggregators authenticate the data produced by sensors before aggregating that data. In the downstream stage, two techniques are involved: (1) One-way functions and (2) μTESLA. The upstream stage requires that a pairwise key is shared between an aggregator and its sensor nodes.

Cam et al. propose an "energy-efficient secure pattern-based data aggregation" (ESPDA) protocol for WSNs [148,149]. It is noted that ESPDA is applicable for hierarchy-based sensor networks. In ESPDA, a cluster-head node first requests sensor nodes to send the corresponding pattern code for the sensed data. If multiple sensor nodes send the same pattern code to the cluster-head node, only one of them is permitted to send the data to the cluster-head node. The ESPDA is secure because it does not require encrypted data to be decrypted by cluster-head nodes to perform data aggregation.

Cam et al. have introduced another scheme, the "secure differential data aggregation" (SDDA) scheme, based on pattern codes [150]. The SDDA prevents redundant data transmission from sensor nodes by implementing the following: The SDDA transmits differential data rather than raw data, performs data aggregation on pattern codes representing the main characteristics of the sensed data, and employs a sleep protocol to coordinate the activation of sensing units in such a way that only one of the sensor nodes capable of sensing the data is activated at a given time. In the SDDA data transmission scheme, raw data from sensor nodes is compared with reference data with the difference data being transmitted. The reference data is obtained by taking the average of previously transmitted data.

Du et al. proposed a "witness-based data aggregation" (WDA) scheme for WSNs to ensure the validation of data fusion nodes to the base station [151]. To prove the validity of the fusion results, the fusion node has to provide proofs from several witnesses. A witness is one who conducts data fusion similar to a data fusion node but does not forward its result to the base station. Instead, each witness computes the MAC of the result and then provides it to the data fusion node, which must forward the proofs to the base station.

Wagner studied secure data aggregation in sensor networks and proposed a mathematical framework for formally evaluating their security [152]. The robustness of an aggregation operator against malicious data is quantified in this framework. Ye et al. propose a statistical en route filtering mechanism to detect the sending of any forged data from sensor nodes to the base station of a WSN using multiple MACs along the path from the aggregator to the base station [11].

15.5.10.2 Secure Aggregation on Ciphertext Data

The SA of ciphertext data in WSNs is required to preserve the privacy of sensor nodes. Efficient in-network data aggregation with preservation of data privacy is an important requirement in many WSN applications [153–157]. As a key approach to fulfilling this requirement of private data aggregation of sensor nodes, concealed data aggregation (CDA) schemes have been proposed in which multiple source nodes send encrypted data to a sink along a convergecast tree with aggregation of ciphertext being performed over the route [153–155,157,158]. Two ciphertext-based secure data aggregation schemes have been proposed by Catelluccia et al. [154], Peter et al. [162], and Girao et al. [155]. The propositions are based on a particular encryption transformation, that is, a "privacy homomorphism" (PH). A PH is an encryption transformation that allows direct computation on encrypted data. Let Q and R denote two rings, and "+" denote addition and "×" denote multiplication on the two rings. Let K be the key space. We denote an encryption transformation $E : K \times Q \Diamond R$ and the corresponding decryption transformation $D : K \times R \Diamond Q$. Given a, $b \, \varepsilon \, Q$ and $k \, \varepsilon \, K$, the following operation is termed as "additively homomorphic":

$$a + b = D_k(E_k(a) + E_k(b)) \tag{15.5}$$

Similarly, the following operation is termed as "multiplicatively homomorphic" [159]:

$$a \times b = D_k(E_k(a) \times E_k(b)) \tag{15.6}$$

The CDA scheme proposed by Girao et al. [155] is based on PH, which was proposed by Domingo-Ferrer [160]. Although the study by Wagner [161] showed that the proposed PH [160] is insecure against chosen plaintext attacks for some parameter settings, Girao et al. [155] claim that for the WSN data aggregation scenario the security level is still adequate and the proposed PH method [160] can be employed for encryption. The CDA can be used to calculate sum and average in a hierarchical WSN. To calculate average, an aggregator needs to know the number of sensor nodes, n.

Castelluccia et al. proposed a simple and provable secure additively homomorphic stream cipher (HSC) that allows the efficient aggregation of encrypted data [154]. The new cipher uses modular addition and, therefore, is very well suited for CPU-constrained devices such as those used in WSNs. The aggregation based on this cipher can be used to efficiently compute statistical values such as mean, variance, and standard deviation of sensed data while achieving significant bandwidth gain.

Secure data aggregation is an extremely important issue in WSNs. Several secure data aggregation protocols have been proposed by researchers. However, no comparisons have been conducted on such proposed protocols. Further evaluations are required to get an idea about the performance of such protocols. The performance metrics for evaluation might include security, processing overhead, communication overhead, energy consumption, and data compression. Moreover, new data aggregation protocols are needed to address higher scalability and higher reliability requirements against aggregator and sensor node cheating [103].

15.5.11 Defense against Physical Attacks

To protect against a possible physical attack, sensor nodes may be equipped with special hardware. The sensor nodes in a WSN may be protected against tampering by tamperproofing the physical packages of the sensors [29]. Researchers have also proposed mechanisms that focus on building tamper-resistant hardware in order to make the memory contents on the sensor chip inaccessible to a potential external attacker [20,21,24]. Special-purpose software and hardware may also be deployed outside the sensor nodes to detect physical tampering. Self-termination of sensor nodes is an effective mechanism to defend against possible data theft in the event of a physical attack. The basic idea in this case is that whenever a sensor senses an attack, it kills itself and destroys all the data and keys stored in its memory. This is particularly feasible in large-scale WSNs where there is much redundancy of information and connectivity among the nodes. However, the main challenge is to accurately identify a physical attack. A simple solution is to periodically verify the neighborhood information for each node. In case of a mobile sensor network, this is an open problem.

A number of techniques have been discussed for extracting protected data from card processors [20–24]. These techniques include manual microprobing, laser cutting, focused ion-beam manipulation, glitch attacks, and power analysis. Most of these techniques may be used to launch physical attacks on sensor nodes in a WSN.

Anderson et al. have proposed countermeasures for each of these attacks [21]. In some studies [20–24], the authors describe techniques for extracting protected software and data from smart card processors. This includes manual microprobing, laser cutting, focused ion-beam manipulation, glitch attacks, and power analysis, most of which are also possible physical attacks on sensors in a WSN. Based on an analysis of these attacks, Andersen et al. give examples of low-cost protection measures that make such attacks considerably more difficult [21].

Deng et al. have proposed various approaches for protecting sensors by deploying components outside them [163]. Sastry et al. have presented the ECHO protocol for secure and reliable location verification of sensor nodes in a WSN [25]. The scheme is based on the physical properties of sound and RF signal propagation from sensor nodes. It is not possible for an adversary to cheat and falsely claim a shorter distance from the base station by transmitting its ultrasonic sound response early, because it will not be able to produce the required nonce for verification.

In a study [2], Deng et al. present defense mechanisms against search-based physical attacks. The authors also discuss a systematic modeling framework for "blind" physical attacks [27]. The defense mechanism against physical attacks as proposed by the authors involves two phases: In the first phase, the sensors detect the attacker and send out attack notification messages in the network. In the second phase, the sensors that receive the notification messages schedule their states to switch off mode. Seshadri et al. have proposed a mechanism called "software-based attestation for embedded devices" (SWATT) to detect a sudden and abrupt change in the memory content of a sensor node [26]. An abrupt change in the memory content of a sensor indicates possibility of a physical attack.

15.5.12 Trust Management

Application of trust- and reputation-based frameworks for enforcing a high level of security in WSNs is another approach for defending networks against attacks. In fact, trust-based schemes can protect against attacks that are beyond the capabilities of cryptographic security. For example, issues such as judging the quality and reliability of sensor nodes and wireless links, data aggregation reliability and correctness of aggregator nodes, and timeliness in packet forwarding of sensors can be addressed effectively in a systematic manner with the help of a trust-based framework. However, trust-based models usually involve high computational overhead and building an efficient scheme for resource-constrained WSNs is a challenging task. A comprehensive discussion on basic concepts of trust and reputation and various security mechanisms based on these concepts for WSNs are given in the literature [164].

Pirzada et al. [165] have proposed an approach for building trust relationship between the nodes of an ad hoc network. It is assumed that the nodes in the network passively monitor the packets that are received and forwarded by other nodes. The receiving and forwarding activities of the nodes are termed as events. Events are observed and given weights depending on the type of application requiring a trust relationship with other nodes. The weights reflect the significance of the observed events for the corresponding application. The trust values for all events from a node are combined using weights to compute an aggregate trust level for the node. The computed trust values are used as link weights for the computation of routes. Links that connect more trustworthy nodes have smaller weights. A shortest-path routing algorithm can compute the most trustworthy paths in a network.

Oram [166] describes methods of finding paths from a source node to a designated target node in a peer-to-peer computing paradigm. Extending this approach, Zhu et al. [19] provide a practical approach for computing trust in wireless networks by treating individual mobile devices as nodes of a delegation graph G and mapping a delegation path from a source node S to a target node T into an edge in the corresponding transitive closure of the graph G. From the edges of the transitive closure of the graph G, the trust values of the wireless links are computed. In the proposed trust-based framework, an undirected transitive signature scheme is used within the authenticated transitive graphs.

In a study by Sen [167], a secure and efficient searching scheme for peer-to-peer networks has been proposed that utilizes topology adaptation by constructing an overlay of trusted peers in which neighbors are selected based on their trust ratings and content similarities. Using a robust trust management mechanism, the scheme provides a highly reliable framework for protecting the privacy of users and data in the network.

Yan et al. have proposed a security solution based on trust framework to ensure data protection, secure routing, and other security features in an ad hoc network [18]. Mechanisms of logical and computational trust analysis and evaluation are applied on the nodes. Each node evaluates the trust of its peers based on factors such as experience statistics, data value, intrusion detection results, and recommendations from its other neighbors. Ren et al. have presented a technique to establish trust relationships among nodes in an ad hoc network [16]. The proposed framework is a probabilistic solution based on a distributed trust model. A secret dealer is introduced only in the system bootstrapping phase to initiate trust propagation. Shorter and more robust trust chains are subsequently developed among the nodes. A fully self-organized trust establishment approach is then adopted to conform to dynamic changes in membership.

Ganeriwal et al. have proposed a reputation-based framework for high-integrity sensor networks [12]. The framework employs a beta distribution for reputation representation, updates, and integration. Using beta distribution for reputation computation and exploiting statitistical theory of estimation, a secure and robust data aggregation scheme for WSNs was presented by Sen [168]. Tanachaiwiwat et al. [17] have proposed a mechanism of location-centric isolation of nodes exhibiting misbehavior and trust-based routing among nodes in a sensor network. The trust value of a node is computed based on the cryptographic suite applied, availability statistics, and the packet-forwarding information of the node. If the computed trust associated with a node falls below a threshold, the node's location is considered insecure and it is avoided in the routing process. The robust reputation computation model allows accurate detection of node misbehavior.

In one study [169], a reputation- and trust-based security framework for ad hoc networks is proposed for detecting malicious packet-dropping attacks. The mechanism is based on a trust model that computes reputation values for the nodes in a network. A similar scheme based on cooperation of nodes in a neighborhood and a distributed algorithm for reputation computation has been presented in one study [170].

Liang et al. have carried out extensive work on the development of models and evaluating robustness and security of various aggregation algorithms in open and untrusted environments [13,14]. These models may be adapted for deployment of trust frameworks in WSNs. In one study [15], Liang et al. propose a model called "personalized trust" (PET) for nodes in a WSN. In another study [14], for aggregation of various ratings received by a node from its peer sensor nodes, a comprehensive analytical and inference model of trust has been presented. The authors have identified two types of uncertainties in a rating system in an open computing environment: (1) Uncertainties associated with rating aggregation algorithms and (2) uncertainties resulting from other algorithm-independent design factors. The authors have shown that complex aggregation algorithms are not suitable in many cases due to memory limitations in the sensor nodes for storing knowledge related to computation of a trust-based framework. Simulation results show that it is better to treat ratings received from different evaluators (i.e., nodes) with equal weight and simply compute the average to arrive at the final trust value. This approach not only has a very low computational overhead but also gives very satisfactory results in practice. The authors also observe that for a trust model, the most important and critical issue is how to adaptively adjust the parameters of the model based on a change in environment.

15.6 Conclusions and Future Trends

Although research efforts are underway on cryptography, key management, secure routing, secure data aggregation, and intrusion detection in WSNs, there are still some challenges left to be addressed. First, the selection of appropriate cryptographic methods depends on the processing capability of sensor nodes, indicating that there is no unified solution for all sensor networks. Instead, the security mechanisms in existence today are highly application specific. Second, sensors are characterized by constraints on energy, computation capability, memory, and communication bandwidth. The design of security services in WSNs must satisfy these constraints. Third, most of the current protocols assume that the sensor nodes and the base station are stationary. However, there are situations, such as battlefield environments, where the base station and possibly the sensors need to be mobile. The mobility of sensor nodes has a great influence on sensor network topology and thus raises many issues in secure routing protocols. The following research issues on security in WSNs are particularly important:

- *Exploit the availability of private key operations on sensor nodes*: Recent studies on public key cryptography have shown that public key operations may be practical in sensor nodes. However, private key operations are still very expensive to realize in sensor nodes. As public key cryptography can greatly ease the design of security in WSNs, improving the efficiency of private key operations on sensor nodes is highly desirable.
- *Secure routing protocols for mobile sensor networks*: The mobility of sensor nodes has a great influence on sensor network topology and thus on the routing protocols used. Mobility may

be required at the base station, the sensor nodes, or both. Current protocols assume a sensor network is stationary. New secure routing protocols for mobile sensor networks need to be developed.

■ *Time synchronization issues*: Current broadcast authentication schemes such as μTESLA and its extensions require sensor networks to be loosely time synchronized. This requirement is often hard to meet, and new techniques that do not have such requirements are in great demand.

■ *Scalability and efficiency in broadcast authentication protocols*: Novel schemes with higher scalability and efficiency than existing ones need to be developed for authenticated broadcast protocols. The recent progress in public key cryptography may facilitate the design of authenticated broadcast protocols.

■ *Defending DoS attacks:* Defending DoS attacks is a great challenge. In the simplest form of this attack, an adversary attempts to disrupt communication by transmitting a broadcast signal of high strength. The adversary can also inhibit communication by violating the MAC protocol by transmitting frames while a neighbor is also transmitting or by continuously requesting channel access with a request-to-send (RTS). New techniques for dealing with these attacks are needed.

■ *Continuous stream security in WSNs*: Current work on security in sensor networks focuses on discrete events such as temperature and humidity. Continuous stream events such as video and images are not discussed. Video and image sensors for WSNs might not be widely available now, but they are likely to be available in the future. Substantial differences in authentication and encryption exist between discrete events and continuous events, indicating that there will be distinctions between continuous stream security and the current protocols in WSNs.

■ *Quality of service (QoS) and security*: Performance is generally degraded with the addition of security services in WSNs. Current studies on security in WSNs focus on individual topics such as key management, secure routing, secure data aggregation, and intrusion detection. It is noted that QoS and security services need to be evaluated together in WSNs.

References

1. Akyildiz, F., W. Su, Y. Sankarasubramaniam, and E. Cayirci. August 2002. "A Survey on Sensor Networks." *IEEE Communications Magazine* 40 (8): 102–14.
2. Deng, J., R. Han, and S. Mishra. November 2002. "INSENS: Intrusion-Tolerant Routing in Wireless Sensor Networks." Technical Report CU-CS-939-02, Department of Computer Science, University of Colorado at Boulder.
3. Karp, B., and H. T. Kung. 2000. "GPSR: Greedy Perimeter Stateless Routing for Wireless Networks." In *Proceedings of the 6th Annual International Conference on Mobile Computing and Networking*, 243–54. New York: ACM Press.
4. Papadimitratos, P., and Z. J. Haas. January 2002. "Secure Routing for Mobile Ad hoc Networks." In *Proceedings of the SCS Communication Networks and Distributed System Modeling and Simulation Conference (CNDS2002)*, 27–31. San Antonio, TX.
5. Tanachaiwiwat, S., P. Dave, R. Bhindwale, and A. Helmy. 2003. " Routing on Trust and Isolating Compromised Sensors in Location-Aware Sensor Networks." In *Proceedings of the 1st International Conference on Embedded Networked Sensor Systems*, 324–25. New York: ACM Press.
6. Estrin, D., R. Govindan, J. S. Heidemann, and S. Kumar. 1999. "Next Century Challenges: Scalable Coordination in Sensor Networks." In *Proceedings of ACM International Conference on Mobile Computing and Networking (MOBICOM'99)*, 263–70. Seattle, Washington.

7. Hu, L., and D. Evans. 2003. "Secure Aggregation for Wireless Networks." In *Proceedings of the International Symposium on Applications and the Internet (SAINT'03) Workshops*, 384, Orlando, Florida: IEEE Computer Society. January 2003.

8. Madden, S., M. J. Franklin, J. M. Hellerstein, and W. Hong. 2002. "TAG: A Tiny Aggregation Service for Ad-hoc Sensor Networks." *ACM SIGOPS Operating Systems Review* (Special Issue): 131–46.

9. Przydatek, B., D. Song, and A. Perrig. 2003. "SIA: Secure Information Aggregation in Sensor Networks." In *Proceedings of the 1st International Conference on Embedded Networked Systems (SenSys'03)*, 255–65. New York: ACM Press.

10. Shrivastava, N., C. Buragohain, D. Agrawal, and S. Suri. 2004. "Medians and Beyond: New Aggregation Techniques for Sensor Networks." In *Proceedings of the 2nd International Conference on Embedded Networked Sensor Systems (ACM SenSys'04)*, 239–49. Baltimore, Maryland, November 2004.

11. Ye, F., L. H. Luo, and S. Lu. 2004. "Statistical En-Route Filtering of Injected False Data in Sensor Networks." In *Proceedings of the 23rd IEEE Joint Annual Conference of Computer and Communication Societies (IEEE INFOCOM'04)*, vol. 4, 2446–57. Hong Kong, China: IEEE Communications Society Press. March 2004.

12. Ganeriwal, S., and M. Srivastava. 2004. "Reputation-Based Framework for High Integrity Sensor Networks." In *Proceedings of the 2nd ACM Workshop on Security on Ad Hoc and Sensor Networks (SASN'04)*, 66–77. Washington DC.

13. Liang, Z., and W. Shi. 2005. "Enforcing Cooperative Resource Sharing in Untrusted Peer-to-Peer Environment." *ACM Journal of Mobile Networks and Applications (MONET)* 10 (6): 771–83.

14. Liang, Z., and W. Shi. February 2005. "Analysis of Ratings on Trust Inference in the Open Environment." Technical report MIST-TR-2005-002, Department of Computer Science, Wayne State University.

15. Liang, Z., and W. Shi. January 2005. "PET: A Personalized Trust Model with Reputation and Risk Evaluation for P2P Resource Sharing." In *Proceedings of the 38th Annual Hawaii International Conference on System Sciences (HICSS)*, 201–02. Hawaii.

16. Ren, K., T. Li, Z. Wan, F. Bao, R. H. Deng, and K. Kim. August 2004. "Highly Reliable Trust Establishment Scheme in Ad hoc Networks." *Computer Networks: The International Journal of Computer and Telecommunications Networking* 45: 687–99.

17. Tanachaiwiwat, S., P. Dave, R. Bhindwale, and A. Helmy. 2004. "Location-Centric Isolation of Misbehavior and Trust Routing in Energy-Constrained Sensor Networks." In *Proceedings of the IEEE International Conference on Performance, Computing, and Communications (IPCCC'04)*, 463–69. Phoenix, Arizona, April 2004.

18. Yan, Z., P. Zhang, and T. Virtanen. 2003. "Trust Evaluation Based Security Solution in Ad hoc Networks." In *Proceedings of the 7th Nordic Workshop on Secure IT Systems (NordSec'03)*, Gjovik, Norway, October 2003.

19. Zhu, H., F. Bao, R. H. Deng, and K. Kim. 2004. "Computing of Trust in Wireless Networks." In *Proceedings of the 60th IEEE Vehicular Technology Conference (VTC'04-Fall)*, vol. 4, 2621–2624. Singapore, September 2004.

20. Anderson, R., and M. Kuhn. 1996. "Tamper Resistance- A Cautionary Note." In *Proceedings of the 2nd USENIX Workshop on Electronic Commerce (WOEC'96)*, 1–11, Oakland, California, November 1996.

21. Anderson, R., and M. Kuhn. 1997. "Low Cost Attacks on Tamper Resistant Devices." In *Proceedings of the5th International Workshop on Security Protocols (IWSP)*, LNCS vol. 1361, 125–36.

22. Hartung, C., J. Balasalle, and R. Han. 2004. "Node Compromise in Sensor Networks: The Need for Secure Systems." Technical Report CU-CS-988-04, Department of Computer Science, University of Colorado at Boulder.

23. Hu, L., and D. Evans. 2004. "Using Directional Antennas to Prevent Wormhole Attacks." In *Proceedings of the 11th Annual Network and Distributed System Security Symposium (NDSS'04)*, 131–141. San Diego, California, February 2004.

24. Komerling, O., and M. G. Kuhn. May 1999. "Design Principles for Tamper-Resistant Smart Card Processors." In *Proceedings of USENIX Workshop on Smartcard Technology*, 9–20. Chicago, IL.

25. Sastry, N., U. Shankar, and D. Wagner. September 2003. "Secure Verification of Location Claims." In *Proceedings of the 2nd ACM Workshop on Wireless Security*, 1–10. Sandiego, CA.

26. Seshadri, A., A. Perrig, L. Van Doorn, and P. Khosla. 2004. "SWATT: Software-Based Attestation for Embedded Devices." In *Proceedings of the IEEE Symposium on Security and Privacy*, 272–82. Oakland, California, May 2004.

27. Wang, X., W. Gu, S. Chellappan, K. Schoseck, and D. Xuan. 2005. "Lifetime Optimization of Sensor Networks under Physical Attacks." In *Proceedings of the IEEE International Conference on Communications (ICC)*, vol. 5, 3295–301. Seoul, Korea, May 2005.

28. Wang, X., W. Gu, S. Chellappan, D. Xuan, and T. H. Laii. February 2005. "Search-Based Physical Attacks in Sensor Networks: Modeling and Defense." Technical Report, Department of Computer Science and Engineering, Ohio State University.

29. Wood, A. D., and J. A. Stankovic. 2002. "Denial of Service in Sensor Networks." *IEEE Computer* 35 (10): 54–62.

30. Carman, D. W., P. S. Krus, and B. J. Matt. 2000. "Constraints and Approaches for Distributed Sensor Network Security." Technical Report 00-010, NAI Labs, Network Associates Inc., Glenwood, MD.

31. Hill, J., R. Szewczyk, A. Woo, S. Hollar, D. E. Culler, and K. Pister. 2000. "System Architecture Directions for Networked Sensors." In *Proceedings of the 9th International Conference on Architectural Support for Programming Languages and Operating Systems*, 93–104. New York: ACM Press.

32. Slijepcevic, S., M. Potkonjak, V. Tsiatsis, S. Zimbeck, and M. B. Srivastava. 2002. "On Communication Security in Wireless Ad-hoc Sensor Networks." In *Proceedings of the 11th IEEE International Workshop on Enabling Technologies: Infrastructure for Collaborative Enterprises (WETICE'04)*, 139–44. Pittsburg, Pennsylvania, June 2002.

33. Yuan, L., and G. Qu. 2002. "Design Space Exploration for Energy-Efficient Secure Sensor Networks." In *Proceedings of the IEEE International Conference on Application-Specific Systems (ASAP'02)*, 88–100. San Jose, California, July 2002.

34. URL: http://www.willow.co.uk/html/telosb_mote_platform.html, 2010. Accessed on July11, 2012.

35. Perrig, A., R. Szewczyk, V. Wen, D. E. Culler, and J. D. Tygar. September 2002. "SPINS: Security Protocols for Sensor Networks." *Wireless Networks* 8 (5): 521–34.

36. Stankovic, J. A., T. Abdelzaher, C. Lu, L. Sha, and J. Hou. July 2003. "Real-Time Communication and Coordination in Embedded Sensor Networks." *Proceedings of the IEEE* 91 (7): 1002–22.

37. Eschenauer, L., and V. D. Gligor. 2002. "A Key-Management Scheme for Distributed Sensor Networks." In *Proceedings of the 9th ACM Conference on Computer and Communications Security (CCS'02)*, 41–47. Washington DC: ACM Press. November 2002.

38. Chan, H., A. Perrig, and D. Song. 2003. "Random Key Pre-Distribution Schemes for Sensor Networks." In *Proceedings of the IEEE Symposium on Security and Privacy (S&P'03)*, 197. Berkeley, California: IEEE Computer Society. May 2003.

39. Hwang, J., and Y. Kim. 2004. "Revisiting Random Key Pre-Distribution Schemes for Wireless Sensor Networks." In *Proceedings of the 2nd ACM Workshop on Security of Ad Hoc and Sensor Networks (SASN'04)*, 43–52. New York: ACM Press.

40. Liu, D., P. Ning, and R. Li. 2005. *ACM Transactions on Information Systems Security* 8 (1): 41–77.

41. Capkun, S., and J.-P. Hubaux. 2006. "Secure Positioning in Wireless Networks." *IEEE Journal on Selected Areas in Communications* 24 (2): 221–32.

42. Lazos, L., and R. Poovendran. 2005. "SERLOC: Robust Localization for Wireless Sensor Networks." *ACM Transactions on Sensor Networks* 1 (1): 73–100.

43. Ganeriwal, S., S. Capkun, C.-C. Han, and M. B. Srivastava. 2005. "Secure Time Synchronization Service for Sensor Networks." In *Proceedings of the 4th ACM Workshop on Wireless Security*, 97–106. New York: ACM Press.

44. Shi, E., and A. Perrig. December 2004. "Designing Secure Sensor Networks." *Wireless Communication Magazine* 11 (6): 38–43.

45. Wang, X., W. Gu, K. Schosek, S. Chellappan, and D. Xuan. July 2004. "Sensor Network Configuration Under Physical Attacks." Technical report (OSU-CISRC-7/04-TR45), Department of Computer Science and Engineering, Ohio State University.

46. Karlof, C., and D. Wagner. 2003. "Secure Routing in Wireless Sensor Networks: Attacks and Countermeasures." In *Proceedings of the 1st IEEE International Workshop on Sensor Network Protocols and Applications*, 113–27. Anchorage, Alaska, May 2003.

47. Newsome, J., E. Shi, D. Song, and A. Perrig. 2004. "The Sybil Attack in Sensor Networks: Analysis and Defenses." In *Proceedings of the 3rd International Symposium on Information Processing in Sensor Networks*, 259–68. ACM Press.

48. Douceur, J. February 2002. "The Sybil Attack." In *Proceedings of the 1st International Workshop on Peer-to-Peer Systems (IPTPS'02)*, vol. 2429. 251–60, Cambridge, Massachusetts: Springer LNCS. March 2002.

49. Awerbuch, B., D. Holmer, C. Nita-Rotaru, and H. Rubens, "An On-Demand Secure Routing Protocol Resilient to Byzantine Failures." In *Proceedings of the 1st ACM Workshop on Wireless Security (WiSe'02)*, 21–30. Atlanta, Georgia: ACM Press. September 2002.

50. Sen, J. December 2010. "Routing Security Issues in Wireless sensor Networks: Attacks and Defense." In *Sustainable Wireless Sensor Networks*, edited by Y. K. Tan, Chapter 12, 279–309. Croatia: INTECH Publishers.

51. Parno, B., A. Perrig, and V. Gligor. 2005. "Distributed Detection of Node Replication Attacks in Sensor Networks." In *Proceedings of the IEEE Symposium on Security and Privacy (S&P'05)*, 49–63. Oakland, California: IEEE Computer Society. May 2005.

52. Gruteser, M., G. Schelle, A. Jain, R. Han, and D. Grunwald. 2003. "Privacy-Aware Location Sensor Networks." In *Proceedings of the 9th USENIX Workshop on Hot Topics in Operating Systems (Hot OS IX)*, vol. 9, 28. Lihue, Hawaii, May 2003.

53. Ozturk, C., Y. Zhang, and W. Trappe. 2004. "Source-Location Privacy in Energy-Constrained Sensor Network Routing." In *Proceedings of the 2nd ACM Workshop on Security of Ad Hoc and Sensor Networks (SASN'04)*, 88–93. Washington DC: ACM Press. October 2004.

54. Chan, H., and A. Perrig. 2003. "Security and Privacy in Sensor Networks." *IEEE Computer Magazine* 36 (10): 103–05.

55. Deng, J., R. Han, and S. Mishra. 2004. "Countermeasures Against Traffic Analysis in Wireless Sensor Networks." Technical Report CU-CS-987-04, University of Colorado at Boulder.

56. Perrig, A., J. Stankovic, and D. Wagner. 2004. "Security in Wireless Sensor Networks." *Communications of ACM* 47 (6): 53–57.

57. Malan, D. J., M. Welsh, and M. D. Smith. October 2004. "A Public-Key Infrastructure for Key Distribution in TinyOS based on Elliptic Curve Cryptography." In *Proceedings of the 1st IEEE International Conference on Sensor and Ad hoc Communications and Networks*. Santa Clara, CA.

58. Rivest, R. L., A. Shamir, and L. Adleman. 1983. "A Method for Obtaining Digital Signatures and Public-Key Cryptosystems." *Communications of the ACM* 26 (1): 96–99.

59. Brown, M., D. Cheung, D. Hankerson, J. L. Hernandez, M. Kirkup, and A. Menezes. 2000. "PGP in Constrained Wireless Devices." In *Proceedings of the 9th USENIX Security Symposium (SSYM'00)*, vol. 9, 19. Denver, Colorado, August 2000.

60. Gura, N., A. Patel, A. Wander, H. Eberle, and S. Shantz. 2004. "Comparing Elliptic Curve Cryptography and RSA on 8-bit CPUs." In *Proceedings of the 6th International Workshop on Cryptographic Hardware and Embedded Systems (CHES'04)*, vol. 3156. 119–32. Cambridge, Massachusetts: Springer LNCS. August 2004.

61. Gaubatz, G., J. P. Kaps, and B. Sunar. 2004. "Public Key Cryptography in Sensor Networks-Revisited." In *proceedings of the 1st European Workshop on Security in Ad-hoc and Sensor Networks (ESAS'04)*, vol. 3313, 2–18. Heidelberg, Germany: Springer LNCS. August 2004.

62. Wander, A. S., N. Gura, H. Eberle, V. Gupta, and S. C. Shantz. 2005. "Energy Analysis of Public-Key Cryptography for Wireless Sensor Networks." In *proceedings of the 3rd IEEE International Conference on Pervasive Computing and Communication (PerCpm'05)*, 324–28. Kauai Island, Hawaii, March 2005.

63. Rabin, M. O. 1979. "Digitalized Signatures and Public-Key Functions as Intractable as Factorization." Cambridge, MA, Technical Report.

64. Hoffstein, J., J. Pipher, and J. H. Silverman. 1998. "NTRU: A Ring-Based Public Key Cryptosystem." In *Proceedings of the 3rd International Symposium on Algorithmic Number Theory (ANTS'98)*, vol. 1423, 267–88. Portland, Oregon: Springer LNCS. June 1998.

65. Miller, V. S. 1986. "Use of Elliptic Curves in Cryptography." In *Proceedings of the Advances in Cryptology—CRYPTO'85*, edited by Y. K. Tan, vol. 218, 417–26. Santa Barbara, California: Springer LNCS. August 1985.

66. Kobiltz, N. 1987. "Elliptic Curve Cryptosystems." *Mathematics of Computation* 48: 203–09.

67. Elliptic Curve Cryptography, SECG Std. SEC1. 2000. http://www.secg.org/collateral/sec1.pdf. (Accessed on July 11, 2012).

68. Kaliski, B. May 2003. TWIRL and RSA Key Size, RSA Laboratories, Technical Note.

69. Recommended Elliptic Curve Domain Parameters, SECG Std. SEC 2. 2000. http://www.secg.org/collateral/sec2_final.pdf. (Accessed on July 11, 2012).

70. Hankerson, D., A. Menezes, and S. Vanstone. 2004. *Guide to Elliptic Curve Cryptography*. New York: Springer-Verlag.

71. Freier, A., P. Karlton, and P. Kocher. *The SSL Protocol* (version 3.0). http://www.mozilla.org/projects/security/pki/nss/ssl/draft302.txt (Accessed on Jul 11, 2012).

72. Watro, R., D. Kong, S. Cuti, C. Gardiner, C. Lynn, and P. Kruus. 2004. "TinyPK: Securing Sensor Networks with Public Key Technology." *In Proceedings of the 2nd ACM Workshop on Security of Ad hoc and Sensor Networks (SASN'04)*, 59–64. New York: ACM Press.

73. Liu, A. and P. Ning. 2008. "TinyECC: A Configurable Library for Elliptic Curve Cryptography in Wireless Sensor Networks." In *Proceedings of the 7th International Conference on Information Processing in Sensor Networks (IPSN'08)*, SPOTS Track, 245–256. St. Louis, Missouri, April 2008. http://discovery.csc.ncsu.edu/software/TinyECC/ (Accessed on July 11, 2012).

74. Karlof, C., N. Sastry, and D. Wagner. November 2004. "TinySec: A Link Layer Security Architecture for Wireless Sensor Networks." In *Proceedings of the 2nd ACM Conference on Embedded Networked Sensor Systems (SenSys'04)*, 162–75. Baltimore, MD.

75. U.S. National Institute of Standards and Technology (NIST). June 1998. SKIPJACK and KEA algorithm specifications, Federal Information Processing Standards Publications 185 (FIPS PUB 185).

76. Rivest, R. L. 1995. "The RC5 Encryption Algorithm." In *Proceedings of the International Workshop on Fast Software Encryption*, LNCS vol. 1008, 86–96.

77. Eastlake, D., and P. Jones. September 2001. "U.S. Secure Hash algorithm 1 (SHA1)." RFC 3174 (Informational).

78. Daemen, J., and V. Rijmen. 1998. "AES proposal: Rijndael." In *Proceedings of the 1st AES Candidate Conference (AES1)*, Ventura, California, August 1998.

79. Menezes, A. J., S. A. Vanstone, and P. C. V. Oorschot. 1996. *Handbook of Applied Cryptography*. Boca Raton, FL: CRC Press.

80. Rivest, R. L. April 1992. "The MD5 Message-Digest Algorithm." RFC 1321.

81. Ganesan, P., R. Venugopalan, P. Peddabachagari, A. Dean, F. Mueller, and M. Sichitiu. 2003. "Analyzing and Modeling Encryption Overhead for Sensor Network Nodes." In *Proceedings of the 2nd ACM International Conference on Wireless Sensor Networks and Applications*, 151–59. New York: ACM Press.

82. Law, Y. W., J. M. Doumen, and P. H. Hartel. "Benchmarking Block Ciphers for Wireless Sensor Networks (Extended Abstract)." In *Proceedings of the 1st IEEE International Conference on Mobile Ad-hoc and Sensor Systems (MASS'04)*, 447–56. Fort Lauderdale, Florida, October 2004.

83. Wheeler, D. J., and R. M. Needham. 1994. "TEA: A Tiny Encryption Algorithm." In *Proceedings of Fast Software Encryption: 2nd International Workshop*, edited by B. Preneel. vol. 1008, 363–366. Leuven, Belgium: Springer LNCS. December 1994.

84. Rivest, R. L., M. J. B. Robshaw, R. Sidney, and Y. L. Yin. "The RC6 Block Cipher." Publisher ftp://ftp.rsasecurity.com/pub/rsalabs/rc6/rc6v11.pdf (Accessed on July 11, 2012)

85. Matsui, M. 1997. "New Block Encryption Algorithm MISTY." In *Proceedings of the 4th International Workshop on Fast Software Encryption (FSE'97)*, edited by E. Biham. vol. 1267, 54–68. Haifa, Israel: Springer LNCS. January 1997.

86. 3GPP Specification Detail 2011: 3G Security: Specification of the 3GPP Confidentiality and Integrity Algorithms: Document 2: KASUMI Specification. Available in URL: http://www.3gpp.org/ftp/Specs/html-info/35202.htm

87. Aoki, K., T. Ichikawa, M. Matsui, S. Moriai, J. Nakajima, and T. Tokita. 2001. Specification of Camellia-A 128-bit Block Cipher, Specification (version 2.0). Nippon Telegraph and Telephone Corporation and Mitsubishi Electric Corporation.

88. Di Pietro, R., L. V. Mancini, Y. W. Law, S. Etalle, and P. Havinga. 2003. "LKHW: A Directed Diffusion-Based Secure Multi-Cast Scheme for Wireless Sensor Networks." In *Proceedings of the 32nd International Conference on Parallel Processing Workshops (ICPPW'03)*, 397–406. Kaohsiung, Taiwan: IEEE Computer Society Press. October 2003.

89. Zhu, S., S. Setia, and S. Jajodia. 2003. "LEAP: Efficient Security Mechanism for Large–Scale Distributed Sensor Networks." In *Proceedings of the 10th ACM Conference on Computer and Communications Security*, 62–72. New York: ACM Press.

90. Lai, B., S. Kim, and I. Verbauwhede. 2002. "Scalable Session Key Construction Protocols for Wireless Sensor Networks." In *Proceedings of the IEEE Workshop on Large Scale Real-Time and Embedded Systems (LATES'02)*, 1–6. Austin, Texas, December 2002.

91. Cametepe, S. A., and B. Yener. 2007. "Combinatorial Design of Key Distribution Mechanisms for Wireless Sensor Networks." *IEEE/ACM Transactions on Networking (TON)* 15 (2): 346–58.

92. Lee, J., and D. R. Stinson. 2004. "Deterministic Key Pre-Distribution Schemes for Distributed Sensor Networks." In *Proceedings of the 11th International Workshop on Selected Areas in Cryptography (SAC'04)*, edited by H.Handschuh and M. A. Hasan, vol. 3357, 294–307. Waterloo, Canada: Springer LNCS. August 2004.

93. Lee, J., and D. R. Stinson. 2005. "A Combinatorial Approach to Key Pre-Distribution for Distributed Sensor Networks." In *Proceedings of the IEEE Wireless Communications and Networking Conference (WCNC'05)*, vol. 2, 1200–05. New Orleans, Los Angeles, March 2005.

94. Chan, H., and A. Perrig. 2005. "PIKE: Peer Intermediaries for Key Establishment in Sensor Networks." In *Proceedings of the 25th IEEE Annual Conference on Computer and Communications (INFOCOM'05)*, 524–35. Miami, Florida, March 2005.

95. Huang, Q., J. Cukier, H. Kobayashi, B. Liu, and J. Zhang. 2003. "Fast Authenticated Key Establishment Protocols for Self-Organizing Sensor Networks." In *Proceedings of the 2nd ACM International Conference on Wireless Sensor Networks and Applications (WSNA'03)*, 141–50. San Diego, CA: ACM Press.

96. Zhou, Y., and Y. Fang. 2006. "A Scalable Key Agreement Scheme for Large Scale Networks." In *Proceedings of IEEE International Conference on Networking, Sensing and Control (ICNSC'06)*, 631–36. Fort Lauderdale, FL, April 23–25, 2006.

97. Du, W., J. Deng, Y. S. Han, and P. K. Varshney. 2003. "A Pair-Wise Key Pre-Distribution Scheme for Wireless Sensor Networks." In *Proceedings of the 10th ACM Conference on Computer and Communications Security*, 42–51. New York: ACM Press.

98. Pietro, R. D., L. V. Mancini, and A. Mei. 2003. "Random Key-Assignment for Secure Wireless Sensor Networks." In *Proceedings of the 1st ACM Workshop on Security of Ad hoc and Sensor Networks*, 62–71. New York: ACM Press.

99. Du, W., J. Deng, Y. S. Han, S. Chen, and P. K. Varshney. 2004. "A Key Management Scheme for Wireless Sensor Networks using Deployment Knowledge." In *Proceedings of IEEE INFOCOM*, 586–97. Hong Kong.

100. Hwang, D. D., B. Lai, and I. Verbauwhede. 2004. "Energy-Memory-Security Trade-offs in Distributed Sensor Networks." In *Proceedings of the 3rd International Conference on Ad-hoc Networks and Wireless (ADHOC-NOW)*, vol. 3158, 70–81. Lecture Notes in Computer Science (LNCS), Springer. July 2004.

101. Blundo, C., A. D. Santis, A. Herzberg, S. Kutten, U. Vaccaro, and M. Yung. 1998. "Perfectly-Secure Key Distribution for Dynamic Conferences." *Information and Computation* 146 (1): 1–23.

102. Liu, D., and P. Ning. October 2003. "Location-Based Pair-Wise Key Establishments for Static Sensor Networks." In *Proceedings of the ACM Workshop on Security in Ad hoc and Sensor Networks*, 72–82.

103. Zhang, Y., J. Zheng, and H. Hu. 2008. *Security in Wireless Mesh Networks*. Taylor & Francis Group, Boca Raton, FL: CRC Press.

104. Chan, H., V. Gligor, A. Perrig, and G. Muralidharan. July–Sept, 2005. "On the Distribution and Revocation of Cryptographic Keys in Sensor Networks." *IEEE Transactions on Dependable and Secure Computing* 2 (3): 233–47.

105. Sen, J., M. G. Chandra, P. Balamuralidhar, S. G. Harihara, and H. Reddy. May 2007. "A Distributed Protocol for Detection of Packet Dropping Attack in Mobile Ad hoc Networks." In *Proceedings of the International Conference on Telecommunications and Malaysian International Conference on Communications (ICT-MICC'07)*. Penang, Malaysia.

106. Hu, Y., A. Perrig, and D. B. Jonson. 2003. "Packet Leashes: A Defense Against Worm-Hole Attacks." In *Proceedings of the 22nd Annual Joint Conference of the IEEE Computer and Communications Societies (INFOCOM'03)*, vol. 3, 1976–86. San Francisco, California, March–April, 2003.

107. Wang, W., and B. Bhargava. 2004. "Visualization of Wormholes in Sensor Networks." In *Proceedings of the 2004 ACM Workshop on Wireless Security*, 51–60. New York: ACM Press.

108. Sen, J., M. G. Chandra, P. Balamuralidhar, S. G. Harihara, and H. Reddy. December, 2007. "A Mechanism for Detection of Gray Hole Attack in Mobile Ad hoc Networks." In *Proceedings of the 6th International Conference on Information, Communications and Signal Processing (ICICS'07)*. Singapore.

109. Aura, T., P. Nikander, and J. Leiwo. 2001. "DOS-Resistant Authentication with Client Puzzles." In *Proceedings of the 8th International Workshop on Security Protocols*, vol. 2133, 170–77. Cambridge, UK: Springer LNCS. April 2000.

110. Rafaeli, S., and D. Hutchison. 2003. "A Survey of Key Management for Secure Group Communications." *ACM Computing Survey* 35 (3): 309–29.

111. Lazos, L., and R. Poovendran. 2003. "Energy-Aware Secure Multi-Cast Communication in Ad-hoc Networks using Geographic Location Information." In *Proceedings of the IEEE International Conference on Acoustic Speech and Signal Processing (ICASSP)*, vol.4, 201–04. Hong Kong, China, April 2003.

112. Lazos, L., and R. Poovendran. 2002. "Secure Broadcast in Energy-Aware Wireless Sensor Networks." In *Proceedings of the IEEE International Symposium on Advances in Wireless Communications (ISWC'02)*, 1, Invited Paper, Victoria, British Columbia, Canada, September 2002.

113. Intanagonwiwat, C., R. Govindan, and D. Estrin. 2000. "Directed Diffusion: A Scalable and Robust Communication Paradigm for Sensor Networks." In *Proceedings of the 6th ACM Annual International Conference on Mobile Computing and Networking (MobiCom'00)*, 56–67. Boston, Massachusetts: ACM Press. August 2000.

114. Kaya, T., G. Lin, G. Noubir, and A. Yilmaz. 2003. "Secure Multicast Groups on Ad hoc Networks." In *Proceedings of the 1st ACM Workshop on Ad hoc and Sensor Networks (SASN'03)*, 94–102. Fairfax, Virginia: ACM Press. October 2003.

115. Al-Karaki, J. N., and A. E. Kamal. December 2004. "Routing Techniques in Wireless Sensor Networks: A Survey." *IEEE Wireless Communications* 11 (6): 6–28.

116. Du, W., R. Wang, and P. Ning. 2005. "An Efficient Scheme for Authenticating Public Keys in Sensor Networks." In *Proceedings of the 6th ACM International Symposium on Mobile Ad hoc Networking and Computing*, 58–67. New York: ACM Press.

117. Liu, D., and P. Ning. February 2003. "Efficient Distribution of Key Chain Commitments for Broadcast Authentication in Distributed Sensor Networks." In *Proceedings of the 10th Annual Network and Distributed System Security Symposium*, 263–76. San Diego, CA.

118. Liu, D., and P. Ning. 2004. "Multilevel μTESLA: Broadcast Authentication for Distributed Sensor Networks." *ACM Transactions on Embedded Computing Systems (TECS)* 3(4): 800–36.

119. Liu, D., P. Ning, S. Zhu, and S. Jajodia. 2005. "Practical Broadcast Authentication in Sensor Networks." In *Proceedings of the 2nd Annual International Conference on Mobile and Ubiquitous Systems: Networking and Services (MobiQuitous'05)*, 118–29. San Diego, California, July 2005.

120. Public-Key Infrastructure (X.509) (pkix) [online], available at http://www.ietf.org/html.charters/pkix-charter.html. (Accessed on July 11, 2012)

121. Sen, J., and A. Ukil. 2010. "A Secure Routing Protocol for Wireless Sensor Networks." In *Proceedings of the International Conference on Computational Science and its Application (ICCSA)*, Fukuoka, Japan, vol. 3, 277–90. LNCS 6018. Heidelberg, Germany: Springer-Verlag.

122. Braginsky, D., and D. Estrin. 2002. "Rumor Routing Algorithm for Sensor Networks." In *Proceedings of the 1st ACM International Workshop on Wireless Sensor Networks and Applications*, 22–31. New York: ACM Press.

123. Gruteser, M., and D. Grunwald. 2003 "Anonymous Usage of Location-Based Services through Spatial and Temporal Cloaking." In *Proceedings of the 1st International Conference on Mobile Systems, Applications, and Services (MobiSys'03)*, 31–42. San Francisco, California, May 2003.

124. Beresford, A. R., and F. Stajano. 2003. "Location Privacy in Pervasive Computing." *IEEE Pervasive Computing* 2 (1): 46–55.

125. Sen, J. December 2010. "An Efficient and User Privacy-Preserving Routing Protocol for Wireless Mesh Networks." *International Journal on Scalable Computing: Practice and Experience, Special Issue on Network and Distributed Systems* 11 (4): 345–58.

126. Rivest, R., A. Shamir, and Y. Tauman. 2001. "How to Leak a Secret." In *Proceedings of the 7th International Conference on the Theory and Application of Cryptology and Information Security (ASIACRYPT'01)*, edited by C. Boyd, vol. 2249, 552–65. Gold Coast, Australia: Springer LNCS. December 2001.

127. Gruteser, M., and D. Grunwald. 2003. "A Methodological Assessment of Location Privacy Risks in Wireless Hotspot Networks." In *Proceedings of the 1st International Conference on Security in Pervasive Computing (SPC'03)*, vol. 2802, 10–24. Boppard, Germany, Lecture Notes in Computer Science (LNCS), Springer. March 2003.

128. Priyantha, N. B., A. Chakraborty, and H. Balakrishnan. 2000. "The Cricket Location Support System." In *proceedings of the 6th Annual International Conference on Mobile Computing and Networking (MobiCom'00)*, 32–43. Boston, Massachusetts, August 2000.

129. Smailagic, A., D. P. Siewiorek, J. Anhalt, Y. Wang, and D. Kogan. 2001. "Location Sensing and Privacy in a Context Aware Computing Environment." *IEEE Wireless Communications* 9: 10–17.

130. Molnar, D., and D. Wagner. 2004. "Privacy and Security in Library RFID: Issues, Practices, and Architectures." In *Proceedings of the 11th ACM Conference on Computer and Communications Security (CCS'04)*, 204–19. Washington DC: ACM Press. October 2004.

131. Duri, S., M. Gruteser, X. Liu, P. Moskowitz, R. Perez, M. Singh, and J. Tang. 2000. "Framework for Security and Privacy in Automotive Telematics." In *Proceedings of the 2nd ACM International Workshop on Mobile Commerce (WMC'02)*, 25–32. Atlanta, Georgia: ACM Press. September 2002.

132. Snekkenes, E. 2001. "Concepts for Personal Location Privacy Policies." *In Proceedings of the 3rd ACM Conference on Electronic Commerce (ACM-EC'01)*, 48–57, Tampa, Florida: ACM Press. October, 2001.

133. Myles, G., A. Friday, and N. Davies. 2003. "Preserving Privacy in Environments with Location-Based Applications." *IEEE Pervasive Computing* 2 (1): 56–64.

134. Hengartner, U., and P. Steenkiste. "Protecting Access to People Location Information." In *Proceedings of the 1st International Conference on Security in Pervasive Computing (SPC'03)*, vol. 2802, 222–31. Boppard, Germany: Springer LNCS. March 2003.

135. Xi, Y., L. Schwiebert, and W. Shi. 2006. "Preserving Privacy in Monitoring-Based Wireless Sensor Networks." In *Proceedings of the 2nd International Workshop on Security in Systems and Networks* In *Proceedings of the 2nd International Workshop on Security in Systems and Networks (SSN'06)*, Rhode Island, Greece, April 2006.

136. Sato, I., Y. Okazaki, and S. Goto. 2002. "An Improved Intrusion Detection Method Based on Process Profiling." *IPSJ Journal* 43 (11): 3316–26.

137. Marti, S., T. J. Giuli, K. Lai, and M. Baker. 2000. "Mitigating Routing Misbehavior in Mobile Ad hoc Networks." In *Proceedings of the 6th Annual International Conference on Mobile Computing and Networking*, 255–65. New York: ACM Press.

138. Zhang, Y., W. Lee, and Y.-A. Huang. 2003. "Intrusion Detection Techniques for Mobile Wireless Networks." *Wireless Networks* 9 (5): 545–56.

139. Huang, Y., W. Fan, W. Lee, and P. S. Yu. May 2003. "Cross-Feature Analysis for Detecting Ad-hoc Routing Anomalies." In *Proceedings of the 23rd International Conference on Distributed Computing Systems (ICDCS)*, 478. Providence, RI.

140. Huang, Y., and W. Lee. September 2004. "Attack Analysis and Detection for Ad hoc Routing Protocols." In *Proceedings of the 7th International Symposium on Recent Advances in Intrusion Detection*, 125–45. Sophia Antipolis, France.

141. Sen, J., and I. Sengupta. December 2005. "Autonomous Agent-Based Distributed Fault-Tolerant Intrusion Detection System.". In *Proceedings of the 2nd International Conference on Distributed Computing and Internet Technology (ICDCIT'05)*, Bhubaneswar, India, edited by G. Chakraborty, LNCS vol. 3186, 125–31. Heidelberg, Germany: Springer-Verlag.

142. Brutch, P., and C. Ko. 2003. "Challenges in Intrusion Detection for Wireless Ad-hoc Networks." In *Proceedings of the Symposium on Applications and the Internet Workshops (SAINT'03 Workshops)*, 368, Orlando, Florida: IEEE Computer Society. January 2003.

143. Zhu, S., S. Setia, S. Jajodia, and P. Ning. May 2004. "An Interleaved Hop-by-Hop Authentication Scheme for Filtering of Injected False Data in Sensor Networks." In *Proceedings of IEEE Symposium on Security and Privacy*, 259–71. Oakland, CA.

144. Wang, G., W. Zhang, C. Cao, and T. L. Porta. 2003. "On Supporting Distributed Collaboration in Sensor Networks." In *Proceedings of the IEEE Military Communications Conference (MILCOM'03)*, vol. 2, 752–57. Monterey, California, October 2003.

145. Albers, P., and O. Camp. 2002. "Security in Ad hoc Networks: A General Intrusion Detection Architecture Enhancing Trust-Based Approaches." In *Proceedings of the 1st International Workshop on Wireless Information Systems (WIS'02)*, 1–12, Ciudad Real, Spain: ICEIS Press 2002. April 2002.

146. Merkle, R. C. 1980. "Protocols for Public Key Cryptosystems." In *Proceedings of the IEEE Symposium on Security and Privacy (S&P'80)*, 122–134, Oakland, California: IEEE Computer Society Press. April 1980.

147. Deng, J., R. Han, and S. Mishra. 2003. "Security Support for In-Network Processing in Wireless Sensor Networks." In *Proceedings of the 1st ACM Workshop on Security of Ad hoc and Sensor Networks*, 83–93. New York: ACM Press.

148. Cam, H., D. Muthuavinashiappan, and P. Nair. October 2003. "ESPDA: Energy-Efficient and Secure Pattern-Based Data Aggregation for Wireless Sensor Networks." In *Proceedings of IEEE Sensors*, 732–36. Toronto, Canada.

149. Cam, H., D. Muthuavinashiappan, and P. Nair. October 2005. "Energy-Efficient Security Protocol for Wireless Sensor Networks." In *Proceedings of IEEE VTC Conference*, 2981–84. Orlando, FL.

150. Cam, H., S. Ozdemir, H. O. Sanli, and P. Nair. 2004. "Secure Differential Data Aggregation for Wireless Sensor Networks". *Sensor Network Operations*, edited by S. Phoha, T. F. La Porta, and C. Griffin, Willy-IEEE Press, May 2006.

151. Du, W., J. Deng, Y. S. Han, and P. K. Varshney. December 2003. "A Witness-Based Approach for Data Fusion Assurance in Wireless Sensor Networks." In *Proceedings of IEEE Global Telecommunications Conference*, 1435–39. San Francisco.

152. Wagner, D. 2004. "Resilient Aggregation in Sensor Networks." In *Proceedings of the 2nd ACM Workshop on Security of Ad Hoc and Sensor Networks (SASN'04)*, 78–87. New York: ACM Press.

153. Acharya, M., J. Girao, and D. Westhoff. 2005. "Secure Comparison of Encrypted Data in Wireless Sensor Networks." In *Proceedings of the 3rd International Symposium on Modeling and Optimization in Mobile, Ad Hoc and Wireless Networks (WIOPT)*, 47–53. Washington, DC.

154. Catelluccia, C., A. C-F. Chan, E. Mykletun, and G. Tsudik. 2009. "Efficient and Provably Secure Aggregation of Encrypted Data in Wireless Sensor Networks." *ACM Transactions on Sensor Networks (TOSN)*, 5(3), Article No: 20, May 2009.

155. Girao, J., D. Westhoff, and M. Schneider. 2005. "CDA: Concealed Data Aggregation for Reverse Multicast Traffic in Wireless Sensor Networks." In *Proceedings of the IEEE International Conference on Communications (ICC'05)*, vol. 5, 3044–49. Seoul, Korea, May 2005.

156. He, W., X. Liu, H. Ngyyen, K. Nahrstedt, and T. Abdelzaher. 2007. "PDA: Privacy-Preserving Data Aggregation in Wireless Sensor Networks." In *Proceedings of the 26th IEEE International Conference on Computer Communications (IEEE INFOCOM'07)*, 2045–53. Anchorage, Alaska, May 2007.

157. Westhoff, D., J. Girao, and M. Acharya. 2006. "Concealed Data Aggregation for Reverse Multicast Traffic in Sensor Networks: Encryption, Key Distribution, and Routing Adaptation." *IEEE Transactions on Mobile Computing* 5 (10): 1417–31.

158. Armknecht, F., D. Westhoff, J. Girao, and A. Hessler. 2008. "A Lifetime-Optimized End-to-End Encryption Scheme for Sensor Networks Allowing in-Network Processing." *Computer Communications* 31 (4): 734–49.

159. Rivest, R. L., L. Adleman, and M. L. Dertouzos. 1978. "On Data Banks and Privacy Homomorphisms." In *Foundations of Secure Computation*, Workshop, Georgia Institute of Technology, Atlanta, 1977, 169–79. New York: Academic Press.

160. Domingo-Ferrer, J. 2002. "A Provably Secure Additive and Multiplicative Privacy Homomorphism." In *Proceedings of the 5th International Conference on Information Security (ISC'02)*, vol. 2433, 471–83. Sao Paulo, Brazil: LNCS. September–October 2002.

161. Wagner, D. October 2003. "Cryptanalysis of an Algebraic Privacy Homomorphism." In *Proceedings of the 6th Information Security Conference*, LNCS, vol. 2851, 234–39. Bristol, UK.

162. Peter, S., D. Westhoff, and C. Castelluccia. 2010. "A Survey on the Encryption of Converge-Cast Traffic with in-Network Processing." *IEEE Transactions on Dependable and Secure Computing* 7 (1): 20–34.

163. Deng, J., R. Han, and S. Mishra. August 2005. "Security, Privacy, and Fault-Tolerance in Wireless Sensor Networks." 215–234, *Wireless Sensor Networks: A Systems Perspective*, edited by N. Bulusu and S. Jha, Artech House, July 2005.

164. Sen, J. 2010. "Reputation- and Trust-Based Systems for Wireless Self-Organizing Networks." In *Security of Self-Organizing Networks: MANET, WSN, WMN, VANET*, edited by Al-Sakib Khan Pathan, Aurbach, Book Chapter No. 5, 91–92. Taylor & Francis Group, Boca Raton, FL: CRC Press.

165. Pirzada, A., and C. McDonald. 2004. "Establishing Trust in Pure Ad Hoc Networks." In *Proceedings of the 27th Australian Conference on Computer Science*, 47–54. Dunedin, New Zealand.

166. Oram, A. March 2001. *Peer-to-Peer: Harnessing the Power of Disruptive Technologies*. O'Reilly & Associates.

167. Sen, J. 2012. "Secure and Privacy-Aware Searching in Peer-to-Peer Networks." In *Proceedings of the 6th International Workshop on Data Privacy Management (DPM2011)*, edited by J. Garcia-Alfaro et al., Leuven, Belgium, LNCS vol. 7122, 72–89. Heidelberg, Germany: Springer-Verlag.

168. Sen, J. January 2011. "A Robust and Secure Aggregation Protocol for Wireless Sensor Networks." In *Proceedings of the 6th IEEE International Symposium on Electronic Design, Test and Applications (DELTA 2011)*, 222–27. Queenstown, New Zealand.

169. Sen, J. October 2010. "A Distributed Trust Management Framework for Detecting Malicious Packet Dropping Nodes in a Mobile Ad Hoc Network." *International Journal of Network Security and its Applications (IJNSA)* 2 (4): 92–104.

170. Sen, J. July 2010. "A Distributed Trust and Reputatation Framework for Mobile Ad Hoc Networks." In *Proceedings of the 1st International Conference on Network Security and its Applications (CNSA 2010)*, edited by Meghanathan et al., Chennai, India. *CCIS*, vol. 89, 538–47. Heidelberg, Germany: Springer-Verlag.

171. Wang, Y., G. Attebury, and B. Ramamurthy. 2006. "A Survey of Security Issues in Wireless Sensor Networks." *IEEE Communications Surveys and Tutorials*, 8(2), 2–23.

172. Blom, R. 1985. "An Optimal Class of Symmetric Key Generation System." In *Proceedings of the EUROCRYPT'84*, edired by T. Beth et al., Paris, France, vol. 209, 335–38. Springer LNCS. April 1984.

Chapter 16

Security in Wireless Video Sensor Networks Based on Watermarking Techniques

Noreen Imran, Boon-Chong Seet, and A. C. M. Fong

Contents

16.1 Introduction

Wireless video sensor networks (WVSNs) have gained considerable attention in recent years due to their vast application domains, flexible deployment structures, and most of all, the availability of low-cost complementary metal–oxide–semiconductor (CMOS) sensor modules. The application domain for WVSN spans from surveillance and monitoring to health care, traffic, and industrial control sectors. The self-organizing, flexible, and easily scalable infrastructure of WVSN is one key factor for its widespread popularity. Secure transmission of video data over radio links is highly desirable for many applications that require data to be shielded from unwanted access, tempering, and loss. These requirements motivate the need for new security solutions to be designed for WVSNs as most existing solutions for wireless sensor networks (WSNs) cannot be straightforwardly adapted to WVSNs (which will be discussed in Section 16.2) [1,2].

Unlike WSNs where sensor nodes capture and transmit only simple scalar data such as temperature, pressure, and humidity, video data is based on rich streaming media generated at a higher rate and thus requires more complex processing, memory storage, higher network bandwidth, and energy for transmission. At the same time, WVSNs have to deal with optimization of performance parameters such as delay, throughput, network lifetime, and quality of service (QoS).

A WVSN generally comprises spatially distributed sensor nodes, each equipped with a miniaturized camera and a transceiver that capture, compress, and transmit visual information about its surroundings to a sink or base station for further content analysis, verification, and distribution. The basic architecture of a WVSN is shown in Figure 16.1.

Sensor nodes are typically deployed in open and unattended areas. Therefore, they are vulnerable to various forms of physical or logical attacks and their transmissions are subjected to radio propagation effects such as fading and shadowing [3]. A wireless video sensor node is basically composed of sensing, processing, transceiver, and power modules as shown in Figure 16.2.

The mobilizer and location-finding modules are optional and deployed in situations where the location of sensor node is not fixed. In such scenarios, these modules are used to track and

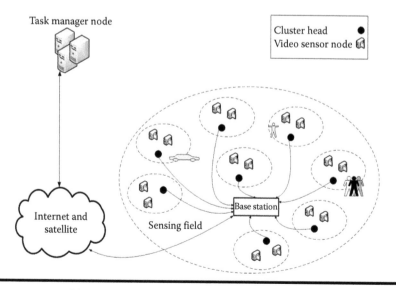

Figure 16.1 WVSN architecture.

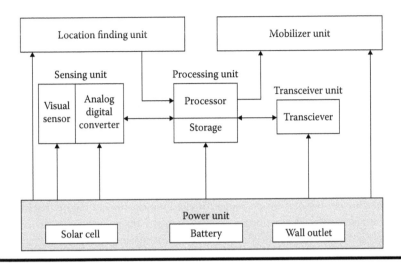

Figure 16.2 Functional modules of wireless video sensor node. (Data from Akyildiz, F. et al., *IEEE Communications Magazine,* **40 (8): 102–114, 2002.)**

provide location information using Global Positioning System (GPS) or other positioning technologies. This information is further used to compute and update routing table entries at neighboring nodes [4].

In contrast to a sensor node that gathers scalar data, a video sensor node requires

- More power to process and compress video data.
- Employ digital signal processors (DSPs) rather than microcontrollers (MCUs).
- Higher storage capacity (for both random access and external memory).
- More bandwidth due to large volume of video data.

The remainder of the chapter is organized as follows. Section 16.2 discusses the motivation for security mechanisms in WVSNs. Section 16.3 overviews the general design challenges in WVSNs. Section 16.4 discusses various security issues and their solutions in the context of WVSN. Section 16.5 presents the technical background of digital video watermarking. Section 16.6 reviews the security mechanisms for WVSNs based on digital watermarking. Finally, Sections 16.7 and 16.8 conclude the chapter and discuss some directions for future work respectively.

16.2 Motivation

A wireless communication channel is available to any entity that tunes its radio interface to the frequency of the channel for transmission or reception. This provision offers substantial ease to attackers for disrupting or misleading communication between legitimate parties. Existing strong security solutions for wireless networks may not be easily adapted to the WVSN environment due to their complex processing that is inappropriate for resource-limited nodes [6]. From our survey of existing literature, it is evident that security in WVSNs is still a relatively emerging research area with limited number of works found so far [7]. WVSNs require low complexity, robust, and scalable security mechanisms, which will not severely impact the overall lifetime of an individual node or the entire network.

When WVSN is deployed in an open and hostile environment with possibilities of encountering attacks, ensuring data integrity and authenticity is an important requirement. For example, a malicious user may intercept the video transmission between a sender and receiver pair and inject manipulated frames to subvert or disrupt their communication. On the contrary, some critical decision making has to be done on basis of the received video. The application needs to ensure that the given video has not been tampered on its way to the receiver and is indeed coming from the intended source. Such validation of video integrity and authenticity is almost obligatory in scenarios such as war zone monitoring and border protection.

Generally, in wireless video surveillance applications, the digital video captured by sensor nodes is used to detect and track some unusual events or activities within the surveillance area, so that necessary actions can be taken when any alarming situation is raised. Moreover, the video can be examined and provided as an evidence for the investigation of that particular event. To ascertain that the given video is the same video captured by the surveillance camera and which has not been manipulated, security mechanisms for validating both integrity and authenticity of the video must be required.

For surveillance cameras deployed in public places such as airports, train stations, malls, or office buildings, the issue of privacy protection may not be a very major concern. However, in more personal environments such as homes or patient rooms in hospital, users will be more concerned about their privacy if known that they are being continuously watched. WVSNs are also more vulnerable to denial-of-service attacks due to high volume of data transmission involved in video streaming applications. The strict node energy, storage, and processing constraints make the implementation of advanced antijamming mechanisms such as frequency hopping and physical tamperproofing an impractical solution due to their complex design and high-energy requirements [3].

16.3 Design Challenges for WVSNs

The WVSNs have a number of requirements that are unique to its architecture such as high bandwidth for video streaming, low-complexity video encoding techniques, application-specific QoS demands, and resource constraints [8]. These requirements should be taken into account when designing not only the security mechanisms but also other lower/higher-layer protocol mechanisms. A brief discussion of these requirements is given in the following.

16.3.1 Bandwidth

Transmitting video to the sink through several intermediate nodes over radio links requires much higher bandwidth than one required for transmitting image or scalar data. Most commercially available sensor motes that comply with IEEE 802.15.4 standard such as TelosB [9] and MICAz [10] have a maximum data rate of only up to 250 kbps, which inevitably limits (depending on actual codec used) the resolution and frame rate of the video that can be transmitted.

16.3.2 Video Encoding Techniques

Video encoding is often a complex and processing intensive operation due to which considerable energy drain can occur at resource-limited sensor nodes. Conventional video codecs such as MPEG-x have a complex encoding and simple decoding architecture, where raw video data output from a video source, for example, video camera, is not encoded by the source itself, but by a more resourceful machine that can perform complex encoding operations such as motion compensation and estimation. These codecs are intended for applications such as broadcasting or streaming video-on-demand (VoD) where video data is encoded once and decoded many times by end user devices. The ability to decode (playback) the encoded video in a timely manner by end user devices would be made possible by having a simple decoding process. In contrast, the resource-limited video sensor nodes in WVSN need to encode their own raw video data before transmitting them to the sink, which makes it necessary for sensor nodes to have a reversed video coding paradigm than conventional video codecs, that is, simple encoding and complex decoding. The video coding frameworks based on this reversed paradigm are as follows:

- Distributed video coding (DVC)
  - DVC is an emerging video coding paradigm for applications that have restricted resources available at encoders. It reverses the conventional video coding paradigm by shifting the complexity of the encoder entirely or partially to the decoder, assumed to be a more resourceful machine than the encoder [11–14]. Therefore, DVC-based encoders are much simpler than conventional encoders, which make them a promising candidate for video encoding in WVSNs.
  - Wyner–Ziv is one of the most well-known video coding architectures based on DVC approach where intraframe encoding is used to encode video frames, which in turn are conditionally decoded using interframe decoder that exploits side information obtained by interpolation or extrapolation of previously decoded frames. So far, several different architectures supporting the concept of evolving DVC paradigm [15–33] have been proposed in literature.
- Distributed compressive video sensing (DVCS)
  - Another low-complexity video coding paradigm that came to existence recently integrates the core principles of DVC and compressive sensing/sampling (CS) [34,35]. CS is a relatively new theory that shows how a sparse signal can be reconstructed from measurements far fewer than required by traditional methods based on Nyquist's sampling theory [36,37].
  - The process of CS video encoding comprises two primary steps: obtaining CS measurements of the sparse video signal and quantizing the compressed measurements to generate the bitstream. On the contrary, the decoder involves complex operations such as decoding of bitstream, dequantization and ℓ_1 regularization of CS measurements [38,39]. This simple encoder–complex decoder architecture of DVCS thus suits well to WVSN environment. Currently, three relevant works support the DCVS architecture [37,40,41].

16.3.3 Application-Specific QoS

Different WVSN applications may have different requirements in terms of bandwidth, processing, and compression, among others. Video streaming involves continuous capturing and delivery of data that requires optimized encoding and compression algorithms in addition to efficient hardware to meet the QoS demands pertinent to particular applications [42]. In the context of WVSNs, the approaches for network-layer QoS can be based on reliability or timeliness of video delivery [43]. For example, some applications are delay-tolerant but require reliable and error-free data transmission, involving packet retransmissions and multipath routing through which sensor node can inject multiple copies of same packet into different paths so that at least one copy is able to make it to the sink.

16.3.4 Resource Constraints

Unlike WSN, WVSN requires more resources in terms of processing capability, memory storage (on board and external), operating energy (battery), and transmission bandwidth. Due to real-time nature of video data, WVSN has to employ mechanisms to deal with jitter, frame loss rate, and end-to-end delay, which subsequently require more resources. Considering a small battery-operated video sensor node, performing operations such as video capturing, encoding, and transmission along with error and recovery mechanisms can be a challenging task [8].

16.4 WVSN Security

As discussed in Section 16.3.4, the resource limitations of WVSNs have made it vulnerable to various attacks that vary from video intercepting, tampering, replaying previously stored frames, and injecting fake frames into the network to disrupting the entire transmission. The fundamental goals of designing security mechanisms for WVSN can be outlined as follows:

- The video should be received from an authentic source (authentication).
- The video should be received from its intended source without any alteration (integrity).
- In certain applications, the video should only capture the behavior or action of the monitored subjects and not their identity (privacy).

Since sensor nodes can be considered as a soft target for attacks due to their limited resources, it is essential to implement at least some basic level of security at sensor nodes to counteract those attacks. A substantial amount of work has been done in regard to security in WSN, comprising mainly cryptographic (symmetric and asymmetric key cryptography) techniques, hashing, certification, trust management, and other advanced techniques [39]. In contrast to WSNs, which mainly process scalar data, addressing security in WVSNs may require a different approach. We now rephrase the privacy, integrity, and authentication in the context of WVSN as follows:

- Privacy protection in video processing environments is defined as a process that requires masking, blurring, or confiscation procedures to hide sensitive information in some regions of the video frame such as human face or vehicle number plate to protect the subject's identity. Therefore, an eavesdropper cannot extract meaningful information by accessing/intercepting the video frames transmitted between the sender and the receiver. Thus far, we have not found many works that deal with privacy issues in WVSNs or wireless multimedia sensor networks (WMSNs). Kundur et al. [1] proposed a cryptography-based process called PICO (privacy through invertible cryptographic obscuration), which applies a face recognition algorithm to an image and subsequently encrypts the identified facial region using symmetric key cryptography. Chessa et al. [44]presented a low-complexity privacy protection mechanism for automated video surveillance systems that primarily comprise two steps. The first step selects the information-sensitive regions from the video frame and the second step is a scrambling technique that is applied to selected regions to obscure the information.

- Integrity: To ensure the receiver that the given video sequence has not been altered during transit by any malicious user. Ju and Jonathan [45] devised a watermarking-based video integrity verification method, which overcomes the inability of conventional verification mechanisms such as digital signatures to distinguish between attacks and regular modifications.

- Authentication: To ensure the receiver that the given video sequence is indeed originated from the intended sender rather than an impersonator of that sender. Following the low-complexity privacy protection mechanism [38], which scrambles the frame region containing sensitive information, each frame can further undergo a blind watermarking process (which is explained in Section 16.4.1) to normalize the scrambling map and embed the authentication information.

Conventionally, cryptosystems used digital signature methodology to address sender authentication and data integrity issues. A digital signature is composed of a stream of bits computed on the basis of entire image/frame to be delivered and a private key. For each frame, a unique frame-dependent digital signature is computed to guarantee that an eavesdropper is unable to compute the same signature from two different frames [44]. In WVSNs, the encoders will be the tiny battery-operated video sensors, which have serious computation and energy constraints. Thus, it is practically not possible for the video sensors to perform the processing intensive computation of digital signature for each individual frame and bear associated transmission overhead in addition to the original video. Additionally, the existing security mechanisms based on traditional video codecs cannot be applied directly to WVSN environment due to the use of a totally different video coding paradigm (as explained in Section 16.3.2) [46]. Moreover, WVSNs also suffer from multipath fading effect, path loss, channel interference, and other environment factors. Therefore, video security mechanisms pertinent to WVSN have to deal with not only resource limitations and a different video coding paradigm but also the environment factors to address issues such as video data integrity and authentication [44].

16.4.1 Cryptography versus Watermarking

As discussed, the problem of maintaining privacy, integrity, and authenticity of video data transmitted over radio links is of vital importance. Several analogies have been found in literature [47–49] to define the relationship between cryptography and watermarking. However, it is evident that no single solution is sufficient to deal with all types of security threats. Therefore, a combination of security mechanisms in the same or different protocol layers has been a common solution approach [50]. Before making the case for which one (cryptography or watermarking) is more suitable for WVSNs, we will first clearly explain each of them along with their anticipated goals.

- Cryptography deals with secure transmission of a message from sender to receiver through an insecure communication channel, with secure transmission being characterized by three aspects, namely, privacy, integrity, and authenticity of the message. It is meant to keep the communication secret primarily by the use of symmetric and asymmetric cryptography and hash primitives, accompanied by various key distribution and trust management techniques. Only those who have the key(s) are able to access the hidden content.

■ Watermarking, on the contrary, is a multidisciplinary field that coalesces diverse areas such as signal processing, cryptography, communication theory, coding, compression, human visual system (HVS), and video quality requirements [44]. It is the branch of information-hiding that is applied to embed watermark (or digital signature) into the digital data signal such that it is hard to remove from the signal without the extraction algorithm.

Solutions to all three security problems (privacy, integrity, and authenticity) have been provided based on cryptography, watermarking, and sometimes a combination of both. Rahman et al. [51] presented a privacy protection mechanism to hide sensitive information from the video while providing efficient surveillance. They used a cryptography-based data scrambling approach to render regions of video frame that contain privacy-breaching information. Multiple levels of abstraction are used to meet the privacy privileges for various user roles. The scheme is claimed to be computationally efficient and intended for real-time video transmission. Some related works are also proposed [52–56].

Wang and Smith [57] proposed a watermarking-based framework to address privacy and authenticity issues in a video surveillance system. The idea is based on monitoring only the unauthorized persons in a given area so that privacy of authorized persons can be ensured. Radio frequency identification (RFID) sensors are used to distinguish between an authorized and an unauthorized person. From a given video frame, the regions representing the authorized persons can be removed (and restored again if necessary in some circumstances) using a secret key. Watermarking is used to embed and hide the privacy information (associated with the authorized person) into the video with minimal perceptual changes, while a digital signature is embedded in the packet header for authentication of the watermarked video. There are some other works that also address the issues of privacy, integrity, and authenticity of video based on the watermarking approach [45,58–64].

Firstly, video authentication using cryptographic approach commonly employs a digital signature or message authentication code (MAC) for transmission along with the message to receiver. However, this additional information introduces transmission overhead and there is a probability that the MAC may be corrupted due to format conversion [47]. Secondly, the authentication process using cryptography is content-dependent since digital signature/MAC is computed on the basis of entire content, which makes it a computation-intensive task. On the contrary, this approach provides robust authentication since a few bit inversions/errors will declare the message as corrupted. However, in practice, bit inversions/errors are common in WVSN environment due to the lossy nature of the wireless channel.

Therefore, watermarking is deemed a more flexible and lower complexity solution than cryptography for addressing security issues. It is able to ensure that the semantic meaning of the digital content has not been modified by illegitimate sources, while being sustainable against wireless channel errors, lossy compression, and other signal processing primitives [47].

In contrast to cryptography, watermarking can be used to embed watermark into the content without additional transmission overhead. In general, the least significant bits (LSBs) of the transformed coefficients for an image/video frame are replaced by watermark bits at particular locations/pixels specified by algorithms such as block, edge, and corner detection [65].

Nevertheless, watermark security can be complemented by using cryptographic approach, such that an attacker is not able to detect the watermark within the intercepted frame sequence. Only the legitimate receiver who successfully reconstructed the "encrypted" watermark using a detection algorithm can decrypt it and compare against the one stored at its own site.

Two primary branches of cryptography are (1) symmetric key cryptography and (2) asymmetric key cryptography, briefly explained as follows:

1. Using symmetric/private key cryptography, the same key is used to encrypt and decrypt the content at the sender and the receiver respectively.
2. Using asymmetric/public key cryptography, two-key pair, namely public–private key, is exploited. Public key is known to everyone and is used to encrypt the message. Only the authorized receiver who has the secret/private key will be able to decrypt it. Therefore, no secret channel is required for key sharing between the sender and the receiver.

There have been a number of works that recommend the use of symmetric cryptography since its processing requirements are relatively low. However, on the contrary, it requires strong trust management for the distribution of keys, which may be feasible for scalar data–based WSN applications, but impractical for those with multimedia transmission [54]. Compared to symmetric cryptography, asymmetric cryptography is a better solution against eavesdropping and compromised nodes attacks. However, it requires some nodes to have higher processing capability to execute signature verification operations [54]. Using a cryptographic approach to privacy, integrity, and authenticity of video data in WVSN may have the following drawbacks:

- Public key cryptography is a computation intensive approach that is not feasible for resource-limited, battery-powered video sensor nodes, which also need to perform video processing and transmission operations along with video capturing. Mechanisms such as RSA [66] and digital signatures, which are primarily based on public key cryptography, are complex with significant storage and energy requirements, making them inappropriate for WVSNs [67].
- Modern crypto systems mainly rely on the distribution and management of keys. Using the public key cryptography, key management in a network of *n* nodes is performed by distributing *n* public–private key pairs to each of the nodes through secure communication channel. Thus, for large-scale distributed WVSNs, scalability of the system may become an issue.
- There are situations when a node is physically compromised or under a form of Sybil attack [68,69] in which nodes have dual identities. In such cases, key revocation process will be necessary, which itself is also a complex mechanism.
- The distribution of unique key to each sensor node pair may require considerable storage space. On the contrary, employing shared key may have low computational and storage costs, but it is a less secure solution because it may take only one compromised node to jeopardize the security of the entire network.

We conclude this section with the remark that the security mechanisms based on cryptography are not suitable for WVSNs, and there still exists a considerable research gap in designing security solutions for WMSNs/WVSNs based on watermarking [2]. Thus, in Section 16.5, we only focus on lightweight digital video watermarking and their applications in secure communications.

16.5 Digital Video Watermarking

16.5.1 Generic Watermarking Framework

Hui-Yu et al. [70] suggested that watermarking techniques can be made practical for WSN by exploiting the principles of distributed source coding, which utilize the duality that lies between data hiding and channel coding with side information. Digital video watermarking techniques have been widely studied and implemented in various application domains but hardly for WVSN [71]. The basic idea is to hide some metadata about the video into the video itself. The video watermarking framework generally comprises three modules: the embedding, detection, and comparator functions (CFs) [72,73]. We briefly discuss the working of each of the modules as follows:

■ Embedding function (EF)
   – The EF resides within the source video sensor, and typically, it has three inputs: (1) the information to be embedded as a watermark, (2) the video sequence for which the watermark is to be embedded, and (3) a key that is optional and employed to provide an additional level of security to the watermark information.
■ Detection function (DF)
   – The DF is applied at the sink after the video decoder decodes the received video. The DF extracts the watermark information using the key (if employed) and passes it on to the CF to verify its successful recovery and to ensure the authenticity and integrity of the received video (Figure 16.3a and b).
■ Comarator function (CF)
   – This function generates a real value that indicates the degree of watermark reconstruction by comparing the original watermark W with the reconstructed watermark. If the value is equal to or greater than the predefined distortion threshold measure, the watermark is taken to be successfully reconstructed and the authenticity of the video is

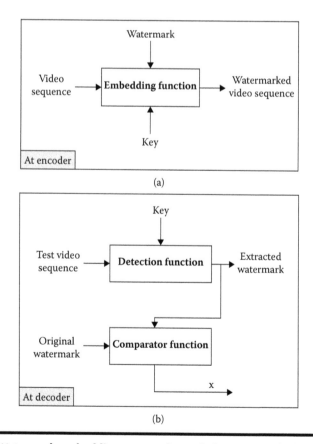

(a)

(b)

Figure 16.3 (a) Watermark embedding at encoder and (b) watermark detection and verification at decoder.

verified. Otherwise, the video frame is considered suspicious and discarded. The distortion threshold measure is usually set up in accordance with the channel conditions. It provides flexibility in situations where a significant portion of the watermark has been recovered but not exactly 100% since data loss is very likely in error-prone wireless environment.

16.5.2 Design Goals for Video Watermarking

Besides the underlying approach and structure of video watermarking algorithm, Hartung and Kutter proposed the following design goals for multimedia watermarking schemes [74]:

- *Robustness*: refers to the degree of survival of watermark against modification in watermarked video due to attacks such as frame dropping, reordering and averaging, compression, noise, cropping, and recoding at lower bit rate, among others.
- *Invisibility*: refers to the degree to which the watermark seems to be unperceptual in the given watermarked video. A relative measure to the HVS.
- *Capacity*: refers to the tolerable amount of information in the form of watermark bits that can be hidden in the multimedia content.
- *Security*: refers to the security of the watermarking algorithm using some cryptographic approach. Usually, the embedding algorithm of watermark scheme is declared to be *Public* while the key used for embedding is kept *Private*. Therefore, the security of the whole watermarking scheme relies on some carefully chosen key *K* from a massive pool of keys [75].

16.5.3 Classification of Video Watermarking Schemes

It is a widely believed that the majority of the image watermarking schemes can be applied to video with little or no modification since video can be seen as periodically spaced sequence of still images. However, the reality is different from this perception due to reasons such as real-time requirements and complexity in video watermarking, excessive redundancy among frames, and un-normalized balance between active and nonactive regions. We may classify watermarking schemes as follows:

- Implementation domain
  Majority of watermarking schemes fall into two categories with reference to their implementation domain, namely spatial domain and transform domain.
  - *Spatial domain video watermarking*: refers to schemes that exploit the correlation among pixels of video frames to embed watermark. Embedding is performed by carrying out simple operations such as changing pixel location, intensity, and pixel replacement at specific regions of video frames to incorporate watermark into original content. Spatial domain schemes have low complexity in terms of implementation and processing, but more perceptible to HVS and less robust to attacks. Some spatial domain watermarking schemes are proposed [76–78].
  - *Transform domain video watermarking*: refers to schemes in which the watermark is inserted into transformed coefficients of video, which provides more information-hiding capacity and more robustness against watermarking attacks. Watermarking in transform domain is more robust than in spatial domain because information can be spread

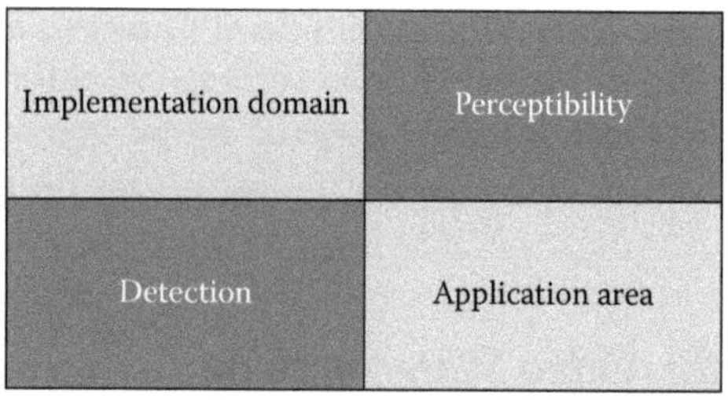

Figure 16.4 Classification parameters of video watermarking schemes.

out to entire video frame (Figure 16.4). The most common transformation methods are discrete cosine transform (DCT) [70,79], discrete wavelet transform (DWT) [72,80], and discrete Fourier transform (DFT) [73].

■ Perceptibility

Video watermarking can be categorized into perceptible and imperceptible schemes having HVS as a classification parameter.

- *Perceptible video watermarking*: refers to schemes in which the watermark is embedded in a visible manner into host video signal [81,82].
- *Imperceptible video watermarking*: refers to schemes in which the watermark is embedded in an invisible manner into host video signal [77,83].

■ Detection

Watermarking schemes can also be classified on the basis of the information required by the watermark detection algorithm at decoder to detect the watermark from reconstructed video.

- *Blind/oblivious video watermarking*: refers to schemes that do not require the original video at the decoder. However, the secret key and watermark bit pattern are required for extraction of watermark from video frames [72,84].
- *Nonblind/nonoblivious video watermarking*: refers to schemes that require the secret key, watermark bit pattern, and the original video at receiver site to detect the watermark from the watermarked video [85].

■ Application area

Video watermarking schemes can also be classified into robust, semifragile, and fragile in reference to application goals and objectives:

- *Robust video watermarking schemes*: refers to watermarking schemes that can endure the majority of video compression and processing operations. Applications such as copyright protection and ownership authentication entail that the watermark have to retain adequate robustness in the context of security against malicious attacks and error-prone network transmission [77,78].
- *Semifragile video watermarking*: refers to schemes that are particularly premeditated to cope with certain levels of distortion in the form of either noise or attacks [61,62,64].
- *Fragile video watermarking*: refers to schemes that are sensitive to all potential modifications in video frames. Generally, fragile watermarking schemes are intended for applications such as content authentication and video integrity verification. Since these schemes are sheltered under cryptographic security along with strong localization characteristics, even minor changes to the frame can be detected [86,87].

16.6 Review of WVSN Security Mechanisms Based on Digital Watermarking

There exists numerous works regarding video watermarking on a variety of codecs and platforms, but those techniques cannot be applied to the sensor network domain due to their major architectural differences. In literature, digital watermarking has been applied mostly to address copyrights protection of digital content, digital fingerprinting, tamper detection, broadcast monitoring, and metadata insertion [88]. Few applications of this technique have been found in wireless multimedia, video, or even scalar data sensor networks. In the context of WMSNs, security issues such as privacy, trust management, and authentication have a high degree of correlation, and digital watermarking alone is not sufficient to address all of them [71]. Thus, an eight-stage conceptual framework that provides guidelines for secure communication in WMSN based on watermarking is presented [2], as shown in Figure 16.5.

A brief description of key stages is as follows:

- *Application scenario extraction*: This stage identifies the application-specific QoS requirements such as timeliness or reliability. For example, surveillance systems deployed in a battlefield, traffic, or hospital monitoring environments have different types of risks, concerns, and resources. Hence, the application settings in terms of network size, node density, topology, and software and hardware resources must be known to design the security mechanism for it.
- *Privacy protection*: This stage deals with providing privacy protection to multimedia content. The conceptual framework proposed that privacy protection mechanism should be incorporated into the design of watermarking.
- *Authentication*: This stage develops a mechanism to identify and authenticate the originator of multimedia content. The framework proposed the use of digital watermarking to ensure correctness and confidentially of multimedia content. Watermarking-based authentication is proposed for WMSNs [60].

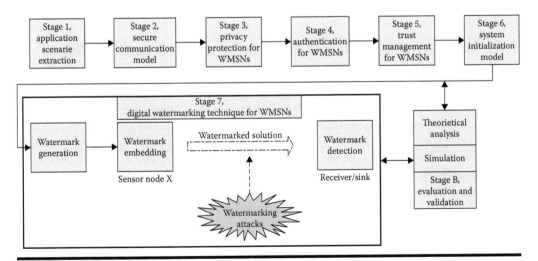

Figure 16.5 Conceptual framework for secure communication in WMSNs. (Data from Harjito, B. et al., *Digital Ecosystems and Technologies (DEST)*, Dubai, UAE, IEEE, 2010.)

Table 16.1 Summary of Reviewed Video Watermarking Techniques

Reference	Classification Parameters			
	Implementation Domain	Detection	Perceptibility	Application Area
[85]	Transform	Blind/nonblind	Invisible	Robust
[45]	Transform	Blind	Invisible	Robust
[93]	Transform	Blind	Invisible	Robust

■ *Trust management*: This stage deals with the establishment of trust management to enhance the security and reliability of nodes in WMSN. Since the network is deployed in an uncontrolled environment, there is a high probability that a node may get compromised.

■ *Watermarking technique for WMSNs*: This stage comprises three substages, namely watermark generation, embedding, and detection. Linear feedback shift register (LFBSR) [89] is used to generate watermark bit sequence that is further transformed using some constraints under Kolmogorov complexity rule [90] to enhance watermark security. For embedding purpose, multimodal fusion of watermark sequence bits and multimedia data (captured from surrounding) is performed at sensor node. Afterwards, watermarked solution is obtained by solving the nonlinear system equation formed using atomic trilateration process on multimodal data. Finally, the watermarked solution is transmitted to the receiver through wireless communication. Blind watermark detection is then performed at the receiver to verify the watermark existence.

■ *Theoretical analysis and simulation*: This stage determines compatibility, benefits, and drawbacks of the proposed architecture in the context of WMSNs.

Overall, the theoretical framework discussed security issues spanning privacy and authentication to trust management in WMSNs. In summary, this section provides an in-depth review of the most relevant video watermarking techniques for WVSNs in literature, which is also summarized as shown in Table 16.1, based on the four classification parameters discussed in Section 16.2.

16.6.1 Watermarking Technique Based on Distributed Video Coding

Zhu et al. [85] proposed the first watermarking scheme for a DVC-based Wyner–Ziv video codec. The distributed source coding theorems was established in 1970s by Slepian and Wolf [91] for distributed lossless coding and by Wyner and Ziv [33] for lossy coding with decoder side information. These information-theoretic results revealed that it is possible to encode the frames of a video sequence independently and still achieve efficient compression as long as decoding is performed jointly. This idea formed the basis of simple video encoder at the cost of increased complexity at decoder. Incoming video frames at the encoder are split into *key* frames, which are compressed using conventional H.264 intra encoder, and *non-key* frames, which are compressed using a Wyner–Ziv encoder. A configurable setting GOP (Group of Pictures) determines the distance between key frames in the video sequence, and thus the coding efficiency. For example, if GOP = m, there will be one key frame for every $m - 1$ non-key frames (also referred to as WZ frames). At the decoder side, key frames are decompressed using H.264 intra decoder, from which side information is generated for decoding of the WZ frames.

To enhance the security and invisibility of watermarking scheme, the Arnold's image transformation [92] is applied on the watermark image. Arnold's transformation stretches pixels of an

($n \times n$) image and wraps the stretched portions to restore the original dimensions. After certain number of iterations, an image will transform back to the original. Following the transformation, a combined corner and edge detection algorithm [65] is applied to the video frames to identify the points of interest (potential regions to invisibly embed watermark image bits). It is possible that the points of interest identified by the key frames and Wyner–Ziv frames do not exactly match with each other. Therefore, the resulting redundancy of the embedded watermark information leads to robustness of the watermarking scheme.

The watermark embedding process is as follows:

1. Perform Arnold's transformation on watermark image and encrypt it with private key $K1$.
2. Identify the points of interest using Harris corner and edge detector in key frames.
3. Locate a block around each interest point to embed whole watermark image.
4. For every 8-bit pixel in a block, replace the bit plane with 1 bit pixel's value in the watermark image matrix.

As illustrated in Figure 16.6, the same watermark can be detected at the decoder from both key frames (K′) and Wyner–Ziv (WZ′) frames, with the difference in that the detection from K′ is blind, while that from WZ′ is nonblind (requires original video frames). Detection from WZ′ frames requires original frames at decoder because the difference between original WZ frame and reconstructed WZ′ frame will enable us to detect the watermark information.

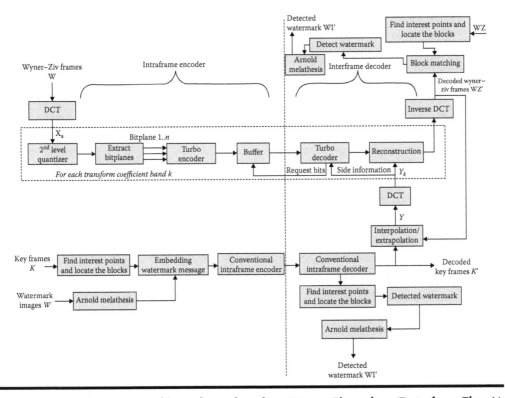

Figure 16.6 Video watermarking scheme based on Wyner–Ziv codec. (Data from Zhu, N. et al., *Proceedings of the 2008 International Conference on Intelligent Information Hiding and Multimedia Signal Processing*, 857–60, Harbin, China, IEEE Computer Society, 2008.)

The watermark detection process from key frames (blind) is as follows:

1. Identify the points of interest from the given frame.
2. Locate the blocks around each point of interest and find the bits in the pixels that were replaced with watermark bits. One complete block around point of interest constructs the watermark image matrix.
3. Average values calculated from all blocks to obtain the watermark image matrix.
4. Perform Arnold transformation on reconstructed watermark to obtain the real watermark.

The watermark detection process from Wyner–Ziv frames (nonblind) is as follows:

1. Identify the points of interest from the given frame.
2. Locate the blocks around each point of interest and find the bits in the pixels that were replaced with watermark bits.
3. Locate the bits corresponding to pixels of the same blocks in the original Wyner–Ziv frame.

```
If (Reconstructed WZ' + CV < Original WZ) then WM = 0
Else if (Reconstructed WZ' > Original WZ + CV) then WM = 1
Else detect from Key frame
```

Where CV is a criterion measure chosen for measuring the difference between original and reconstructed Wyner–Ziv frame, $0 < CV < 2^{BitPlane}$.
4. Average values calculated from all block to obtain the watermark image matrix.
5. Perform Arnold transformation on reconstructed watermark to obtain real watermark.

This DVC-based watermarking scheme is distinctive since it is the only video watermarking scheme based on Wyner–Ziv video codec, which suits the WSN environment. The simulation results showed that the scheme works well in terms of invisibility, correctness, and robustness. However, it is still not ready for application in WVSN because at receiver site, the watermark extraction from Wyner–Ziv frame is nonblind and thus requires the original Wyner–Ziv frame, which is not possible. Although watermark can be extracted from key frames in a blind manner, this alone will not be sufficient to authenticate the video. Therefore, the watermarking scheme needs further enhancements to make it operational for WVSN environment.

16.6.2 Video Watermarking Technique against Correlation Attack Analysis in WSN

A blind video watermarking scheme for WVSN using F-modulation method is presented by Ju [45,57] to deal specifically with content analysis–based correlation attacks. The proposed design focused on a cross-layer methodology for addressing the authentication issues for video transmission in WVSN. Traditional watermarking schemes are designed to be resilient to various image processing attacks that attempt to remove or weaken the embedded watermark (i.e., watermark robustness against unintended signal processing is desired). In this scheme, this property is no longer a priority since a video frame without the expected watermark will be considered as a counterfeit. The following problem scenario is used as a case study: A wireless video surveillance system deployed to monitor an area of high security, and a malicious user or attacker accessed the radio communication channel and tried to invisibly destroy/manipulate the original video. An attack is made on the integrity of video data by suppressing the authentic transmission and intercepting the video for the intended receiver using highly directional and powerful antennas.

An invisible, low-complexity signature-based authentication mechanism is proposed to thwart these attacks, in which any receiver can validate the integrity of video data, provided that the signatures have been known previously. A test bed is set up along with a device called VCU mote that comprises a CMOS camera and two ZigBee transceivers. A blind watermarking embedding process, available in both simple and enhanced versions, is presented. The embedding process replaces the LSB of the selected DCT coefficient with the watermark bits, making the scheme invisible, while the receiver copies the LSB from predefined fixed locations from the frames of received video. The watermark is then embedded at the predefined locations, making the verification process easier. However, it is noted that such an approach does not work well against collusion, playback, and middleman attacks.

- *Collusion attack*: Refers to attacks in which the attacker uses statistical methods to identify potential watermark locations. Such attack requires the attacker to have access to a large number of watermarked video frames. For example, the attacker has multiple watermarked video frames where the same watermark is embedded in the same locations. Then, for each watermarked video frame, the attacker derives the LSB plane where the LSB of each DCT coefficient is extracted and forms an image of the same resolution as the original one. The attacker next calculates the average of all LSB planes to form a grayscale image. After processing a sufficient number of video frames, the color of watermarked locations becomes either strong white or black, while other pixels approach a grayscale intensity level of 0.5. Therefore, to counter collusion attack, the watermarking scheme must involve nondeterministic watermark.
- *Playback attacks*: Refers to attacks in which the attacker playbacks or uses an old (expired) video frame to fool the system.
- *Middleman attacks*: Refers to attacks in which the attacker exploits certain loopholes in the verification process to make fraudulent video frames.

Improvements are made by embedding into each frame a distinct watermark and using characteristic bit derived from DCT coefficients to modulate the watermark bit. Firstly, the watermark is considered a time-variant pseudorandom variable that generates a distinct signature for every frame. Two DCT coefficients are selected to embed each watermark bit. The fixed predefined embedding locations for watermark are vulnerable for correlation and playback attacks. Therefore, the scheme proposed dynamic and random selection of DCT blocks for embedding watermark so that the location of watermark varies over time from one frame to another. Secondly, the communication protocol needs to integrate the intrusion detection mechanism for the watermark security to thwart the channel capturing attacks over a period of time. Additionally, this scheme embeds some information as noise by varying DCT coefficient values to act as a safeguard against content and correlation analysis.

Tsong-Yi et al. [64] also suggested secure implementation of block selection algorithm at encoder/embedder and decoder/detector. The location generator algorithm identifies the same embedding location at both source and receiver sites by providing identical frame numbers. However, the reliability of the algorithm solely depends on the synchronization of the frame number between embedder and detector.

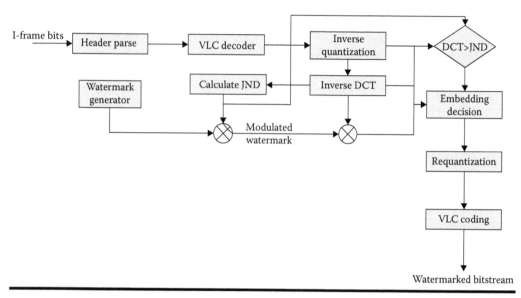

Figure 16.7 **Block diagram for embedding watermark into intraframe sequence. (Data from Wang., J., and G. L. Smith, *International Journal of Security and Networks*, 5 (1): 63–76, 2010.)**

The watermark detection process comprises the following steps (Figure 16.7):

1. Locate the character bit from DCT coefficients of the 8 × 8 DCT block.
2. Extract the watermark bits based on the value of character bit.
3. Calculate the detection statistic metrics.
4. Compare with the threshold value to decide the authenticity of the frame.

The watermarking scheme discussed earlier has the following security properties:

■ *Resistant against correlation attack*: A distinct watermark is generated for every video frame on the basis of pseudorandom sequence, which makes it difficult for the attacker to judge the watermark location by averaging the LSB plane values.
■ *Resistant against playback attack*: The playback attack can be detected by maintaining the record of the frame numbers received at the receiver site. A playback attack may be in progress if the receiver notices a duplicate frame number. Request for resetting the frame sequence number from receiver to video sensor node may help to overcome the playback attack.
■ *Resistant against middleman attack*: Since the location of modulation bit is unknown to the attacker, it would not be able to produce the same watermark embedding process by modulating watermark bit. Even if the attacker copies the entire LSBs to the injected video frame, the likelihood to pass the watermark verification is negligible because the watermark bits are modulated using characteristic bit.
■ *Resistant against content induction attack*: Such attacks attempt to discover the watermark location from the video frame by disrupting the communication link between the authentic video transmitters and receivers. The attacker can then feed the video sensor with "bait" images and analyzes the output watermarked video frames for watermark signature and locations.

Correlation analysis is performed between the F-modulation bit and DCT coefficients, with and without block selection algorithm at embedder and detector. The results show that with dynamic block selection, the correlation does not have much variation, which means it is almost impossible to find the exact correlation among the watermark and F-modulation bits.

While the proposed watermarking scheme is shown to be robust against correlation analysis and playback attacks, it does not address the energy consumption issue pertinent to WVSNs, as the scheme is built on MPEG-2, a conventional video codec with high complexity encoder and simple decoder architecture. Moreover, the evaluation only demonstrates the resilience of the proposed scheme against correlation and playback attacks. It does not provide any idea about the amount of distortion/noise introduced to the original video due to watermark embedding.

16.6.3 Wavelet-Based Resource-Aware, Adaptive Watermarking for WMSN

Wang [93] presented a communication resource-aware, adaptive watermarking scheme for multimedia authentication in WMSNs. The primary challenges addressed by the scheme were to embed/protect/extract watermark efficiently in low-cost sensors and to transmit authenticated multimedia in an energy-efficient manner. The transmission quality of watermark is maintained by embedding watermark with adaptive coding redundancies and allocating network resources adaptively to protect the multimedia packets containing watermark information. Since the watermark is adaptive to network conditions and the processing delay is also reduced by exploiting interframe correlation, the proposed scheme is shown to achieve reasonable communication energy efficiency and real-time performance.

When watermarked multimedia content is transmitted over WMSN, the multimedia quality could be degraded significantly due to packet loss caused by channel error and in the worst case makes the watermark undetectable at the receiver site. Therefore, a high-quality and efficient watermarking system in WMSN should be robust against transmission errors along with efficient resource utilization. The scheme embeds the watermark to selective coefficients of the three-level DWT middle-frequency bands of an image frame, based on the network conditions.

The embedding algorithm [93] exploits qualified significant wavelet tree (QSWT) [94] to separate the frequency bands of the multimedia and embeds the watermark into the frame's middle-frequency bands. The reason behind using the middle-frequency bands is that after wavelet transform, low-frequency bands contain the most energy of the image frame, while high-frequency bands contain noise and the information in those bands is often lost during compression. The watermark is embedded into host multimedia at chosen locations of LH3/HL3 (middle-frequency bands) coefficients. Figure 16.8 shows the arrows from the parent sub-band to its children sub-bands. The lowest frequency sub-band is at the top left and the highest frequency sub-band is at the bottom right.

Two thresholds t_1 and t_2 are selected adaptively based on network conditions and parameters such as packet loss ratio to identify the optimal number of QSWTs to embed the watermark. Although embedding in lesser number of QSWTs maintains the invisibility of scheme, it will be difficult to detect watermark in case of packet losses. On the contrary, embedding watermark redundantly makes use of large number of QSWTs, which lead to greater distortion in the multimedia content as well as higher energy consumption at the sensor node (Figure 16.9).

The multimedia packets are classified as either WM packets (with watermark information) or NWM packet (without watermark information). The WM packets are protected more by allocating extra network resource to improve watermark transmission quality to facilitate watermark detection

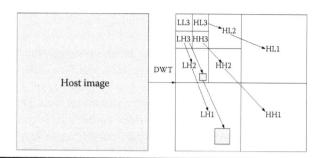

Figure 16.8 DWT decomposition of an image. (Data from Wang, H., *Journal of Supercomputing*, 1–15, 2010.)

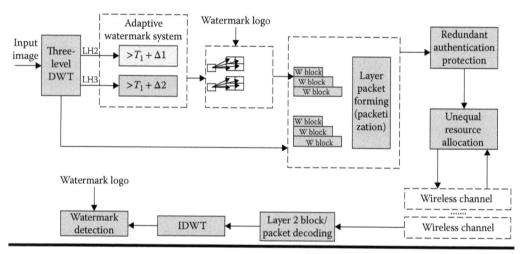

Figure 16.9 Resource-aware adaptive watermarking system for WMSN. (Data from Wang, H., *Journal of Supercomputing*, 1–15, 2010.)

and authentication. The network resource is referred to as the average total energy required, computed as a function of the desirable bit error rate (BER), frame size, packet transmission rate, transmission power, and retransmission limit. The primary goal of the scheme is to improve the watermark authentication while at the same time maintaining received multimedia quality.

The watermarking algorithm of the proposed adaptive wavelet based scheme is as follows:

1. Apply three-level DWT on host multimedia frame to obtain wavelet sub-bands (LH2 and LH3).
2. Decompose LH2 and LH3 into sub-blocks of size M.
3. Compute mean of sub-blocks in LH3 and LH2 and store them in arrays T_1 and T_2 respectively.
4. Find the optimal number of QSWTs for every sub-block. For each sub-block m, compare coefficient location at LH3 (i, j, m) with threshold $T_1 (m) + \Delta_1$, where Δ_1 is selected adaptively based on network conditions.
   a. If LH3 (i, j, m) is greater than threshold $T_1 (m) + \Delta_1$, then look for at least three child coefficients that must be greater than threshold $T_2 (m) + \Delta_2$, where Δ_2 is also an adaptive parameter.

      i. If at least three child coefficients are greater than $T_2(m) + \Delta_2$, then LH3 (i, j) is selected as one of the QSWT for sub-block m.

      ii. Sum the coefficient values from parents to all its children.

   b. After finding sufficient number of QSWTs for each block, sort them in the descending order.

5. Watermark embedding decision in current frame is based on QSWT location. If the difference between LH3 coefficients of consecutive frames is beyond some predetermined limit, the watermark will not embedded in current frame.

6. Else embed watermark in each QSWT for each sub-block m.

The simulation results show that the proposed wavelet-based watermarking process achieves a good performance in terms of invisibility and can resist JPEG lossy compression provided that thresholds T_1 and T_2 are chosen carefully. However, the scheme is designed for images and thus cannot be applied directly to watermark video content, which has much greater requirements in terms of frame rate, inherent redundancy in consecutive frames, and complexity than image watermarking as discussed in Section 16.3.

16.7 Conclusion

In this chapter, a comprehensive review on the state-of-the-art video watermarking techniques for WVSNs has been presented. Each technique is shown to have its pros and cons, and open research issues for this relatively unexplored area are discussed. We summarize the two primary design requirements of watermarking schemes for WVSNs as follows:

1. *Energy efficiency*: Practically, a network of wireless video sensors inherently suffers from constraints such as *limited power/energy supply and computational capabilities*, which lead to the requirement for careful design of the watermarking scheme with appropriate selection of compression and signal processing algorithms (preferably lightweight) that efficiently utilize the power not only during video processing but also during their transmission.

2. *Robustness*: Data loss or corruption inherent in error-prone wireless and open operating environment raises the need for more robust watermarking schemes that should survive packets loss or corruption pertinent to the wireless environment and attacks by malicious nodes.

16.8 Future Research Directions

It is expected that most real-life WVSN applications will require adequate security mechanisms, and watermarking is a potential solution as traditional approaches such as cryptography and stenography have much higher computational requirements. There is a clear need for video watermarking techniques that deal with not only the security issues but also the resource constraints of WVSNs. Achieving optimal use of resources such as node processing capability, memory storage, power, and network bandwidth while providing levelheaded protection against various attacks at the same time is thus the fundamental design goal of security mechanisms in WVSNs.

   Resilience of video data against channel errors can also be exploited in designing cost-effective security mechanisms because in contrast to scalar data, minor loss of video data that cannot be perceptually visible to HVS can be afforded. Thus far, research on addressing issues such as privacy, confidentiality, and authentication in WVSNs is still at a preliminary stage and, therefore, represents an open problem space with much room for innovation.

References

1. Kundur, D. et al. 2008. "Security and Privacy for Distributed Multimedia Sensor Networks." *Proceedings of the IEEE* 96 (1): 112–30.
2. Harjito, B. et al. 2010. "Secure Communication in Wireless Multimedia Sensor Networks Using Watermarking." In *Digital Ecosystems and Technologies (DEST), 2010 4th IEEE International Conference on*. Dubai, UAE: IEEE.
3. Harjito, B., and H. Song. 2010. "Wireless Multimedia Sensor Networks Applications and Security Challenges." In *Broadband, Wireless Computing, Communication and Applications (BWCCA), 2010 International Conference on*. Fukuoka, Japan: IEEE.
4. Lee, I., W. Shaw, and X. Fan. 2009. *Wireless Multimedia Sensor Networks Guide to Wireless Sensor Networks*, 561–82. London: Springer.
5. Akyildiz, I. F. et al. 2002. "A Survey on Sensor Networks." *IEEE Communications Magazine* 40 (8): 102–14.
6. Yun, Z., F. Yuguang, and Z. Yanchao. 2008. "Securing Wireless Sensor Networks: A Survey." *IEEE Communications Surveys & Tutorials* 10 (3): 6–28.
7. Guerrero-Zapata, M. et al. 2010. "The Future of Security in Wireless Multimedia Sensor Networks." *Telecommunication Systems* 45 (1): 77–91.
8. Al Nuaimi, M., F. Sallabi, and K. Shuaib. 2011. "A Survey of Wireless Multimedia Sensor Networks Challenges and Solutions." In *Innovations in Information Technology (IIT), 2011 International Conference on*. Abu Dhabi, UAE: IEEE.
9. *MEMSIC TelosB Mote Specifications*. http://www.memsic.com/. Accessed January 1, 2012, http://www.memsic.com/products/wireless-sensor-networks/wireless-modules.html.
10. *MEMSIC MICAz Mote Specifications*. http://www.memsic.com/. Accessed January 1, 2012, http://www.memsic.com/products/wireless-sensor-networks/wireless-modules.html.
11. Dufaux, F. et al. 2009. "Distributed Video Coding: Trends and Perspectives." *EURASIP Journal on Image and Video Processing*, 508167, 13 pp.
12. Girod, B. et al. 2005. "Distributed Video Coding." *Proceedings of the IEEE* 93(1): 71–83.
13. Pereira, F. et al. 2009. "Distributed Video Coding: Selecting the Most Promising Application Scenarios." *Image Communication* 23 (5): 339–52.
14. Puri, R. et al. 2006. "Distributed Video Coding in Wireless Sensor Networks." *IEEE Signal Processing Magazine* 23 (4): 94–106.
15. Aaron, A., S. Rane, and B. Girod. 2004. "Wyner–Ziv Video Coding with Hash-Based Motion Compensation at the Receiver." In *Image Processing, 2004. ICIP '04. 2004 International Conference on*. Singapore: IEEE.
16. Aaron, A. et al. 2004. "Transform-Domain Wyner–Ziv Codec for Video." In *Proceedings of the SPIE* 5308: 520–8.
17. Aaron, A., Z. Rui, and B. Girod. 2002. "Wyner–Ziv Coding of Motion Video." In *Signals, Systems and Computers, 2002. Conference Record of the Thirty-Sixth Asilomar Conference on*. Pacific Grove, CA: IEEE.
18. Aiguo, Y. et al. 2010. "A Fast Video Transcoder from Wyner–Ziv to AVS." In *Proceedings of the Advances in Multimedia Information Processing, and 11th Pacific Rim Conference on Multimedia: Part II*, 328–39. Shanghai, China: Springer-Verlag.
19. Aiguo, Y. et al. 2011. "A Fast Video Transcoder from Wyner–Ziv to AVS." In *Advances in Multimedia Information Processing - PCM 2010*, edited by G. Qiu et al., 328–39. Berlin/Heidelberg: Springer.
20. Anne, A. 2003. "Wyner–Ziv Coding for Video: Applications to Compression and Error Resilience." In *Proceedings of the IEEE Data Compression Conference*. Snowbird, UT.
21. Ascenso, J., and F. Pereira. 2007. "Adaptive Hash-Based Side Information Exploitation for Efficient Wyner–Ziv Video Coding." In *Image Processing, 2007. ICIP 2007. IEEE International Conference on*. San Antonio, TX.
22. Brites, C., J. Ascenso, and F. Pereira. 2006. "Studying Temporal Correlation Noise Modeling for Pixel Based Wyner–Ziv Video Coding." In *Image Processing, 2006 IEEE International Conference on*. Atlanta, GA: IEEE.

23. Kubasov, D., J. Nayak, and C. Guillemot. 2007. "Optimal Reconstruction in Wyner–Ziv Video Coding with Multiple Side Information." In *Multimedia Signal Processing, 2007. MMSP 2007, IEEE 9th Workshop on*. Crete, Greece: IEEE.

24. Martinez, J. L. et al. 2009. "Wyner–Ziv to H.264 Video Transcoder." In *Image Processing (ICIP), 2009 16th IEEE International Conference on*. Cairo, Egypt: IEEE.

25. Pedro, J., 2007. "Studying Error Resilience Performance for a Feedback Channel Based Transform Domain Wyner–Ziv Video Codec." In *Picture Coding Symposium*, Lisbon, Portugal.

26. Peixoto, E., R. L. de Queiroz, and D. Mukherjee. 2010. "A Wyner–Ziv Video Transcoder." *Circuits and Systems for Video Technology, IEEE Transactions on* 20 (2): 189–200. Hanover, Germany: IEEE.

27. Pereira, F. et al. 2008. "Wyner–Ziv Video Coding: A Review of the Early Architectures and Further Developments." In *Multimedia and Expo, 2008 IEEE International Conference on*. Hanover, Germany: IEEE.

28. Puri, R., A. Majumdar, and K. Ramchandran. 2007. "PRISM: A Video Coding Paradigm with Motion Estimation at the Decoder." *Image Processing, IEEE Transactions on* 16 (10): 2436–48.

29. Qian, X., and X. Zixiang. 2006. "Layered Wyner–Ziv Video Coding." *Image Processing, IEEE Transactions on* 15 (12): 3791–803.

30. Sehgal, A., A. Jagmohan, and N. Ahuja. 2004. "Wyner–Ziv Coding of Video: An Error-Resilient Compression Framework." *Multimedia, IEEE Transactions on* 6 (2): 249–58.

31. Tagliasacchi, M. et al. 2006. "Intra Mode Decision Based on Spatio-Temporal Cues in Pixel Domain Wyner-ZIV Video Coding." In *Acoustics, Speech and Signal Processing, 2006. ICASSP 2006 Proceedings, IEEE International Conference on*. Toulouse, France: IEEE.

32. Wang, H., N.-M. Cheung, and A. Ortega. 2006. "A Framework for Adaptive Scalable Video Coding Using Wyner-Ziv Techniques." *EURASIP Journal on Applied Signal Processing* 60971: 1–18.

33. Wyner, A., and J. Ziv. 1976. "The Rate-Distortion Function for Source Coding with Side Information at the Decoder." *Information Theory, IEEE Transactions on* 22 (1): 1–10.

34. Candes, E. J., and M. B. Wakin. 2008. "An Introduction To Compressive Sampling." *Signal Processing Magazine, IEEE* 25 (2): 21–30.

35. Duarte, M. F. et al. 2008. "Single-Pixel Imaging via Compressive Sampling." *Signal Processing Magazine, IEEE* 25 (2): 83–91.

36. Do, T. T. et al. 2009. "Distributed Compressed Video Sensing." In *Information Sciences and Systems, CISS 2009. 43rd Annual Conference on*. Baltimore, MD: IEEE.

37. Zhang, C., and J. Leng. 2011. "Distributed Video Coding Based on Compressive Sensing. In *Multimedia Technology (ICMT), 2011 International Conference on*. Hangzhou, China: IEEE.

38. Goldstein, T., and S. Osher. 2009. "The Split Bregman Method for L1-Regularized Problems." *SIAM Journal on Imaging Sciences* 2 (2): 323–43.

39. Yin, W. et al. 2008. "Bregman Iterative Algorithms for L1-Minimization with Applications to Compressed Sensing." *SIAM Journal on Imaging Sciences* 1 (1): 143–68.

40. Chen, H. W., K. Li-Wei, and L. Chun-Shein. 2010. "Dynamic Measurement Rate Allocation for Distributed Compressive Video Sensing." In *Proceedings of the SPIE*. Anhui, China: IEEE.

41. Kang, L.-W., and C.-S. Lu. 2009. "Distributed Compressive Video Sensing." In *IEEE International Conference on Acoustics, Speech and SP*, Taiwan.

42. Akyildiz, I. F., T. Melodia, and K. R. Chowdhury. 2007. "A Survey on Wireless Multimedia Sensor Networks." *Computer Networks* 51 (4): 921–60.

43. Misra, S., M. Reisslein, and X. Guoliang. 2008. "A Survey of Multimedia Streaming in Wireless Sensor Networks." *Communications Surveys & Tutorials, IEEE* 10 (4): 18–39.

44. Chessa, S. et al. 2007. "Mobile Application Security for Video Streaming Authentication and Data Integrity Combining Digital Signature and Watermarking Techniques." In *Vehicular Technology Conference, 2007. VTC2007-Spring. IEEE 65th*. Dublin, Ireland: IEEE.

45. Ju, W., and L. Jonathan. 2008. "Video Authentication against Correlation Analysis Attack in Wireless Network." In *Multimedia, 2008*.

46. Imran, N., B.-C. Seet, and A. C. M. Fong. 2011. "Performance Analysis of Video Encoders for Wireless Video Sensor Networks.' In *Communications, Computers and Signal Processing (PacRim), 2011 IEEE Pacific Rim Conference on*. Victoria, British Columbia, Canada: IEEE.

47. Cox, I., G. Doërr, and T. Furon. 2006. "Watermarking is Not Cryptography" In *Digital Watermarking*, edited by Y. Shi and B. Jeon, 1–15. Berlin/Heidelberg: Springer.

48. Katzenbeisser, S. 2004. "On the Integration of Watermarks and Cryptography" In *Digital Watermarking*, edited by T. Kalker, I. Cox, and Y. Ro, 267–8. Berlin/Heidelberg: Springer.

49. Sadeghi, A.-R. 2008. "The Marriage of Cryptography and Watermarking—Beneficial and Challenging for Secure Watermarking and Detection" In *Digital* Watermarking, edited by Y. Shi, H.-J. Kim, and S. Katzenbeisser, 2–18. Berlin/Heidelberg: Springer.

50. Haque, M. M. et al. 2007. "An Efficient PKC-Based Security Architecture for Wireless Sensor Networks." In *Military Communications Conference, 2007. MILCOM 2007*. Orlando, FL: IEEE.

51. Rahman, S. et al. "Chaos-Cryptography Based Privacy Preservation Technique for Video Surveillance." *Multimedia Systems* 18 (2): 145–155.

52. Rahman, S. M. M. et al. 2010. "A Real-Time Privacy-Sensitive Data Hiding Approach Based on Chaos Cryptography." In *Multimedia and Expo (ICME), 2010 IEEE International Conference on*. Singapore: IEEE

53. Wander, A. S. et al. 2005. "Energy Analysis of Public-Key Cryptography for Wireless Sensor Networks." In *Pervasive Computing and Communications, 2005. PerCom 2005. Third IEEE International Conference on*. Kauai Island, HI: IEEE.

54. Almalkawi, I. T. et al. 2010. "Wireless Multimedia Sensor Networks: Current Trends and Future Directions. *Sensors* 10 (7): 6662–717.

55. Cheung, S. C. S., J. K. Paruchuri, and T. P. Nguyen. 2008. "Managing Privacy Data in Pervasive Camera Networks." In *Image Processing, ICIP 2008. 15th IEEE International Conference on*. San Diego, CA: IEEE.

56. Saini, M. et al. 2011. "Anonymous Surveillance." In *Multimedia and Expo (ICME), 2011 IEEE International Conference on*.

57. Wang, J., and G. L. Smith. 2010. "A Cross-Layer Authentication Design for Secure Video Transportation in Wireless Sensor Network." *International Journal of Security and Networks* 5 (1): 63–76.

58. Shen, J., and X. Zheng. 2010. "Security for Video Surveillance with Privacy." In *Internet Technology and Applications, 2010 International Conference on*. Wuhan, China: IEEE.

59. Fakhari, P., E. Vahedi, and C. Lucas. 2011. "Protecting Patient Privacy from Unauthorized Release of Medical Images Using a Bio-Inspired Wavelet-Based Watermarking Approach." *Digital Signal Processing* 21 (3): 433–46.

60. Honggang, W. et al. 2008. "Energy-Aware Adaptive Watermarking for Real-Time Image Delivery in Wireless Sensor Networks." In *Communications, 2008. ICC '08. IEEE International Conference on*. Beijing, China: IEEE.

61. Jiang, H., H. Wang, and X. Wang. 2009. "A Solution of Video Semi-Fragile Watermarking of Authentication Based on Binary Characteristic Strings." In *Multimedia Information Networking and Security, MINES '09. International Conference on*. Hubel, China: IEEE.

62. Kitanovski, V., D. Taskovski, and S. Bogdanova. 2007. "Semi-Fragile Watermarking Scheme for Authentication of MPEG-1/2 Coded Videos." In *Systems, Signals and Image Processing, 2007 and 6th EURASIP Conference focused on Speech and Image Processing, Multimedia Communications and Services. 14th International Workshop on*. Maribor, Slovenia: IEEE.

63. Rey, C., and J.-L. Dugelay. 2002. "A Survey of Watermarking Algorithms for Image Authentication." *EURASIP Journal on Applied Signal Processing* 2002 (1): 613–21.

64. Tsong-Yi, C. et al. 2008. "H.264 Video Authentication Based on Semi-Fragile Watermarking." In *Intelligent Information Hiding and Multimedia Signal Processing, 2008. IIHMSP '08 International Conference on*. Harbin, China: IEEE.

65. Harris C., and M. Stephens. 1988. "A Combined Corner and Edge Detector." In *Proceeding of the Fourth Alvey Vision Conference*. Manchester, UK: IEEE.

66. Jonsson, J., and B. Kaliski. 2003. Public-Key Cryptography Standards (PKCS) #1: RSA Cryptography Specifications Version 2.1, in RFC 3447.

67. Kamel, I., and H. Juma. 2011. 'A Lightweight Data Integrity Scheme for Sensor Networks." *Sensors* 11 (4): 4118–36.

68. Newsome, J. et al. 2004. "The Sybil Attack in Sensor Networks: Analysis & Defenses." In *Proceedings of the 3rd International Symposium on Information Processing in Sensor Networks*, 259–68. Berkeley, California: ACM.

69. Shaohe, L. et al. 2008. "Detecting the Sybil Attack Cooperatively in Wireless Sensor Networks." In *Computational Intelligence and Security, 2008. CIS '08. International Conference on*. Suzhou, China: IEEE.

70. Hui-Yu, H., Y. Cheng-Han, and H. Wen-Hsing. 2009. "A video watermarking algorithm based on pseudo 3D DCT." In *Computational Intelligence for Image Processing, 2009. CIIP '09. IEEE Symposium on*. IEEE.

71. Grieco, L. A. et al. 2009. "Secure Wireless Multimedia Sensor Networks: A Survey." In *Mobile Ubiquitous Computing, Systems, Services and Technologies, 2009. UBICOMM '09. Third International Conference on*. Sliema, Malta: IEEE.

72. Chun-Xing, W. 2009. "A Blind Video Watermarking Scheme Based on DWT." In *Intelligent Information Hiding and Multimedia Signal Processing. IIH-MSP'09*. 5th International Conference on. Kyoto, Japan.

73. Liu, Y. and J. Zhao. 2010. "A New Video Watermarking Algorithm based on 1D DFT and Radon Transform." *Signal Processing* 90 (2): 626–39.

74. Hartung, F., and M. Kutter. 1999. "Multimedia Watermarking Techniques." *Proceedings of the IEEE* 87 (7): 1079–107.

75. Bhattacharya, S., T. Chattopadhyay, and P. Arpan. 2006. "A Survey on Different Video Watermarking Techniques and Comparative Analysis with Reference to H.264/AVC." In *Consumer Electronics, 2006. ISCE '06. 2006 IEEE Tenth International Symposium on*.

76. Xu, D., R. Wang, and J. Wang. 2009. "Video Watermarking Based on Spatio-temporal JND Profile Digital Watermarking." In *The Smithsonian/NASA Astrophysics Data System*, edited by H.-J. Kim, S. Katzenbeisser, and A. Ho, 327–41. Berlin/Heidelberg: Springer.

77. Lancini, R., F. Mapelli, and S. Tubaro. 2002. "A Robust Video Watermarking Technique in the Spatial Domain." In *Video/Image Processing and Multimedia Communications 4th EURASIP-IEEE Region 8 International Symposium on VIPromCom*. Zadar, Croatia: IEEE.

78. Ling, H. et al., 2011. "Robust video watermarking based on affine invariant regions in the compressed domain." *Signal Processing* 91 (8): 1863–75.

79. Cheng-Han, Y., H. Hui-Yu, and H. Wen-Hsing. 2008. "An adaptive video watermarking technique based on DCT domain." In *Computer and Information Technology, 2008. CIT 2008. 8th IEEE International Conference on*. Sydney, NSW, Australia: IEEE.

80. Ahmed, K., I. El-Henawy, and A. Atwan. 2009. "Novel DWT video watermarking schema." *Machina Graphics and Vision* 18 (3): 363–80.

81. Campisi, P., and A. Neri. 2005. "Perceptual Video Watermarking in the 3D-DWT Domain Using a Multiplicative Approach" In *Digitial Watermarking*, edited by M. Barni et al., 432–443. Berlin/Heidelberg: Springer.

82. Mohanty, S. P. et al. 2009. "VLSI Architectures of Perceptual Based Video Watermarking for Real-Time Copyright Protection." In *Quality of Electronic Design, ISQED 2009. Quality Electronic Design*. San Jose, CA: IEEE.

83. Zhi, L., and C. Xiaowei. 2008. "The Imperceptible Video Watermarking Based on the Model of Entropy." In *Audio, Language and Image Processing, ICALIP 2008. International Conference on*. Shanghai, China: IEEE.

84. Li, J. 2009. "A Novel Scheme of Robust and Blind Video Watermarking." In *Information Technology and Applications, 2009. IFITA '09. International Forum on*. Chengdu, China: IEEE.

85. Zhu, N. et al. 2008. "A Novel Watermarking Method For Wyner-Ziv Video Coding" In *Proceedings of the 2008 International Conference on Intelligent Information Hiding and Multimedia Signal Processing*, 857–60. Harbin, China: IEEE Computer Society.

86. Gui, F., and W. Guo-Zheng. 2011. "Motion Vector and Mode Selection Based Fragile Video Watermarking Algorithm." In *Anti-Counterfeiting, Security and Identification (ASID), 2011 IEEE International Conference on*. Xiamen, China: IEEE.

87. Tien-Ying, K., L. Yi-Chung, and I. L. Chen. 2008. "Fragile Video Watermarking Technique by Motion Field Embedding with Rate-Distortion Minimization." In *Intelligent Information Hiding and Multimedia Signal Processing, 2008. IIHMSP '08 International Conference on.* Washington, DC: IEEE.

88. Wong, J. L. et al. 2004. "Security in Sensor Networks: Watermarking Techniques." In *Wireless Sensor Networks*, edited by C. S. Raghavendra, K. M. Sivalingam, and T. Znati, 305–23. Norwell, MA: Kluwer Academic Publishers.

89. Harjito, B. 2003. "Watermarking Technique based on Linear Feed Back Shift Register (LFSR)." *Seminar Konferda at National Association of Regional Mathematics*, Faculty Central Java and Yogyakarta UNS. Indonesia.

90. Li, M., Vitanyi., and M. B. Paul. 2008. *An Introduction to Kolmogorov Complexity and Its Applications.* 3rd ed. Texts in Computer Science. Springer, USA.

91. Slepian, D., and J. Wolf. 1973. "Noiseless Coding of Correlated Information Sources." *Information Theory, IEEE Transactions on* 19 (4): 471–80.

92. He., M. 2007. Adaptive Image Digital Watermarking Algorithm Based on Best Scramble." Master's dissertation. Wuhan, China: South-Central University for Nationalities.

93. Wang, H. 2010. "Communication-Resource-Aware Adaptive Watermarking for Multimedia Authentication in Wireless Multimedia Sensor Networks." *The Journal of Supercomputing* 1–15.

94. Ming-Shing, H., T. Din-Chang, and H. Yong-Huai. 2001. "Hiding Digital Watermarks Using Multiresolution Wavelet Transform." *Industrial Electronics, IEEE Transactions on* 48 (5) 875–82.

Chapter 17

Intrusion Detection and Prevention in Wireless Sensor Networks

Abror Abduvaliyev, Al-Sakib Khan Pathan, Jianying Zhou, Rodrigo Roman, and Wai-Choong Wong

Contents

17.1 Introduction

In many wireless sensor network (WSN) application scenarios, security is a vital concern, especially the applications designed for WSNs deployed in hostile environments and commercial applications. With the level of importance of security in a WSN application, ensuring it to the expected level also becomes relatively more difficult than its other wireless network counterparts. In fact, security in WSN has a great number of challenges that may not be seen in other types of wireless networks. This is due to many reasons such as the broadcast nature of wireless communications, limited resources of the sensor nodes, unattended environment where sensor nodes might be susceptible to physical attacks, and so on [1–3]. Security solutions such as authentication, cryptography, or key management can enhance the security of WSNs. Nevertheless, these solutions alone cannot prevent all possible attacks. As a wide range of attacks can be launched by compromised nodes in a WSN (i.e., nodes that appear to be legitimate in the network but not working for other party [4,5]), a second line of defense such as intrusion detection system (IDS) [6,7] is needed.

There are mainly three angles of looking at the security in WSNs. These angles could cover all the security requirements and issues that we should consider. Figure 17.1 shows these aspects so as to reveal where exactly IDS could be located in WSNs.

17.1.1 WSN Security Viewing Angle 1

The first angle is based on the mechanism used to deal with security in WSNs. These mechanisms include: (a) key management, (b) security routing, (c) secure services, and (d) IDSs.

17.1.2 WSN Security Viewing Angle 2

The second angle could be based on where the security is employed. This angle includes the following:

■ *Physical security:* The physical protection of the sensors in a network; tamper-proof methods, self-destruction method if cracked by attacker, shielding and camouflaging of sensors, and so on.

■ *Deployment security:* Dependent on whether the network is sparsely deployed or densely deployed. A densely deployed sensor network may have redundancy in a small area, which

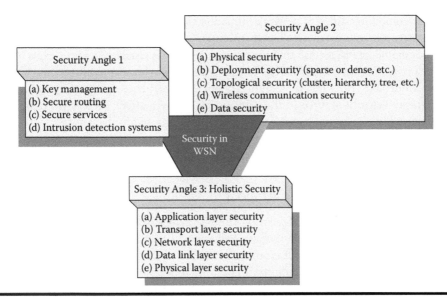

Figure 17.1 Angles of WSN security.

could find out alternative ways of protecting the traffic flow, if attacked by attackers in one way or the other. Besides, based on the deployment types or the method of deployment of sensors, the security measures may need different types of prior works. If the network is uniformly distributed, the security schemes may be installed uniformly among nodes, while a random deployment may require installing security components in key nodes in the network that cover the entire network.

■ *Topological security:* Based on the network structure or network formation, the security could be different. There are mainly three types of network structures: cluster, tree, and hierarchy. In a cluster structure, there is a cluster head in each cluster and some subordinate nodes under the cluster head. In this formation, instead of installing security schemes in each node, cluster heads could be the most suitable entities. This is because the cluster heads collect data from the other subordinate nodes and process them before forwarding toward the base station. If the cluster heads with higher computing and energy resources are used in a network, the task becomes easier as they can take the load of processing and forwarding secure packets. If the network formation is tree based, the nodes have parent–child relationships among themselves from the leaf toward the sink node or vice versa. In such a case, each individual node may include security measures and along a path in a tree, the packets could be checked before forwarding to the next hop or to the sink node or base station. The third type of network formation is hierarchical, where there are several hierarchical levels of the nodes in the network. For example, in one level, there are several clusters with cluster heads and subordinate nodes. The cluster heads of this level could be considered the subordinate nodes in another bigger scale cluster (another level), which might have a higher powered cluster head and it could be repeated for several levels. A well-scalable and large WSN with some strategically positioned high-power nodes with higher transmission ranges could have such a structure. So, such a network formation needs security measures in a different way than the other two types of formations. The thematic diagrams of all these types of network formations are shown later in this chapter when discussing these in relation to our work (see Section 17.5). Other than these categories, there might be a hybrid topology in the network combining different network formation styles, for example, a network with partly cluster structure and partly tree-based structure. So, it becomes crucial where the security schemes should be installed so that the network security is ensured up to the expected level.

■ *Wireless communication security:* Due to the nature of wireless communications, a WSN is always vulnerable. The wireless medium is of open nature, hence the signaling and reception mechanisms must be secured in the best possible way. An attack such as jamming [8], for example, could disrupt the natural wireless transmissions within the network.

■ *Data security:* This includes the encryption, decryption of data packets, efficient packet authentication techniques, hop-by-hop checking, and so on.

17.1.3 WSN Security Viewing Angle 3

The third angle takes a holistic view of security. This brings forward the concept of layer-wise security in such type of network. Based on the very well-known Open Systems Interconnection (OSI) reference model, we could think about ensuring security in each layer. Especially for WSNs, five layers are relevant: application layer, transport layer, network layer, link layer, and physical layer. Lack of security in any of these levels weakens the overall security of the network. A full working solution where different mechanisms could work in cooperation is still an open area of research, which would take huge effort to develop.

17.1.4 Basic Information and Contributions of this Work

After knowing all these views and angles of security in a WSN, in Section 17.3, we will explore the major types of threats and attacks against such type of network. However, before that, let us get a clear picture of our main focus of this work. From the earlier discussion, it is clear that IDS is one of the major mechanisms to be considered for WSN security. An IDS that has been successfully implemented in wired networks can detect the misbehavior of participating nodes and notify other nodes in the network to take appropriate countermeasures. However, an IDS scheme designed for wired networks cannot be applied directly to WSNs because of their specific network characteristics. In a WSN, an IDS is an important security mechanism against both insider and outsider attacks [9]. It focuses on the detection of misbehaving or malicious nodes. When the IDS detects a sensor node misbehaving, it tries to isolate that malicious node from the network.

In the recent years, many IDSs for WSNs have been proposed. However, there is still a great need for a comprehensive survey on the recent developments in this particular area. In spite of the presence of some partial works [10–13], to date there has not been any comprehensive survey paper that collects all the significant IDSs and gives overviews of those works in terms of the techniques they use along with important observations and obtained results. Thus, the purpose of this chapter is to analyze the vital signs of IDS in WSN from various angles. We critically analyze works that have been proposed over the last decade and discuss the current state of the art in this research area. We classify the IDSs based on their used detection techniques and analyze the works considering the specific features of WSNs. We analyze all of the discussed IDS mechanisms with respect to network architecture. In addition, we present the well-known attacks for a quick look and then introduce some less-known security attacks that need to be detected and prevented as well.

The remainder of this chapter is organized as follows: Section 17.2 explains the background of IDSs in WSN. The major security threats and attacks against WSNs are explored in Section 17.3. Section 17.4 reviews the significant IDS approaches proposed for WSNs. In Section 17.5, we discuss a few key issues and finally, Section 17.6 concludes the chapter based on our findings and analysis.

17.2 Intrusion Detection Systems in WSN

In reality, it is extremely difficult to design a network where attackers cannot find some ways to break it. In fact, networks should seriously consider the integration of self-awareness and fault-tolerance capabilities, that is, not only to assume that problems will appear in one way or the other, but also to provide some mechanisms that will detect and reduce the impact of a particular threat. Therefore, we need a second line of defense that can detect attackers or intruder nodes. An IDS is able to detect misbehaving nodes and inform neighbor nodes to take proper countermeasures [14]. Although some type of IDS is used as a major prevention mechanism in wired networks, it is infeasible to apply that directly in WSNs because of the vast difference in network characteristics. Sensor networks inherit all aspects of wireless networks and have their own distinct characteristics that make the design of a security measure more complex than that of wireless ad hoc networks [15]. It is a fact that the computing and power resources of sensor nodes are more constrained than that of ad hoc nodes. Moreover, in most application scenarios, sensor nodes are stationary and more application oriented. Thus, WSNs demand a more novel and lightweight design of IDS.

There are three main approaches that an IDS can use to classify the attacks:

1. *Anomaly detection:* This detection technique compares the behaviors of observed nodes with normal behaviors with the intent to detect anomalies from expected normal behaviors rather than malicious attacks. This approach first describes normal behaviors that are established by automated training and then flags any activities varying from these behaviors as intrusions. If a sensor node does not act according to the defined specification of a particular protocol, the IDS would have high confidence to decide that there is an anomaly and the node may also be potentially malicious. The wrong decisions made by the system results in false-positive and false-negative rates, thus reducing the accuracy of detection. Hence, the disadvantage of this methodology is that the system can exhibit legitimate but unseen behavior that could lead to a substantial false alarm rate. Also, an intrusion that does not exhibit anomalous behavior may not be detected, resulting in false negatives.

2. *Misuse detection:* The actions or behaviors of nodes are compared with well-known attack patterns. In this case, attack patterns must be defined and given to the system. The disadvantages are that this technique needs knowledge to build attack patterns and they are not able to detect novel attacks. In addition, there is a need to frequently update the database of attack signatures. These drawbacks significantly reduce the efficiency of this approach in terms of system management where the administrator of the network always has to provide IDS agents with up-to-date attack profiles. At the current stage, most of the known attacks are only the results of some assumptions or imitated from other classic networks. Whether these well-known attacks or any unknown security attack would be a serious problem for WSNs remains still unclear.

3. *Specification-based detection:* This technique combines the aims of misuse and anomaly detection. Note that, in particular cases, misuse and anomaly based detection techniques are used side by side, resulting in hybrid detection mechanisms. This model is based on deviations from normal behaviors that are defined neither by machine learning techniques nor by training data. The attack specifications are defined manually, which describe what normal behavior is. Any action is monitored with respect to these specifications. The drawback of this approach is the manual development of attack specifications, which is a huge time-consuming process for security personnel. Another disadvantage of this technique is that it cannot detect malicious behaviors that do not violate defined specifications of the IDS protocol.

Details of particular IDS models and techniques are discussed later in this chapter.

17.3 Security Threats and Types of Attacks in WSN

There are several well-known and a few less-known security attacks that exist in WSNs. In this section, we discuss these security attacks in brief with respect to their countermeasures. Almost all of the attacks described in Sections 17.3.1 through 17.3.7 focus on the limitations of routing protocols in WSNs [16]. However, some unknown attacks that are launched considering other security constraints of the network are presented as well. Table 17.1 introduces a brief summary of well-known and less-known (or less studied) security attacks and their characteristics in terms of attack behaviors and techniques. In addition, the relevant detection techniques for the attacks are highlighted in the table. In Section 17.4, we will discuss some of these techniques in terms of their benefits and drawbacks.

Table 17.1 Security Attacks in WSNs

Well-known		Less-known (or Less Studied)	
Name	*Characteristics*	*Name*	*Characteristics*
DoS attacks in different layers [8,17,18]	Flooding, jamming, and misdirection	Fabrication during reprogramming [19]	Unsecure reprogramming process with bogus messages
Sinkhole/blackhole [1,5,20,21]	Shortest path, drop the packets	External stimuli [22]	Use external physical stimuli to create a large number of packets
Selective forwarding [23–27]	Selectively drop the packets	Homing [22]	Hamper the normal functioning of cluster heads
The node replication [28–30]	Add extra node to the network with the same cryptographic secrets	Neglect and greed [31]	Deny transmission of legitimate packets and give higher priority to own packets
HELLO flood [32]	Flood with HELLO packets	Unfairness [31]	Unfair resource allocation on MAC protocols
Wormhole [33–39]	Offer less number of hops and less delay which is fake	Forced delay [40]	A node delays packets within its forwarding component
Sybil [26,41–45]	A malicious node pretends to be more than one node		

17.3.1 Denial of Service Attacks

We consider any type of intentional activity that can disrupt, subvert, or even destroy the network as a denial of service (DoS) attack. Basically, DoS attacks can be categorized into three types:

1. Consumption of scarce, limited, or nonrenewable resources
2. Destruction or alteration of configuration information
3. Physical destruction or alteration of network resources

These types of DoS attacks are the most significant for WSNs as the sensors in the network suffer from limited resources. Also, DoS can be categorized according to the layers of the network architecture. An attacker can use different tools at different layers to stop proper functioning of the entire network or some sensor nodes. Even though, it has been said that it is too difficult to know whether any particular DoS situation is caused intentionally or unintentionally, there are some detection methods that exist to thwart each type of DoS attack [46]. In general, DoS means any situation that prevents providing proper service that is expected from the network and "DoS attack" means any deliberate activity by an entity (or, someone) that causes DoS in the network.

Jamming and *tampering attacks* that exist in physical layer of WSNs are also considered as kinds of DoS attacks. Jamming is the deliberate injection of interference to disrupt radio reception to deny a target's use of a communication channel. Owing to their unpredictable nature, WSNs are highly vulnerable to "radio channel jamming"-based DoS attack [17]. Tampering is actually any type of physical attack on sensors in the network. They might involve physical damage or replacing the sensors, parts of computational or sensitive hardware. These types of attacks cannot be defended by some system or base station, and only accurate and effective designer of the network and regular maintenance can handle it.

17.3.2 Sinkhole/Blackhole Attacks

In this attack, a malicious node acts as a blackhole [47] to pull in all the traffic in the network. The attacker listens to the route requests and then replies to the target node informing that it has the shortest path to the base station. A victim node is enticed to select it as a forwarder for its packets. Once a malicious node is able to put itself between the base station and the sensor node, it is able to do whatever it wants (drop packets, change the content, etc.) with the packets that pass through it. This type of attack can be very harmful for sensor nodes that are deployed considerably far from the base station. We have to keep in mind that blackhole and sinkhole attacks are basically the same by definition. Some recent works have addressed this attack and possible IDSs have been proposed [5,8,20,48].

17.3.3 Selective Forwarding

Multihop networks such as WSNs rely on an important assumption that all nodes in the network will faithfully forward the received messages to the base station. In these attacks, a malicious node acts as a normal node by forwarding only certain messages, but selectively drops sensitive packets that are hard to detect by the system. The specific form of this attack is the sinkhole/blackhole attack with which a node might drop all messages that it receives. As possible solutions to detect this type of attack, some secure routing algorithms and IDSs using different techniques have been proposed [48,23,24,25,27].

17.3.4 Node Replication Attacks

Owing to the resource constraints of sensor nodes and often unattended environment of WSNs, an attacker can easily capture the nodes, analyze, and replicate them. In this attack, an attacker attempts to add one or more nodes in a network that use the same cryptographic secrets as any other legitimate node in that network. One can even extract cryptographic keys to gain unrestricted access to higher communication layers. A node sends advertising information that is not consistent with the state of the network (i.e., feature advertising [49]) to perform bad actions at the base station or neighborhood level. This kind of attack may have severe consequences such as corruption of data by an adversary or even disconnection of some critical parts of the network. Some centralized detection schemes with one point of failure, neighborhood-voting protocols with the lack of detecting distributed node replications, and some successful distributed detection techniques have been proposed [28,29]. Zhu et al. [30] discuss existing approaches on the detection of node replication attacks in mobile sensor networks and propose a new detection technique that is based on node statistics.

17.3.5 HELLO Flood Attacks

This attack uses HELLO packets as a tool for convincing the sensors in the network. Many of the routing protocols require broadcasting of HELLO packets to discover the neighbors. An attacker uses this assumption as a weapon to attract the sensor nodes. A node that receives such a packet may assume that it is within the normal radio range of the sender node. Hence, an attacker with a large radio range and enough processing power can send HELLO packets to a large number of sensor nodes by flooding the entire network. Thus, the sensor nodes could be persuaded that the adversary is their neighbor. Possible solutions to detect this type of attacks could be the use of bidirectional verification of links before using them, secure multipath routing, and the use of multiple base stations [32].

17.3.6 Wormhole Attacks

In this attack, an attacker records the packets at one location in the network and tunnels those to another location with the help of a long-range wireless or optical link. Wormhole attack is another significant and serious threat to WSNs, because this is possible even if the attacker has not compromised any node, and even if all communications provide authenticity and confidentiality. Owing to the nature of wireless transmissions, the attacker can still create a wormhole for the packets that are not addressed to itself since it can overhear them. Attackers offer less number of hops and less delay than other normal routing paths, which lead to attract the sensor nodes to send data through them. While forwarding packets, the attackers can arbitrarily drop sensitive packets. In a recent work, Sharif and Leckie propose three types of wormhole attacks: Energy Depleting Wormhole Attack (EDWA), indirect Wormhole Attack (IBA), and Targeted Energy Depleting Wormhole Attack (TEDWA) [35]. Also, IDS using connectivity information to detect the wormhole attacks has been proposed [37]. In another work, Hu and Evans [38] propose wormhole detection technique using directional antennas, which is, in most of the cases, infeasible for sensor networks due to their limited resources.

17.3.7 Sybil Attacks

In some applications, the sensor might need to work collaboratively to accomplish a certain task; hence, management policy of the network can use distribution of subtasks or redundancy of information. In this case, a malicious node can pretend to be more than one node at the same time using the identities of other legitimate nodes. This is known as a Sybil attack, and has been studied by Newsome et al. [41]. By using this attack, a malicious node tries to degrade the integrity of data, level of security, routing mechanism, data aggregation, and even misbehavior detection techniques. As possible countermeasures, we can use a logically centralized authority (base station or cluster head) in the network. Some other recent IDSs could be found [26,42–45,50].

17.3.8 Other Security Attacks in WSNs

There are a few less-known (or commonly unknown or less-studied) security threats that exist in WSNs. These attacks mostly concentrate on service availability (i.e., DoS) of the networks in different layers. We briefly describe them as follows:

> *Fabrication during reprogramming:* This attack could be launched in the application layer if a WSN application allows reprogramming of the network. Reprogramming of the network may be needed for scope selection, encoding–decoding, completion validation, code acquisition, or network management purposes [19]. If the reprogramming process is not secure enough, the attackers can effectively cut off a portion of the network by using bogus messages.
>
> *External stimuli:* A possible attack against WSNs in the application layer could be launched by using some external physical stimuli. The attacker uses the external stimuli to stimulate the nodes with a huge number of events to be sent directly to the base station. However, this attack is not effective when packets are sent with predefined regular intervals. The possible solution might be using an IDS that detects attackers in the network if a particular region creates a large number of packets within a short period of time [22].
>
> *Homing:* Depending on WSN application, some nodes (e.g., cluster heads) are given special responsibilities such as managing keys, maintaining a local group, and so on. The adversaries try to handle and eavesdrop on the activities of those leader nodes. In this attack, the attackers hamper the normal functioning (i.e., receiving and sending packets, periodic advertisement and/or information update, etc.) of leader nodes within a WSN application [22].
>
> *Neglect and greed:* If a sensor node drops packets or denies transmitting legitimate packets or if a node is very greedy to give undue priority to its own messages, then it could be considered neglecting a node. This attack is a special case of selective forwarding attack where the node may still acknowledge the received packets to the sender, but it drops them randomly and gives excessive priority to its own packets. The protocols based on dynamic source routing (DSR) are the most vulnerable to this type of attack [31].
>
> *Unfairness:* This attack is a weaker form of DoS attack at the link layer. This attack could degrade service for real-time media access control (MAC) protocols by using unfair resource allocations (e.g., an attacker causes nodes to miss their transmission deadline). In fact, providing fairness in WSNs is often viewed as a separate research issue [31].
>
> *Forced delay:* A sensor node deliberately delays packets within its forwarding component to delay messages [40]. This attack can be effectively used to degrade the quality of service in systems with near-real-time requirements.

So far, we have discussed various types of security threats in WSNs. These attacks can be tackled by using some successful and efficient countermeasures. Most of the research works basically rely on some statistical assumptions and simulation results. At the time of the implementation of those mechanisms in real environments, they might face lots of difficulties due to the unpredictable nature of WSNs.

17.4 Taxonomy of IDS Approaches in WSN

As mentioned earlier, an IDS is based on the assumption that there exists a noticeable difference in the behavior of an attacker and a legitimate node in the network such that the IDS can match those preprogrammed or learned rules. Based on the analysis model used for analyzing the audit data to detect intrusions, we classify IDSs into three detection techniques: (1) misuse, (2) anomaly, and (3) specification based. The misuse detection systems are used to detect known patterns of intrusions, while anomaly detection techniques are used to detect new or unknown intrusions. Specification-based detection is based on some deviations from normal behaviors. Figure 17.2 shows the overall taxonomy of various IDSs applied to WSNs.

17.4.1 Anomaly Detection Schemes

There are many IDS mechanisms that use anomaly detection techniques in WSNs. These types of systems usually rely on specific normal behaviors of sensor nodes. Thus, most of the researchers take this approach as a main method to detect intrusions and they have found it easier than misuse- or specification-based detections. The first attempts to apply anomaly detection in sensor networks are partially inherited by misuse-detection techniques such as the watchdog approach. However, the main difference is that the agents monitor and compare normal behaviors of sensor nodes with abnormal activities rather than predefined attack patterns.

Basically, these techniques involve simple assumptions to define normal behaviors of the sensor nodes, such as

- Payload of a packet should not be altered or modified.
- Retransmission of a packet must occur in a certain time threshold.
- Same packet can be resubmitted a limited number of times.
- Packet sending rate must be within some limits, and so on.

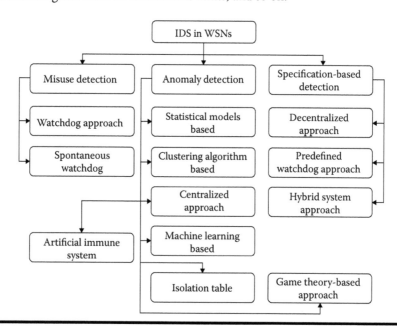

Figure 17.2 Taxonomy of IDSs in WSNs.

Table 17.2 Comparison of Anomaly Based Detection Techniques

IDS	Statistical Models Based	Clustering Algorithm Based	Centralized	Artificial Immune System	Isolation Table	Game Theory Based	Machine Learning
Accuracy	**Medium**	**High**	**High**	**High/ Medium**	**Low**	**High/ Medium**	**High**
Energy efficiency	No detail	Yes	No	No detail	No	No	Yes
Memory requirement	No detail	High	Low	No detail	Medium	Medium	High
Network structure	Normal	Clustered	Normal	Normal	Clustered	Normal/ distributed	Normal

Table 17.2 gives an overall comparison of anomaly based detection techniques in terms of their energy efficiency, accuracy, and memory requirements.

17.4.1.1 Statistical Model-Based Approach

Onat and Miri [6] proposed an anomaly detection-based security scheme for WSNs. In their method, each sensor node builds a simple statistical model of its neighbor's behavior, and these statistics are used to detect node impersonation and resource depletion changes. The system features that analyze anomalies are the average of received power and packet arrival rate. At every node, only the last N packets received from each neighbor are used to calculate the statistics for that neighbor node and each arriving packet is then compared with those values. If the packet conforms to the statistics of the neighbor, it is accepted as a normal behavior. *Drawbacks:* The authors do not present how the experimental setup was designed. The information about the used routing protocol and simulator is missing. The system cannot detect selective forwarding and wormhole attacks due to their simple statistical features.

The same main idea of anomaly detection with different evaluation metrics is presented by Onat and Miri [51]. Instead of previously implemented interarrival times, mean and standard deviation (in the buffers) metrics are used. A packet is identified as anomalous if the absolute value of difference between the mean of received packet buffer and the mean of intrusion buffer is greater than the standard deviation of the received packet buffer. *Drawbacks:* Again, no information is given about how nodes were tested, the number of nodes, the analysis of cost of communications, and computations.

17.4.1.2 Clustering Algorithm-Based Approach

Loo et al. [14] developed an intrusion detection scheme for routing attacks, which uses a fixed-width clustering algorithm to build a model of normal behavior. Note that the clustering algorithm relates to unsupervised learning algorithms, not to cluster-based network structure. They use this model to detect anomalous traffic patterns. The IDS module is implemented on each sensor node and 12 network traffic patterns are identified. These features are used in the training and testing stages. In the training stage, a fixed-width clustering algorithm is used to build a set of clusters with predefined radius in the feature space. Clusters that contain less than a specific threshold of the total set of points are identified as anomalous. During the testing stage, each traffic sample is compared to the cluster set to determine whether it is anomalous or not. *Drawbacks:* Their method puts too much computation on the sensor node. The authors claim that since the proposed IDS do not require communication between the sensor nodes, it significantly reduces the power consumption. However, a statistical analysis of reduction in power consumption compared to other existing IDSs is not provided.

An almost similar approach to Loo et al.'s [14] has been presented [52]. The main difference between the two systems is the clustering technique where an Apriori algorithm is used [52]. Therefore, selected network features might slightly be different at different time intervals. For simulation purposes, five training data sets with normal traffic and two testing data sets with DoS and selective forwarding attack instances are used. *Benefits:* The results show that the algorithm is able to detect both attacks with a high detection rate. The algorithm is adaptive in the sense that each node might have different detection models. *Drawbacks:* Providing each node with local training data set would be infeasible in large WSNs where the sensor nodes should receive and forward a huge amount of packets in addition to their heavy packet-processing duty, or at least would require the sensor nodes that have higher computing capabilities. This issue makes the algorithm implementation inapplicable in practical environments.

Wang and Zhang [53] propose an anomaly detection system based on the arriving order of different packets. The system is based on the assumption that all sensor nodes can be cluster heads and usually only communicate with a limited number of nodes and should follow corresponding protocol specifications. The IDS has two stages: profile learning and anomaly detection. In profile learning stage, the source and destination address in the network layer and packet type are extracted from the packet payload and compared against previously received packet's information. *Drawbacks:* The limitation of this work is that the algorithm has not been evaluated and performance results are not provided.

17.4.1.3 Centralized Approach

A centralized, active anomaly detection system called ANDES is proposed by Gupta et al. [54]. This IDS is made up of collection and analysis of application data, collection and analysis of management information, and the detection algorithm. It is stated in the paper that the usage of management information and application data (node's ID, hops toward the sink, total transmitted packets, and total number of failures to route a packet) make the algorithm a novel approach of IDSs in WSNs. ANDES is implemented at the base node and it maintains the list of active and connected nodes. The sensor nodes can be in normal, unavailable, duplicated and abnormal state based on the packet count and the timestamp of application packets. Application and management data are then combined to identify the type of anomaly. *Benefits:* This system is implemented in TinyOS [55] on Tmote Sky sensor nodes. Although the management information might impose overhead to gather additional management traffic, the results obtained from experiments are shown to be positive.

17.4.1.4 Artificial Immune System

A totally different approach from traditional anomaly detection techniques by Kim et al. [56] introduces a biologically inspired algorithm, namely artificial immune system (AIS). The authors in this work show the similarities between the properties of WSNs and a biological immune

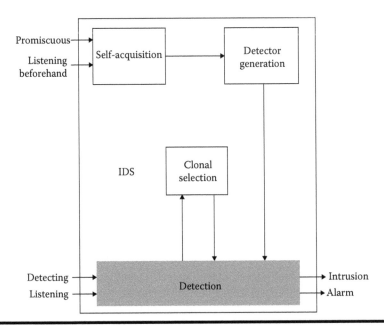

Figure 17.3 Architecture of immunity-based IDS.

system. Analogies between sensor network with directed diffusion routing and biological tissues are presented. The concept of a danger theory–based AIS employing dendritic cell algorithm (DCA) for anomaly detection of interest cache poisoning attack is also proposed. A sensor node employing directed diffusion maintains an interest cache and a data cache tables. When a node receives a packet, directed diffusion updates both caches and extracts the signals and antigens (e.g., bogus interest packets) from the received packets and caches, and then passes to the DCA. It evaluates whether antigens are benign or malicious. The algorithm is implemented in J-Sim and also is tested in TOSSIM [57]. *Drawbacks:* There is no information available about performance and statistical analysis that might prove the effectiveness of the approach.

With a similar approach, Liu and Yu [58] propose an anomaly detection scheme based on immunology. The proposed algorithm is divided into four phases: (1) self-acquisition, (2) generation, (3) detection, and (4) clonal selection. In the self-acquisition phase, beacon packets are collected from neighbor nodes and key parameters are extracted for further processing before being stored in the self-pool. After these processes, negative selection starts to generate detectors that are stored in the detector pool. The clonal selection phase, which is based on a costimulation mechanism, is used for reducing false-positive rates. The algorithm is simulated in TOSSIM. Figure 17.3 shows the architecture of this scheme, for a quick reference. The necessity of AISs in anomaly detection is discussed [59], which advocates the immune system as a possible solution to detect misbehavior in WSNs.

17.4.1.5 Isolation Table

In Chen et al. [60], anomaly detection method for hierarchical WSNs based on an isolation table is proposed. This method has four stages: First, the administrator of the network does the predefinition of IDS. Second, the secondary cluster heads monitor the sensor nodes where they gather data from sensor nodes and record malicious nodes into their isolation table. Third, the primary cluster head will be monitored by secondary cluster heads and sensor nodes. In this stage, the primary cluster head integrates sensing data and isolation table records from secondary cluster heads and forwards them to the base station. At the final stage, IDS backs up the isolation table in the base station. The method is simulated in the ns-2 simulator. *Drawbacks:* The results of simulations show that the method has disadvantages in terms of high energy consumption when the number of nodes is increased. In addition, the authors have not considered node failures and node tampering problems that lead to an increase of the false-negative rate. That work is further extended in terms of the detailed algorithm explanation but not the improvement on energy consumption [61].

17.4.1.6 Machine Learning-Based Approaches

There are some IDSs, which basically rely on machine learning techniques that are available for anomaly detection in WSNs. For example, introduce machine learning and automata-based learning approaches are introduced as anomaly detection tool for WSNs [62–65].

The learning automata-based approach (which is commonly used in optimization problems) has been used for detecting misbehaving nodes [62]. This approach relies on packet sampling where some parts of the packets are sampled to identify whether they are malicious nodes or not. *Benefits:* Decisions are made depending on the feedback of the environment to the automaton in partially favorable or partially unfavorable cases. The results obtained from analytical analysis show that the detection rate is high and the energy consumption is favorable for WSNs. The extended version of the work is presented [66].

Doumit and Agrawal [64] introduce an anomaly approach based on the structure of naturally occurring events. This approach takes advantage of the self-organized criticality of a certain location based on an environment variable and uses it to detect future anomalies by comparing new data with older ones. This method uses the hidden Markov model to characterize and learn from past patterns.

A recent work [67] uses one class of support vector machines (SVMs) to detect network anomalies. The paper proposes two SVM-based approaches called centered hyperellipsoidal support vector machine (CESVM) and quarter-sphere support vector machine (QSSVM), respectively. Even though CESVM has advantages in terms of parameter selection flexibility and computational complexity, it faces certain limitations in distributed WSNs, that is, it uses a centralized approach. On the other hand, QSSVM works well in a distributed environment. The results from real and simulated data sets show that both approaches achieve high detection accuracy.

17.4.1.7 Game Theory-Based Approaches

Another approach for anomaly detection in WSNs is a game theory-based approach. A few works have been introduced related to this approach.

Agah et al. [4] introduce a non-cooperative game approach to detect misbehaving nodes in the network. The authors use three different approaches: (1) a non-cooperative game approach, (2) Markov decision process, and (3) intuitive metric technique. A non-cooperative game approach is used for the formulation of attack-defense game as a two-sum, nonzero game, and non-cooperative game. Markov decision process is used for predicting the most vulnerable nodes in the network that can be attacked. This learning scheme informs the IDS which nodes are the most vulnerable to attack. For defending the sensor nodes with maximum activity load, they use "intuitive metric" technique that defines traffic of each cluster. The IDS defends the cluster that has the highest value of traffic, because that cluster is the most vulnerable in the network and should be defended. *Drawbacks:* The authors claim that the IDS can improve the detection rate.

However, providing each node with a heavy IDS module and learning mechanism will raise the problem of high energy consumption and communication overheads.

Similar approaches are presented [68–72]. Game theory-based models can be excellent solutions for wired networks in terms of computation and level of security, but for WSNs, it is necessary to prove the applicability of these algorithms. The sensors are equipped with constrained energy sources. The performances of these methods decrease when the scale of the network is large. In addition, some of these works, for instance [68], needs to prove the validity of the method either by experimentation or by real implementation.

17.4.2 Misuse Detection Schemes

In this section, we provide details of systems that use rule-based or misuse detection techniques. One of the main ideas used in these types of detection protocols is called a watchdog approach where packet monitoring takes place in several specific nodes in the network [73].

17.4.2.1 Watchdog Approach

This approach relies on the broadcast nature of the wireless communications and the assumption that sensors are usually densely deployed. Each packet broadcasted in the network is not only received by the receiver, but also by a set of neighboring nodes within the sender's radio range. In normal cases, neighbor nodes should discard the packet, since they are not actual receivers, but for intrusion detection, this can be used as a valuable audit data. Hence, a node can activate its IDS agent and monitor the packets sent by its neighbors by overhearing them. Furthermore, to detect attacks with high accuracy, it is not enough to monitor only one node; the system involves more information from other neighbor nodes as well. For instance, to detect selective forwarding attack, a watchdog should overhear packets arriving at a node and transmitted by that node.

If we want to see whether node B forwards packets sent by node A, we have to activate watchdogs that reside within the intersection of the radio ranges of A and B. For instance, in Figure 17.4, nodes C, D, and E can be watchdogs for the link between A and B.

Roman et al. [15] have proposed a novel technique for optimal monitoring of neighbors called "spontaneous watchdog" that extends the watchdog monitoring mechanism proposed [73]. The mechanism uses local agents in every sensor node to monitor local activities (i.e., information sent and received by the sensor node) and global agents in order to overhear the communications of neighbors. *Drawbacks:* The problem with this approach is that the authors fail to consider the selection of a global agent. Another drawback of the work is that it does not deal with the collision of packets, which is highly likely due to the high density of nodes in WSNs.

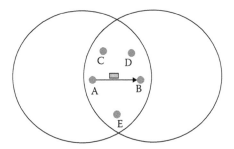

Figure 17.4   Nodes C, D, and E are watchdogs of the link A→B.

Some researchers may argue that watchdog approaches incur more energy consumption on the sensor nodes as the nodes should overhear every packet that is not addressed to them. However, each node receives packets sent by the neighbor nodes anyway. Furthermore, the nodes are not able to know if a packet is destined to them unless they receive it and check the packet. Additional overhead on the nodes can be imposed by further packet processing, which is based on stored attack signatures.

In practice, it is difficult to think exactly as an attacker or to know the motive of the attacker. The administrator of the network has to model attack patterns according to attacks that might occur in future. Moreover, the severe memory constraints of WSNs make misuse-detection-based IDSs that need to store attack signatures relatively difficult to implement and less likely to be effective [10]. Thus, there are very few papers that study misuse-detection technique for WSNs. Hence, based on the number of previously proposed works in this area and current standings, it can be construed that there are opportunities to place greater emphasis and concentration on developing misuse-detection-based IDSs.

17.4.3 Specification-Based Schemes

Some specification-based schemes have been proposed as IDS solutions for WSNs. As noted earlier, the main disadvantage of this approach is that the development of attack or protocol specifications is done by human beings. In this case, the administrator or the designer of the network has to manually define the specifications that describe what a correct operation is and monitor any behavior with respect to those constraints.

17.4.3.1 Decentralized Approach

One of the first works in this research track is introduced by da Silva et al. [74]. They propose a decentralized IDS based on several predefined rules.

The method has three phases: (1) data acquisition phase where packets are collected in a promiscuous mode to filter out the important data before being stored, (2) rule application phase where the rules are applied to the stored data, and (3) an IDS phase where the number of raised failures are compared with the expected amount of occasional failures that defines whether intrusion has occurred or not. Figure 17.5 illustrates the architecture of a monitor node, which has an IDS function in addition to sensing and message transmission capabilities. *Drawbacks:* The algorithm is simulated by a proprietary WSN simulator, which is not well known and the technical details of the simulator are unknown. This makes it difficult to revalidate the results presented as a simplified WSN model may not be something that could be used in practice and in that case, other types of analyses (numerical or probabilistic or logical) may have been added alongside the presented outputs. The algorithm is tested for different types of simulated attacks such as jamming, blackhole, and wormhole. The results obtained from the simulations show that the method performs well for simulated attacks. However, the algorithm has no information about how to select the monitor nodes in the application.

There are many other works in this topic [75–81,90] that use different techniques (e.g., group based and collaborative) to specify intrusion detection patterns and attack signatures.

For instance, Bhuse et al. [78] introduce a specification-based approach for detecting masquerade attacks. They propose two techniques that complement each other when used concurrently. The first one is mutual guarding where the sensor nodes check the source *id*s of received packets

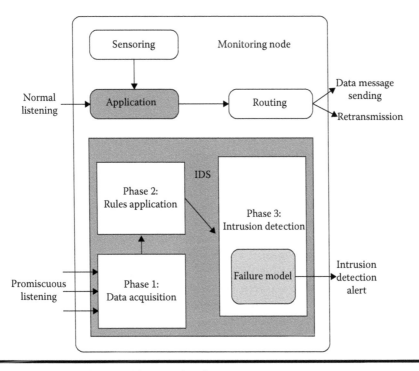

Figure 17.5   Detection phases of decentralized IDS.

for intrusion. The second technique is send receive packets (SRP): verification of the number of packets sent and received for masquerade detection. *Drawbacks:* Simulation results show that the mutual guard method has considerable overhead and it fails to protect nodes when the attacker has shorter communication range than the sensor nodes.

17.4.3.2 Predefined Watchdog Approach

Krontiris et al. have proposed various specification-based IDS to detect blackhole [10], selective forwarding [10], and sinkhole [5,82] attacks in WSNs. Their approach is based on watchdogs, which have predefined rules that raise an intrusion alert. One of those rules is stated as following: "If more than half of the watchdog nodes have raised an alert, then the target node is considered compromised and should be revoked, or the base station should be notified." In defining the threshold value, the authors also take into consideration the loss of messages arising from other network-related reasons. The method has three common modules: (1) local monitoring and detection engine for collecting and analyzing data according to the rules, (2) cooperative detection engine for making accurate decisions collaboratively, and (3) local response module for taking appropriate actions if an intrusion is verified by the network. *Drawbacks:* The method produces very low false-negative and false-positive rates. However, it is not clear from the work which simulator or experimental setting is used.

In a more recent work [9], the above authors propose a cooperative IDS scheme that has been tested in a real environment. *Benefits:* In fact, to the best of our knowledge, this paper is one of the few works that contains the results from practical implementation of IDS on TelosB motes in the real environment. The method inherits extended modules from the authors' previous works. The algorithm is based on defined intrusion detection conditions (IDC). The authors argue that these conditions are necessary and sufficient for intrusion detection problem. The results show that the proposed algorithm is lightweight enough to run on resource-constrained sensor nodes. Note that the conclusion of this work is exactly the same as the conclusion by Krontiris et al. [82].

17.4.3.3 Hybrid System Approach

As stated in Section 2, specification-based approach integrates the aims of misuse and anomaly detection techniques. However, some specific IDSs allow both detection techniques to coexist and interact in one single detection agent. That is, such agents will make use of automated training-based anomaly detection techniques and human-made rule-based misuse detection techniques. These approaches are known as hybrid systems.

Hai et al. [83] propose a hybrid IDS that integrates both anomaly and misuse techniques. The method is designed to detect routing attacks in WSNs. For energy efficiency, they use hierarchical WSNs. In the misuse detection module, predefined rules such as packet interval rule, integrity rule, packet delay rule, and radio transmission range rule are used. *Drawbacks:* Unfortunately, there is no proper and full explanation of the anomaly detection techniques used in this paper, that is, how to effectively analyze the collected data and how to make decisions on the existence of intrusions.

Later, the extended versions of the above-mentioned work were published by the same leading author (along with others) [24,84]. The methods use two-hop neighbor knowledge to prevent routing attacks. Two-hop neighbor knowledge is basically used in broadcasting protocols to reduce the number of packet transmissions such as source-based protocol and dominant pruning [85]. The two-hop neighbor list is established in each sensor node via a single phase, by modifying the HELLO packet. Other parts of this work consist of local and global agents and predefined rules. The global agents use the two-hop neighbors' list and predefined rules to monitor transmissions in their neighborhood. The method performs well for routing attacks. However, it needs to be tested in different attack scenarios in order to have a more comprehensive understanding of its effectiveness.

Yan et al. [86] introduce a similar hybrid approach. The algorithm contains the misuse detection model, anomaly detection model, and decision-making model. *Benefits:* The novelty of their method is the use of a back propagation network (BPN) for the anomaly detection module. First, the packet records are given to the anomaly detection model to check for abnormal activities. If the activity is determined as "abnormal," then it will be forwarded to both the misuse detection model and the decision-making model. The misuse detection model then analyzes the received data with the help of the BPN and sends them to the decision-making model. Finally, the decision-making model combines the outputs of both models to determine whether or not an output is an intrusion, and the category of attack. In case of intrusion, the model reports to the base station. Interested readers could refer to Wang et al. [87] for comprehensive and detailed simulation results.

Dynamic IDS (DIDS) has been proposed by Huo and Wang [88]. This work is similar to that proposed by Techateerawat and Jennings [89] in terms of applied approaches. The method has an event monitor module, rules record base, misuse and anomaly detection modules, and an alert module. The core architecture of the DIDS is shown in Figure 17.6. The method was simulated in ns-2 with 70 nodes. The results obtained by simulations state that their work has some advantages compared to other static IDSs. *Drawbacks:* The authors state that it can detect multiple intruders

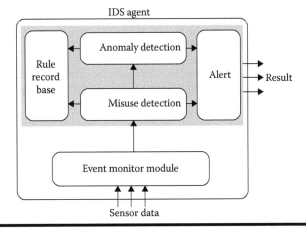

Figure 17.6   Core architecture of DIDS.

with distributed defense, albeit, with increased energy consumptions with the increase in cluster size. However, to have real advantages over the other IDS schemes, the work needs to be implemented in a real environment.

17.5 Discussion on the Vital Issues

We have so far discussed various types of IDSs in WSNs. Furthermore, we have classified them into different types based on the detection techniques used. Despite the fact that IDSs are a well-implemented technology in wired networks, there still remains enough scope of research on IDSs for WSNs. With this motivation, in this section, we will highlight the different vital facets that have been seldom considered by the previously surveyed major schemes.

WSN is a highly application-dependent network. Hence, network structures vastly differ depending on the application types. We have discussed the major structures in the introductory section in a more detailed way. To get maximum effect out of IDS, it is important to define optimal IDS locations in network structures. Before putting the concluding remarks, let us talk about some other possible solutions and categories of future IDSs. Of course, we need considerable efforts to develop an IDS of any of these kinds that could make a real positive impact.

Dynamic IDS: Little work has been done on IDS for mobile WSNs. In fact, applying IDS for mobile nodes or in the presence of dynamic change of network topology would be very challenging. Hence, IDS should take into account autoconfigurability and scalability with respect to dynamic network topology or communication failures.

Internet-enabled IDS: These days, in the era of evolving Internet of Things (IoT), we are talking about every object having its own IP address, which makes them identifiable and inventoried by computers. The next-generation Internet applications using IPv6 would be able to communicate with all devices including sensor nodes. In such a way, WSNs are becoming one strong part of IoT, which motivate experts to build IPv6-enabled WSN applications and protocols. As we have seen earlier, there is no IDS solution that can work in both wired sensor network and WSN. Another strong motivation is that the end users would prefer to have one complete IDS solution rather than having separate IDSs for each type of network they are using. Even though this problem is very challenging, it is worth further studying.

17.6 Conclusions

In this chapter, we performed a comprehensive study on IDSs in WSNs and classified them based on their employed technique. In addition, we briefly introduced the existing security attacks in WSNs and their respective countermeasures. Furthermore, we critically analyzed the IDS mechanisms with respect to network architecture of WSNs and highlighted certain shortcomings of existing works and defined the future research tracks for IDS in WSNs.

The drawbacks of existing works certainly yield to design new IDS mechanisms. We hope that our detailed investigation on various facets of IDS in WSN, their advancements, and the basic information would be beneficial for both the beginners and the active researchers in this area.

Acknowledgments

The work was supported by NDC Lab, KICT, IIUM, Malaysia and project grant NRF2007IDMIDM002-069 on "Life Spaces" from the IDM Project Office, Media Development Authority of Singapore.

References

1. Zhou, Y., Y. Fang, and Y. Zhang. 2008. "Securing Wireless Sensor Networks: A Survey." *IEEE Communications Survey* 10 (3): 6–28.
2. Pathan, A.-S. K., H.-W. Lee, and C. S. Hong. 20–22 February 2006. "Security in Wireless Sensor Networks: Issues and Challenges." In *Proceedings of the 8th International Conference on Advanced Communication Technology (IEEE ICACT 2006)*, II, Phoenix Park, Korea, 1043–48.
3. Akyildiz, I. F., W. Su, Y. Sankarasubramaniam, and E. Cayirci. August 2002. "A Survey on Sensor Networks." *IEEE Communications Magazine* 40 (8): 102–14.
4. Agah, A., S. K. Das, K. Basu, and M. Asadi. September 2004. "Intrusion Detection in Sensor Networks: A Non-Cooperative Game Approach." In *Proceedings of the Network Computing and Applications, Third IEEE International Symposium*, 343–46. Berlin/Heidelberg: Springer-Verlag.
5. Krontiris, I., T. Dimitriou, Th. Giannetsos, and M. Mpasoukos. 2008. "Intrusion Detection of Sinkhole Attacks in Wireless Sensor Networks." LNCS, vol. 4837, 150–61.
6. Onat, I., and A. Miri. 2005. "An Intrusion Detection System for Wireless Sensor Networks." *Wireless and Mobile Computing, Networking and Communications* 3: 253–59.
7. Lin, H. Y., and T. C. Chiang. June 2010. "Intrusion Detection Mechanisms Based on Queuing Theory in Remote Distribution Sensor Networks." *Advanced Materials Research* 121 (122): 58–63.
8. Xu, W., W. Trappe, Y. Zhang, and T. Wood. May 2005. "The Feasibility of Launching and Detecting Jamming Attacks in Wireless Networks." In *Proceedings of 6th ACM International Symposium Mobile Ad Hoc Networking and Computing (Mobi-Hoc'05)*, Urbana-Champaign, IL.
9. Krontiris, I., Z. Benenson, T. Giannetsos, F. C. Freiling, and T. Dimitriou. 2009. "Cooperative Intrusion Detection in Wireless Sensor Networks." *EWSN 2009*, LNCS, vol. 5432, 263–78.
10. Krontiris, I., T. Dimitriou, and F. C. Freiling. 2007. "Towards Intrusion Detection in Wireless Sensor Networks." In *Proceedings of the 13th European Wireless Conference*, Paris, France.
11. Farooqi, A. H., and F. A. Khan. 2009. "Intrusion Detection Systems for Wireless Sensor Networks: A Survey." *FGCN/ACN 2009*, CCIS, vol. 56: 234–41.
12. Zhang, Y., N. Meratnia, and P. Havinga. 2010. "Outlier Detection Techniques for Wireless Sensor Networks: A Survey." IEEE *Communications Surveys & Tutorials* 12 (2): 159–70.
13. Bhattasali, T., and R. Chaki. 2011. "A Survey of Recent Intrusion Detection Systems for Wireless Sensor Network." In *4th International Conference on Network Security and Applications (CNSA-2011)*, 268–80, Springer.

14. Loo, C. E., M. Y. Ng, C. Leckie, and M. Palaniswami. 2006. "Intrusion Detection for Routing Attacks in Sensor Networks." *International Journal of Distributed Sensor Networks* 2: 313–32.

15. Roman, R., J. Zhou, and J. Lopez. 2006. "Applying Intrusion Detection Systems to Wireless Sensor Networks." *Consumer Communications and Networking Conference*, vol. 1, 640–44. Berlin/Heidelberg: Springer-Verlag.

16. Wang, Y., G. Attebury, and B. Ramamurthy. 2006. "A Survey of Security Issues in Wireless Sensor Networks." *IEEE Communication Surveys* 8: 2–23.

17. Cagalj, M., S. Capkun, and J.-P. Hubaux. 2007. "Wormhole-Based Anti Jamming Techniques in Sensor Networks." *IEEE Transactions on Mobile Computing* 6 (1): 100–14.

18. Chen, H., P. Han, X. Zhou, and C. Gao. 2007. "Lightweight Anomaly intrusion Detection in Wireless Sensor Networks." *PAISI 2007*, LNCS 4430, 105–16.

19. Wang. Q., Y. Zhu, and L. Cheng. May 2006. "Reprogramming wireless sensor networks: Challenges and Approaches." *IEEE Network* 20 (3): 48–55.

20. Krontiris, I., T. Dimitriou, T. Giannetsos, and M. Mpasoukos. 2007. "Intrusion Detection of Sinkhole Attacks in Wireless Sensor Networks." In *3rd International Workshop on Algorithmic Aspects of Wireless Sensor Networks (AlgoSensors'07)*, Wroclaw, Poland.

21. Ngai, E. C. H., J. Liu, and M. R. Lyu. 2007. "An Efficient Intruder Detection Algorithm Against Sinkhole Attacks in Wireless Sensor Networks." *Computer Communication* 30: 2353–64.

22. Raymond, D. R., and S. F. Midkiff. March 2008. "Denial of Service in Wireless Sensor Network: Attacks and Defenses." *IEEE Pervasive Computing* 7 (1): 74–81

23. Kaplantzis, S., A. Shilton, N. Mani, Y.A. S.Kaplantzis, A. Shilton, N. Mani, and Y.A. Sekercioglu. 2007. "Detecting Selective Forwarding Attacks in Wireless Sensor networks using Support Vector Machines." *ISSNIP 2007*, Melbourne, Australia, 335–40.

24. Hai, T. H., and E. N. Huh. 2008. "Detecting Selective Forwarding Attacks in Wireless Sensor Networks Using Two-hops Neighbor Knowledge." *In Proceedings of the 2008 Seventh IEEE International Symposium on Network Computing and Applications*, 325–31.

25. Karlof, C., and D. Wagner. 2003. "Secure Routing in Wireless Sensor Networks: Attacks and Countermeasures." *Elsevier's Ad Hoc Network Journal, Special Issue on Sensor Network Applications and Protocols*, 293–315.

26. Demirbas, M., and Y. Song. 2006. "An RSSI-based Scheme for Sybil Attack Detection in Wireless Sensor Networks." In *Proceedings of IEEE WoWMoM*, 564–70.

27. Loo, C.E., M.Y. Ng, C. Leckie, and M. Palaniswami. 2006. "Intrusion Detection for Routing Attacks in Sensor Networks." *International Journal of Distributed Sensor Networks* 2 (4): 313–32.

28. Zhou, J., T.K. Das, and J. Lopez. 2008. "An Asynchronous Node Replication Attack in Wireless Sensor Networks." In *Proceedings of the IFIP TC 11 23rd International Information Security Conference*, vol. 278, 125–39, Boston Springer.

29. Parno, B., A. Perrig, and V. Gligor. 2005. "Distributed Detection of Node Replication Attack in Sensor Networks." *IEEE SP'05*, 49–63.

30. Zhu, W. T., J. Zhou, R. H. Deng, and F. Bao. 2011. "Detecting Node Replication Attacks in Mobile Sensor Networks: Theory and Approaches." *Security and Communication Networks* 5 (5): 496–507.

31. Wang, Y., G. Attebury, and B. Ramamurthy. 2nd Quarter 2006. "A Survey Of Security Issues In Wireless Sensor Networks." *IEEE Communications Surveys & Tutorials* 8 (2): 2–23.

32. Hamid, M. A., M. Mamun-Or-Rashid, and C. S. Hong. 2–4 January 2006. "Routing Security in Sensor Network: HELLO Flood Attack and Defense." In *Proceedings of IEEE ICNEWS 2006*, Dhaka, Bangladesh, 77–81.

33. Azer, M., Sh. El-Kassas, A. Hassan, and M. El-Soudani. 2008. "Intrusion Detection for Wormhole Attacks in Ad hoc Networks a Survey and a proposed Decentralized Scheme." In *Proceedings of the 3rd International Conference on Availability, Reliability and Security*, 636–41.

34. Hu, L., and D. Evans. Feb. 2004. "Using Directional Antennas to Prevent Wormhole Attacks." In *Proceedings of the 11th Annual Network and Distributed System Security Symposium (NDSS'04)*, San Diego, CA.

35. Sharif, W., and C. Leckie. 2006. "New variants of Wormhole Attacks for Sensor Networks." In *Proceedings of the Australian Telecommunication Networks and Applications Conference*, 26–30. Berlin/Heidelberg: Springer-Verlag.

36. Hu, C. Y., and A. Perrig. 2006. "Wormhole Attacks in Wireless Networks." *IEEE Journal on Selected Areas in Communications* 24 (2): 370–80.

37. Maheshwari, R., J. Gao, and S. R. Das. 2007. "Detecting Wormhole Attacks in Wireless Sensor Networks Using Connectivity Information." In *Proceedings of INFOCOM*, 107–15. Berlin/Heidelberg: Springer-Verlag.

38. Hu, L., and D. Evans. 2003. "Using Directional Antennas to Prevent Wormhole Attacks." In *Proceedings of the 11th Network and Distributed System Security Symposium*, 131–41. Berlin/Heidelberg: Springer-Verlag.

39. Graaf, R. D., I. Hegazy, J. Horton, and R. Safavi-Naini. 2010. "Distributed Detection of Wormhole attacks in Wireless Sensor Networks." *Ad Hoc Networks,* LNCS, vol. 28(1), 208–23

40. Lopez, J., R. Roman, and C. Alcaraz. August 2009. "Analysis of Security Threats, Requirements, Technologies and Standards in Wireless Sensor Networks." In *Foundations of Security Analysis and Design 2009,* LNCS 56705, 289–338.

41. Newsome, J., E. Shi, D. Song, and A. Perrig. 2004. "The Sybil Attack in Sensor Networks: Analysis & Defense." In *Proceedings of ACM IPSN'04*, 259–68. Berlin/Heidelberg: Springer-Verlag.

42. Yu, H., M. Kaminsky, P. B. Gibbons, and A. Flaxman. 2006. "SybilGuard: Defending Against Sybil Attacks via Social Networks." In *Proceedings of ACM SIGCOMM*, 267–78. Berlin/Heidelberg: Springer-Verlag.

43. Jiangtao, W., Y. Geng, S. Yuan, and C. Shengshou. 2007. "Sybil Attack Detection Based on RSSI for Wireless Sensor Networks." In *Proceedings of WiCom*, 2684–87. Berlin/Heidelberg: Springer-Verlag.

44. Mukhopadhyay, D., and I. Saha. 2006. "Location Verification Based Defense Against Sybil Attack in Sensor Networks." *ICDCN 2006.* LNCS 4308, 509–21. Berlin/Heidelberg: Springer-Verlag.

45. Chen, R. C., Y. F. Haung, and Ch. F. Hsieh. 2010. "Ranger Intrusion Detection System for Wireless Sensor Networks with Sybil Attack Based on Ontology." *AIC'10*, 176–80.

46. Kong, J., Z. Ji, W. Wang, M. Gerla, R. Bagrodia and B. Bhargava. 2005. "Low-Cost Attacks Against Packet Delivery, Localization and Time Synchronization Services in Underwater Sensor Networks." In. *Proceedings of the Fourth ACM Workshop on Wireless Security*, 87–96. Berlin/Heidelberg: Springer-Verlag.

47. Ahmed, N., S. Kanhere, and S. Jha. 2005. "The holes problem in wireless sensor networks: A Survey." *ACM SIGMOBILE Mobile Computing and Communications Review* 9 (2): 4–18.

48. Yu, B., and B. Xiao. April 2006. "Detecting selective forwarding attacks in wireless sensor networks." In *Proceedings of the 20th International Parallel and Distributed Processing Symposium (SSN2006 Workshop)*, Rhodes, Greece, 1–8.

49. Roman, R., J. Lopez, and S. Gritzalis. April 2008. "Situation Awareness Mechanisms for Wireless Sensor Networks." In *IEEE Communications Magazine* 46 (4): 102–07.

50. Pathan, A.-S. K. 2010. *Security of Self-Organizing Networks: MANET, WSN, WMN, VANET.* Auerbach Publications, CRC Press, Taylor & Francis Group. Boca Raton, FL. ISBN: 978-1-4398-1919-7.

51. Onat, I., and A. Miri. 2005. "A Real-Time Node-Based Traffic Anomaly Detection Algorithm for Wireless Sensor Networks." In *Proceedings of the ICW*, 422–27. Berlin/Heidelberg: Springer-Verlag.

52. Jian-hua, S., and M. Chuan-Xiang. 22–24 Aug. 2007. "Anomaly Detection Based on Data-Mining for Routing Attacks in Wireless Sensor Networks." In *Proceedings of CHINACOM '07*, 296–300. Berlin/Heidelberg: Springer-Verlag.

53. Wang, Q., and T. Zhang. 2007. "Detecting Anomaly Node Behavior in Wireless Sensor Networks." *AINAW*, 451–56.

54. Gupta, S., R. Zheng, and A. Cheng. 2007. "ANDES: An Anomaly Detection System for Wireless Sensor Networks." *MASS'2007*, 1–9.

55. Tiny OS. http://www.tinyos.net/. Accessed November 15, 2011.

56. Kim, J., P. Bentley, C. Wallenta, M. Ahmed, and S. Hailes. 2006. "Danger is Ubiquitous: Detecting Malicious Activities in Sensor Networks using the Dendritic Cell Algorithm" In *Proceedings of the ICARIS*, LNCS 4163. Berlin/Heidelberg: Springer-Verlag.

57. Levis, P., N. Lee, M. Welsh, and D. Culler. 2003. "TOSSIM: Accurate and Scalable Simulation of Entire TinyOS Applications." In *Proceedings of the 1st International Conference on Embedded Networked Sensor System*, 126–37. Berlin/Heidelberg: Springer-Verlag.

58. Liu, Y., and F. Yu. 2008. "Immunity-Based Intrusion Detection for Wireless Sensor Networks." In *Proceedings of the 2008 International Joint Conference on Neural Networks*, 439–44. Berlin/Heidelberg: Springer-Verlag.

59. Shaust, S., and H. Szczerbicka. 2007. "Misbehavior Detection for Wireless Sensor Networks— Necessary or Not?" In *Proceedings of the 6. Fachgespräch "Drahtlose Sensornetze" der GI/ITG-Fachgruppe "Kommunikation und Verteilte Systeme,"* Germany, 51–54.

60. Chen, R., Ch. Hsieh, and Y. Huang. January, 2009. "A New Method for Intrusion Detection on Hierarchical Wireless Sensor Networks." In *Proceedings of the ICUIMC-09*, Suwon, Korea, 238–45.

61. Chen, R. C., Ch. F. Hsieh, and Y. F. Haung. 2010. "An Isolation Intrusion Detection System for Hierarchical Wireless Sensor Networks." *Journal of Networks* 5 (3): 335–42.

62. Misra, S., K. Abraham, Md. Obaidat, and P. V. Krishna. 2008. "LAID: A Learning Automata-Based Scheme for Intrusion Detection in Wireless Sensor Networks." *Secure Communication Networks* 2: 105–15.

63. Yu, Z., and J. Tsai. 2008. "A Framework of Machine Learning Based Intrusion Detection for Wireless Sensor Networks." In *Proceedings of the SUTC'08*, 272–79. Berlin/Heidelberg: Springer-Verlag.

64. Doumit, S., and D. P. Agrawal. 2003. "Self-Organized Criticality & Stochastic Learning Based Intrusion Detection System for Wireless Sensor Network." *MILCOM*, 609–14.

65. Banerjee, S., C. Grosan, A. Abraham, and P. Mahanti. 2005. "Intrusion Detection on Sensor Networks Using Emotional Ants." *International Journal of Applied Science and Computations* 12 (3): 152–73.

66. Misra, S., P. V. Krishna, and K. I. Abraham. 2011. "A simple Learning Automata-Based Solution for Intrusion Detection in Wireless Sensor Networks." *Wireless Communications and Mobile Computing, Special Issue on Architectures and Protocols for Wireless Mesh, Ad Hoc, and Sensor Networks* 11 (3): 426–41.

67. Rajasegarar, S., C. Leckie, J. C. Bezdek, and M. Palaniswami. 2010. "Centered Hyperspherical and Hyperellipsoidal One-Class Support Vector Machines for Anomaly Detection in Sensor Networks." *IEEE Transactions on Information Forensics and Security* 5 (3): 518–33.

68. Ma, Y., H. Cao, and J. Ma. 2008. "The Intrusion Detection Method Based on Game Theory in Wireless Sensor Network." In *Proceedings of the IEEE Ubi-Media Computing*, 326–31. Berlin/Heidelberg: Springer-Verlag.

69. Agah, A., and S. K. Das. 2006. "Preventing DoS Attacks in Wireless Sensor Networks: A Repeated Game Theory Approach." *International Journal of Network Security (IJNS)* 5 (2): 145–53.

70. Krishnan, M. "Intrusion Detection in Wireless Sensor Networks." Project Paper, University of California at Berkeley, Unpublished.

71. Reddy, Y. B. 2009. "A Game Theory Approach to Detect Malicious Nodes in Wireless Sensor Networks." In *Proceedings of the SENSORCOMM'09*, Greece.

72. Reddy, Y. B., and S. Srivathsan. 2009. "Game Theory Model for Selective Forward Attacks in Wireless Sensor Networks." In *17th Mediterranean Conference on Control and Automat*. Berlin/Heidelberg: Springer-Verlag.

73. Marti, S., T. J. Giuli, K. Lai, and M. Baker. 2000. "Mitigating Routing Misbehavior in Mobile Ad hoc Networks." *MobiCom'00*, 255–65.

74. da Silva, A.P.R., M.H.T.Martins, B.P.S. Rocha, A.A.F. Loureiro, L. B. Ruiz, and H. C. Wong. October 2005. "Decentralized Intrusion Detection in Wireless Sensor Networks." In *Proceedings of the 1st ACM International Workshop on Quality of Service & Security in Wireless and Mobile Networks*, Montreal, Quebec, Canada.

75. Mostarda, L., and A. Navarra. 2008. "Distributed Intrusion Detection Systems for Enhancing Security in Mobile Wireless Sensor Networks." *International Journal of Distributed Sensor Networks* 4 (2): 83–109.

76. Wang, Y., X. Wang, B. Xie, D. Wang, and P. Agrawal. 2008. "Intrusion Detection in Homogeneous and Heterogeneous Wireless Sensor Networks." *IEEE Transactions on Mobile Computing*, 8 (6): 698–711.

77. Guorui, L., H. Jingsha, and F. Yingfang. 2008. "Group-Based Intrusion Detection System in Wireless Sensor Networks." *Computer Communications* 32 (18): 4324–32.

78. Bhuse, V., A. Gupta, and Ala Al-Fuqaha. 2007. "Detection of Masquerade Attacks on Wireless Sensor Networks." In *Proceedings of the ICC'07*, 1142–47. Berlin/Heidelberg: Springer-Verlag.

79. de Sousa Lemos, M. V., L. B. Leal, and R. H. Filho. 2010. "A New Collaborative Approach for Intrusion Detection System on Wireless Sensor Networks." *Novel Algorithms and Techniques*. Springer.

80. Shin, S., T. Kwon, G. Y. Jo, Y. Park, and H. Rhee. 2010. "An Experimental Study of Hierarchical Intrusion Detection for Wireless Industrial Sensor Networks." *IEEE Transactions on Industrial Informatics* 6 (4): 744–57.
81. Mubarak, T. M., S. A. Sattar, A. Rao, and M. Sajitha. 2011. "A Collaborative, Secure and Energy Efficient Intrusion Detection Method for Homogeneous WSN." In *International Conference on Advances in Computing and Communications (ACC-2011)*, Springer. Berlin/Heidelberg: Springer-Verlag.
82. Krontiris, I., T. Dimitriou, and Th. Giannetsos. 2008. "LIDeA: A Distributed Lightweight Intrusion Detection Architecture for Sensor Networks." In *Proceedings of the 4th International Conference on Security and Privacy in Communication Networks*, Istanbul, Turkey.
83. Hai, T. H., F. Khan, and E.-N. Huh. 2007. "Hybrid Intrusion Detection System for Wireless Sensor Networks." In *Proceedings of the ICCSA 2007*, LNCS 4706, 383–96. Berlin/Heidelberg: Springer-Verlag.
84. Hai, T. H., E.-N. Huh, and Minho Jo. 2007. "A Lightweight Intrusion Detection Framework for Wireless Sensor Networks." *Wireless Communication Mobile Computing* 10 (4): 559–72.
85. Durresi, A., V. Parucheri, S. Iyengar, and R. Kannan. 2005. "Optimized Broadcast Protocol for Sensor Networks." *IEEE Transactions on Computers* 54 (8): 1013–24.
86. Yan, K. Q., S. C. Wang, and C. W. Liu. 2009. "A Hybrid Intrusion Detection System of Cluster-based Wireless Sensor Networks." In *Proceedings of the IMECS 2009*, Hong Kong, 411–16.
87. Wang, S. S., K. Q. Yan, S. C. Wang, and C. W. Liu. 2011. "An Integrated Intrusion Detection System for Cluster-based Wireless Sensor Networks." *Elsevier's Expert Systems and Applications* 38: 15234–243.
88. Huo, G., and X. Wang. 2008. "DIDS: A Dynamic Model of Intrusion Detection System in Wireless Sensor Networks." In *Proceedings of the 2008 IEEE ICIA*, 374–78. Berlin/Heidelberg: Springer-Verlag.
89. Techateerawat, P., and A. Jennings. "Energy Efficiency of Intrusion Detection Systems in Wireless Sensor Networks." In *Proceedings of the WI-IATW'06*. Berlin/Heidelberg: Springer-Verlag.
90. Bankovic, Z., J. M. Moya, A. Araujo, D. Fraga, J. C. Vallejo, and J. M. de Goyeneche. 2010. "Distributed Intrusion Detection System for Wireless Sensor Networks Based on a Reputation System Coupled with Kernel Self-Organizing Maps." *Integrated Computer-Aided Engineering* 17 (2): 87–102.

Index